An Introduction to Envelopes

WILEY SERIES IN PROBABILITY AND STATISTICS

The **Wiley Series in Probability and Statistics** is well established and authoritative. It covers many topics of current research interest in both pure and applied statistics and probability theory. Written by leading statisticians and institutions, the titles span both state-of-the-art developments in the field and classical methods.

Reflecting the wide range of current research in statistics, the series encompasses applied, methodological and theoretical statistics, ranging from applications and new techniques made possible by advances in computerized practice to rigorous treatment of theoretical approaches.

This series provides essential and invaluable reading for all statisticians, whether in academia, industry, government, or research.

A complete list of titles in this series can be found at http://www.wiley.com/go/wsps

An Introduction to Envelopes

Dimension Reduction for Efficient Estimation in Multivariate Statistics

R. Dennis Cook
School of Statistics
University of Minnesota
U.S.A.

WILEY

This edition first published 2018

Registered Office(s)
John Wiley & Sons, Inc., 111 River Street, Hoboken, NJ 07030, USA

Editorial Office
111 River Street, Hoboken, NJ 07030, USA

For details of our global editorial offices, customer services, and more information about Wiley products visit us at www.wiley.com.

Wiley also publishes its books in a variety of electronic formats and by print-on-demand. Some content that appears in standard print versions of this book may not be available in other formats.

Library of Congress Cataloging-in-Publication Data

Names: Cook, R. Dennis, author.
Title: An introduction to envelopes : dimension reduction for efficient
estimation in multivariate statistics / R. Dennis Cook.
Description: 1st edition. | Hoboken, NJ : John Wiley & Sons, 2018. | Series:
Wiley series in probability and statistics |
Identifiers: LCCN 2018023695 (print) | LCCN 2018036057 (ebook) | ISBN
9781119422952 (Adobe PDF) | ISBN 9781119422969 (ePub) | ISBN 9781119422938
(hardcover)
Subjects: LCSH: Multivariate analysis. | Dimension reduction (Statistics)
Classification: LCC QA278 (ebook) | LCC QA278 .C648 2018 (print) | DDC
519.5/35 – dc23
LC record available at https://lccn.loc.gov/2018023695

Cover Design: Wiley
Cover Image: © teekid/iStockphoto and Courtesy of R. Dennis Cook

Set in 10/12pt WarnockPro by SPi Global, Chennai, India

Printed in the United States of America

V10004269_090418

For Sandra
Sister and Artist
1949–2017

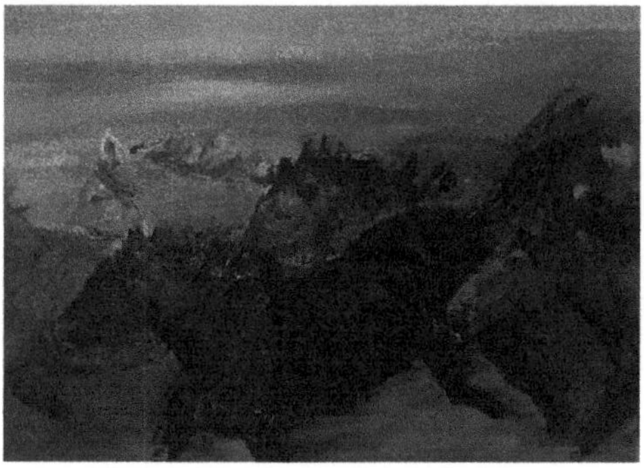

Contents

Preface *xv*
Notation and Definitions *xix*

1 **Response Envelopes** *1*
1.1 The Multivariate Linear Model *2*
1.1.1 Partitioned Models and Added Variable Plots *5*
1.1.2 Alternative Model Forms *6*
1.2 Envelope Model for Response Reduction *6*
1.3 Illustrations *10*
1.3.1 A Schematic Example *10*
1.3.2 Compound Symmetry *13*
1.3.3 Wheat Protein: Introductory Illustration *13*
1.3.4 Cattle Weights: Initial Fit *14*
1.4 More on the Envelope Model *19*
1.4.1 Relationship with Sufficiency *19*
1.4.2 Parameter Count *19*
1.4.3 Potential Gains *20*
1.5 Maximum Likelihood Estimation *21*
1.5.1 Derivation *21*
1.5.2 Cattle Weights: Variation of the X-Variant Parts of Y *23*
1.5.3 Insights into $\widehat{\mathcal{E}}_{\Sigma}(\mathcal{B})$ *24*
1.5.4 Scaling the Responses *25*
1.6 Asymptotic Distributions *25*
1.7 Fitted Values and Predictions *28*
1.8 Testing the Responses *29*
1.8.1 Test Development *29*
1.8.2 Testing Individual Responses *32*
1.8.3 Testing Containment Only *34*
1.9 Nonnormal Errors *34*
1.10 Selecting the Envelope Dimension, u *36*
1.10.1 Selection Methods *36*

1.10.1.1 Likelihood Ratio Testing *36*
1.10.1.2 Information Criteria *37*
1.10.1.3 Cross-validation *37*
1.10.2 Inferring About rank(β) *38*
1.10.3 Asymptotic Considerations *38*
1.10.4 Overestimation Versus Underestimation of u *41*
1.10.5 Cattle Weights: Influence of u *43*
1.11 Bootstrap and Uncertainty in the Envelope Dimension *45*
1.11.1 Bootstrap for Envelope Models *45*
1.11.2 Wheat Protein: Bootstrap and Asymptotic Standard Errors, u Fixed *46*
1.11.3 Cattle Weights: Bootstrapping u *47*
1.11.4 Bootstrap Smoothing *48*
1.11.5 Cattle Data: Bootstrap Smoothing *49*

2 Illustrative Analyses Using Response Envelopes *51*
2.1 Wheat Protein: Full Data *51*
2.2 Berkeley Guidance Study *51*
2.3 Banknotes *54*
2.4 Egyptian Skulls *55*
2.5 Australian Institute of Sport: Response Envelopes *58*
2.6 Air Pollution *59*
2.7 Multivariate Bioassay *63*
2.8 Brain Volumes *65*
2.9 Reducing Lead Levels in Children *67*

3 Partial Response Envelopes *69*
3.1 Partial Envelope Model *69*
3.2 Estimation *71*
3.2.1 Asymptotic Distribution of $\widehat{\beta}_1$ *72*
3.2.2 Selecting u_1 *73*
3.3 Illustrations *74*
3.3.1 Cattle Weight: Incorporating Basal Weight *74*
3.3.2 Mens' Urine *74*
3.4 Partial Envelopes for Prediction *77*
3.4.1 Rationale *77*
3.4.2 Pulp Fibers: Partial Envelopes and Prediction *78*
3.5 Reducing Part of the Response *79*

4 Predictor Envelopes *81*
4.1 Model Formulations *81*
4.1.1 Linear Predictor Reduction *81*
4.1.1.1 Predictor Envelope Model *83*

4.1.1.2 Expository Example 83
4.1.2 Latent Variable Formulation of Partial Least Squares Regression 84
4.1.3 Potential Advantages 86
4.2 SIMPLS 88
4.2.1 SIMPLS Algorithm 88
4.2.2 SIMPLS When $n < p$ 90
4.2.2.1 Behavior of the SIMPLS Algorithm 90
4.2.2.2 Asymptotic Properties of SIMPLS 91
4.3 Likelihood-Based Predictor Envelopes 94
4.3.1 Estimation 95
4.3.2 Comparisions with SIMPLS and Principal Component Regression 97
4.3.2.1 Principal Component Regression 98
4.3.2.2 SIMPLS 98
4.3.3 Asymptotic Properties 98
4.3.4 Fitted Values and Prediction 100
4.3.5 Choice of Dimension 101
4.3.6 Relevant Components 101
4.4 Illustrations 102
4.4.1 Expository Example, Continued 102
4.4.2 Australian Institute of Sport: Predictor Envelopes 103
4.4.3 Wheat Protein: Predicting Protein Content 105
4.4.4 Mussels' Muscles: Predictor Envelopes 106
4.4.5 Meat Properties 109
4.5 Simultaneous Predictor–Response Envelopes 109
4.5.1 Model Formulation 109
4.5.2 Potential Gain 110
4.5.3 Estimation 113

5 Enveloping Multivariate Means 117
5.1 Enveloping a Single Mean 117
5.1.1 Envelope Structure 117
5.1.2 Envelope Model 119
5.1.3 Estimation 120
5.1.3.1 Maximum Likelihood Estimation 120
5.1.3.2 Asymptotic Variance of $\widehat{\mu}$ 121
5.1.3.3 Selecting $u = \dim(\mathcal{E}_{\Sigma}(\mathcal{M}))$ 122
5.1.4 Minneapolis Schools 122
5.1.4.1 Two Transformed Responses 123
5.1.4.2 Four Untransformed Responses 124
5.1.5 Functional Data 126
5.2 Enveloping Multiple Means with Heteroscedastic Errors 126

5.2.1 Heteroscedastic Envelopes *126*
5.2.2 Estimation *128*
5.2.3 Cattle Weights: Heteroscedastic Envelope Fit *129*
5.3 Extension to Heteroscedastic Regressions *130*

6 Envelope Algorithms *133*
6.1 Likelihood-Based Envelope Estimation *133*
6.2 Starting Values *135*
6.2.1 Choosing the Starting Value from the Eigenvectors of $\widehat{\mathrm{M}}$ *135*
6.2.2 Choosing the Starting Value from the Eigenvectors of $\widehat{\mathrm{M}} + \widehat{\mathrm{U}}$ *137*
6.2.3 Summary *138*
6.3 A Non-Grassmann Algorithm for Estimating $\mathcal{E}_{\mathrm{M}}(\mathcal{U})$ *139*
6.4 Sequential Likelihood-Based Envelope Estimation *141*
6.4.1 The 1D Algorithm *141*
6.4.2 Envelope Component Screening *142*
6.4.2.1 ECS Algorithm *143*
6.4.2.2 Alternative ECS Algorithm *144*
6.5 Sequential Moment-Based Envelope Estimation *145*
6.5.1 Basic Algorithm *145*
6.5.2 Krylov Matrices and $\dim(\mathcal{U}) = 1$ *147*
6.5.3 Variations on the Basic Algorithm *147*

7 Envelope Extensions *149*
7.1 Envelopes for Vector-Valued Parameters *149*
7.1.1 Illustrations *151*
7.1.2 Estimation Based on a Complete Likelihood *154*
7.1.2.1 Likelihood Construction *154*
7.1.2.2 Aster Models *156*
7.2 Envelopes for Matrix-Valued Parameters *157*
7.3 Envelopes for Matrix-Valued Responses *160*
7.3.1 Initial Modeling *161*
7.3.2 Models with Kronecker Structure *163*
7.3.3 Envelope Models with Kronecker Structure *164*
7.4 Spatial Envelopes *166*
7.5 Sparse Response Envelopes *168*
7.5.1 Sparse Response Envelopes when $r \ll n$ *168*
7.5.2 Cattle Weights and Brain Volumes: Sparse Fits *169*
7.5.3 Sparse Envelopes when $r > n$ *170*
7.6 Bayesian Response Envelopes *171*

8 Inner and Scaled Envelopes *173*
8.1 Inner Envelopes *173*
8.1.1 Definition and Properties of Inner Envelopes *174*
8.1.2 Inner Response Envelopes *175*
8.1.3 Maximum Likelihood Estimators *176*
8.1.4 Race Times: Inner Envelopes *179*
8.2 Scaled Response Envelopes *182*
8.2.1 Scaled Response Model *183*
8.2.2 Estimation *184*
8.2.3 Race Times: Scaled Response Envelopes *185*
8.3 Scaled Predictor Envelopes *186*
8.3.1 Scaled Predictor Model *187*
8.3.2 Estimation *188*
8.3.3 Scaled SIMPLS Algorithm *189*

9 Connections and Adaptations *191*
9.1 Canonical Correlations *191*
9.1.1 Construction of Canonical Variates and Correlations *191*
9.1.2 Derivation of Canonical Variates *193*
9.1.3 Connection to Envelopes *194*
9.2 Reduced-Rank Regression *195*
9.2.1 Reduced-Rank Model and Estimation *195*
9.2.2 Contrasts with Envelopes *196*
9.2.3 Reduced-Rank Response Envelopes *197*
9.2.4 Reduced-Rank Predictor Envelopes *199*
9.3 Supervised Singular Value Decomposition *199*
9.4 Sufficient Dimension Reduction *202*
9.5 Sliced Inverse Regression *204*
9.5.1 SIR Methodology *204*
9.5.2 Mussels' Muscles: Sliced Inverse Regression *205*
9.5.3 The "Envelope Method" *206*
9.5.4 Envelopes and SIR *207*
9.6 Dimension Reduction for the Conditional Mean *207*
9.6.1 Estimating One Vector in $\mathcal{S}_{\mathrm{E}(Y|\mathrm{X})}$ *208*
9.6.2 Estimating $\mathcal{S}_{\mathrm{E}(Y|\mathrm{X})}$ *209*
9.7 Functional Envelopes for SDR *211*
9.7.1 Functional SDR *211*
9.7.2 Functional Predictor Envelopes *211*
9.8 Comparing Covariance Matrices *212*
9.8.1 SDR for Covariance Matrices *213*
9.8.2 Connections with Envelopes *215*

9.8.3 Illustrations *216*
9.8.4 SDR for Means and Covariance Matrices *217*
9.9 Principal Components *217*
9.9.1 Introduction *217*
9.9.2 Random Latent Variables *219*
9.9.2.1 Envelopes *220*
9.9.2.2 Envelopes with Isotropic Intrinsic and Extrinsic Variation *222*
9.9.2.3 Envelopes with Isotropic Intrinsic Variation *223*
9.9.2.4 Selection of the Dimension u *225*
9.9.3 Fixed Latent Variables and Isotropic Errors *225*
9.9.4 Numerical Illustrations *226*
9.10 Principal Fitted Components *229*
9.10.1 Isotropic Errors, $\Sigma_{X|Y} = \sigma^2 I_p$ *230*
9.10.2 Anisotropic Errors, $\Sigma_{X|Y} > 0$ *231*
9.10.3 Nonnormal Errors and the Choice of f *232*
9.10.3.1 Graphical Choices *232*
9.10.3.2 Basis Functions *232*
9.10.3.3 Categorical Response *232*
9.10.3.4 Sliced Inverse Regression *233*
9.10.4 High-Dimensional PFC *233*

Appendix A Envelope Algebra *235*
A.1 Invariant and Reducing Subspaces *235*
A.2 M-Envelopes *240*
A.3 Relationships Between Envelopes *241*
A.3.1 Invariance and Equivariance *241*
A.3.2 Direct Sums of Envelopes *244*
A.3.3 Coordinate Reduction *244*
A.4 Kronecker Products, vec and vech *246*
A.5 Commutation, Expansion, and Contraction Matrices *248*
A.6 Derivatives *249*
A.6.1 Derivatives for η, Ω, and Ω_0 *249*
A.6.2 Derivatives with Respect to Γ *250*
A.6.3 Derivatives of Grassmann Objective Functions *251*
A.7 Miscellaneous Results *252*
A.8 Matrix Normal Distribution *255*
A.9 Literature Notes *256*

Appendix B Proofs for Envelope Algorithms *257*
B.1 The 1D Algorithm *257*
B.2 Sequential Moment-Based Algorithm *262*
B.2.1 First Direction Vector w_1 *263*
B.2.2 Second Direction Vector w_2 *263*

B.2.3 $(q + 1)$st Direction Vector w_{q+1}, $q < u$ *264*
B.2.4 Termination *265*

Appendix C Grassmann Manifold Optimization *267*
C.1 Gradient Algorithm *268*
C.2 Construction of B *269*
C.3 Construction of $\exp\{\delta A(B)\}$ *271*
C.4 Starting and Stopping *272*

Bibliography *273*

Author Index *283*

Subject Index *287*

Preface

> ... the objective of statistical methods is the reduction of data. A quantity of data... is to be replaced by relatively few quantities which shall adequately represent... the relevant information contained in the original data.
> Since the number of independent facts supplied in the data is usually far greater than the number of facts sought, much of the information supplied by an actual sample is irrelevant. It is the object of the statistical process employed in the reduction of data to exclude this irrelevant information, and to isolate the whole of the relevant information contained in the data.
>
> Fisher (1922): *On the Mathematical Foundations of Theoretical Statistics.*

Dimension reduction has always been a leitmotif of statistical thought, as illustrated by the title of a nineteenth century article by Edgeworth (1884) – "On the reduction of observations" – and by the foundations of theoretical statistics established by Fisher (1922). Dimension reduction interpreted loosely is today a huge area. Many ad hoc methods have been developed by the Computer Science and Machine Learning communities (Burges, 2009), and it seems that new methods are being proposed every day. The material in this book reflects a subset of the dimension reduction literature that can be motivated in a manner that is reminiscent of Fisher's original motivation of sufficiency.

Sufficient dimension reduction for regression strives to find a low-dimensional set of proxy predictors that can be used in place of the full set of predictors without the loss of regression information and without a prespecified parametric model. It has expanded substantially over the past three decades, and today is a widely recognized statistical paradigm that can be particularly helpful in the initial stages of an analysis, prior to model selection. However, the formal paradigm stops typically at the estimation of the proxy predictors, leaving the analyst to carry on with other context-dependent methods, and has little to offer when a traditional model is available.

Envelopes, the topic of this book, can be seen as a descendent of sufficient dimension reduction that can be applied in model-based contexts without altering traditional objectives, and they have also the potential to offer advances in model-free contexts as well. *Most importantly, envelope methods can result in massive efficiency gains relative to standard methods, gains that are equivalent to increasing the sample size many times over.*

Outline

We begin in Chapter 1 by using response envelopes to improve estimative efficiency in the context of the usual multivariate (multiresponse) linear model, following a brief review of the model itself. This chapter was written to emphasize the fundamental rationale underlying envelopes, providing arguments that hopefully appeal to intuition along with selected mathematical details. These foundations are then adapted to other contexts in subsequent chapters. I have tried to achieve a useful balance between foundations and methodology by integrating illustrative examples throughout the book, but particularly in the first few chapters. Most proofs are provided in the appendices, or the reader is referred to the literature. But proofs are presented occasionally in the main text when they seem useful to show how various ideas integrate to produce the desired results. Some of the essential development in Chapter 1 is revisited in subsequent chapters because some readers may prefer to skip around, and a reminder from time to time could be useful to show how arguments carry from one context to another.

Still in the context of response envelopes, Chapter 2 consists of a number of small examples to show the potential advantages of envelopes and to illustrate how they might be used in practice, and in Chapter 3, we discuss how to use envelopes to target selected coefficients. Chapter 4 is devoted to predictor envelopes and their connection with partial least squares regression, and Chapter 5 shows the potential for envelope methodology to improve estimation of a multivariate mean. We turn to objective function optimization in Chapter 6 where we discuss several algorithms, including relatively slow likelihood-based methods and relatively fast moment-based methods. The discussion to this point is in terms of the multivariate linear model introduced in Chapter 1. In Chapter 7, we sketch how envelope methodology can be applied to improve an asymptotically normal vector or matrix estimator. We also discuss how envelopes can be used in regressions with a matrix-valued response and sketch work on sparse and Bayesian response envelopes. Inner and scaled envelopes, which are discussed in Chapter 8, are envelope forms that allow for efficiency gains in ways that are different from that introduced in Chapter 1. We discuss in Chapter 9 the relationships between envelopes and other dimension reduction methods, including canonical correlations, reduced-rank regression, supervised singular value decomposition, sufficient dimension reduction, principal components, and principal fitted components.

Computing

R or MatLab computer programs are available for nearly all of the methods discussed in this book and many have been implemented in both languages. Most of the methods are available in integrated packages, but some come as standalone programs. These programs and packages are not discussed in this book, but descriptions of and links to them can be found at z.umn.edu/envelopes. This format will allow for updates and links to developments following this book.

Acknowledgments

Earlier versions of this book were used as lecture notes for a one-semester course at the University of Minnesota. Most students who attended this course had background in mathematical statistics and multivariate analysis. Students in this course and my collaborators contributed to the ideas and flavor of the book. In particular, I would like to thank Austin Brown, Jami Cook, Shanshan Ding, Daniel Eck, Liliana Forzani, Ming Gao, Inga Helland, Zhihua Su, Xin (Henry) Zhang, and Xinyu Zhang for many helpful discussions. Daniel Eck wrote the first draft of the discussion of Aster models in Section 7.1.2, and Liliana Forzani and Zhihua Su collaborated on Appendix C. Zhihua Su and her students wrote nearly all of the R and MatLab programs for implementing envelopes. Much of this book reflects the many stimulating conversations I had with Bing Li and Francesca Chiaromonte during the genesis of envelopes.

Nearly all of my work on envelopes was supported by grants from the National Science Foundation's Division of Mathematical Sciences.

St. Paul, Minnesota
December 2017

R. Dennis Cook

Notation and Definitions

Matrices. For positive integers r and p, $\mathbb{R}^{r\times p}$ stands for the class of all real matrices of dimension $r \times p$, and $\mathbb{S}^{r\times r}$ denotes the class of all symmetric $r \times r$ matrices. For $\mathbf{A} \in \mathbb{R}^{r\times p}$, $\mathbf{A}^{\dagger}$ indicates the Moore–Penrose inverse of $\mathbf{A}$. Vectors and matrices will typically be written in bold face, while scalars are not bold. The $r \times r$ identity matrix is denoted as $\mathbf{I}_r$, and $\mathbf{1}_n$ denotes the $n \times 1$ vector of 1's. The ijth element of a matrix $\mathbf{A}$ is often denoted as $(\mathbf{A})_{ij}$.

For $\mathbf{M} \in \mathbb{S}^{r\times r}$, the notation $\mathbf{M} > 0$ means that $\mathbf{M}$ is positive definite, $\mathbf{M} \geq 0$ means that $\mathbf{M}$ is positive semi-definite, $|\mathbf{M}|$ denotes the determinant of $\mathbf{M}$, and $|\mathbf{M}|_0$ denotes the product of the nonzero eigenvalues of $\mathbf{M}$. The spectral norm of $\mathbf{M} \in \mathbb{R}^{r\times p}$ is denoted as $\|\mathbf{M}\|$.

The *vector* operator vec : $\mathbb{R}^{r\times p} \mapsto \mathbb{R}^{rp}$ stacks the columns of the argument matrix. On the symmetric matrices, we use the related *vector-half* operator vech : $\mathbb{S}^{r\times r} \mapsto \mathbb{R}^{r(r+1)/2}$, which stacks only the unique part of each column that lies on or below the diagonal. The operators vec and vech are related through a *contraction* matrix $\mathbf{C}_r \in \mathbb{R}^{r(r+1)/2\times r^2}$ and an *expansion* matrix $\mathbf{E}_r \in \mathbb{R}^{r^2\times r(r+1)/2}$, which are defined so that $\text{vech}(\mathbf{A}) = \mathbf{C}_r\text{vec}(\mathbf{A})$ and $\text{vec}(\mathbf{A}) = \mathbf{E}_r\text{vech}(\mathbf{A})$ for any $\mathbf{A} \in \mathbb{S}^{r\times r}$. These relations uniquely define $\mathbf{C}_r$ and $\mathbf{E}_r$ and imply $\mathbf{C}_r\mathbf{E}_r = \mathbf{I}_{r(r+1)/2}$. The $pr \times pr$ commutation matrix that maps vec($\mathbf{A}$) to vec($\mathbf{A}^T$) is denoted by $\mathbf{K}_{rp}$: $\text{vec}(\mathbf{A}^T) = \mathbf{K}_{rp}\text{vec}(\mathbf{A})$. For further background on these operators, see Appendix A.5, Henderson and Searle (1979) and Magnus and Neudecker (1979).

Let $\mathbf{A} = (a_{ij}) \in \mathbb{R}^{r\times s}$ and $\mathbf{B} \in \mathbb{R}^{t\times u}$. The Kronecker product of two matrices $\otimes : \mathbb{R}^{r\times s} \times \mathbb{R}^{t\times u} \mapsto \mathbb{R}^{rt\times su}$ can be defined block-wise as $\mathbf{A} \otimes \mathbf{B} = (a_{ij}\mathbf{B})$, $i = 1, \ldots, r, j = 1, \ldots, s$.

We often denote the eigenvalues of $\mathbf{A} \in \mathbb{S}^{p\times p}$ as $\varphi_1(\mathbf{A}) \geq \ldots \geq \varphi_p(\mathbf{A})$ with corresponding ordered eigenvector $\boldsymbol{\ell}_1(\mathbf{A}), \cdots, \boldsymbol{\ell}_p(\mathbf{A})$. The arguments to φ_j and $\boldsymbol{\ell}_j$ may be suppressed when they are expected to be clear from context.

A block diagonal matrix with diagonal blocks $\mathbf{A}_j$, $j = 1, \ldots, q$, is represented as bdiag($\mathbf{A}_1, \ldots, \mathbf{A}_q$). The direct sum $\oplus$ of two matrices $\mathbf{A} \in \mathbb{R}^{m\times n}$ and $\mathbf{B} \in \mathbb{R}^{p\times q}$ is defined as $(m + p) \times (n + q)$ block diagonal matrix bdiag($\mathbf{A}, \mathbf{B}$): $\mathbf{A} \oplus \mathbf{B} = \text{bdiag}(\mathbf{A}, \mathbf{B})$.

Subspaces. For $\mathbf{A} \in \mathbb{R}^{p\times r}$ and a subspace $\mathcal{S} \subseteq \mathbb{R}^r$, $\mathbf{A}\mathcal{S} := \{\mathbf{A}\mathbf{x} : \mathbf{x} \in \mathcal{S}\}$. For $\mathbf{B} \in \mathbb{R}^{r\times p}$, span($\mathbf{B}$) denotes the subspace of $\mathbb{R}^r$ spanned by the columns of $\mathbf{B}$. We occasionally use $\mathcal{S}_\mathbf{B}$ as shorthand for span($\mathbf{B}$) when $\mathbf{B}$ has been defined. Subscripts $\mathcal{S}_{(\cdot)}$ will also be used to name commonly occurring subspaces. A *basis matrix* for a subspace $\mathcal{S}$ is any matrix whose columns form a basis for $\mathcal{S}$. A *semi-orthogonal matrix* $\mathbf{A} \in \mathbb{R}^{r\times p}$ has orthogonal columns, $\mathbf{A}^T\mathbf{A} = \mathbf{I}_p$. We will frequently refer to semi-orthogonal basis matrices.

Let $\mathbf{A} \in \mathbb{S}^{r\times r}$ and $\mathbf{B} \in \mathbb{S}^{r\times r}$ with $\mathbf{A} > 0$. Then $\mathcal{S}_d(\mathbf{A},\mathbf{B})$ equals $\mathbf{A}^{-1/2}$ times the span of the first $d \leq r$ eigenvectors of $\mathbf{A}^{-1/2}\mathbf{B}\mathbf{A}^{-1/2}$. This subspace can also be described as the span of the first d eigenvectors of $\mathbf{B}$ relative to $\mathbf{A}$.

A sum of subspaces of $\mathbb{R}^r$ is defined as $\mathcal{S}_1 + \mathcal{S}_2 = \{\mathbf{x}_1 + \mathbf{x}_2 : \mathbf{x}_1 \in \mathcal{S}_1, \mathbf{x}_2 \in \mathcal{S}_2\}$. We use $\mathcal{S}_1 \subset \mathcal{S}_2$ to indicate that the subspace $\mathcal{S}_1$ is a proper subset of $\mathcal{S}_2$, while $\mathcal{S}_1 \subseteq \mathcal{S}_2$ allows $\mathcal{S}_1 = \mathcal{S}_2$.

For a positive definite matrix $\mathbf{\Sigma} \in \mathbb{S}^{r\times r}$, the inner product in $\mathbb{R}^r$ defined by $\langle \mathbf{x}_1, \mathbf{x}_2\rangle_\mathbf{\Sigma} = \mathbf{x}_1^T\mathbf{\Sigma}\mathbf{x}_2$ is referred to as the $\mathbf{\Sigma}$ inner product; when $\mathbf{\Sigma} = \mathbf{I}_r$, the r by r identity matrix, this inner product is called the usual inner product. A projection relative to the $\mathbf{\Sigma}$ inner product is the projection operator in the inner product space $\{\mathbb{R}^r, \langle\cdot,\cdot\rangle_\mathbf{\Sigma}\}$. If $\mathbf{B} \in \mathbb{R}^{r\times p}$, then the projection onto span($\mathbf{B}$) relative to $\mathbf{\Sigma}$ has the matrix representation $\mathbf{P}_{\mathbf{B}(\mathbf{\Sigma})} := \mathbf{B}(\mathbf{B}^\mathbf{T}\mathbf{\Sigma}\mathbf{B})^\dagger\mathbf{B}^\mathbf{T}\mathbf{\Sigma}$. The projection onto the orthogonal complement of span ($\mathbf{B}$) relative to the $\mathbf{\Sigma}$ inner product, $\mathbf{I}_r - \mathbf{P}_{\mathbf{B}(\mathbf{\Sigma})}$, is denoted by $\mathbf{Q}_{\mathbf{B}(\mathbf{\Sigma})}$. Projection operators employing the usual inner product are written with a single subscript argument $\mathbf{P}_{(\cdot)}$, where the subscript describes the subspace, and $\mathbf{Q}_{(\cdot)} = \mathbf{I}_r - \mathbf{P}_{(\cdot)}$. The orthogonal complement $\mathcal{S}^\perp$ of a subspace $\mathcal{S}$ is constructed with respect to the usual inner product, unless indicated otherwise.

Let $\mathcal{S}_1 \subseteq \mathcal{S}$ be two nested subspaces of $\mathbb{R}^r$. We write $\mathcal{S} \setminus \mathcal{S}_1$ for the part of $\mathcal{S}$ that is orthogonal to $\mathcal{S}_1$. That is, $\mathcal{S} \setminus \mathcal{S}_1 = \text{span}(\mathbf{P}_\mathcal{S} - \mathbf{P}_{\mathcal{S}_1})$.

If $\mathcal{S}_1 \in \mathbb{R}^p$ and $\mathcal{S}_2 \in \mathbb{R}^q$, then their direct sum is defined as $\mathcal{S}_1 \oplus \mathcal{S}_2 = \text{span}(\mathbf{B}_1 \oplus \mathbf{B}_2)$, where $\mathbf{B}_j$ is a basis matrix for $\mathcal{S}_j$, $j = 1, 2$.

The set of all u-dimensional subspaces of $\mathbb{R}^r$ is called a Grassmann manifold or Grassmannian and denoted as $\mathcal{G}(u, r)$. A Grassmannian is a compact topological manifold named for Hermann Grassmann (1809–1877).

Envelopes. Let $\mathbf{M} \in \mathbb{S}^{r\times r}$, and let $\mathcal{S} \subseteq (\mathbf{M})$. The $\mathbf{M}$-envelope of $\mathcal{S}$, to be written as $\mathcal{E}_\mathbf{M}(\mathcal{S})$, is the intersection of all reducing subspaces of $\mathbf{M}$ that contain $\mathcal{S}$ (see Definition 1.2). A variety of envelopes will be used in some chapters, and to avoid proliferation of notation, we may use a matrix as the argument of $\mathcal{E}_\mathbf{M}(\cdot)$: If $\mathbf{B} \in \mathbb{R}^{r\times d}$ and span($\mathbf{B}$) $= \mathcal{S}$, then $\mathcal{E}_\mathbf{M}(\mathbf{B}) = \mathcal{E}_\mathbf{M}(\text{span}(\mathbf{B})) = \mathcal{E}_\mathbf{M}(\mathcal{S})$.

Random vectors and their distributions. The notation $\mathbf{Y} \sim \mathbf{X}$ means that the random vectors $\mathbf{Y}$ and $\mathbf{X}$ have the same distribution, and $\mathbf{Y} \perp\!\!\!\perp \mathbf{X}$ means that they are independent. The conditional distribution of $\mathbf{Y}$ given $\mathbf{X} = \mathbf{x}$, or equivalently of $\mathbf{Y} \mid (\mathbf{X} = \mathbf{x})$, varies with the value $\mathbf{x}$ assumed by $\mathbf{X}$. When there is no ambiguity, we will write $\mathbf{Y} \mid (\mathbf{X} = \mathbf{x})$ as $\mathbf{Y} \mid \mathbf{X}$, understanding that $\mathbf{X}$ is fixed at some value. The notation $\mathbf{Y} \perp\!\!\!\perp \mathbf{X} \mid \mathbf{V}$ means that $\mathbf{Y}$ and $\mathbf{X}$ are independent given any value for the random vector $\mathbf{V}$.

We often use $\boldsymbol{\Sigma}_{\mathbf{V}} = \mathrm{var}(\mathbf{V}) \in \mathbb{S}^{r \times r}$ to denote the covariance matrix of the random vector $\mathbf{V} \in \mathbb{R}^r$ and $\boldsymbol{\Sigma}_{\mathbf{U},\mathbf{V}} = \mathrm{cov}(\mathbf{U}, \mathbf{V})$ to denote the matrix of covariances between the element of $\mathbf{U}$ and $\mathbf{V}$. Similarly, $\boldsymbol{\Sigma}_{\mathbf{V}|\mathbf{U}}$ denotes the covariance matrix of $\mathbf{V}$ given $\mathbf{U}$. Sample covariance matrices are denoted as $\mathbf{S}_{(\cdot)}$ with the subscript indicating the random vectors involved. All sample covariances use the sample size as the denominator, unless indicated otherwise. We use $\mathbf{S}_{\mathbf{U},\mathbf{V}}$ to denote the sample covariance matrix for random vectors $\mathbf{U}$ and $\mathbf{V}$. The sample cross correlations between the standardized vectors $\mathbf{S}_{\mathbf{U}}^{-1/2}\mathbf{U}$ and $\mathbf{S}_{\mathbf{V}}^{-1/2}\mathbf{V}$ are denoted as $\mathbf{C}_{\mathbf{U},\mathbf{V}} = \mathbf{S}_{\mathbf{U}}^{-1/2}\mathbf{S}_{\mathbf{U},\mathbf{V}}\mathbf{S}_{\mathbf{V}}^{-1/2}$.

We write an asymptotic covariance matrix as $\mathrm{avar}(\cdot)$; that is, if $\sqrt{n}(\mathbf{T} - \boldsymbol{\theta}) \xrightarrow{D} N(\mathbf{0}, \mathbf{A})$, then $\mathrm{avar}(\sqrt{n}\,\mathbf{T}) = \mathbf{A}$.

1

Response Envelopes

Envelopes, which were introduced by Cook et al. (2007) and developed for the multivariate linear model by Cook et al. (2010), encompass a class of methods for increasing efficiency in multivariate analyses without altering traditional objectives. They serve to reshape classical methods by exploiting response–predictor relationships that affect the accuracy of the results but are not recognized by classical methods. Multivariate data are often modeled by combining a selected structural component to be estimated with an error component to account for the remaining unexplained variation. Capturing the desired signal and only that signal in the structural component can be an elusive task with the consequence that, in an effort to avoid missing important information, there may be a tendency to overparameterize, leading to overfitting and relatively soft inferences and interpretations. Essentially a type of targeted dimension reduction that can result in substantial gains in efficiency, envelopes operate by enveloping the signal and thereby account for extraneous variation that might otherwise be present in the structural component.

In this chapter, we consider multivariate (multiresponse) linear regression allowing for the presence of "immaterial variation" (described herein) in the response vector. The possibility of such variation being present in the predictors is considered in Chapter 4, where we develop a connection with partial least squares regression. Section 1.1 contains a very brief review of the multivariate linear model, with an emphasis on aspects that will play a role in later developments. Additional background is available from Muirhead (2005). The envelope model for response reduction is introduced in Section 1.2. Introductory illustrations are given in Section 1.3 to provide intuition, to set the tone for later developments, and to provide running examples. In later sections, we discuss additional properties of the envelope model, maximum likelihood estimation, and the asymptotic variance of the envelope estimator of the coefficient matrix. Most of the technical materials used in this chapter are taken from Cook et al. (2010). Some algebraic details are presented without justification. The missing development is given extensively in Appendix A, which covers the linear algebra of envelopes.

An Introduction to Envelopes: Dimension Reduction for Efficient Estimation in Multivariate Statistics, First Edition. R. Dennis Cook.

1.1 The Multivariate Linear Model

Consider the multivariate regression of a response vector $\mathbf{Y} \in \mathbb{R}^r$ on a vector of nonstochastic predictors $\mathbf{X} \in \mathbb{R}^p$. The standard linear model for describing a sample $(\mathbf{Y}_i, \mathbf{X}_i)$ can be represented in vector form as

$$\mathbf{Y}_i = \boldsymbol{\alpha} + \boldsymbol{\beta}\mathbf{X}_i + \boldsymbol{\varepsilon}_i, \quad i = 1, \ldots, n, \tag{1.1}$$

where the predictors are centered in the sample $\sum_{i=1}^n \mathbf{X}_i = 0$, the error vectors $\boldsymbol{\varepsilon}_i \in \mathbb{R}^r$ are independently and identically distributed normal vectors with mean 0 and covariance matrix $\boldsymbol{\Sigma} > 0$, $\boldsymbol{\alpha} \in \mathbb{R}^r$ is an unknown vector of intercepts, and $\boldsymbol{\beta} \in \mathbb{R}^{r \times p}$ is an unknown matrix of regression coefficients. Centering the predictors facilitates discussion and presentation of some results, but is technically unnecessary. If $\mathbf{X}$ is stochastic, so $\mathbf{X}$ and $\mathbf{Y}$ have a joint distribution, we still condition on the observed values of $\mathbf{X}$ since the predictors are ancillary under model (1.1). The normality requirement for $\boldsymbol{\varepsilon}$ is not essential, as discussed in Section 1.9 and in later chapters.

Let $\mathbb{Y}$ denote the $n \times r$ centered matrix with rows $(\mathbf{Y}_i - \bar{\mathbf{Y}})^T$, let $\mathbb{Y}_0$ denote the $n \times r$ uncentered matrix with rows $\mathbf{Y}_i^T$, and let $\mathbb{X}$ denote the $n \times p$ matrix with rows $\mathbf{X}_i^T$, $i = 1, \ldots, n$. Also, let $\mathbf{S}_{\mathbf{Y},\mathbf{X}} = \mathbb{Y}^T\mathbb{X}/n$ and

$$\mathbf{S}_{\mathbf{X}} = \mathbb{X}^T\mathbb{X}/n = n^{-1}\sum_{i=1}^n \mathbf{X}_i\mathbf{X}_i^T.$$

Then the maximum likelihood estimator of $\boldsymbol{\alpha}$ is $\bar{\mathbf{Y}}$, and the maximum likelihood estimator of $\boldsymbol{\beta}$, which is also the ordinary least squares estimator, is

$$\mathbf{B} = \mathbb{Y}^T\mathbb{X}(\mathbb{X}^T\mathbb{X})^{-1} = \mathbb{Y}_0^T\mathbb{X}(\mathbb{X}^T\mathbb{X})^{-1} = \mathbf{S}_{\mathbf{Y},\mathbf{X}}\mathbf{S}_{\mathbf{X}}^{-1}, \tag{1.2}$$

where the second equality follows because the predictors are centered. To see this result, let $\widehat{\mathbf{Y}}_i = \bar{\mathbf{Y}} + \mathbf{B}\mathbf{X}_i$ and $\mathbf{r}_i = \mathbf{Y}_i - \widehat{\mathbf{Y}}_i$ denote the ith vectors of fitted values and residuals, $i = 1, \ldots, n$, and let $\mathbf{D} = \boldsymbol{\beta} - \mathbf{B}$. Then after substituting $\bar{\mathbf{Y}}$ for $\boldsymbol{\alpha}$, the remaining log-likelihood $L(\boldsymbol{\beta}, \boldsymbol{\Sigma})$ to be maximized can be expressed as

$$\begin{aligned}
(2/n)L(\boldsymbol{\beta}, \boldsymbol{\Sigma}) &= c - \log|\boldsymbol{\Sigma}| - n^{-1}\sum_{i=1}^n (\mathbf{Y}_i - \bar{\mathbf{Y}} - \boldsymbol{\beta}\mathbf{X}_i)^T\boldsymbol{\Sigma}^{-1}(\mathbf{Y}_i - \bar{\mathbf{Y}} - \boldsymbol{\beta}\mathbf{X}_i) \\
&= c - \log|\boldsymbol{\Sigma}| - n^{-1}\sum_{i=1}^n (\mathbf{r}_i - \mathbf{D}\mathbf{X}_i)^T\boldsymbol{\Sigma}^{-1}(\mathbf{r}_i - \mathbf{D}\mathbf{X}_i) \\
&= c - \log|\boldsymbol{\Sigma}| - n^{-1}\,\mathrm{tr}\left(\sum_{i=1}^n \mathbf{r}_i\mathbf{r}_i^T\boldsymbol{\Sigma}^{-1}\right) \\
&\quad - n^{-1}\,\mathrm{tr}\left(\mathbf{D}\sum_{i=1}^n \mathbf{X}_i\mathbf{X}_i^T\mathbf{D}^T\boldsymbol{\Sigma}^{-1}\right) \\
&= c - \log|\boldsymbol{\Sigma}| - n^{-1}\,\mathrm{tr}\left(\sum_{i=1}^n \mathbf{r}_i\mathbf{r}_i^T\boldsymbol{\Sigma}^{-1}\right) - \mathrm{tr}(\mathbf{D}\mathbf{S}_{\mathbf{X}}\mathbf{D}^T\boldsymbol{\Sigma}^{-1}),
\end{aligned}$$

where $c = -r\log(2\pi)$ and the last step follows because $\sum_{i=1}^{n} \mathbf{r}_i\mathbf{X}_i^T = 0$. Consequently, $L(\boldsymbol{\beta}, \boldsymbol{\Sigma})$ is maximized over $\boldsymbol{\beta}$ by setting $\boldsymbol{\beta} = \mathbf{B}$ so $\mathbf{D} = 0$, leaving the partially maximized log-likelihood

$$(2/n)L(\boldsymbol{\Sigma}) = -r\log(2\pi) - \log|\boldsymbol{\Sigma}| - n^{-1}\operatorname{tr}\left(\sum_{i=1}^{n} \mathbf{r}_i\mathbf{r}_i^T\boldsymbol{\Sigma}^{-1}\right).$$

It follows that the maximum likelihood estimator of $\boldsymbol{\Sigma}$ is $\mathbf{S}_{\mathrm{Y}|\mathrm{X}} := n^{-1}\sum_{i=1}^{n} \mathbf{r}_i\mathbf{r}_i^T$ and that the fully maximized log-likelihood is

$$\hat{L} = -(nr/2)\log(2\pi) - nr/2 - (n/2)\log|\mathbf{S}_{\mathrm{Y}|\mathrm{X}}|.$$

We notice from (1.2) that $\mathbf{B}$ can be constructed by doing r separate univariate linear regressions, one for each element of $\mathbf{Y}$ on $\mathbf{X}$. The coefficients from the jth regression then form the jth row of $\mathbf{B}$, $j = 1, \dots, r$. The stochastic relationships among the elements of $\mathbf{Y}$ are not used in forming these estimators. However, as will be seen later, relationships among the elements of $\mathbf{Y}$ play a central role in envelope estimation. Standard inference on $(\boldsymbol{\beta})_{jk}$, the (j,k) th element of $\boldsymbol{\beta}$, under model (1.1) is the same as inference obtained under the univariate linear regression of Y_j, the jth element of $\mathbf{Y}$, on $\mathbf{X}$. Model (1.1) becomes operational as a multivariate construction when inferring simultaneously about elements in different rows of $\boldsymbol{\beta}$ or when predicting elements of $\mathbf{Y}$ jointly.

The sample covariance matrices of $\mathbf{Y}$, $\hat{\mathbf{Y}}$, and $\mathbf{r}$ can be expressed as

$$\mathbf{S}_{\mathrm{Y}} = n^{-1}\mathbb{Y}^T\mathbb{Y} = \mathbf{S}_{\mathrm{Y}\circ\mathrm{X}} + \mathbf{S}_{\mathrm{Y}|\mathrm{X}}, \tag{1.3}$$

$$\mathbf{S}_{\mathrm{Y}\circ\mathrm{X}} = n^{-1}\mathbb{Y}^T\mathbf{P}_{\mathbb{X}}\mathbb{Y} = \mathbf{S}_{\mathrm{Y,X}}\mathbf{S}_{\mathrm{X}}^{-1}\mathbf{S}_{\mathrm{X,Y}}, \tag{1.4}$$

$$\begin{aligned} \mathbf{S}_{\mathrm{Y}|\mathrm{X}} &= n^{-1}\sum_{i=1}^{n} \mathbf{r}_i\mathbf{r}_i^T = n^{-1}\mathbb{Y}^T\mathbf{Q}_{\mathbb{X}}\mathbb{Y}, \\ &= \mathbf{S}_{\mathrm{Y}} - \mathbf{S}_{\mathrm{Y,X}}\mathbf{S}_{\mathrm{X}}^{-1}\mathbf{S}_{\mathrm{Y,X}}^T, \\ &= \mathbf{S}_{\mathrm{Y}} - \mathbf{S}_{\mathrm{Y}\circ\mathrm{X}}, \end{aligned} \tag{1.5}$$

where $\mathbf{S}_{\mathrm{X}}$ is nonstochastic, $\mathbf{P}_{\mathbb{X}} = \mathbb{X}(\mathbb{X}^T\mathbb{X})^{-1}\mathbb{X}^T$ denotes the projection onto the column space of $\mathbb{X}$, $\mathbf{Q}_{\mathbb{X}} = \mathbf{I}_n - \mathbf{P}_{\mathbb{X}}$, $\mathbf{S}_{\mathrm{Y}\circ\mathrm{X}}$ is the sample covariance matrix of the fitted vectors $\hat{\mathbf{Y}}_i$, and $\mathbf{S}_{\mathrm{Y}|\mathrm{X}}$ is the sample covariance matrix of the residuals $\mathbf{r}_i$.

We will occasionally encounter the standardized version of $\mathbf{B}$,

$$\tilde{\mathbf{B}} = \mathbf{S}_{\mathrm{Y}|\mathrm{X}}^{-1/2}\mathbf{B}\mathbf{S}_{\mathrm{X}}^{1/2}, \tag{1.6}$$

which corresponds to the estimated coefficient matrix from the ordinary least squares fit of the standardized responses $\mathbf{S}_{\mathrm{Y}|\mathrm{X}}^{-1/2}\mathbf{Y}$ on the standardized predictors $\mathbf{S}_{\mathrm{X}}^{-1/2}\mathbf{X}$.

The joint distribution of the elements of $\mathbf{B}$ can be found by using the vec operator to stack the columns of $\mathbf{B}$: $\operatorname{vec}(\mathbf{B}) = \{(\mathbb{X}^T\mathbb{X})^{-1}\mathbb{X}^T \otimes \mathbf{I}_r\}\operatorname{vec}(\mathbb{Y}_0^T)$, where $\otimes$ denotes the Kronecker product. Since $\operatorname{vec}(\mathbb{Y}_0^T)$ is normally distributed with mean $\mathbf{1}_n \otimes \boldsymbol{\alpha} + (\mathbb{X} \otimes \mathbf{I}_r)\operatorname{vec}(\boldsymbol{\beta})$ and variance $\mathbf{I}_n \otimes \boldsymbol{\Sigma}$, it follows that $\operatorname{vec}(\mathbf{B})$ is normally distributed with mean and variance

$$\mathrm{E}\{\mathrm{vec}(\mathbf{B})\} = \mathrm{vec}(\boldsymbol{\beta}), \tag{1.7}$$

$$\mathrm{var}\{\mathrm{vec}(\mathbf{B})\} = (\mathbb{X}^T\mathbb{X})^{-1} \otimes \boldsymbol{\Sigma} = n^{-1}\mathbf{S}_{\mathbf{X}}^{-1} \otimes \boldsymbol{\Sigma}. \tag{1.8}$$

The covariance matrix can also be represented in terms of $\mathbf{B}^T$ by using the $rp \times rp$ commutation matrix $\mathbf{K}_{rp}$ to convert $\mathrm{vec}(\mathbf{B})$ to $\mathrm{vec}(\mathbf{B}^T)$: $\mathrm{vec}(\mathbf{B}^T) = \mathbf{K}_{rp}\,\mathrm{vec}(\mathbf{B})$ and

$$\mathrm{var}\{\mathrm{vec}(\mathbf{B}^T)\} = n^{-1}\mathbf{K}_{rp}(\mathbf{S}_{\mathbf{X}}^{-1} \otimes \boldsymbol{\Sigma})\mathbf{K}_{rp}^T = n^{-1}\boldsymbol{\Sigma} \otimes \mathbf{S}_{\mathbf{X}}^{-1}.$$

Background on the commutation matrix, vec and related operators is available in Appendix A. The variance $\mathrm{var}\{\mathrm{vec}(\mathbf{B})\}$ is typically estimated by substituting the residual covariance matrix for $\boldsymbol{\Sigma}$,

$$\widehat{\mathrm{var}}\{\mathrm{vec}(\mathbf{B})\} = n^{-1}\mathbf{S}_{\mathbf{X}}^{-1} \otimes \mathbf{S}_{\mathbf{Y}|\mathbf{X}}. \tag{1.9}$$

Let $\mathbf{e}_i \in \mathbb{R}^r$ denote the indicator vector with a 1 in the ith position and 0s elsewhere. Then the covariance matrix for the ith row of $\mathbf{B}$ is

$$\mathrm{var}\{\mathrm{vec}(\mathbf{e}_i^T\mathbf{B})\} = (\mathbf{I}_p \otimes \mathbf{e}_i^T)\mathrm{var}\{\mathrm{vec}(\mathbf{B})\}(\mathbf{I}_p \otimes \mathbf{e}_i) = n^{-1}\mathbf{S}_{\mathbf{X}}^{-1} \otimes (\boldsymbol{\Sigma})_{ii}.$$

We see from this that the covariance matrix for the ith row of $\mathbf{B}$ is the same as that from the marginal linear regression of Y_i on $\mathbf{X}$. We refer to the estimator $(\mathbf{B})_{ij}$ divided by its standard error $\{n^{-1}(\mathbf{S}_{\mathbf{X}}^{-1})_{jj}(\mathbf{S}_{\mathbf{Y}|\mathbf{X}})_{ii}\}^{1/2}$ as a Z-score:

$$Z = \frac{(\mathbf{B})_{ij}}{\{n^{-1}(\mathbf{S}_{\mathbf{X}}^{-1})_{jj}(\mathbf{S}_{\mathbf{Y}|\mathbf{X}})_{ii}\}^{1/2}}. \tag{1.10}$$

This statistic will be used from time to time for assessing the magnitude of $(\mathbf{B})_{ij}$, sometimes converting to a p-value using the standard normal distribution.

We will occasionally encounter a conditional variate of the form $\mathbf{N} \mid \mathbf{C}^T\mathbf{N}$, where $\mathbf{N} \in \mathbb{R}^r$ is a normal vector with mean $\boldsymbol{\mu}$ and variance $\boldsymbol{\Delta}$, and $\mathbf{C} \in \mathbb{R}^{r \times q}$ is a nonstochastic matrix with $q \leq r$. The mean and variance of this conditional form are as follows:

$$\mathrm{E}(\mathbf{N} \mid \mathbf{C}^T\mathbf{N}) = \boldsymbol{\mu} + \mathbf{P}_{\mathbf{C}(\boldsymbol{\Delta})}^T(\mathbf{N} - \boldsymbol{\mu}), \tag{1.11}$$

$$\begin{aligned} \mathrm{var}(\mathbf{N} \mid \mathbf{C}^T\mathbf{N}) &= \boldsymbol{\Delta} - \boldsymbol{\Delta}\mathbf{C}(\mathbf{C}^T\boldsymbol{\Delta}\mathbf{C})^{-1}\mathbf{C}^T\boldsymbol{\Delta} \\ &= \boldsymbol{\Delta}\mathbf{Q}_{\mathbf{C}(\boldsymbol{\Delta})} \\ &= \mathbf{Q}_{\mathbf{C}(\boldsymbol{\Delta})}^T\boldsymbol{\Delta}\mathbf{Q}_{\mathbf{C}(\boldsymbol{\Delta})}. \end{aligned} \tag{1.12}$$

The usual log-likelihood ratio statistic for testing that $\boldsymbol{\beta} = 0$ is

$$\Lambda = n \log \frac{|\mathbf{S}_{\mathbf{Y}}|}{|\mathbf{S}_{\mathbf{Y}|\mathbf{X}}|}, \tag{1.13}$$

which is asymptotically distributed under the null hypothesis as a chi-square random variable with pr degrees of freedom. We will occasionally use this statistic in illustrations to assess the presence of any detectable dependence of $\mathbf{Y}$ on $\mathbf{X}$. This statistic is sometimes reported with an adjustment that is useful when n is not large relative to r and p (Muirhead 2005, Section 10.5.2).

The Fisher information $\mathbf{J}$ for $(\text{vec}^T(\boldsymbol{\beta}), \text{vech}^T(\boldsymbol{\Sigma}))^T$ in model (1.1) is

$$\mathbf{J} = \begin{pmatrix} \boldsymbol{\Sigma}_{\mathbf{X}} \otimes \boldsymbol{\Sigma}^{-1} & \mathbf{0} \\ \mathbf{0} & \frac{1}{2}\mathbf{E}_r^T(\boldsymbol{\Sigma}^{-1} \otimes \boldsymbol{\Sigma}^{-1})\mathbf{E}_r \end{pmatrix}, \tag{1.14}$$

where $\mathbf{E}_r$ is the expansion matrix that satisfies $\text{vec}(\mathbf{A}) = \mathbf{E}_r \, \text{vech}(\mathbf{A})$ for $\mathbf{A} \in \mathbb{S}^{r\times r}$, and $\boldsymbol{\Sigma}_{\mathbf{X}} = \lim_{n\to\infty} \sum_{i=1}^{n} \mathbf{X}_i\mathbf{X}_i^T/n > 0$. It follows from standard likelihood theory that $\sqrt{n}(\text{vec}(\mathbf{B}) - \text{vec}(\boldsymbol{\beta}))$ is asymptotically normal with mean 0 and variance given by the upper left block of $\mathbf{J}^{-1}$,

$$\text{avar}(\sqrt{n}\,\text{vec}(\mathbf{B})) = \boldsymbol{\Sigma}_{\mathbf{X}}^{-1} \otimes \boldsymbol{\Sigma}. \tag{1.15}$$

Asymptotic normality holds also without normal errors but with some technical conditions: if the errors have finite fourth moments and $\max_{1\le i\le n}(\mathbf{P}_{\mathbb{X}})_{ii} \to 0$, then $\sqrt{n}(\text{vec}(\mathbf{B}) - \text{vec}(\boldsymbol{\beta}))$ converges in distribution to a normal vector with mean 0 (e.g. Su and Cook 2012, Theorem 2).

1.1.1 Partitioned Models and Added Variable Plots

A subset of the predictors may occasionally be of special interest in multivariate regression. Partition $\mathbf{X}$ into two sets of predictors $\mathbf{X}_1 \in \mathbb{R}^{p_1}$ and $\mathbf{X}_2 \in \mathbb{R}^{p_1}$, $p_1 + p_2 = p$, and conformably partition the columns of $\boldsymbol{\beta}$ into $\boldsymbol{\beta}_1$ and $\boldsymbol{\beta}_2$. Then model (1.1) can be rewritten as

$$\mathbf{Y} = \boldsymbol{\mu} + \boldsymbol{\beta}_1\mathbf{X}_1 + \boldsymbol{\beta}_2\mathbf{X}_2 + \boldsymbol{\varepsilon}, \tag{1.16}$$

where $\boldsymbol{\beta}_1$ holds the coefficients of interest. We next reparameterize this model to force the new predictors to be uncorrelated in the sample and to focus attention on $\boldsymbol{\beta}_1$.

Recalling that $\bar{\mathbf{X}} = 0$, let $\widehat{\mathbf{R}}_{1|2} = \mathbf{X}_1 - \mathbf{S}_{\mathbf{X}_1,\mathbf{X}_2}\mathbf{S}_{\mathbf{X}_2}^{-1}\mathbf{X}_2$ denote a typical residual from the ordinary least squares fit of $\mathbf{X}_1$ on $\mathbf{X}_2$, and let $\boldsymbol{\beta}_2^* = \boldsymbol{\beta}_1\mathbf{S}_{\mathbf{X}_1,\mathbf{X}_2}\mathbf{S}_{\mathbf{X}_2}^{-1} + \boldsymbol{\beta}_2$. Then the partitioned model can be reexpressed as

$$\mathbf{Y} = \boldsymbol{\mu} + \boldsymbol{\beta}_1\widehat{\mathbf{R}}_{1|2} + \boldsymbol{\beta}_2^*\mathbf{X}_2 + \boldsymbol{\varepsilon}. \tag{1.17}$$

In this version of the partitioned model, the parameter vector $\boldsymbol{\beta}_1$ is the same as that in (1.16), while $\boldsymbol{\beta}_2 \neq \boldsymbol{\beta}_2^*$ unless $\mathbf{S}_{\mathbf{X}_1,\mathbf{X}_2} = 0$. The predictors – $\widehat{\mathbf{R}}_{1|2}$ and $\mathbf{X}_2$ – in (1.17) are uncorrelated in the sample $\mathbf{S}_{\widehat{\mathbf{R}}_{1|2},\mathbf{X}_2} = 0$, and consequently the maximum likelihood estimator of $\boldsymbol{\beta}_1$ is obtained by regressing $\mathbf{Y}$ on $\widehat{\mathbf{R}}_{1|2}$. The maximum likelihood estimator of $\boldsymbol{\beta}_1$ can also be obtained by regressing $\widehat{\mathbf{R}}_{\mathbf{Y}|2}$, the residuals from the regression of $\mathbf{Y}$ on $\mathbf{X}_2$, on $\widehat{\mathbf{R}}_{1|2}$. A plot of $\widehat{\mathbf{R}}_{\mathbf{Y}|2}$ versus $\widehat{\mathbf{R}}_{1|2}$ is called an added variable plot (Cook and Weisberg 1982). These plots are often used in univariate linear regression ($r = 1$) as general graphical diagnostics for visualizing how hard the data are working to fit individual coefficients.

Added variable plots and the partitioned forms of the multivariate linear model (1.16) and (1.17) will be used in this book from time to time, particularly in Chapter 3.

1.1.2 Alternative Model Forms

Because the elements of $\boldsymbol{\beta}$ and $\boldsymbol{\Sigma} > 0$ are unconstrained, model (1.1) allows each coordinate of $\mathbf{Y}$ to have a different linear regression on $\mathbf{X}$. It could be necessary in some applications to restrict the elements of $\boldsymbol{\beta}$ so that they are linked over rows and then iterations may be required because closed-form expressions for the maximum likelihood estimators of $\boldsymbol{\beta}$ and $\boldsymbol{\Sigma}$ will not normally be possible. The maximum likelihood estimator of $\boldsymbol{\beta}$ will be in the form of a weighted least squares estimator with weight matrix that depends on $\boldsymbol{\Sigma}$. For instance, suppose that we wish to restrict the coordinate regressions $\mathrm{E}(Y_j \mid \mathbf{X})$ to be parallel. This can be accomplished by requiring $\boldsymbol{\beta}$ to be a rank 1 matrix of the form $\boldsymbol{\beta} = \mathbf{1}_r\mathbf{b}^T$, where $\mathbf{b} \in \mathbb{R}^p$. Then $\boldsymbol{\beta}\mathbf{X} = \mathbf{1}_r\mathbf{X}^T\mathbf{b}$, and the model becomes $\mathbf{Y}_i = \boldsymbol{\alpha} + \mathbf{W}_i\mathbf{b} + \boldsymbol{\varepsilon}_i$, $i = 1, \dots, n$, where $\mathbf{W}_i = \mathbf{1}_r\mathbf{X}_i^T$.

For a second instance, consider a longitudinal study where n independent subjects are each observed at r times t_j, $j = 1, \dots, r$, with the elements of $\mathbf{Y}_i$ corresponding to the ordered sequence of r observations on subject i. Suppose further that $\mathrm{E}((\mathbf{Y}_i)_j \mid t_j) = \alpha + \mathbf{f}^T(t_j)\mathbf{b}$, where $\mathbf{f}(t) \in \mathbb{R}^p$ is a vector-valued user-specified function of time and $\mathbf{b} \in \mathbb{R}^p$. The elements of $\mathrm{E}(\mathbf{Y} \mid t_1, \dots, t_r)$ then correspond to points along the curve $\alpha + \mathbf{f}(t)^T\mathbf{b}$, and the linear model becomes

$$\mathbf{Y}_i = \alpha\mathbf{1} + \mathbf{W}\mathbf{b} + \boldsymbol{\varepsilon}_i,$$

where $\mathbf{W} \in \mathbb{R}^{r\times p}$ with rows $\mathbf{f}^T(t_j), j = 1, \dots, r$. Many variations of this model are available in the literature on longitudinal data (e.g. Weiss 2005). Again, the point here is that the maximum likelihood estimator of $\mathbf{b}$ will in general no longer be an ordinary least squares estimator because the parameters in the individual coordinate regressions are linked. Instead, the maximum likelihood estimator of $\mathbf{b}$ will be in the form of a weighted least squares estimator that depends on $\boldsymbol{\Sigma}$.

We will employ model (1.1) in this book, unless stated otherwise, although as discussed in Chapter 7, envelopes can apply regardless of the form of the model.

1.2 Envelope Model for Response Reduction

The motivation for response envelopes comes from allowing for the possibility that there are linear combinations of the response vector whose distribution is invariant to changes in the nonstochastic predictor vector. We refer to such linear combinations of $\mathbf{Y}$ as X-invariants. If X-invariants exist, then allowing for them in model (1.1) can result in substantial reduction in estimative variation. The linear transformation $\mathbf{G}^T\mathbf{Y}$, where $\mathbf{G} \in \mathbb{R}^{r\times q}$ with $q \leq r$, is X-invariant if and only if $\mathbf{A}^T\mathbf{G}^T\mathbf{Y}$ is X-invariant for any nonstochastic full-rank matrix $\mathbf{A} \in \mathbb{R}^{q\times q}$. Consequently, a specific transformation $\mathbf{G}$ is not identifiable but span($\mathbf{G}$) is identifiable, which leads us to consider subspaces rather than individual coordinates. The envelope model arises by parameterizing the multivariate linear model (1.1) in terms of the smallest subspace $\mathcal{E}$ of $\mathbb{R}^r$ with the properties that, for all relevant $\mathbf{x}_1$ and $\mathbf{x}_2$,

$$\text{(i) } \mathbf{Q}_{\mathcal{E}}\mathbf{Y} \mid (\mathbf{X} = \mathbf{x}_1) \sim \mathbf{Q}_{\mathcal{E}}\mathbf{Y} \mid (\mathbf{X} = \mathbf{x}_2) \quad \text{and} \quad \text{(ii) } \mathbf{P}_{\mathcal{E}}\mathbf{Y} \perp\!\!\!\perp \mathbf{Q}_{\mathcal{E}}\mathbf{Y} \mid \mathbf{X}, \tag{1.18}$$

where $\mathbf{P}_{\mathcal{E}}$ is the projection onto $\mathcal{E}$ and $\mathbf{Q}_{\mathcal{E}} = \mathbf{I}_r - \mathbf{P}_{\mathcal{E}}$. These properties serve to identify parametrically the X-invariant part of $\mathbf{Y}$, $\mathbf{Q}_{\mathcal{E}}\mathbf{Y}$. Condition (i) stipulates that the marginal distribution of $\mathbf{Q}_{\mathcal{E}}\mathbf{Y}$ must be unaffected by changes in $\mathbf{X}$. It holds if and only if span($\boldsymbol{\beta}$) $\subseteq \mathcal{E}$, because then

$$\mathbf{Q}_{\mathcal{E}}\mathbf{Y} = \mathbf{Q}_{\mathcal{E}}\boldsymbol{\alpha} + \mathbf{Q}_{\mathcal{E}}\boldsymbol{\beta}\mathbf{X} + \mathbf{Q}_{\mathcal{E}}\boldsymbol{\varepsilon} = \mathbf{Q}_{\mathcal{E}}\boldsymbol{\alpha} + \mathbf{Q}_{\mathcal{E}}\boldsymbol{\varepsilon}.$$

Condition (ii) requires that $\mathbf{Q}_{\mathcal{E}}\mathbf{Y}$ be unaffected by changes in $\mathbf{X}$ through an association with $\mathbf{P}_{\mathcal{E}}\mathbf{Y}$, and it holds if and only if

$$\text{cov}(\mathbf{P}_{\mathcal{E}}\mathbf{Y}, \mathbf{Q}_{\mathcal{E}}\mathbf{Y} \mid \mathbf{X}) = \mathbf{P}_{\mathcal{E}}\boldsymbol{\Sigma}\mathbf{Q}_{\mathcal{E}} = 0.$$

Conditions (i) and (ii) together imply that any dependence of $\mathbf{Y}$ on $\mathbf{X}$ must be concentrated in $\mathbf{P}_{\mathcal{E}}\mathbf{Y}$, the X-variant part of $\mathbf{Y}$ that is material to the regression, while $\mathbf{Q}_{\mathcal{E}}\mathbf{Y}$ is X-invariant and thus immaterial. The next two definitions, which do not require model (1.1), formalize the construction of an envelope in general.

Definition 1.1 A subspace $\mathcal{R} \subseteq \mathbb{R}^r$ is said to be a reducing subspace of $\mathbf{M} \in \mathbb{S}^{r\times r}$ if $\mathcal{R}$ decomposes $\mathbf{M}$ as $\mathbf{M} = \mathbf{P}_{\mathcal{R}}\mathbf{M}\mathbf{P}_{\mathcal{R}} + \mathbf{Q}_{\mathcal{R}}\mathbf{M}\mathbf{Q}_{\mathcal{R}}$. If $\mathcal{R}$ is a reducing subspace of $\mathbf{M}$, we say that $\mathcal{R}$ reduces $\mathbf{M}$.

This definition of a reducing subspace is equivalent to that used by Cook et al. (2010), as described in Appendix A. It is common in the literature on invariant subspaces and functional analysis, although the underlying notion of "reduction" differs from the usual understanding in statistics. Here it is used to guarantee condition (ii) of (1.18) since the decomposition holds if and only if $\mathbf{P}_{\mathcal{R}}\mathbf{M}\mathbf{Q}_{\mathcal{R}} = 0$. The following definition makes use of reducing subspaces.

Definition 1.2 Let $\mathbf{M} \in \mathbb{S}^{r\times r}$ and let $\mathcal{B} \subseteq \text{span}(\mathbf{M})$. Then the $\mathbf{M}$-envelope of $\mathcal{B}$, denoted by $\mathcal{E}_{\mathbf{M}}(\mathcal{B})$, is the intersection of all reducing subspaces of $\mathbf{M}$ that contain $\mathcal{B}$.

Definition 1.2 is the definition of an envelope introduced by Cook et al. (2007, 2010). It formalizes the construction of the smallest subspace that satisfies conditions (1.18) by asking for the intersection of all subspaces that envelop span($\boldsymbol{\beta}$), and thus that satisfy condition (i), from among those that satisfy condition (ii). Further discussion of this definition is available in Section A.2. We will often identify the subspace $\mathcal{B}$ as the span of a specified matrix U: $\mathcal{B} = \text{span}(\mathbf{U})$. To avoid proliferation of notation in such cases, we will occasionally use the matrix as the argument to $\mathcal{E}_{\mathbf{M}}$: $\mathcal{E}_{\mathbf{M}}(\mathbf{U}) := \mathcal{E}_{\mathbf{M}}(\text{span}(\mathbf{U}))$.

The actual construction of $\mathcal{E}_{\mathbf{M}}(\mathcal{B})$ can be characterized in terms of $\mathcal{B}$ and the spectral structure of $\mathbf{M}$. Suppose that $\mathbf{M}$ has $q \leq r$ distinct eigenvalues with projections onto the corresponding eigenspaces represented by $\mathbf{P}_k$, $k = 1, \ldots, q$. Then, as shown by Cook et al. (2010, see also Proposition A.2),

$$\mathcal{E}_{\boldsymbol{\Sigma}}(\mathcal{B}) = \sum_{k=1}^{q} \mathbf{P}_k \mathcal{B}. \tag{1.19}$$

This result shows that $\mathcal{B}$ is enveloped by using the eigenspaces of $\mathbf{M}$. Figure 1.1 gives a schematic representation of (1.19) when $r = 3$. The three axes of the plot represent three eigenvectors $\boldsymbol{\ell}_k$ of $\mathbf{M}$ with corresponding eigenvalues φ_k, $k = 1, 2, 3$. Three different possibilities for a one-dimensional $\mathcal{B}$ are represented by the spanning vectors $\boldsymbol{\beta}_j$, so $\mathcal{B}_j = \text{span}(\boldsymbol{\beta}_j)$, $j = 1, 2, 3$. The ordering of the eigenvalues is irrelevant for this discussion, but equality among the eigenvalues is relevant.

- β_1: Since $\boldsymbol{\beta}_1$ aligns with $\boldsymbol{\ell}_1$, $\text{span}(\boldsymbol{\beta}_1) = \text{span}(\boldsymbol{\ell}_1)$ and thus $\mathcal{E}_{\boldsymbol{\Sigma}}(\mathcal{B}_1) = \mathcal{B}_1$ regardless of any equalities among the eigenvalues.
- β_2: Since $\boldsymbol{\beta}_2$ falls in the $(\boldsymbol{\ell}_1, \boldsymbol{\ell}_2)$-plane, $\mathcal{E}_{\boldsymbol{\Sigma}}(\mathcal{B}_2)$ depends on the corresponding eigenvalues. If $\varphi_1 = \varphi_2$, then regardless of the value for φ_3, $\mathcal{E}_{\boldsymbol{\Sigma}}(\mathcal{B}_2) = \mathcal{B}_2$. But if $\varphi_1 \neq \varphi_2$, then $\mathcal{E}_{\boldsymbol{\Sigma}}(\mathcal{B}_2) = \text{span}(\boldsymbol{\ell}_1, \boldsymbol{\ell}_2)$.
- β_3: In the final case represented in Figure 1.1, $\boldsymbol{\beta}_3$ does not fall in any subspace spanned by a subset of the eigenvectors represented in the figure, except for the full space $\mathbb{R}^3 = \text{span}(\boldsymbol{\ell}_1, \boldsymbol{\ell}_2, \boldsymbol{\ell}_3)$. Consequently, if the eigenvalues are distinct, then $\mathcal{E}_{\boldsymbol{\Sigma}}(\mathcal{B}_3) = \mathbb{R}^3$. If $\varphi_1 \neq \varphi_2 = \varphi_3$, then there are two eigenspaces $\text{span}(\boldsymbol{\ell}_1)$ and $\text{span}(\boldsymbol{\ell}_2, \boldsymbol{\ell}_3)$, and $\mathcal{E}_{\boldsymbol{\Sigma}}(\mathcal{B}_3) = \text{span}(\boldsymbol{\ell}_1, \mathbf{P}_{2,3}\beta_3)$, where $\mathbf{P}_{2,3}$ is the projection onto $\text{span}(\boldsymbol{\ell}_2, \boldsymbol{\ell}_3)$. If $\varphi_1 = \varphi_2 = \varphi_3$, then there is only one eigenspace and $\mathcal{E}_{\boldsymbol{\Sigma}}(\mathcal{B}_3) = \mathcal{B}_3$.

Back to model (1.1), let $\mathcal{B} = \text{span}(\boldsymbol{\beta})$. The response projection $\mathbf{P}_{\mathcal{E}}$ is then defined as the projection onto $\mathcal{E}_{\boldsymbol{\Sigma}}(\mathcal{B})$, which by construction is the smallest reducing subspace of $\boldsymbol{\Sigma}$ that contains $\mathcal{B}$. Model (1.1) can be parameterized in terms of $\mathcal{E}_{\boldsymbol{\Sigma}}(\mathcal{B})$ by using a basis. Let $u = \dim(\mathcal{E}_{\boldsymbol{\Sigma}}(\mathcal{B}))$, and let $(\boldsymbol{\Gamma}, \boldsymbol{\Gamma}_0) \in \mathbb{R}^{r \times r}$ be an orthogonal matrix with $\boldsymbol{\Gamma} \in \mathbb{R}^{r \times u}$, $\text{span}(\boldsymbol{\Gamma}) = \mathcal{E}_{\boldsymbol{\Sigma}}(\mathcal{B})$, and $\text{span}(\boldsymbol{\Gamma}_0) = \mathcal{E}_{\boldsymbol{\Sigma}}^{\perp}(\mathcal{B})$. Then the envelope model can be written as

$$\mathbf{Y} = \boldsymbol{\alpha} + \boldsymbol{\Gamma}\boldsymbol{\eta}\mathbf{X} + \boldsymbol{\varepsilon}, \text{ with } \boldsymbol{\Sigma} = \boldsymbol{\Gamma}\boldsymbol{\Omega}\boldsymbol{\Gamma}^T + \boldsymbol{\Gamma}_0\boldsymbol{\Omega}_0\boldsymbol{\Gamma}_0^T. \tag{1.20}$$

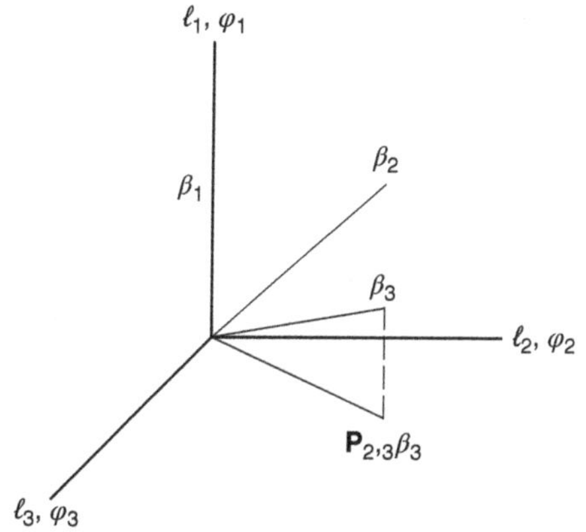

Figure 1.1 Schematic representation of an envelope.

The coefficient vector $\boldsymbol{\beta} = \boldsymbol{\Gamma}\boldsymbol{\eta}$, where $\boldsymbol{\eta} \in \mathbb{R}^{u \times p}$ carries the coordinates of $\boldsymbol{\beta}$ relative to the basis matrix $\boldsymbol{\Gamma}$, and $\boldsymbol{\Omega} \in \mathbb{R}^{u \times u}$ and $\boldsymbol{\Omega}_0 \in \mathbb{R}^{(r-u) \times (r-u)}$ are positive definite matrices. To see how this model reflects the X-invariant part of $\mathbf{Y}$, multiply both sides by $\boldsymbol{\Gamma}_0^T$ to get $\boldsymbol{\Gamma}_0^T\mathbf{Y} = \boldsymbol{\Gamma}_0^T\boldsymbol{\alpha} + \boldsymbol{\Gamma}_0^T\boldsymbol{\varepsilon}$, where $\text{var}(\boldsymbol{\Gamma}_0^T\boldsymbol{\varepsilon}) = \boldsymbol{\Omega}_0$. We see from this representation that the marginal distribution of $\boldsymbol{\Gamma}_0^T\mathbf{Y}$ does not depend on $\mathbf{X}$, so condition (i) of (1.18) is met. Further, since $\text{cov}(\boldsymbol{\Gamma}^T\mathbf{Y}, \boldsymbol{\Gamma}_0^T\mathbf{Y} \mid \mathbf{X}) = \boldsymbol{\Gamma}^T\boldsymbol{\Sigma}\boldsymbol{\Gamma}_0 = 0$, we see also that condition (ii) of (1.18) holds.

The parameterization of (1.20) is intended to facilitate the estimation of $\boldsymbol{\beta}$ and $\boldsymbol{\Sigma}$. We normally do not attempt to reify or infer about the constituent parameters $\boldsymbol{\Gamma}$, $\boldsymbol{\eta}$, $\boldsymbol{\Omega}$, and $\boldsymbol{\Omega}_0$. The values of $\boldsymbol{\eta}$, $\boldsymbol{\Omega}$, and $\boldsymbol{\Omega}_0$ depend on the choice of $\boldsymbol{\Gamma}$ to represent $\mathcal{E}_{\boldsymbol{\Sigma}}(\mathcal{B})$, and so may be difficult to interpret even if there is a desire to do so, while $\boldsymbol{\beta}$ and $\boldsymbol{\Sigma}$ depend on $\mathcal{E}_{\boldsymbol{\Sigma}}(\mathcal{B})$ but not on the particular basis. The basis matrix $\boldsymbol{\Gamma}$ is not identifiable in model (1.20) since, for any orthogonal matrix $\mathbf{O} \in \mathbb{R}^{u \times u}$, replacing $\boldsymbol{\Gamma}$ with $\boldsymbol{\Gamma}\mathbf{O}$ leads to an equivalent model. For example, $\boldsymbol{\beta} = \boldsymbol{\Gamma}\boldsymbol{\eta} = (\boldsymbol{\Gamma}\mathbf{O})(\mathbf{O}^T\boldsymbol{\eta})$, so replacing $\boldsymbol{\Gamma}$ with $\boldsymbol{\Gamma}\mathbf{O}$ and $\boldsymbol{\eta}$ with $\mathbf{O}^T\boldsymbol{\eta}$ leads to an equivalent expression for $\boldsymbol{\beta}$. However, the envelope $\mathcal{E}_{\boldsymbol{\Sigma}}(\mathcal{B}) = \text{span}(\boldsymbol{\Gamma})$ is identifiable, which allows us to estimate $\boldsymbol{\beta}$. Additional properties of model (1.20) are discussed in Section 1.4.2.

Separate application of either of the two conditions in (1.18) does not necessarily lead to progress, but see the discussion of reduced-rank regression in Section 9.2.1. A subspace that reduces $\boldsymbol{\Sigma}$ may have no useful connection with $\mathcal{B}$. If $\mathcal{B}$ is a proper subspace of $\mathbb{R}^r$, then there can be many subspaces $\mathcal{S}$ that contain $\mathcal{B}$, while any particular subspace may not reduce $\boldsymbol{\Sigma}$. It is typically when the two conditions of (1.18) are used in concert that we obtain substantial gains in efficiency. The X-invariant component $\boldsymbol{\Gamma}_0^T\mathbf{Y}$ induces extraneous variation into the maximum likelihood estimator of $\boldsymbol{\beta}$ under model (1.1). Envelopes distinguish between $\mathbf{P}_{\mathcal{E}}\mathbf{Y}$ and $\mathbf{Q}_{\mathcal{E}}\mathbf{Y}$ in the estimation process and, as discussed in Section 1.5, the envelope estimator of $\boldsymbol{\beta}$ may then be more efficient than $\mathbf{B}$, as the variation from the X-invariant part of $\mathbf{Y}$ is effectively removed. Given the dimension u of the envelope and assuming normal errors, the envelope model can be fitted by maximum likelihood. The details of this are described later in Section 1.5; for now, we note that the maximum likelihood estimator of $\boldsymbol{\beta}$ is just the projection of $\mathbf{B}$ onto the estimated envelope, $\widehat{\boldsymbol{\beta}} = \mathbf{P}_{\widehat{\mathcal{E}}}\mathbf{B}$. The asymptotic covariance matrix of $\widehat{\boldsymbol{\beta}}$ is described in Section 1.6.

Definitions 1.1 and 1.2 are quite general – they do not require normality or even a linear model. Consequently, the envelope is still well defined if the linear model holds, but the errors are not normal. In that case, the maximum likelihood estimators under normality are still $\sqrt{n}$-consistent (Section 1.9), and the variance of $\widehat{\boldsymbol{\beta}}$ can be estimated by using the residual bootstrap (Section 1.11).

Figure 1.1 is applicable in the context of model (1.20) with $r = 3$ responses, $p = 1$ predictor, the $\boldsymbol{\ell}_j$s and φ_js interpreted as the eigenvectors and values of $\boldsymbol{\Sigma}$, and the $\boldsymbol{\beta}_j$s interpreted as three different possibilities for the coefficient vector. Those three configurations represent three ways in which an X-invariant can occur. Conversely, if part of $\mathbf{Y}$ is X-invariant, then a setting similar to those represented in the figure must hold.

The dimension $u \in \{0, 1, \dots, r\}$ can be selected based on any of the standard methods, including information criteria such as Akaike's information criterion (AIC) and Bayes information criterion (BIC), likelihood ratio testing, cross-validation, or a hold out sample, as described later in Section 1.10. If $u = 0$, then $\boldsymbol{\beta} = 0$ and there is no dependence of $\mathbf{Y}$ on $\mathbf{X}$, a setting that is often tested in practice. On the other extreme, if $u = r$, then $\mathcal{E}_{\boldsymbol{\Sigma}}(\mathcal{B}) = \mathbb{R}^r$, the envelope model reduces to the usual model (1.1), and the distribution of every linear combination of $\mathbf{Y}$ responds to changes in $\mathbf{X}$, a conclusion that might also be useful in some applications.

1.3 Illustrations

Before expanding our discussion of the envelope model (1.20), we give in this section illustrations to provide intuition about the working mechanism of envelope estimation and its potential advantages over the standard estimators. These will also be used as running examples to illustrate various phases of an envelope analysis as they are developed in later sections.

1.3.1 A Schematic Example

Consider comparing the means $\boldsymbol{\mu}_1$ and $\boldsymbol{\mu}_2$ of two bivariate normal populations, $N_2(\boldsymbol{\mu}_1, \boldsymbol{\Sigma})$ and $N_2(\boldsymbol{\mu}_2, \boldsymbol{\Sigma})$. This problem can be cast into the framework of model (1.1) by letting $\mathbf{Y} = (Y_1, Y_2)^T$ denote the bivariate response vector and letting X be an indicator variable taking value $X = 0$ in population 1 and $X = 1$ in population 2. Parameterizing so that $\boldsymbol{\alpha} = \boldsymbol{\mu}_1$ is the mean of the first population and $\boldsymbol{\beta} = \boldsymbol{\mu}_2 - \boldsymbol{\mu}_1$ is the mean difference, we have the multivariate linear model

$$\mathbf{Y} = \boldsymbol{\mu}_1 + (\boldsymbol{\mu}_2 - \boldsymbol{\mu}_1)X + \boldsymbol{\varepsilon} = \boldsymbol{\alpha} + \boldsymbol{\beta}X + \boldsymbol{\varepsilon}.$$

Starting with a sample, $(\mathbf{Y}_i, X_i)$, $i = 1, \ldots, n$, having independent response vectors, the standard estimator of $\boldsymbol{\beta}$ is just the difference in the sample means for the two populations $\mathbf{B} = \bar{\mathbf{Y}}_2 - \bar{\mathbf{Y}}_1$ and the corresponding estimator of $\boldsymbol{\Sigma}$ is the pooled intra-sample covariance matrix. As in the general multivariate normal model (1.1), this estimator of $\boldsymbol{\beta}$ does not make use of the dependence between the responses and is equivalent to performing two univariate regressions of Y_j on X, $j = 1, 2$. Bringing envelopes into play can lead to a very different estimator of $\boldsymbol{\beta}$, one with substantially smaller variation than the maximum likelihood estimator $\mathbf{B}$ under model (1.1). In the remainder of this section, we illustrate one way this can happen.

A standard analysis will likely be sufficient when the populations are well separated, as illustrated in Figure 1.2, even with a larger number of responses. The two ellipses in that figure represent the two normal populations indicated by the predictor values $X = 0$ and $X = 1$. However, a standard analysis may not do as well when the populations are close, as illustrated in Figure 1.3. Without loss of generality, we set $\boldsymbol{\mu}_1 + \boldsymbol{\mu}_2 = 0$, so that $\mathcal{B} = \text{span}(\boldsymbol{\mu}_1, \boldsymbol{\mu}_2) = \text{span}(\boldsymbol{\mu}_1 - \boldsymbol{\mu}_2) = \text{span}(\boldsymbol{\beta})$. The left panel represents a standard likelihood analysis under model (1.1). For inference on β_2, the second element of $\boldsymbol{\beta}$, a standard analysis directly projects the data "y" onto the Y_2 axis following the dashed line marked A and then proceeds with inference based on the resulting univariate samples. The curves along the horizontal axis in the left panel stand for the projected distributions from the two populations. A standard analysis might involve constructing a two-sample t-test on samples drawn from these populations. There is a considerable overlap between the two projected distributions, so it may take a large sample size to infer that $\beta_2 \neq 0$ in a standard analysis. This illustration is based on β_2 to facilitate visualization; the same conclusions could be reached using a different linear combination of the elements of $\boldsymbol{\beta}$.

The two populations depicted in Figure 1.3 have the same eigenvectors, as they must because they have equal covariance matrices. We saw in (1.19) that an envelope can be constructed as $\mathcal{E}_{\boldsymbol{\Sigma}}(\mathcal{B}) = \sum_{i=1}^{q} \mathbf{P}_i \mathcal{B}$. There are only two

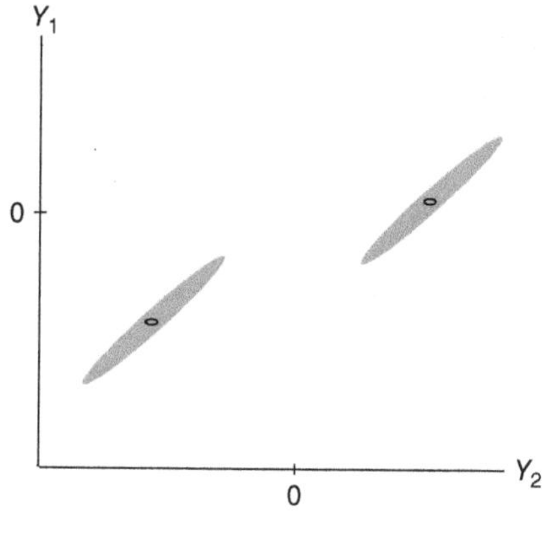

Figure 1.2 Graphical illustration of a relatively uncomplicated scenario. The axes are centered responses.

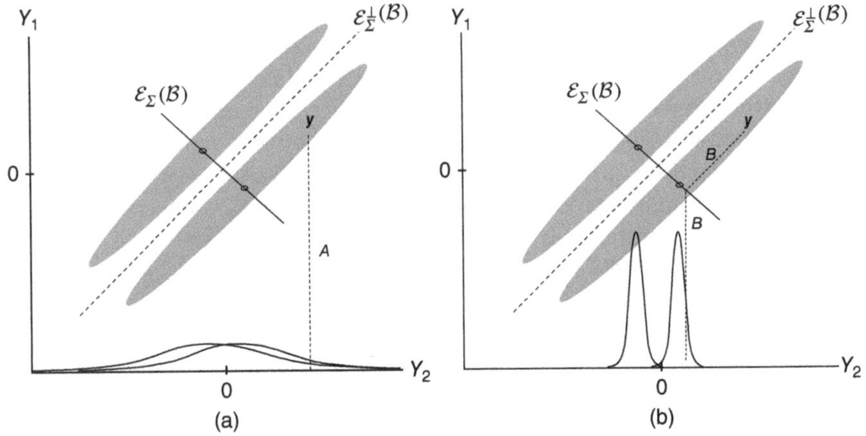

Figure 1.3 Graphical illustration envelope estimation. (a) Standard analysis; (b) envelope analysis.

eigenspaces in Figure 1.3, so the envelope must have dimension $u = 0$, $u = 1$, or $u = 2$. Since $\mathcal{B}$ equals the second eigenspace, we have $u = 1$ and $\mathcal{B} = \mathcal{E}_{\mathbf{\Sigma}}(\mathcal{B})$, although in higher dimensions only $\mathcal{B} \subseteq \mathcal{E}_{\mathbf{\Sigma}}(\mathcal{B})$ is required. Accordingly, the eigenvector corresponding to the smaller eigenvalue is marked by the notation for the envelope $\mathcal{E}_{\mathbf{\Sigma}}(\mathcal{B})$, and the first eigenvector is marked by notation for the orthogonal complement of the envelope $\mathcal{E}_{\mathbf{\Sigma}}^{\perp}(\mathcal{B})$. Condition (i) of (1.18) holds because the two populations have equal distributions when projected onto $\mathcal{E}_{\mathbf{\Sigma}}^{\perp}(\mathcal{B})$; that is, $\mathbf{Q}_{\mathcal{E}}\mathbf{Y} \mid (X = 0) \sim \mathbf{Q}_{\mathcal{E}}\mathbf{Y} \mid (X = 1)$, where $\mathcal{E}_{\mathbf{\Sigma}}(\mathcal{B})$ is shortened to $\mathcal{E}$ for use in subscripts. Since the populations are normal and $\mathcal{E}_{\mathbf{\Sigma}}(\mathcal{B})$ and $\mathcal{E}_{\mathbf{\Sigma}}^{\perp}(\mathcal{B})$ are spanned by eigenvectors, we also have condition (ii) of (1.18), $\mathbf{P}_{\mathcal{E}}\mathbf{Y} \perp\!\!\!\perp \mathbf{Q}_{\mathcal{E}}\mathbf{Y} \mid X$.

The maximum likelihood envelope estimator of β_2 (see Section 1.5) can be formed by first projecting the data onto $\mathcal{E}_{\mathbf{\Sigma}}(\mathcal{B})$ to remove the X-invariant component $\mathbf{Q}_{\mathcal{E}}\mathbf{Y}$ and extract the X-variant part of the response $\mathbf{P}_{\mathcal{E}}\mathbf{Y}$, and then projecting onto the horizontal axis, as illustrated by the paths marked "B" in Figure 1.3b. Figure 1.3b also shows the resulting projected distributions corresponding to the projected distributions in Figure 1.3a. Now the projected distributions are well separated, and the envelope estimator of β_2 should be much more efficient than the estimator represented in Figure 1.3a. The estimative gain represented by passing from the standard estimator in Figure 1.3a to the envelope estimator in Figure 1.3b is reflected by the difference in the magnitude of the variances $\text{var}(\mathbf{\Gamma}^T\mathbf{Y} \mid X) = \mathbf{\Omega}$ and $\text{var}(\mathbf{\Gamma}_0^T\mathbf{Y} \mid X) = \mathbf{\Omega}_0$. The distributions in Figure 1.3 were constructed so that $\mathbf{\Omega} \ll \mathbf{\Omega}_0$, and consequently we anticipate substantial estimative gains.

There are two noteworthy caveats to our discussion of Figure 1.3. First, because the response is two-dimensional, the only nontrivial envelope must have dimension 1 and thus must align with one of the eigenvectors of $\mathbf{\Sigma}$. This is not required in higher dimensions, as shown by (1.19). Second, the illustration is based on the true envelope. The envelope needs to be estimated in practice, which has the effect of causing it to wobble in Figure 1.3b. That wobble produces increased variation in the envelope distributions of Figure 1.3b. As shown in Proposition 1.1, regardless of the degree of wobble, the asymptotic variance of the envelope estimator will not exceed the asymptotic variance of the standard estimator, which is reflected by the distributions in Figure 1.3a (Cook et al. 2010). In other words, the envelope estimator will always do at least as well as the standard maximum likelihood estimator asymptotically.

1.3.2 Compound Symmetry

Suppose that the elements $(\mathbf{\Sigma})_{ij}$ of the residual covariance matrix for envelope model (1.20) have the form $(\mathbf{\Sigma})_{ii} = \sigma^2$ and $(\mathbf{\Sigma})_{ij} = \sigma^2 \rho$ for $-(r-1)^{-1} < \rho < 1$ and $i \neq j$. Then $\mathbf{\Sigma}$ can be decomposed as

$$\begin{aligned}\mathbf{\Sigma}\sigma^{-2} &= \rho \mathbf{1}_r \mathbf{1}_r^T + (1-\rho)\mathbf{I}_r \\ &= (1+(r-1)\rho)\mathbf{P}_{\mathbf{1}_r} + (1-\rho)\mathbf{Q}_{\mathbf{1}_r}.\end{aligned}$$

In consequence, $\mathbf{\Sigma}$ has two eigenspaces $\text{span}(\mathbf{1}_r)$ and $\text{span}^{\perp}(\mathbf{1}_r)$ with corresponding eigenvalues $\varphi_1 = \sigma^2(1+(r-1)\rho)$ and $\varphi_2 = \sigma^2(1-\rho)$. It follows from (1.19) that the envelope is the sum of the projections of $\mathcal{B}$ onto $\text{span}(\mathbf{1}_r)$ and $\text{span}^{\perp}(\mathbf{1}_r)$: $\mathcal{E}_{\mathbf{\Sigma}}(\mathcal{B}) = \text{span}(\mathbf{P}_{\mathbf{1}_r}\boldsymbol{\beta}) + \text{span}(\mathbf{Q}_{\mathbf{1}_r}\boldsymbol{\beta})$, which has dimension at most $1 + \min(p, r-1)$.

Compound symmetry, which is a common name given to the covariance structure considered in this section, is frequently used for analysis of longitudinal data where the response vector is made up of repeated measures on a single individual over time (Weiss 2005), as in the cattle weights illustration introduced in Section 1.3.4.

1.3.3 Wheat Protein: Introductory Illustration

The classic wheat protein data (Fearn 1983) contain measurements on protein content and the logarithms of near-infrared reflectance at six wavelengths across the range 1680–2310 nm measured on each of $n = 50$ samples of ground wheat. To illustrate the ideas associated with Figure 1.3 in data analysis, we use $r = 2$ wavelengths as responses $\mathbf{Y} = (Y_1, Y_2)^T$ and convert the continuous measure of protein content into a categorical predictor X, indicating low and high protein (24 and 26 samples, respectively).

The mean difference $\boldsymbol{\mu}_2 - \boldsymbol{\mu}_1$ corresponds to the parameter vector $\boldsymbol{\beta}$ in model (1.1), with X representing a binary indicator: $X = 0$ for high protein and $X = 1$ for low protein wheat. For these data, which are shown in Figure 1.4, $\mathbf{B}^T = (7.5, -2.1)$ with standard errors (SEs) 8.6 and 9.5 (Figure 1.3a). There is

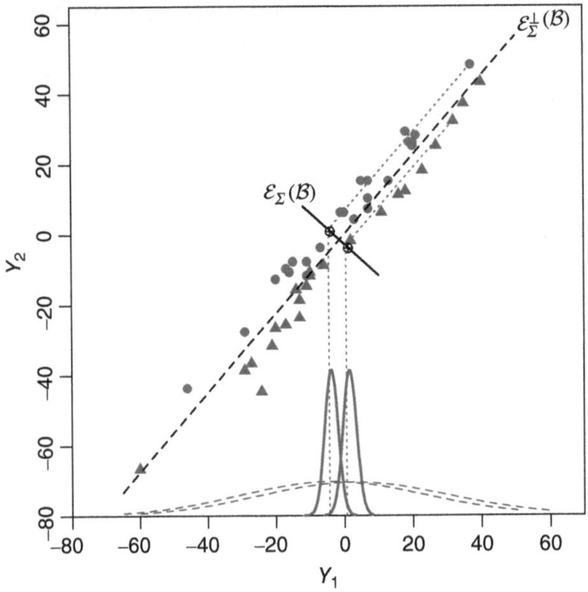

Figure 1.4 Wheat protein data with the estimated envelope superimposed.

no indication from these marginal results that $\mathbf{Y}$ depends on X, while the likelihood ratio test statistic (1.13) has the value 27.5 on 2 degrees of freedom and thus indicates otherwise. The simultaneous occurrence in a standard analysis of relatively small Z-scores and a relatively large likelihood ratio statistic is often a clue that an envelope analysis will be advantageous, although these conditions are certainly not necessary.

The envelope estimate is $\widehat{\boldsymbol{\beta}}^T = (5.1, -4.7)$ with asymptotic standard errors of 0.51 and 0.46 (Figure 1.3b). To more fully appreciate the magnitude of this drop in standard errors, we would need for a standard analysis a sample size of $n \sim 20\,000$ to reduce the standard error from 9.4 to 0.46. Figure 1.4 shows the projected distributions of the data like those from Figure 1.3.

1.3.4 Cattle Weights: Initial Fit

Roundworm is an intestinal parasite that takes nutrients from animals and lowers their resistance to disease, resulting in relatively low body weights. Kenward (1987) presented data[1] from a randomized experiment to compare two treatments for controlling roundworm in cattle. The two treatments were randomly assigned to 60 cows, with 30 cows per treatment. Their weights in

1 The data were obtained from the website for Weiss (2005) http://rem.ph.ucla.edu/~rob/mld/data.html.

kilograms were recorded at the beginning of the study prior to treatment application and at 10 times during the study corresponding to weeks $2, 4, 6, \ldots, 18$ and 19; that is, at two-week intervals except the last that was over a one-week interval. The goal of the experiment was to determine if the treatments have a differential effect on weight and, if so, to estimate the time point at which the difference was first manifested. The experimenter evidently anticipated a lag between the time of treatment application and manifestation of their effects.

The basal weights, which were taken before treatment, could be used as covariates in an effort to control some of the variations between animals. In this illustration, we take a simpler approach and neglect the basal weights in our analysis. (Basal weights are included as covariates in an illustrative analysis in Chapter 3.) Since treatments were assigned randomly this should not introduce a bias, although we might incur variation that could otherwise be removed. Figure 1.5a shows a profile plot, also called a parallel coordinate plot, of animal weight against week, beginning with the first posttreatment measurements in week 2. Each line traces the weight of an animal over time, with the two treatments represented by different colors. Profile plots are useful for seeing clusters over time, but for these data no clusters or treatment effects seem apparent visually. Figure 1.5b shows a profile plot of mean weight by week and treatment on the same scale as Figure 1.5a. The two mean profiles are roughly parallel until week 12, cross between weeks 12 and 14, and then cross back between weeks 18 and 19. Judged against the variation of the individual profiles in Figure 1.5a and recognizing that the intra-animal weights are surely positively correlated, these visual representations hint that there is no detectable difference between the two treatments from these data.

Let $\mathbf{Y}_i \in \mathbb{R}^{10}$, $i = 1, \ldots, 60$, be the vector of weight measurements for each animal over time, and let $X_i = -0.5$ or 0.5 indicate the two treatments, so $\sum_{i=1}^{60} X_i = 0$. Our interest lies in the regression coefficient $\boldsymbol{\beta}$ from the multivariate linear regression $\mathbf{Y} = \boldsymbol{\alpha} + \boldsymbol{\beta} X + \boldsymbol{\varepsilon}$, where it is still assumed that $\boldsymbol{\varepsilon} \sim N_{10}(0, \boldsymbol{\Sigma})$. Recall that $\mathbf{B}$ denotes the ordinary least squares estimator of $\boldsymbol{\beta}$, which is also the maximum likelihood estimator under normality. For these data $\mathbf{B}$ is simply the difference in the mean vectors for the two treatments,

$$\mathbf{B} = \bar{\mathbf{Y}} \mid (X = 0.5) - \bar{\mathbf{Y}} \mid (X = -0.5),$$

and the corresponding estimator of $\boldsymbol{\alpha}$ is the grand mean $\bar{\mathbf{Y}}$. The plot in Figure 1.5b can also be described as a profile plot of the fitted weights $\hat{\mathbf{Y}} = \bar{\mathbf{Y}} + \mathbf{B}X$. The coefficient estimates $\mathbf{B}$ and their Z-scores (1.10) are given in the second and third columns of Table 1.1. This fit supports the visual impression from Figure 1.5, as the largest absolute Z-score is only 1.30. On the other hand, the likelihood ratio statistic (1.13) for the hypothesis $\boldsymbol{\beta} = 0$ has the value 26.9 on 10 degrees of freedom, which suggests differences somewhere that are not manifested by the marginal Z-scores of Table 1.1. As mentioned in the wheat protein illustration, the simultaneous occurrence of relatively small

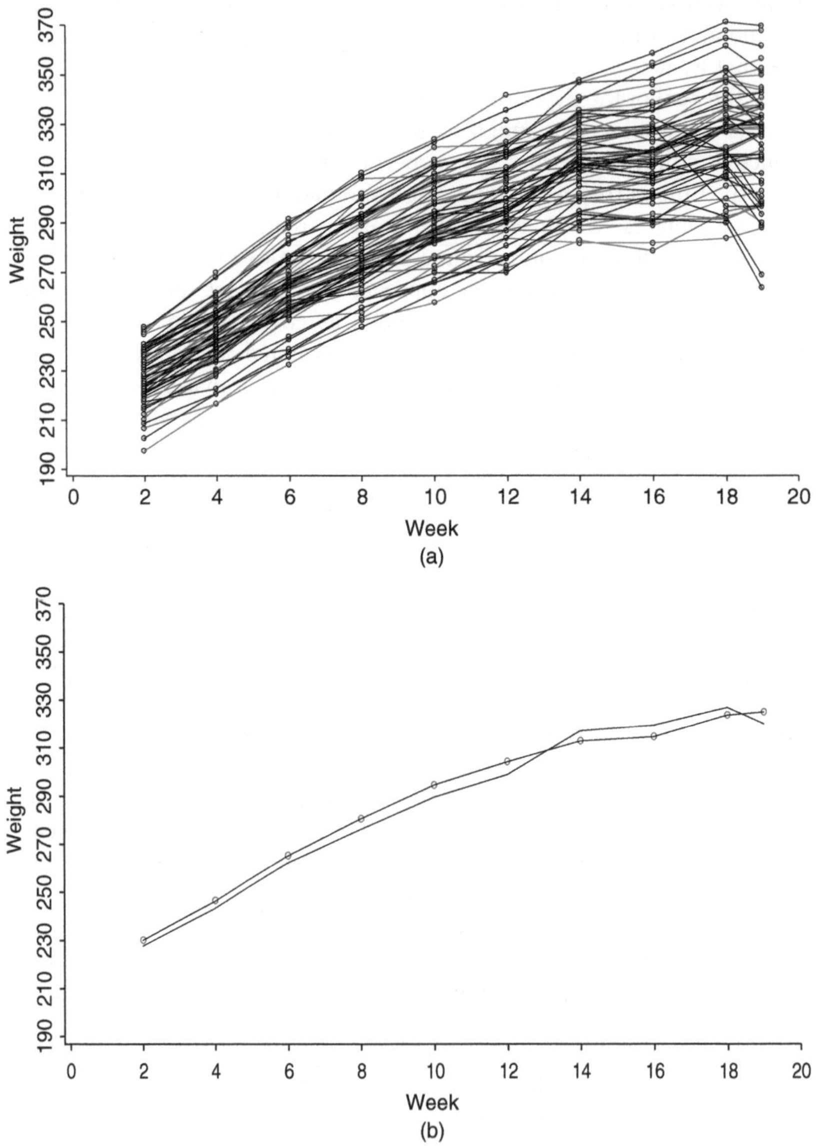

Figure 1.5 Cattle data: profile plots of weight for 60 cows over 10 time periods. Colors designate the two treatments. (a) Profile plots for individual cows by treatment. (b) Profile plots of average weight by treatment.

Table 1.1 Cattle data: Estimated coefficients for the standard model and envelope model with $u = 1$.

Week	B	Z-score	$\widehat{\beta}$	Z-score	R
2	2.43	0.83	−2.17	−2.48	3.32
4	3.33	1.05	−0.48	−0.65	4.27
6	3.13	0.89	0.88	1.23	4.89
8	4.73	1.22	2.38	2.82	4.56
10	4.73	1.14	2.89	4.14	5.94
12	5.50	1.30	5.40	5.30	4.15
14	−4.80	−1.11	−5.09	−5.55	4.69
16	−4.53	−0.97	−4.62	−5.36	5.40
18	−2.87	−0.54	−3.67	−4.06	5.86
19	5.00	0.86	4.21	4.92	6.78

The Z-score is the estimate divided by its standard error (se(·)) as defined in (1.10), and $R = \text{se}\{(\mathbf{B})_j\}/\text{se}\{(\widehat{\beta})_j\}$.

Z-scores and a relatively large likelihood ratio statistic is often a good clue that an envelope analysis will offer advantages, although again these conditions are not necessary. We next turn to an envelope analysis.

Using the LRT(.05) method of Section 1.10, we first estimated that $u = 1$, so it was inferred that the treatment difference is manifested in only one linear combination $\boldsymbol{\Gamma}^T\mathbf{Y}$ of the response vector. Then with $u = 1$, we estimated a basis $\widehat{\boldsymbol{\Gamma}}$ for $\mathcal{E}_{\boldsymbol{\Sigma}}(\mathcal{B})$ using the method of Section 1.5. The fourth and fifth columns of Table 1.1 give the envelope coefficient estimates $\widehat{\boldsymbol{\beta}} = \mathbf{P}_{\widehat{\boldsymbol{\Gamma}}}\mathbf{B}$ and their Z-scores determined by using the standard errors described in (1.33). Making use of a Bonferroni inequality at level 0.05 to adjust for multiple testing, differential treatment effects are indicated by at least week 10 and persist thereafter. The final column of Table 1.1 gives the ratios of the standard errors for the elements of $\mathbf{B}$ to those of $\widehat{\boldsymbol{\beta}}$. We see from these results that the standard errors for the elements of $\mathbf{B}$ were 3.32–6.78 times those of $\widehat{\boldsymbol{\beta}}$. These ratios represent a substantial increase in efficiency of the envelope estimator relative to the usual maximum likelihood estimator. To more fully appreciate the magnitude of this drop in standard errors, we would need $n \sim 1500$ for a standard analysis to achieve the standard errors from an envelope analysis with $n = 60$.

Figure 1.6 shows a profile plot on the same scale as Figure 1.5 of the fitted weights, $\widehat{\mathbf{Y}} = \widehat{\boldsymbol{\alpha}} + \widehat{\boldsymbol{\Gamma}}\widehat{\eta}X$ from the envelope model with $u = 1$. Comparing to Figure 1.5b, the corresponding profile plot of the fitted weights from the standard model, we see that the two plots agree well after about week 10, but before then the fitted weights for the two treatments are closer from the

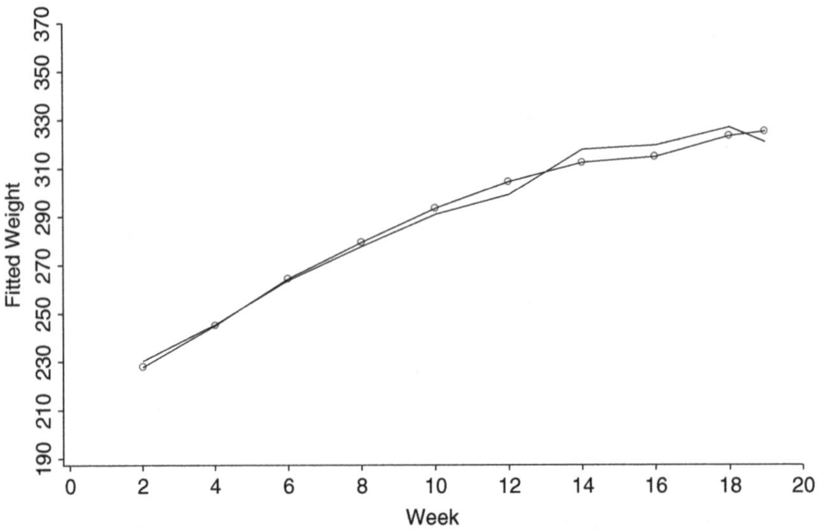

Figure 1.6 Cattle data: profile plot of fitted weights from the envelope model with $u = 1$. Colors designate the two treatments.

envelope fit than from the standard fit. This supports the prior notion of a lag between treatment application and effect.

This example has 10 responses, so an overall graphical construction like that shown in Figure 1.3 is not possible. However, marked plots of the weights for week w versus the weights for week $w + 2$ provide some intuition on the structure of the data. For instance, the plot for week 12 weight versus week 14 weight given in Figure 1.7 suggests a clear difference in weights and exhibits the

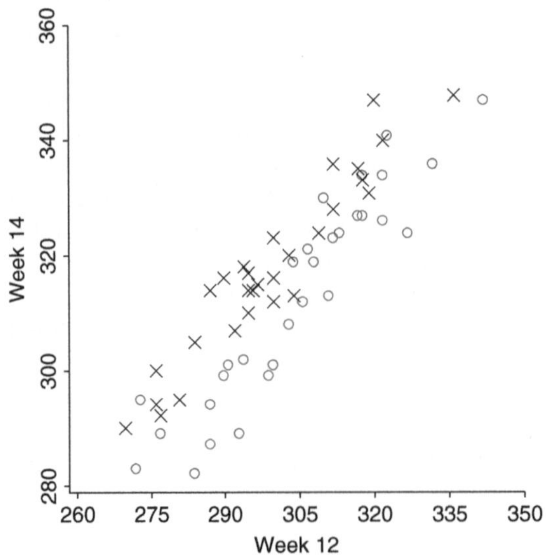

Figure 1.7 Cattle data: scatterplot of week 12 weight versus week 14 weight with treatments marked.

envelope structure represented schematically in Figure 1.3. A formal envelope analysis of these bivariate data indicates that $u = 1$ (Section 1.10), leading to envelope standard errors that are about 5.7 times smaller than those from the standard model and highly significant coefficient estimates. This level of reduction is commensurate with those shown previously in Table 1.1.

1.4 More on the Envelope Model

In this section, we revisit and expand upon the envelope model (1.20) introduced in Section 1.2.

1.4.1 Relationship with Sufficiency

Suppose we know a subspace $\mathcal{S}$ that contains $\mathcal{B}$, so that condition (i) of (1.18) holds; we are not yet enforcing condition (ii) of (1.18). The estimator of $\boldsymbol{\beta}$ based on $\mathbf{P}_{\mathcal{S}}\mathbf{Y}$ is $\mathbf{P}_{\mathcal{S}}\mathbf{B} = \mathbf{S}_{\mathbf{P}_{\mathcal{S}}\mathbf{Y},\mathbf{X}}\mathbf{S}_{\mathbf{X}}^{-1}$. If $\mathbf{P}_{\mathcal{S}}\mathbf{Y}$ is to capture the part of $\mathbf{Y}$ that is material to the estimation of $\boldsymbol{\beta}$, then it is reasonable to require that the conditional distribution of $\text{vec}(\mathbf{B}) \mid \text{vec}(\mathbf{P}_{\mathcal{S}}\mathbf{B})$ not depend on $\boldsymbol{\beta}$.

Since $\text{vec}(\mathbf{B}) \mid \text{vec}(\mathbf{P}_{\mathcal{S}}\mathbf{B})$ is normally distributed, we need to consider only its mean and variance. Its variance does not depend on $\boldsymbol{\beta}$, but its mean may do so. Writing $\text{vec}(\mathbf{P}_{\mathcal{S}}\mathbf{B}) = (\mathbf{I}_p \otimes \mathbf{P}_{\mathcal{S}})\text{vec}(\mathbf{B})$, we have from (1.11) that

$$\text{E}\{\text{vec}(\mathbf{B}) \mid \text{vec}(\mathbf{P}_{\mathcal{S}}\mathbf{B})\} = \text{vec}(\boldsymbol{\beta}) + \mathbf{P}^T_{\mathbf{I}_p\otimes\mathbf{P}_{\mathcal{S}}(\mathbf{V})}\{\text{vec}(\mathbf{B}) - \text{vec}(\boldsymbol{\beta})\},$$

where $\mathbf{V} = \text{var}\{\text{vec}(\mathbf{B})\}$ as given in (1.8). Because $\mathbf{P}_{\mathbf{I}_p\otimes\mathbf{P}_{\mathcal{S}}(\mathbf{V})}(\mathbf{I}_p \otimes \mathbf{P}_{\mathcal{S}}) = \mathbf{I}_p \otimes \mathbf{P}_{\mathcal{S}}$ and $\mathcal{B} \subseteq \mathcal{S}$, we have

$$\begin{aligned}\text{E}\{\text{vec}(\mathbf{B}) \mid \text{vec}(\mathbf{P}_{\mathcal{S}}\mathbf{B})\} &= (\mathbf{I}_p \otimes \mathbf{P}_{\mathcal{S}} + \mathbf{I}_p \otimes \mathbf{Q}_{\mathcal{S}})\text{E}\{\text{vec}(\mathbf{B}) \mid \text{vec}(\mathbf{P}_{\mathcal{S}}\mathbf{B})\} \\ &= \text{vec}(\boldsymbol{\beta}) + (\mathbf{I}_p \otimes \mathbf{P}_{\mathcal{S}})\mathbf{P}^T_{\mathbf{I}_p\otimes\mathbf{P}_{\mathcal{S}}(\mathbf{V})}\{\text{vec}(\mathbf{B}) - \text{vec}(\boldsymbol{\beta})\} \\ &\quad +(\mathbf{I}_p \otimes \mathbf{Q}_{\mathcal{S}})\mathbf{P}^T_{\mathbf{I}_p\otimes\mathbf{P}_{\mathcal{S}}(\mathbf{V})}\{\text{vec}(\mathbf{B}) - \text{vec}(\boldsymbol{\beta})\} \\ &= \text{vec}(\mathbf{P}_{\mathcal{S}}\mathbf{B}) + (\mathbf{I}_p \otimes \mathbf{Q}_{\mathcal{S}})\mathbf{P}^T_{\mathbf{I}_p\otimes\mathbf{P}_{\mathcal{S}}(\mathbf{V})}\{\text{vec}(\mathbf{B}) - \text{vec}(\boldsymbol{\beta})\}.\end{aligned}$$

The second term on the right-hand side will be free of $\boldsymbol{\beta}$ if and only if $\mathbf{P}_{\mathbf{I}_p\otimes\mathbf{P}_{\mathcal{S}}(\mathbf{V})}(\mathbf{I}_p \otimes \mathbf{Q}_{\mathcal{S}}) = 0$; that is, if and only if $\mathbf{I}_p \otimes \mathbf{P}_{\mathcal{S}}$ is orthogonal to $\mathbf{I}_p \otimes \mathbf{Q}_{\mathcal{S}}$ in the $\mathbf{V}$ inner product. This holds if and only if $\mathbf{P}_{\mathcal{S}}\boldsymbol{\Sigma}\mathbf{Q}_{\mathcal{S}} = 0$, so $\mathcal{S}$ must reduce $\boldsymbol{\Sigma}$. Consequently, we are led back to condition (ii) of (1.18).

The general point here is that conditions (i) and (ii) of (1.18) are designed to insure that $\mathbf{P}_{\mathcal{S}}\mathbf{B}$ is sufficient for $\boldsymbol{\beta}$ when $\mathcal{S}$ is known; that is, $\mathbf{P}_{\mathcal{S}}\mathbf{Y}$ is a sufficient reduction of $\mathbf{Y}$. The context here is distinct from classical sufficiency because $\mathcal{S}$ is unknown and must be estimated.

1.4.2 Parameter Count

The total number of real parameters required for the envelope model (1.20) is

$$\begin{aligned}N_u &= r + pu + u(r-u) + u(u+1)/2 + (r-u)(r-u+1)/2 \qquad (1.21) \\ &= r + pu + r(r+1)/2.\end{aligned}$$

This count arises as follows. The first addend on the right-hand side of (1.21) corresponds to the intercept $\boldsymbol{\alpha} \in \mathbb{R}^r$, and the second addend corresponds to the unconstrained coordinate matrix $\boldsymbol{\eta} \in \mathbb{R}^{u \times p}$. The last two addends correspond to $\boldsymbol{\Omega}$ and $\boldsymbol{\Omega}_0$. Their parameter counts arise because, for any integer $k > 0$, it takes $k(k+1)/2$ numbers to specify a nonsingular $k \times k$ symmetric matrix. The third addend $u(r-u)$, which gives the parameter count for $\boldsymbol{\Gamma}$, arises as follows. As mentioned previously, the matrix $\boldsymbol{\Gamma}$ is not identifiable since, for any orthogonal matrix $\mathbf{O} \in \mathbb{R}^{u \times u}$, replacing $\boldsymbol{\Gamma}$ with $\boldsymbol{\Gamma}\mathbf{O}$ results in an equivalent model. However, $\text{span}(\boldsymbol{\Gamma}) = \mathcal{E}_{\boldsymbol{\Sigma}}(\mathcal{B})$ is identifiable. The parameter space for $\mathcal{E}_{\boldsymbol{\Sigma}}(\mathcal{B})$ is the Grassmannian $\mathcal{G}(u, r)$ of dimension u in $\mathbb{R}^r$: $\mathcal{G}(u, r)$ is the collection of all u-dimensional subspaces of $\mathbb{R}^r$. From basic properties of Grassmann manifolds it is known that $u(r-u)$ real parameters are needed to specify an element of $\mathcal{G}(u, r)$ uniquely. Once $\mathcal{E}_{\boldsymbol{\Sigma}}(\mathcal{B})$ is determined, so is its orthogonal complement $\text{span}(\boldsymbol{\Gamma}_0)$, and no additional free parameters are required. The difference between the total parameter count for the standard model (1.1) with $r = u$ and the envelope model (1.20) with $u < r$ is therefore $p(r-u)$. Further discussion of Grassmannians is available from Edelman et al. (1998).

The number of real parameters $u(r-u)$ needed to identify a subspace can be seen intuitively as follows. ru parameters are needed to specify uniquely an unconstrained matrix $\boldsymbol{\Gamma} \in \mathbb{R}^{r \times u}$. But when dealing with subspaces, $\text{span}(\boldsymbol{\Gamma}) = \text{span}(\boldsymbol{\Gamma}\mathbf{A})$ for any nonsingular $\mathbf{A} \in \mathbb{R}^{u \times u}$. Since it requires u^2 parameters to determine an $\mathbf{A}$, specifying a subspace takes $ru - u^2 = u(r-u)$ parameters.

1.4.3 Potential Gains

A specific envelope model is identified by the value of u. All envelope models are nested within the standard model (1.1), but two envelope models with different values of u are not necessarily nested. To see this, notice that the number of real parameters needed to specify an element of $\mathcal{G}(1, r)$ is the same as that for $\mathcal{G}(r-1, r)$. If $u = r$, then $\mathcal{E}_{\boldsymbol{\Sigma}}(\mathcal{B}) = \mathbb{R}^r$, the envelope model degenerates to the standard model (1.1) and enveloping offers no gain. If $r \leq p$ and $\dim(\mathcal{B}) = r$, then again the envelope model reduces to the standard model. However, if (i) $r > p$ or (ii) if $\dim(\mathcal{B}) < r \leq p$, then efficiency gains are possible. These gains can arise in two ways. The first is through the parameter count. Since the number of parameters in the envelope model is less than that in the standard model, we can expect some efficiency gains from parsimony. But the second source is where we have the potential to realize massive gains.

To explore envelope gains and in anticipation of maximum likelihood estimation discussed in Section 1.5, consider the maximum likelihood estimator of $\boldsymbol{\beta}$ when $\mathcal{E}_{\boldsymbol{\Sigma}}(\mathcal{B})$ is known and represented by a semi-orthogonal basis matrix $\boldsymbol{\Gamma}$. In this case, the maximum likelihood estimator of $\boldsymbol{\beta}$ under the envelope model

(1.20) is again the projection of the standard estimator onto $\mathcal{E}_{\boldsymbol{\Sigma}}(\mathcal{B})$: $\widehat{\boldsymbol{\beta}}_{\boldsymbol{\Gamma}} = \mathbf{P}_{\boldsymbol{\Gamma}}\mathbf{B}$. Using (1.8), the variance of this estimator is

$$\begin{aligned}\mathrm{var}\{\mathrm{vec}(\widehat{\boldsymbol{\beta}}_{\boldsymbol{\Gamma}})\} &= \mathrm{var}\{\mathrm{vec}(\mathbf{P}_{\boldsymbol{\Gamma}}\mathbf{B})\}\\ &= (\mathbf{I}_p \otimes \mathbf{P}_{\boldsymbol{\Gamma}})\mathrm{var}\{\mathrm{vec}(\mathbf{B})\}(\mathbf{I}_p \otimes \mathbf{P}_{\boldsymbol{\Gamma}})\\ &= n^{-1}(\mathbf{I}_p \otimes \mathbf{P}_{\boldsymbol{\Gamma}})(\mathbf{S}_{\mathbf{X}}^{-1} \otimes \boldsymbol{\Sigma})(\mathbf{I}_p \otimes \mathbf{P}_{\boldsymbol{\Gamma}})\\ &= n^{-1}\mathbf{S}_{\mathbf{X}}^{-1} \otimes \boldsymbol{\Gamma}\boldsymbol{\Omega}\boldsymbol{\Gamma}^T. \end{aligned} \tag{1.22}$$

Comparing this to the variance of the standard estimator $\mathbf{B}$,

$$\mathrm{var}\{\mathrm{vec}(\mathbf{B})\} - \mathrm{var}\{\mathrm{vec}(\widehat{\boldsymbol{\beta}}_{\boldsymbol{\Gamma}})\} = n^{-1}\mathbf{S}_{\mathbf{X}}^{-1} \otimes \boldsymbol{\Gamma}_0\boldsymbol{\Omega}_0\boldsymbol{\Gamma}_0^T \geq 0. \tag{1.23}$$

From this we conclude that if the variance of the X-invariant part of $\mathbf{Y}$, $\mathrm{var}(\boldsymbol{\Gamma}_0^T\mathbf{Y} \mid \mathbf{X}) = \boldsymbol{\Omega}_0$ is large relative to the variance of the X-variant part of $\mathbf{Y}$, $\mathrm{var}(\boldsymbol{\Gamma}^T\mathbf{Y}) = \boldsymbol{\Omega}$, then the gain from the envelope model can be substantial. Using the spectral norm $\|\cdot\|$ as a measure of overall size, the envelope model may be particularly advantageous when $\|\boldsymbol{\Gamma}\boldsymbol{\Omega}\boldsymbol{\Gamma}^T\| = \|\boldsymbol{\Omega}\| \ll \|\boldsymbol{\Gamma}_0\boldsymbol{\Omega}_0\boldsymbol{\Gamma}_0^T\| = \|\boldsymbol{\Omega}_0\|$. The envelope $\mathcal{E}_{\boldsymbol{\Sigma}}(\mathcal{B})$ will normally be estimated in practice, and these results will then be mitigated by the variability in its estimator. Nevertheless, experience has shown that they are a useful indicator of the kinds of regressions in which envelopes offer substantial gains.

1.5 Maximum Likelihood Estimation

1.5.1 Derivation

In this section, we discuss the derivation of the maximum likelihood estimators of the parameters in envelope model (1.20), assuming that the dimension $u = \dim(\mathcal{E}_{\boldsymbol{\Sigma}}(\mathcal{B}))$ of the envelope is known.

The log-likelihood $L_u(\boldsymbol{\alpha}, \boldsymbol{\eta}, \mathcal{E}_{\boldsymbol{\Sigma}}(\mathcal{B}), \boldsymbol{\Omega}, \boldsymbol{\Omega}_0)$ under model (1.20) with known u can be expressed as

$$\begin{aligned}L_u &= -(nr/2)\log(2\pi) - (n/2)\log|\boldsymbol{\Gamma}\boldsymbol{\Omega}\boldsymbol{\Gamma}^T + \boldsymbol{\Gamma}_0\boldsymbol{\Omega}_0\boldsymbol{\Gamma}_0^T|\\ &\quad -(1/2)\sum_{i=1}^{n}(\mathbf{Y}_i - \boldsymbol{\alpha} - \boldsymbol{\Gamma}\boldsymbol{\eta}\mathbf{X}_i)^T(\boldsymbol{\Gamma}\boldsymbol{\Omega}\boldsymbol{\Gamma}^T + \boldsymbol{\Gamma}_0\boldsymbol{\Omega}_0\boldsymbol{\Gamma}_0^T)^{-1}(\mathbf{Y}_i - \boldsymbol{\alpha} - \boldsymbol{\Gamma}\boldsymbol{\eta}\mathbf{X}_i)\\ &= -(nr/2)\log(2\pi) - (n/2)\log|\boldsymbol{\Omega}| - (n/2)\log|\boldsymbol{\Omega}_0|\\ &\quad -(1/2)\sum_{i=1}^{n}(\mathbf{Y}_i - \boldsymbol{\alpha} - \boldsymbol{\Gamma}\boldsymbol{\eta}\mathbf{X}_i)^T(\boldsymbol{\Gamma}\boldsymbol{\Omega}^{-1}\boldsymbol{\Gamma}^T + \boldsymbol{\Gamma}_0\boldsymbol{\Omega}_0^{-1}\boldsymbol{\Gamma}_0^T)(\mathbf{Y}_i - \boldsymbol{\alpha} - \boldsymbol{\Gamma}\boldsymbol{\eta}\mathbf{X}_i),\end{aligned}$$

where the second equality arises by applying the third and fourth conclusions of Corollary A.1. Also, while the likelihood function depends on $\mathcal{E}_{\boldsymbol{\Sigma}}(\mathcal{B})$, we have written it in terms of the semi-orthogonal basis matrix $\boldsymbol{\Gamma}$ to facilitate the derivation. The original derivation by Cook et al. (2010) uses projections instead of bases.

The maximum likelihood estimator of $\boldsymbol{\alpha}$ is $\hat{\boldsymbol{\alpha}} = \bar{\mathbf{Y}}$ because the predictors are centered. Substituting this into the likelihood function and then decomposing $\mathbf{Y}_i - \bar{\mathbf{Y}} = \mathbf{P}_{\boldsymbol{\Gamma}}(\mathbf{Y}_i - \bar{\mathbf{Y}}) + \mathbf{Q}_{\boldsymbol{\Gamma}}(\mathbf{Y}_i - \bar{\mathbf{Y}})$ and simplifying, we arrive at the first partially maximized log-likelihood,

$$L_u^{(1)}(\boldsymbol{\eta}, \mathcal{E}_{\boldsymbol{\Sigma}}(\mathcal{B}), \boldsymbol{\Omega}, \boldsymbol{\Omega}_0) = -(nr/2)\log(2\pi) + L_u^{(11)}(\boldsymbol{\eta}, \mathcal{E}_{\boldsymbol{\Sigma}}(\mathcal{B}), \boldsymbol{\Omega}) + L_u^{(12)}(\mathcal{E}_{\boldsymbol{\Sigma}}(\mathcal{B}), \boldsymbol{\Omega}_0),$$

where

$$L_u^{(11)} = -(n/2)\log|\boldsymbol{\Omega}| - (1/2)\sum_{i=1}^{n}\{\boldsymbol{\Gamma}^T(\mathbf{Y}_i - \bar{\mathbf{Y}}) - \boldsymbol{\eta}\mathbf{X}_i\}^T\boldsymbol{\Omega}^{-1}\{\boldsymbol{\Gamma}^T(\mathbf{Y}_i - \bar{\mathbf{Y}}) - \boldsymbol{\eta}\mathbf{X}_i\},$$

$$L_u^{(12)} = -(n/2)\log|\boldsymbol{\Omega}_0| - (1/2)\sum_{i=1}^{n}(\mathbf{Y}_i - \bar{\mathbf{Y}})^T\boldsymbol{\Gamma}_0\boldsymbol{\Omega}_0^{-1}\boldsymbol{\Gamma}_0^T(\mathbf{Y}_i - \bar{\mathbf{Y}}).$$

Holding $\boldsymbol{\Gamma}$ fixed, $L_u^{(11)}$ can be seen as the log-likelihood for the multivariate regression of $\boldsymbol{\Gamma}^T(\mathbf{Y}_i - \bar{\mathbf{Y}})$ on $\mathbf{X}_i$, and thus $L_u^{(11)}$ is maximized over $\boldsymbol{\eta}$ at the value $\boldsymbol{\eta} = \boldsymbol{\Gamma}^T\mathbf{B}$. Substituting this into $L_u^{(11)}$ and simplifying, we obtain a partially maximized version of $L_u^{(11)}$

$$L_u^{(21)}(\mathcal{E}_{\boldsymbol{\Sigma}}(\mathcal{B}), \boldsymbol{\Omega}) = -(n/2)\log|\boldsymbol{\Omega}| - (1/2)\sum_{i=1}^{n}(\boldsymbol{\Gamma}^T\mathbf{r}_i)^T\boldsymbol{\Omega}^{-1}\boldsymbol{\Gamma}^T\mathbf{r}_i,$$

where, as defined in Section 1.1, $\mathbf{r}_i$ is the ith residual vector from the fit of the standard model (1.1). From this it follows immediately that, still with $\boldsymbol{\Gamma}$ fixed, $L_u^{(21)}$ is maximized over $\boldsymbol{\Omega}$ at $\boldsymbol{\Omega} = \boldsymbol{\Gamma}^T\mathbf{S}_{\mathbf{Y}|\mathbf{X}}\boldsymbol{\Gamma}$. Consequently, we arrive at the third partially maximized log-likelihood $L_u^{(31)}(\mathcal{E}_{\boldsymbol{\Sigma}}(\mathcal{B})) = -(n/2)\log|\boldsymbol{\Gamma}^T\mathbf{S}_{\mathbf{Y}|\mathbf{X}}\boldsymbol{\Gamma}| - nu/2$. By similar reasoning, the value of $\boldsymbol{\Omega}_0$ that maximizes $L_u^{(21)}(\mathcal{E}_{\boldsymbol{\Sigma}}(\mathcal{B}), \boldsymbol{\Omega}_0)$ is $\boldsymbol{\Omega}_0 = \boldsymbol{\Gamma}_0^T\mathbf{S}_{\mathbf{Y}}\boldsymbol{\Gamma}_0$. This leads to the maximization of $L_u^{(22)}(\mathcal{E}_{\boldsymbol{\Sigma}}(\mathcal{B})) = -(n/2)\log|\boldsymbol{\Gamma}_0^T\mathbf{S}_{\mathbf{Y}}\boldsymbol{\Gamma}_0| - n(r-u)/2$.

Combining the above steps, we arrive at the partially maximized form

$$L_u^{(2)}(\mathcal{E}_{\boldsymbol{\Sigma}}(\mathcal{B})) = -(nr/2)\log(2\pi) - nr/2 - (n/2)\log|\boldsymbol{\Gamma}^T\mathbf{S}_{\mathbf{Y}|\mathbf{X}}\boldsymbol{\Gamma}| - (n/2)\log|\boldsymbol{\Gamma}_0^T\mathbf{S}_{\mathbf{Y}}\boldsymbol{\Gamma}_0|.$$

Next, since $\log|\boldsymbol{\Gamma}_0^T\mathbf{S}_{\mathbf{Y}}\boldsymbol{\Gamma}_0| = \log|\mathbf{S}_{\mathbf{Y}}| + \log|\boldsymbol{\Gamma}^T\mathbf{S}_{\mathbf{Y}}^{-1}\boldsymbol{\Gamma}|$ (Lemma A.13), we can express $L_u^{(2)}(\mathcal{E}_{\boldsymbol{\Sigma}}(\mathcal{B}))$ as a function of $\boldsymbol{\Gamma}$ alone:

$$L_u^{(2)}(\mathcal{E}_{\boldsymbol{\Sigma}}(\mathcal{B})) = -(nr/2)\log(2\pi) - nr/2 - (n/2)\log|\mathbf{S}_{\mathbf{Y}}| - (n/2)\log|\boldsymbol{\Gamma}^T\mathbf{S}_{\mathbf{Y}|\mathbf{X}}\boldsymbol{\Gamma}| - (n/2)\log|\boldsymbol{\Gamma}^T\mathbf{S}_{\mathbf{Y}}^{-1}\boldsymbol{\Gamma}|. \tag{1.24}$$

Summarizing, the maximum likelihood estimators $\widehat{\mathcal{E}}_{\boldsymbol{\Sigma}}(\mathcal{B})$ of $\mathcal{E}_{\boldsymbol{\Sigma}}(\mathcal{B})$ and of the remaining parameters are determined as

$$\widehat{\mathcal{E}}_{\boldsymbol{\Sigma}}(\mathcal{B}) = \text{span}\{\arg\min_{\mathbf{G}} (\log |\mathbf{G}^T \mathbf{S}_{\mathbf{Y}|\mathbf{X}} \mathbf{G}| + \log |\mathbf{G}^T \mathbf{S}_{\mathbf{Y}}^{-1} \mathbf{G}|)\}, \tag{1.25}$$

$$\begin{aligned}
\widehat{\eta} &= \widehat{\boldsymbol{\Gamma}}^T \mathbf{B}, \\
\widehat{\boldsymbol{\beta}} &= \widehat{\boldsymbol{\Gamma}}\,\widehat{\eta} = \mathbf{P}_{\widehat{\mathcal{E}}} \mathbf{B}, \qquad (1.26)\\
\widehat{\boldsymbol{\Omega}} &= \widehat{\boldsymbol{\Gamma}}^T \mathbf{S}_{\mathbf{Y}|\mathbf{X}} \widehat{\boldsymbol{\Gamma}}, \\
\widehat{\boldsymbol{\Omega}}_0 &= \widehat{\boldsymbol{\Gamma}}_0^T \mathbf{S}_{\mathbf{Y}} \widehat{\boldsymbol{\Gamma}}_0, \\
\widehat{\boldsymbol{\Sigma}} &= \widehat{\boldsymbol{\Gamma}} \widehat{\boldsymbol{\Omega}} \widehat{\boldsymbol{\Gamma}}^T + \widehat{\boldsymbol{\Gamma}}_0 \widehat{\boldsymbol{\Omega}}_0 \widehat{\boldsymbol{\Gamma}}_0^T,
\end{aligned}$$

where $\min_{\mathbf{G}}$ is over all semi-orthogonal matrices $\mathbf{G} \in \mathbb{R}^{r \times u}$, $\widehat{\boldsymbol{\Gamma}}$ is any semi-orthogonal basis matrix for $\widehat{\mathcal{E}}_{\boldsymbol{\Sigma}}(\mathcal{B})$, and $\widehat{\boldsymbol{\Gamma}}_0$ is any semi-orthogonal basis matrix for the orthogonal complement of $\widehat{\mathcal{E}}_{\boldsymbol{\Sigma}}(\mathcal{B})$. The fully maximized log-likelihood for fixed u is then

$$\begin{aligned}
\widehat{L}_u = &-(nr/2)\log(2\pi) - nr/2 - (n/2)\log|\mathbf{S}_{\mathbf{Y}}| \\
&-(n/2)\log|\widehat{\boldsymbol{\Gamma}}^T \mathbf{S}_{\mathbf{Y}|\mathbf{X}} \widehat{\boldsymbol{\Gamma}}| - (n/2)\log|\widehat{\boldsymbol{\Gamma}}^T \mathbf{S}_{\mathbf{Y}}^{-1} \widehat{\boldsymbol{\Gamma}}|.
\end{aligned} \tag{1.27}$$

The optimization required in (1.25) can be sensitive to starting values. Discussions of starting values and optimization algorithms are given in Chapter 6 within a broader context. The estimators $\widehat{\boldsymbol{\beta}}$ and $\widehat{\boldsymbol{\Sigma}}$ are invariant to the selection of basis $\widehat{\boldsymbol{\Gamma}}$ and thus are unique, but the remaining estimators $\widehat{\eta}$, $\widehat{\boldsymbol{\Omega}}$ and $\widehat{\boldsymbol{\Omega}}_0$ are basis dependent and thus not unique.

1.5.2 Cattle Weights: Variation of the *X*-Variant Parts of Y

The envelope estimates in Table 1.3.4 were constructed based on (1.25) and the corresponding estimates of the other parameters. We commented at the end of Section 1.4.2 that an envelope analysis will be particularly advantageous when $||\boldsymbol{\Omega}||$ is substantially smaller than $||\boldsymbol{\Omega}_0||$. From the fit of the envelope model with $u = 1$ to the cattle data, we obtain $||\widehat{\boldsymbol{\Omega}}|| = 27.8$ and $||\widehat{\boldsymbol{\Omega}}_0|| = 2351.5$, which conforms qualitatively to the gains in Table 1.1 and the structure shown in Figure 1.7.

In an envelope model with $u = 1$, $\boldsymbol{\beta}$ is constrained to lie in a one-dimensional reducing subspace of $\boldsymbol{\Sigma}$. If the eigenvalues of $\boldsymbol{\Sigma}$ are all distinct, then $\boldsymbol{\beta}$ will align with an eigenvector of $\boldsymbol{\Sigma}$, as described previously in (1.19). The estimate $\widehat{\boldsymbol{\Gamma}}$ will not correspond to an eigenvector of $\mathbf{S}_{\mathbf{Y}|\mathbf{X}}$, but informally we might expect $\widehat{\boldsymbol{\Gamma}}$ to fall close to an eigenvector because of the result shown in (1.19). We can often obtain intuition about the fit of an envelope model by considering the correlations between $\widehat{\boldsymbol{\Gamma}}^T\mathbf{Y}$ and the elements of $\mathbf{L}^T\mathbf{Y}$, where the columns of $\mathbf{L} = (\boldsymbol{\ell}_1, \dots, \boldsymbol{\ell}_r)$ are the ordered eigenvectors of $\mathbf{S}_{\mathbf{Y}|\mathbf{X}}$. For the cattle data, the two largest absolute correlations 0.97 and 0.48 occur with the fourth and sixth eigenvectors, giving $\boldsymbol{\ell}_4^T\mathbf{Y}$ and $\boldsymbol{\ell}_6^T\mathbf{Y}$. Evidently, the treatment differences are largely associated with the fourth eigenvector of $\mathbf{S}_{\mathbf{Y}|\mathbf{X}}$.

1.5.3 Insights into $\widehat{\mathcal{E}}_{\boldsymbol{\Sigma}}(\mathcal{B})$

In this section, we provide different forms for $\widehat{\mathcal{E}}_{\boldsymbol{\Sigma}}(\mathcal{B})$ that might aid intuition. First, $\widehat{\mathcal{E}}_{\boldsymbol{\Sigma}}(\mathcal{B})$ can be reexpressed as

$$\widehat{\mathcal{E}}_{\boldsymbol{\Sigma}}(\mathcal{B}) = \arg\min_{\mathcal{S}\in\mathcal{G}(u,r)} \{\log|\mathbf{P}_{\mathcal{S}}\mathbf{S}_{\mathbf{Y}|\mathbf{X}}\mathbf{P}_{\mathcal{S}}|_0 + \log|\mathbf{Q}_{\mathcal{S}}\mathbf{S}_{\mathbf{Y}}\mathbf{Q}_{\mathcal{S}}|_0\}, \tag{1.28}$$

where $|\cdot|_0$ denotes the product of the nonzero eigenvalues of the matrix argument, and minimization $\min_{\mathcal{S}\in\mathcal{G}(u,r)}$ is over the Grassmannian $\mathcal{G}(u,r)$ of dimension u in $\mathbb{R}^r$. This form highlights the fact that only the subspace is being estimated, not a particular basis.

We next reexpress the objective function in (1.28) to show how it reflects the constraints on $\boldsymbol{\beta}$ and $\boldsymbol{\Sigma}$ in envelope model (1.20). Let $\mathbf{G}$ and $\mathbf{G}_0$ be semi-orthogonal basis matrices for a subspace $\mathcal{S}$ and its orthogonal complement $\mathcal{S}^{\perp}$. Then it follows from Lemma A.14 that $|\mathbf{P}_{\mathcal{S}}\mathbf{S}_{\mathbf{Y}|\mathbf{X}}\mathbf{P}_{\mathcal{S}}|_0 = |\mathbf{G}^T\mathbf{S}_{\mathbf{Y}|\mathbf{X}}\mathbf{G}|$ and $|\mathbf{Q}_{\mathcal{S}}\mathbf{S}_{\mathbf{Y}}\mathbf{Q}_{\mathcal{S}}|_0 = |\mathbf{G}_0^T\mathbf{S}_{\mathbf{Y}}\mathbf{G}_0|$. Consequently,

$$\begin{aligned}\log|\mathbf{P}_{\mathcal{S}}\mathbf{S}_{\mathbf{Y}|\mathbf{X}}\mathbf{P}_{\mathcal{S}}|_0 + \log|\mathbf{Q}_{\mathcal{S}}\mathbf{S}_{\mathbf{Y}}\mathbf{Q}_{\mathcal{S}}|_0 &= \log|\mathbf{G}^T\mathbf{S}_{\mathbf{Y}|\mathbf{X}}\mathbf{G}| + \log|\mathbf{G}_0^T\mathbf{S}_{\mathbf{Y}}\mathbf{G}_0| \\ &= \log|\mathbf{S}_{\mathbf{G}^T\mathbf{Y}|\mathbf{X}}| + \log|\mathbf{S}_{\mathbf{G}_0^T\mathbf{Y}}|.\end{aligned}$$

From (1.5), we can express

$$\begin{aligned}\mathbf{S}_{\mathbf{G}_0^T\mathbf{Y}} &= \mathbf{S}_{\mathbf{G}_0^T\mathbf{Y}|\mathbf{X}} + \mathbf{S}_{\mathbf{G}_0^T\mathbf{Y}\circ\mathbf{X}} \\ &= \mathbf{S}_{\mathbf{G}_0^T\mathbf{Y}|\mathbf{X}} + \mathbf{S}_{\mathbf{G}_0^T\mathbf{Y},\mathbf{X}}\mathbf{S}_{\mathbf{X}}^{-1}\mathbf{S}_{\mathbf{G}_0^T\mathbf{Y},\mathbf{X}}^T \\ &= \mathbf{S}_{\mathbf{G}_0^T\mathbf{Y}|\mathbf{X}} + \mathbf{B}_{\mathbf{G}_0^T\mathbf{Y}|\mathbf{X}}\mathbf{S}_{\mathbf{X}}\mathbf{B}_{\mathbf{G}_0^T\mathbf{Y}|\mathbf{X}}^T \\ &= \mathbf{S}_{\mathbf{G}_0^T\mathbf{Y}|\mathbf{X}} + \mathbf{S}_{\mathbf{G}_0^T\mathbf{Y}|\mathbf{X}}^{1/2}\tilde{\mathbf{B}}_{\mathbf{G}_0^T\mathbf{Y}|\mathbf{X}}\tilde{\mathbf{B}}_{\mathbf{G}_0^T\mathbf{Y}|\mathbf{X}}^T\mathbf{S}_{\mathbf{G}_0^T\mathbf{Y}|\mathbf{X}}^{1/2},\end{aligned}$$

where

$$\tilde{\mathbf{B}}_{\mathbf{G}_0^T\mathbf{Y}|\mathbf{X}} = \mathbf{S}_{\mathbf{G}_0^T\mathbf{Y}|\mathbf{X}}^{-1/2}\mathbf{B}_{\mathbf{G}_0^T\mathbf{Y}|\mathbf{X}}\mathbf{S}_{\mathbf{X}}^{1/2}$$

is the standardized version of the estimated coefficient matrix $\mathbf{B}_{\mathbf{G}_0^T\mathbf{Y}|\mathbf{X}}$ from the fit of $\mathbf{G}_0^T\mathbf{Y}$ on $\mathbf{X}$ (1.6). Consequently,

$$|\mathbf{S}_{\mathbf{G}_0^T\mathbf{Y}}| = |\mathbf{S}_{\mathbf{G}_0^T\mathbf{Y}|\mathbf{X}}| \times \left|\mathbf{I}_{r-u} + \tilde{\mathbf{B}}_{\mathbf{G}_0^T\mathbf{Y}|\mathbf{X}}\tilde{\mathbf{B}}_{\mathbf{G}_0^T\mathbf{Y}|\mathbf{X}}^T\right|$$

and optimization (1.28) can now be reexpressed as

$$\begin{aligned}\widehat{\mathcal{E}}_{\boldsymbol{\Sigma}}(\mathcal{B}) = \operatorname{span}\Big\{\arg\min_{\mathbf{G}}\Big(&\log|\mathbf{S}_{\mathbf{G}^T\mathbf{Y}|\mathbf{X}}| + \log|\mathbf{S}_{\mathbf{G}_0^T\mathbf{Y}|\mathbf{X}}| \\ &+ \log\left|\mathbf{I}_{r-u} + \tilde{\mathbf{B}}_{\mathbf{G}_0^T\mathbf{Y}|\mathbf{X}}\tilde{\mathbf{B}}_{\mathbf{G}_0^T\mathbf{Y}|\mathbf{X}}^T\right|\Big)\Big\},\end{aligned} \tag{1.29}$$

where the minimization is over semi-orthogonal matrices $\mathbf{G} \in \mathbb{R}^{r \times u}$. Since $|\mathbf{S}_{\mathbf{G}^T\mathbf{Y}|\mathbf{X}}| \times |\mathbf{S}_{\mathbf{G}_0^T\mathbf{Y}|\mathbf{X}}| \geq |\mathbf{S}_{\mathbf{Y}|\mathbf{X}}|$ with equality if and only if span($\mathbf{G}$) reduces $\mathbf{S}_{\mathbf{Y}|\mathbf{X}}$ (see Lemma A.14), the sum of the first two terms in the objective function of (1.29) is minimized when the columns of $\mathbf{G}$ are any u eigenvectors of $\mathbf{S}_{\mathbf{Y}|\mathbf{X}}$. Thus, the role of these terms is to enforce the constraint on $\boldsymbol{\Sigma}$ in model (1.20). The third term measures bias using the standardized coefficients from the regression of $\mathbf{G}_0^T\mathbf{Y}$ on $\mathbf{X}$, corresponding to the population constraint that $\mathcal{B} \subseteq$ span($\mathbf{G}$). The third term will equal 0 for any subspace span($\mathbf{G}$) that contains span($\mathbf{B}$), since then $\mathbf{G}_0^T\mathbf{B} = 0$. In short, the objective function balances the two constraints on the envelope model (1.20), minimizing bias while insuring that the solution is "close" to a reducing subspace of $\mathbf{S}_{\mathbf{Y}|\mathbf{X}}$.

1.5.4 Scaling the Responses

Like principal component regression, ridge regression, partial least squares, and other methods, the maximum likelihood estimators of $\boldsymbol{\beta}$ and $\boldsymbol{\Sigma}$ are not invariant or equivariant under rescaling of $\mathbf{Y}$. For instance, rescale $\mathbf{Y}$ by a nonsingular diagonal matrix $\mathbf{D}$ with positive diagonal elements, $\mathbf{Y} \mapsto \mathbf{DY}$. Expressed in terms of the parameters from the regression of $\mathbf{Y}$ on $\mathbf{X}$, the envelope for the regression of $\mathbf{DY}$ on $\mathbf{X}$ is $\mathcal{E}_{\mathbf{D\Sigma D}}(\mathbf{D}\mathcal{B})$. Then generally $\mathcal{E}_{\mathbf{D\Sigma D}}(\mathbf{D}\mathcal{B}) \neq \mathcal{E}_{\boldsymbol{\Sigma}}(\mathcal{B})$ with possibly different dimensions $\dim(\mathcal{E}_{\mathbf{D\Sigma D}}(\mathbf{D}\mathcal{B})) \neq \dim(\mathcal{E}_{\boldsymbol{\Sigma}}(\mathcal{B}))$. Let $\widehat{\boldsymbol{\beta}}_{\mathbf{D}}$ be the envelope estimator of $\boldsymbol{\beta}$ from the regression of $\mathbf{DY}$ on $\mathbf{X}$. Then generally we do not have $\widehat{\boldsymbol{\beta}}_{\mathbf{D}} = \boldsymbol{\beta}$ or $\widehat{\boldsymbol{\beta}}_{\mathbf{D}} = \mathbf{D}\widehat{\boldsymbol{\beta}}$. For these reasons, envelope methods based on the constructions of this chapter tend to work best when the responses are in the same or similar scales, although this is not required. Nearly all of the examples in this book are of that type. Similar comments hold for the predictor envelopes discussed in Chapter 4.

In Chapter 8, we extend the envelope model to allow for simultaneous estimation of a rescaling matrix like $\mathbf{D}$. Until then we stay close to the basic envelope construction described in this chapter.

1.6 Asymptotic Distributions

Asymptotic variances of $\widehat{\boldsymbol{\beta}}$ and $\widehat{\boldsymbol{\Sigma}}$ can in principle be determined from the inverse of the Fisher information matrix. However, we encounter complications when attempting to apply this general procedure to model (1.20) because $\boldsymbol{\Gamma}$ is not identifiable due to its overparameterization. Results from Shapiro (1986, Proposition 4.1) allow for overparameterization and show how to find the asymptotic distribution of an estimable function of the parameters. In this section, we sketch the process of determining the asymptotic distribution of the estimable functions $\boldsymbol{\beta} = \boldsymbol{\Gamma}\boldsymbol{\eta}$ and $\boldsymbol{\Sigma} = \boldsymbol{\Gamma}\boldsymbol{\Omega}\boldsymbol{\Gamma}^T + \boldsymbol{\Gamma}_0\boldsymbol{\Omega}_0\boldsymbol{\Gamma}_0^T$ of the parameters in model (1.20). Additional details are available from Cook et al. (2010).

The general procedure is the same for other envelope models described in later chapters.

We begin by defining the vector $\boldsymbol{\phi}$ of parameters associated with model (1.20) along with the estimable functions $\mathbf{h}(\boldsymbol{\phi})$. We do not include the intercept $\boldsymbol{\alpha}$ because it is not typically of interest and its maximum likelihood estimator is asymptotically independent of the others in the model. Let

$$\boldsymbol{\phi} = \begin{pmatrix} \text{vec}(\boldsymbol{\eta}) \\ \text{vec}(\boldsymbol{\Gamma}) \\ \text{vech}(\boldsymbol{\Omega}) \\ \text{vech}(\boldsymbol{\Omega}_0) \end{pmatrix} := \begin{pmatrix} \boldsymbol{\phi}_1 \\ \boldsymbol{\phi}_2 \\ \boldsymbol{\phi}_3 \\ \boldsymbol{\phi}_4 \end{pmatrix} \tag{1.30}$$

and

$$\mathbf{h}(\boldsymbol{\phi}) = \begin{pmatrix} \text{vec}(\boldsymbol{\beta}) \\ \text{vech}(\boldsymbol{\Sigma}) \end{pmatrix} = \begin{pmatrix} \text{vec}(\boldsymbol{\Gamma\eta}) \\ \text{vech}(\boldsymbol{\Gamma\Omega\Gamma}^T + \boldsymbol{\Gamma}_0\boldsymbol{\Omega}_0\boldsymbol{\Gamma}_0^T) \end{pmatrix} := \begin{pmatrix} \mathbf{h}_1(\boldsymbol{\phi}) \\ \mathbf{h}_2(\boldsymbol{\phi}) \end{pmatrix}.$$

We will also require the gradient matrix,

$$\mathbf{H} := \begin{pmatrix} \partial\mathbf{h}_1/\partial\boldsymbol{\phi}_1^T & \cdots & \partial\mathbf{h}_1/\partial\boldsymbol{\phi}_4^T \\ \partial\mathbf{h}_2/\partial\boldsymbol{\phi}_1^T & \cdots & \partial\mathbf{h}_2/\partial\boldsymbol{\phi}_4^T \end{pmatrix}$$

$$= \begin{pmatrix} \mathbf{I}_p \otimes \boldsymbol{\Gamma} & \boldsymbol{\eta}^T \otimes \mathbf{I}_r & \mathbf{0} & \mathbf{0} \\ \mathbf{0} & \mathbf{H}_{22} & \mathbf{C}_r(\boldsymbol{\Gamma}\otimes\boldsymbol{\Gamma})\mathbf{E}_u & \mathbf{C}_r(\boldsymbol{\Gamma}_0\otimes\boldsymbol{\Gamma}_0)\mathbf{E}_{(r-u)} \end{pmatrix},$$

where $\mathbf{H}_{22} = 2\mathbf{C}_r(\boldsymbol{\Gamma\Omega}\otimes\mathbf{I}_r - \boldsymbol{\Gamma}\otimes\boldsymbol{\Gamma}_0\boldsymbol{\Omega}_0\boldsymbol{\Gamma}_0^T)$, and $\mathbf{C}_r$ and $\mathbf{E}_u$ are the contraction and expansion matrices that connect the vec and vech operators. The derivatives needed for $\mathbf{H}$ are also required in other asymptotic calculations. These and related derivatives are summarized in Section A.6 along with basic properties of the contraction and expansion matrices in Appendix A.5.

Because of the overparameterization in $\boldsymbol{\Gamma}$, $\mathbf{H}$ is not of full rank and standard likelihood methods cannot be applied directly. But $\mathbf{h}$ is estimable, allowing us to use Shapiro (1986, Proposition 4.1) to conclude that

Proposition 1.1 *Under the envelope model (1.20) with normal errors and known* $u = \dim\{\mathcal{E}_{\boldsymbol{\Sigma}}(\mathcal{B})\}$, $\sqrt{n}\{\mathbf{h}(\widehat{\boldsymbol{\phi}}) - \mathbf{h}(\boldsymbol{\phi})\}$ *is asymptotically normal with mean* $\mathbf{0}$ *and covariance matrix*

$$\text{avar}\{\sqrt{n}\mathbf{h}(\widehat{\boldsymbol{\phi}})\} = \mathbf{P}_{\mathbf{H}(\mathbf{J})}\mathbf{J}^{-1}\mathbf{P}_{\mathbf{H}(\mathbf{J})}^T = \mathbf{H}(\mathbf{H}^T\mathbf{J}\mathbf{H})^{\dagger}\mathbf{H}^T, \tag{1.31}$$

where $\mathbf{J}$ *is the information matrix (1.14) for* $(\text{vec}^T(\boldsymbol{\beta}), \text{vech}^T(\boldsymbol{\Sigma}))$ *in the model (1.1). Since* $\mathbf{J}^{-1} - \text{avar}(\sqrt{n}\mathbf{h}(\widehat{\boldsymbol{\phi}})) \geq 0$, *the envelope estimator never does worse than the standard estimator.*

Additionally, $\sqrt{n}(\widehat{\boldsymbol{\beta}} - \boldsymbol{\beta})$ *is asymptotically normal with mean 0 and variance*

$$\text{avar}\{\sqrt{n}\,\text{vec}(\widehat{\boldsymbol{\beta}})\} = \boldsymbol{\Sigma}_{\mathbf{X}}^{-1}\otimes\boldsymbol{\Gamma\Omega\Gamma}^T + (\boldsymbol{\eta}^T\otimes\boldsymbol{\Gamma}_0)\mathbf{U}^{\dagger}(\boldsymbol{\eta}\otimes\boldsymbol{\Gamma}_0^T), \tag{1.32}$$

where

$$\begin{aligned}\mathbf{U} &= \boldsymbol{\eta}\boldsymbol{\Sigma}_{\mathbf{X}}\boldsymbol{\eta}^T \otimes \boldsymbol{\Omega}_0^{-1} + \boldsymbol{\Omega} \otimes \boldsymbol{\Omega}_0^{-1} + \boldsymbol{\Omega}^{-1} \otimes \boldsymbol{\Omega}_0 - 2\mathbf{I}_u \otimes \mathbf{I}_{r-u} \\ &= \boldsymbol{\eta}\boldsymbol{\Sigma}_{\mathbf{X}}\boldsymbol{\eta}^T \otimes \boldsymbol{\Omega}_0^{-1} + (\boldsymbol{\Omega}^{1/2} \otimes \boldsymbol{\Omega}_0^{-1/2} - \boldsymbol{\Omega}^{-1/2} \otimes \boldsymbol{\Omega}_0^{1/2})^2.\end{aligned}$$

These results can be used in practice to construct an asymptotic standard error for $(\widehat{\boldsymbol{\beta}})_{ij}$, $i = 1, \dots, r$, $j = 1, \dots, p$, by first substituting estimates for the unknown quantities on the right-hand side of (1.32) to obtain an estimated asymptotic variance $\widehat{\text{avar}}\{\sqrt{n}\,\text{vec}(\widehat{\boldsymbol{\beta}})\}$. The estimated asymptotic variance $\widehat{\text{avar}}\{\sqrt{n}(\widehat{\boldsymbol{\beta}})_{ij}\}$ is then the corresponding diagonal element of $\widehat{\text{avar}}\{\sqrt{n}\,\text{vec}(\widehat{\boldsymbol{\beta}})\}$, and its *asymptotic standard error* is

$$\text{se}\{(\widehat{\boldsymbol{\beta}})_{ij}\} = \frac{[\widehat{\text{avar}}\{\sqrt{n}(\widehat{\boldsymbol{\beta}})_{ij}\}]^{1/2}}{\sqrt{n}}, \tag{1.33}$$

with corresponding Z-score equal to $(\widehat{\boldsymbol{\beta}})_{ij}/\text{se}\{(\widehat{\boldsymbol{\beta}})_{ij}\}$. This is the method that is used to obtain standard errors in the previous illustrations. The bootstrap can also be used, as described in Section 1.11.

If $u = r$, then $\boldsymbol{\Gamma}\boldsymbol{\Omega}\boldsymbol{\Gamma}^T = \boldsymbol{\Sigma}$, and the second addend on the right-hand side of (1.32) does not appear. The first addend on the right-hand side of (1.32) is the asymptotic variance of $\widehat{\boldsymbol{\beta}}_{\boldsymbol{\Gamma}}$, the envelope estimator of $\boldsymbol{\beta}$ when $\boldsymbol{\Gamma}$ is known (cf. (1.22)),

$$\text{avar}\{\sqrt{n}\,\text{vec}(\widehat{\boldsymbol{\beta}}_{\boldsymbol{\Gamma}})\} = \boldsymbol{\Sigma}_{\mathbf{X}}^{-1} \otimes \boldsymbol{\Gamma}\boldsymbol{\Omega}\boldsymbol{\Gamma}^T.$$

The second addend can be interpreted as the "cost" of estimating $\mathcal{E}_{\boldsymbol{\Sigma}}(\mathcal{B})$. The total on the right does not exceed $\text{avar}\{\sqrt{n}\,\text{vec}(\mathbf{B})\}$; that is,

$$\text{avar}\{\sqrt{n}\,\text{vec}(\mathbf{B})\} - \text{avar}\{\sqrt{n}\,\text{vec}(\widehat{\boldsymbol{\beta}})\} \geq 0.$$

The asymptotic variance (1.32) can be reexpressed informatively as

$$\text{avar}\{\sqrt{n}\,\text{vec}(\widehat{\boldsymbol{\beta}})\} = \text{avar}\{\sqrt{n}\,\text{vec}(\widehat{\boldsymbol{\beta}}_{\boldsymbol{\Gamma}})\} + \text{avar}\{\sqrt{n}\,\text{vec}(\mathbf{Q}_{\boldsymbol{\Gamma}}\widehat{\boldsymbol{\beta}}_{\boldsymbol{\eta}})\}. \tag{1.34}$$

As mentioned previously, the first addend $\text{avar}\{\sqrt{n}\,\text{vec}(\widehat{\boldsymbol{\beta}}_{\boldsymbol{\Gamma}})\}$ on the right-hand side is the asymptotic variance of $\widehat{\boldsymbol{\beta}}_{\boldsymbol{\Gamma}}$, and in the second addend $\text{avar}\{\sqrt{n}\,\text{vec}(\widehat{\boldsymbol{\beta}}_{\boldsymbol{\eta}})\}$ is the asymptotic variance of the maximum likelihood estimator $\widehat{\boldsymbol{\beta}}_{\boldsymbol{\eta}}$ of $\boldsymbol{\beta}$ when $\boldsymbol{\eta}$ is known, both corresponding to asymptotic variances in multivariate linear models.

We next consider a special case to gain intuition about the gains offered by envelopes over the standard method. In preparation, write the asymptotic variance of the estimator $\mathbf{B}$ of $\boldsymbol{\beta}$ under the standard model in terms of the envelope parameters as

$$\text{avar}\{\sqrt{n}\,\text{vec}(\mathbf{B})\} = \boldsymbol{\Sigma}_{\mathbf{X}}^{-1} \otimes \boldsymbol{\Sigma} = \boldsymbol{\Sigma}_{\mathbf{X}}^{-1} \otimes \boldsymbol{\Gamma}\boldsymbol{\Omega}\boldsymbol{\Gamma}^T + \boldsymbol{\Sigma}_{\mathbf{X}}^{-1} \otimes \boldsymbol{\Gamma}_0\boldsymbol{\Omega}_0\boldsymbol{\Gamma}_0^T.$$

Let $\mathbf{D} = \text{avar}\{\sqrt{n}\,\text{vec}(\mathbf{B})\} - \text{avar}\{\sqrt{n}\,\text{vec}(\widehat{\boldsymbol{\beta}})\}$ denote the difference in asymptotic variances. We then have

$$\mathbf{D} = \boldsymbol{\Sigma}_{\mathbf{X}}^{-1} \otimes \boldsymbol{\Gamma}_0\boldsymbol{\Omega}_0\boldsymbol{\Gamma}_0^T - (\boldsymbol{\eta}^T \otimes \boldsymbol{\Gamma}_0)\mathbf{U}^{\dagger}(\boldsymbol{\eta} \otimes \boldsymbol{\Gamma}_0^T) \geq 0.$$

Suppose now that $\boldsymbol{\beta}$ has full column rank p, $\boldsymbol{\Omega} = \omega\mathbf{I}_u$ and $\boldsymbol{\Omega}_0 = \omega_0\mathbf{I}_{r-u}$. Compound symmetry is an instance of this structure (see Section 1.3.2). As a consequence, we see that $\boldsymbol{\Sigma}$ has two eigenspaces, one corresponding to ω and the other to ω_0. Since $\mathcal{B}$ is contained in the eigenspace corresponding to ω, we must have $\mathcal{B} = \mathcal{E}_{\boldsymbol{\Sigma}}(\mathcal{B})$, $u = p$, $\boldsymbol{\Gamma} = \boldsymbol{\beta}(\boldsymbol{\beta}^T\boldsymbol{\beta})^{-1/2}$, and $\boldsymbol{\eta} = (\boldsymbol{\beta}^T\boldsymbol{\beta})^{1/2}$. Then after a little algebra, we can write

$$\mathbf{D} = \boldsymbol{\Sigma}_{\mathbf{X}}^{-1} \otimes \boldsymbol{\Gamma}_0\boldsymbol{\Gamma}_0^T\omega_0 - \boldsymbol{\eta}^T\{\boldsymbol{\eta}\boldsymbol{\Sigma}_{\mathbf{X}}\boldsymbol{\eta}^T\omega_0^{-1} + (\omega\omega_0^{-1} + \omega^{-1}\omega_0 - 2)\mathbf{I}_u\}^{-1}\boldsymbol{\eta} \otimes \boldsymbol{\Gamma}_0\boldsymbol{\Gamma}_0^T.$$

Simplifying $\omega\omega_0^{-1} + \omega^{-1}\omega_0 - 2 = (\omega - \omega_0)^2/\omega\omega_0$ and using the fact that $\boldsymbol{\eta} = (\boldsymbol{\beta}^T\boldsymbol{\beta})^{1/2} \in \mathbb{R}^{p\times p}$ is nonsingular,

$$\mathbf{D} = \boldsymbol{\Sigma}_{\mathbf{X}}^{-1} \otimes \boldsymbol{\Gamma}_0\boldsymbol{\Gamma}_0^T\omega_0 - \{\boldsymbol{\Sigma}_{\mathbf{X}} + \omega^{-1}(\omega - \omega_0)^2(\boldsymbol{\beta}^T\boldsymbol{\beta})^{-1}\}^{-1} \otimes \boldsymbol{\Gamma}_0\boldsymbol{\Gamma}_0^T\omega_0 \geq 0.$$

Recall that if $\mathcal{E}_{\boldsymbol{\Sigma}}(\mathcal{B})$ is known, the envelope model will offer substantial gains when $\omega \ll \omega_0$. The second addend on the right-hand side of the last expression shows the cost of estimating $\mathcal{E}_{\boldsymbol{\Sigma}}(\mathcal{B})$. From the expression, we see that the cost will be relatively small when again $\omega \ll \omega_0$. It can also be small if $\omega \neq \omega_0$ and $(\boldsymbol{\beta}^T\boldsymbol{\beta})^{-1}$ is large relative to $\boldsymbol{\Sigma}_{\mathbf{X}}$.

The next corollary summarizes an important special case of this illustration.

Corollary 1.1 *Assume that* $\boldsymbol{\Sigma} = \omega\mathbf{I}_r$ *and that* $\text{rank}(\boldsymbol{\beta}) = p$. *Then the asymptotic variance of the envelope estimator is the same as the asymptotic variance of the usual maximum likelihood estimator,* $\text{avar}\{\sqrt{n}\,\text{vec}(\mathbf{B})\} = \text{avar}\{\sqrt{n}\,\text{vec}(\widehat{\boldsymbol{\beta}})\}$.

Gains are still possible when $\boldsymbol{\Sigma} = \omega\mathbf{I}_r$ if the rank of $\boldsymbol{\beta}$ is less than p. In that case, $\mathcal{B} = \mathcal{E}_{\boldsymbol{\Sigma}}(\mathcal{B})$, $\omega\omega_0^{-1} + \omega^{-1}\omega_0 - 2 = 0$, and

$$\begin{aligned}\mathbf{D} &= \{\boldsymbol{\Sigma}_{\mathbf{X}}^{-1} - \boldsymbol{\eta}^T(\boldsymbol{\eta}\boldsymbol{\Sigma}_{\mathbf{X}}\boldsymbol{\eta}^T)^{-1}\boldsymbol{\eta}\} \otimes \boldsymbol{\Gamma}_0\boldsymbol{\Gamma}_0^T\omega_0 \\ &= \mathbf{Q}_{\boldsymbol{\eta}^T(\boldsymbol{\Sigma}_{\mathbf{X}})} \otimes \boldsymbol{\Gamma}_0\boldsymbol{\Gamma}_0^T\omega_0 \geq 0.\end{aligned}$$

1.7 Fitted Values and Predictions

The previous asymptotic results can be used to derive the asymptotic distribution of the fitted values, as well as the asymptotic prediction variance. The fitted values at a particular $\mathbf{X}$ can be written as $\widehat{\mathbf{Y}} = \bar{\mathbf{Y}} + \widehat{\boldsymbol{\beta}}\mathbf{X} = \bar{\mathbf{Y}} + (\mathbf{X}^T \otimes \mathbf{I}_r)\text{vec}(\widehat{\boldsymbol{\beta}})$. Hence, the fitted value $\widehat{\mathbf{Y}}$ has the following asymptotic distribution:

$$\sqrt{n}(\widehat{\mathbf{Y}} - \text{E}(\widehat{\mathbf{Y}})) \xrightarrow{\mathcal{L}} N_r[0, \text{avar}\{\sqrt{n}\,\text{vec}(\bar{\mathbf{Y}} + \widehat{\boldsymbol{\beta}}\mathbf{X})\}]. \tag{1.35}$$

Using (1.34) and the fact that $\bar{\mathbf{Y}}$ and $\hat{\boldsymbol{\beta}}$ are asymptotically independent, the asymptotic variance in this distribution can be expressed informatively as

$$\begin{aligned}
&\mathrm{avar}\{\sqrt{n}\,\mathrm{vec}(\bar{\mathbf{Y}}+\hat{\boldsymbol{\beta}}\mathbf{X})\}\\
&\quad=\boldsymbol{\Sigma}+(\mathbf{X}^T\otimes\mathbf{I}_r)\mathrm{avar}\{\sqrt{n}\,\mathrm{vec}(\hat{\boldsymbol{\beta}})\}(\mathbf{X}\otimes\mathbf{I}_r))\\
&\quad=\boldsymbol{\Sigma}+(\mathbf{X}^T\otimes\mathbf{I}_r)\mathrm{avar}\{\sqrt{n}\,\mathrm{vec}(\hat{\boldsymbol{\beta}}_{\boldsymbol{\Gamma}})\}(\mathbf{X}\otimes\mathbf{I}_r)\\
&\qquad+(\mathbf{X}^T\otimes\mathbf{I}_r)\mathrm{avar}\{\sqrt{n}\,\mathrm{vec}(\mathbf{Q}_{\boldsymbol{\Gamma}}\hat{\boldsymbol{\beta}}_{\eta})\}(\mathbf{X}\otimes\mathbf{I}_r)\\
&\quad=\boldsymbol{\Sigma}+\mathrm{avar}\{\sqrt{n}\,\mathrm{vec}(\hat{\boldsymbol{\beta}}_{\boldsymbol{\Gamma}}\mathbf{X})\}+\mathrm{avar}\{\sqrt{n}\,\mathrm{vec}(\mathbf{Q}_{\boldsymbol{\Gamma}}\hat{\boldsymbol{\beta}}_{\eta}\mathbf{X})\}.
\end{aligned}$$

Consequently, the variance of a fitted value has the same essential decomposition as the variance of $\hat{\boldsymbol{\beta}}$ discussed previously.

Turning to prediction, suppose that at some value of $\mathbf{X}$, we wish to infer about a new $\mathbf{Y}$, say $\mathbf{Y}_{\text{new}}$, independently of the past observations. Then

$$\begin{aligned}
&\mathrm{E}\{(\hat{\mathbf{Y}}-\mathbf{Y}_{\text{new}})(\hat{\mathbf{Y}}-\mathbf{Y}_{\text{new}})^T\}\\
&\quad=\mathrm{E}\{(\hat{\mathbf{Y}}-\mathrm{E}(\hat{\mathbf{Y}}))(\hat{\mathbf{Y}}-\mathrm{E}(\hat{\mathbf{Y}}))^T\}+\mathrm{E}\{(\mathrm{E}(\hat{\mathbf{Y}})-\mathbf{Y}_{\text{new}})(\mathrm{E}(\hat{\mathbf{Y}})-\mathbf{Y}_{\text{new}})^T\},
\end{aligned}$$

where the cross-product terms vanish because $\mathbf{Y}_{\text{new}}$ and $\hat{\mathbf{Y}}$ are independent. Combining this with expression (1.35), we see that the mean squared error of the prediction is approximated by

$$\begin{aligned}
\mathrm{E}\{(\hat{\mathbf{Y}}-\mathbf{Y}_{\text{new}})(\hat{\mathbf{Y}}-\mathbf{Y}_{\text{new}})^T\}&=n^{-1}\mathrm{avar}\{\sqrt{n}\,\mathrm{vec}(\hat{\boldsymbol{\beta}}\mathbf{X})\}\\
&\quad+(1+n^{-1})\boldsymbol{\Sigma}+o(n^{-1}).
\end{aligned}\tag{1.36}$$

Envelope model (1.20) can be quite effective at reducing $\mathrm{avar}\{\sqrt{n}\,\mathrm{vec}(\hat{\boldsymbol{\beta}}\mathbf{X})\}$, but it has no impact on the underlying variance $\boldsymbol{\Sigma}$ except for the induced structure. Envelopes give greatest estimative gain when $\mathrm{var}(\boldsymbol{\Gamma}_0^T\mathbf{Y}\mid\mathbf{X})=\boldsymbol{\Omega}_0$ is large relative to the material variation $\mathrm{var}(\boldsymbol{\Gamma}^T\mathbf{Y}\mid\mathbf{X})=\boldsymbol{\Omega}$. Nevertheless, the X-invariant variation is still present in $\boldsymbol{\Sigma}$, and consequently the advantages that model (1.20) bring in the estimation of $\boldsymbol{\beta}$ may not be present to the same degree in prediction. This can be seen in the schematic illustration of Figure 1.3, where the distributions represented in the left-hand display contribute to prediction, but not to estimation as shown in the right-hand display. Greater predictive gain might be realized by using partial envelopes for prediction, as discussed in Section 3.4, or by using envelopes for predictor reduction, as discussed in Chapter 4.

1.8 Testing the Responses

1.8.1 Test Development

In some regressions, we may wish to test the hypothesis that the part of $\mathbf{Y}$ represented by the projection $\mathbf{P}_{\mathcal{H}}\mathbf{Y}$ onto a known user-specified subspace $\mathcal{H}\subset\mathbb{R}^r$

holds the entire X-variant part of $\mathbf{Y}$. That is, starting with model (1.20), we may wish to test the hypothesis that

$$\text{(i) } \mathbf{Q}_{\mathcal{H}}\mathbf{Y} \mid (\mathbf{X} = \mathbf{x}_1) \sim \mathbf{Q}_{\mathcal{H}}\mathbf{Y} \mid (\mathbf{X} = \mathbf{x}_2) \text{ and (ii) } \mathbf{P}_{\mathcal{H}}\mathbf{Y} \perp\!\!\!\perp \mathbf{Q}_{\mathcal{H}}\mathbf{Y} \mid \mathbf{X}. \quad (1.37)$$

This is similar to specification (1.18), except here we are not requiring $\mathcal{H}$ to be the smallest subspace. Equivalently, this hypothesis specifies that $\mathbf{Q}_{\mathcal{H}}\mathbf{Y}$ is X-invariant, while allowing for the possibility that $\mathbf{P}_{\mathcal{H}}\mathbf{Y}$ may also contain an X-invariant part of $\mathbf{Y}$. Since $\mathcal{E}_{\mathbf{\Sigma}}(\mathcal{B})$ is defined as the intersection of all subspaces that satisfy (1.37), it follows that $\mathcal{E}_{\mathbf{\Sigma}}(\mathcal{B})$ is contained in any subspace that satisfies (1.37) and thus $\mathcal{E}_{\mathbf{\Sigma}}(\mathcal{B}) \subseteq \mathcal{H}$. Hypothesis (1.37) can be tested by using the likelihood ratio test statistic $\Lambda_u(\mathcal{H}) = 2(\widehat{L}_u - \widehat{L}_u(\mathcal{H}))$, where $\widehat{L}_u$ is the maximized envelope log-likelihood (1.27), and $\widehat{L}_u(\mathcal{H})$ is the maximized log-likelihood under the hypothesis. As in previous sections, we treat u as known in this development. Methods for selecting u are discussed in Section 1.10.

To construct $\widehat{L}_u(\mathcal{H})$, let $v = \dim(\mathcal{H}) \geq u$ denote the known dimension of $\mathcal{H}$, let $\mathbf{H} \in \mathbb{R}^{r \times v}$ be a semi-orthogonal basis matrix for $\mathcal{H}$, and let $(\mathbf{H}, \mathbf{H}_0) \in \mathbb{R}^{r \times r}$ be an orthogonal matrix. Under hypothesis (1.37), $\mathcal{H}$ is a reducing subspace of $\mathbf{\Sigma}$ that contains $\mathcal{B}$. It follows from Proposition A.6 that $\mathcal{E}_{\mathbf{\Sigma}}(\mathcal{B}) = \mathbf{H}\mathcal{E}_{\mathbf{H}^T\mathbf{\Sigma}\mathbf{H}}(\mathbf{H}^T\mathcal{B})$. Let $\mathbf{h} \in \mathbb{R}^{v \times u}$ be a semi-orthogonal basis matrix for $\mathcal{E}_{\mathbf{H}^T\mathbf{\Sigma}\mathbf{H}}(\mathbf{H}^T\mathcal{B})$, which corresponds to the envelope regression of $\mathbf{H}^T\mathbf{Y}$ on $\mathbf{X}$. Then we have that $\mathbf{\Gamma} = \mathbf{H}\mathbf{h}$ is a basis for $\mathcal{E}_{\mathbf{\Sigma}}(\mathcal{B})$. Let $\mathbf{\Gamma}_{01} = \mathbf{H}\mathbf{h}_0$ be a semi-orthogonal basis for the orthogonal complement of $\mathcal{E}_{\mathbf{\Sigma}}(\mathcal{B})$ within $\mathcal{H}$, where $\mathbf{h}_0 \in \mathbb{R}^{v \times (v-u)}$ is semi-orthogonal and $\mathbf{h}^T\mathbf{h}_0 = 0$. Envelope model (1.20) can now be expressed under (1.37) as

$$\mathbf{Y} = \boldsymbol{\alpha} + \mathbf{H}\mathbf{h}\boldsymbol{\eta}\mathbf{X} + \boldsymbol{\varepsilon}, \quad (1.38)$$

$$\mathbf{\Sigma} = \mathbf{H}\mathbf{h}\mathbf{\Omega}\mathbf{h}^T\mathbf{H}^T + \mathbf{H}\mathbf{h}_0\mathbf{\Omega}_{01}\mathbf{h}_0^T\mathbf{H}^T + \mathbf{H}_0\mathbf{\Omega}_{00}\mathbf{H}_0^T, \quad (1.39)$$

where the structure of $\mathbf{\Sigma}$ follows because $\mathbf{h}^T\mathbf{H}^T\mathbf{Y}$, $\mathbf{h}_0^T\mathbf{H}^T\mathbf{Y}$, and $\mathbf{H}_0^T\mathbf{Y}$ are mutually independent given $\mathbf{X}$. The terms $\boldsymbol{\alpha}$, $\mathbf{\Omega}$, and $\boldsymbol{\eta}$ that occur in envelope model (1.20) are the same as those under hypothesis (1.37), while $\boldsymbol{\beta} = \mathbf{H}\mathbf{h}\boldsymbol{\eta}$, $\mathbf{\Gamma}_0 = (\mathbf{H}\mathbf{h}_0, \mathbf{H}_0)$, and $\mathbf{\Omega}_0 = \text{bdiag}(\mathbf{\Omega}_{01}, \mathbf{\Omega}_{00})$. We see from (1.38) and (1.39) that $\mathbf{H}^T\mathbf{Y} \mid \mathbf{X}$ follows an envelope model, which is helpful when studying the likelihood.

The log-likelihood

$$L_u(\mathcal{H}) := L_u(\boldsymbol{\alpha}, \boldsymbol{\eta}, \mathbf{h}, \mathbf{\Omega}, \mathbf{\Omega}_{01}, \mathbf{\Omega}_{00} \mid \mathcal{H})$$

for this model can now be expressed as

$$\begin{aligned} L_u(\mathcal{H}) &= -(nr/2)\log(2\pi) - (n/2)\log|\mathbf{\Sigma}| \\ &\quad -(1/2)\sum_{i=1}^{n}(\mathbf{Y}_i - \boldsymbol{\alpha} - \mathbf{H}\mathbf{h}\boldsymbol{\eta}\mathbf{X}_i)^T\mathbf{\Sigma}^{-1}(\mathbf{Y}_i - \boldsymbol{\alpha} - \mathbf{H}\mathbf{h}\boldsymbol{\eta}\mathbf{X}_i), \end{aligned}$$

where $\boldsymbol{\Sigma}$ is as given in (1.39). Continuing,

$$\begin{aligned}L_u(\mathcal{H}) = &-(nr/2)\log(2\pi) - (n/2)\log|\boldsymbol{\Omega}| \\ &-(n/2)\log|\boldsymbol{\Omega}_{01}| - (n/2)\log|\boldsymbol{\Omega}_{00}| \\ &-(1/2)\sum_{i=1}^{n}(\mathbf{Y}_i - \boldsymbol{\alpha} - \mathbf{H}\mathbf{h}\boldsymbol{\eta}\mathbf{X}_i)^T(\mathbf{H}\mathbf{h}\boldsymbol{\Omega}^{-1}\mathbf{h}^T\mathbf{H}^T)(\mathbf{Y}_i - \boldsymbol{\alpha} - \mathbf{H}\mathbf{h}\boldsymbol{\eta}\mathbf{X}_i) \\ &-(1/2)\sum_{i=1}^{n}(\mathbf{Y}_i - \boldsymbol{\alpha} - \mathbf{H}\mathbf{h}\boldsymbol{\eta}\mathbf{X}_i)^T(\mathbf{H}\mathbf{h}_0\boldsymbol{\Omega}_{01}^{-1}\mathbf{h}_0^T\mathbf{H}^T)(\mathbf{Y}_i - \boldsymbol{\alpha} - \mathbf{H}\mathbf{h}\boldsymbol{\eta}\mathbf{X}_i) \\ &-(1/2)\sum_{i=1}^{n}(\mathbf{Y}_i - \boldsymbol{\alpha} - \mathbf{H}\mathbf{h}\boldsymbol{\eta}\mathbf{X}_i)^T(\mathbf{H}_0\boldsymbol{\Omega}_{00}^{-1}\mathbf{H}_0^T)(\mathbf{Y}_i - \boldsymbol{\alpha} - \mathbf{H}\mathbf{h}\boldsymbol{\eta}\mathbf{X}_i).\end{aligned}$$

Rearranging terms to match regressions of $\mathbf{H}^T\mathbf{Y}$ and $\mathbf{H}_0^T\mathbf{Y}$ on $\mathbf{X}$, we get

$$\begin{aligned}L_u(\mathcal{H}) = &-(nr/2)\log(2\pi) - (n/2)\log|\boldsymbol{\Omega}| - (n/2)\log|\boldsymbol{\Omega}_{01}| \\ &-(1/2)\sum_{i=1}^{n}(\mathbf{H}^T(\mathbf{Y}_i - \boldsymbol{\alpha}) - \mathbf{h}\boldsymbol{\eta}\mathbf{X}_i)^T\mathbf{h}\boldsymbol{\Omega}^{-1}\mathbf{h}^T(\mathbf{H}(\mathbf{Y}_i - \boldsymbol{\alpha}) - \mathbf{h}\boldsymbol{\eta}\mathbf{X}_i) \\ &-(1/2)\sum_{i=1}^{n}(\mathbf{H}^T(\mathbf{Y}_i - \boldsymbol{\alpha}))^T\mathbf{h}_0\boldsymbol{\Omega}_{01}^{-1}\mathbf{h}_0^T(\mathbf{H}^T(\mathbf{Y}_i - \boldsymbol{\alpha})) \\ &-(n/2)\log|\boldsymbol{\Omega}_{00}| - (1/2)\sum_{i=1}^{n}(\mathbf{H}_0^T(\mathbf{Y}_i - \boldsymbol{\alpha}))^T\boldsymbol{\Omega}_{00}^{-1}(\mathbf{H}_0^T(\mathbf{Y}_i - \boldsymbol{\alpha})).\end{aligned}$$

Addends 2–5 plus $-(nv/2)\log(2\pi)$ correspond to an envelope likelihood with v responses, as described in Section 1.5.1, and the last two addends plus $-(n(r-v)/2)\log(2\pi)$ correspond to a mean only regression. The sum $-(nv/2)\log(2\pi) - (n(r-v)/2)\log(2\pi) = -(nr/2)\log(2\pi)$ is the first addend.

Let $L_u^{(1)}(\mathbf{h}, \boldsymbol{\eta}, \boldsymbol{\Omega}, \boldsymbol{\Omega}_{01} \mid \mathcal{H})$ denote the log-likelihood for the envelope regression of $\mathbf{H}^T\mathbf{Y}$ on $\mathbf{X}$:

$$\begin{aligned}L_u^{(1)} = &-(nv/2)\log(2\pi) - (n/2)\log|\boldsymbol{\Omega}| - (n/2)\log|\boldsymbol{\Omega}_{01}| \\ &-(1/2)\sum_{i=1}^{n}(\mathbf{H}^T(\mathbf{Y}_i - \boldsymbol{\alpha}) - \mathbf{h}\boldsymbol{\eta}\mathbf{X}_i)^T\mathbf{h}\boldsymbol{\Omega}^{-1}\mathbf{h}^T(\mathbf{H}(\mathbf{Y}_i - \boldsymbol{\alpha}) - \mathbf{h}\boldsymbol{\eta}\mathbf{X}_i) \\ &-(1/2)\sum_{i=1}^{n}(\mathbf{H}^T(\mathbf{Y}_i - \boldsymbol{\alpha}))^T\mathbf{h}_0\boldsymbol{\Omega}_{01}^{-1}\mathbf{h}_0^T(\mathbf{H}^T(\mathbf{Y}_i - \boldsymbol{\alpha})).\end{aligned}$$

Let $L_u^{(2)}(\boldsymbol{\Omega}_{00} \mid \mathcal{H})$ denote the log-likelihood arising from the mean-only regression $\mathbf{H}_0^T\mathbf{Y} = \mathbf{H}_0^T\boldsymbol{\alpha} + \mathbf{H}_0^T\boldsymbol{\varepsilon}$:

$$\begin{aligned}L_u^{(2)} = &-(n(r-v)/2)\log(2\pi) - (n/2)\log|\boldsymbol{\Omega}_{00}| \\ &-(1/2)\sum_{i=1}^{n}(\mathbf{Y}_i - \boldsymbol{\alpha})^T\mathbf{H}_0\boldsymbol{\Omega}_{00}^{-1}\mathbf{H}_0^T(\mathbf{Y}_i - \boldsymbol{\alpha}).\end{aligned}$$

Then

$$L_u(\mathcal{H}) = L_u^{(1)}(\mathbf{h}, \boldsymbol{\eta}, \boldsymbol{\Omega}, \boldsymbol{\Omega}_{01} \mid \mathcal{H}) + L_u^{(2)}(\boldsymbol{\Omega}_{00} \mid \mathcal{H}).$$

and the estimators of the parameters in (1.38) and (1.39) can be found by following the derivation in Section 1.5.1:

$$\begin{aligned}
\hat{\mathbf{h}} &= \arg\min_{\mathbf{G}}\{\log|\mathbf{G}^T\mathbf{S}_{\mathbf{H}^T\mathbf{Y}|\mathbf{X}}\mathbf{G}| + \log|\mathbf{G}^T\mathbf{S}_{\mathbf{H}^T\mathbf{Y}}^{-1}\mathbf{G}|\},\\
\hat{\boldsymbol{\eta}} &= \hat{\mathbf{h}}^T\mathbf{S}_{\mathbf{H}^T\mathbf{Y},\mathbf{X}}\mathbf{S}_{\mathbf{X}}^{-1},\\
\hat{\boldsymbol{\beta}} &= \mathbf{H}\hat{\mathbf{h}}\hat{\boldsymbol{\eta}},\\
\hat{\boldsymbol{\Omega}} &= \hat{\mathbf{h}}^T\mathbf{S}_{\mathbf{H}^T\mathbf{Y}|\mathbf{X}}\hat{\mathbf{h}},\\
\hat{\boldsymbol{\Omega}}_{01} &= \hat{\mathbf{h}}_0^T\mathbf{S}_{\mathbf{H}^T\mathbf{Y}}\hat{\mathbf{h}}_0,\\
\hat{\boldsymbol{\Omega}}_{00} &= \mathbf{S}_{\mathbf{H}_0^T\mathbf{Y}},\\
\hat{\boldsymbol{\Sigma}} &= \mathbf{H}\hat{\mathbf{h}}\hat{\boldsymbol{\Omega}}\hat{\mathbf{h}}^T\mathbf{H}^T + \mathbf{H}\hat{\mathbf{h}}_0\hat{\boldsymbol{\Omega}}_{01}\hat{\mathbf{h}}_0^T\mathbf{H}^T + \mathbf{H}_0\mathbf{S}_{\mathbf{H}_0^T\mathbf{Y}}\mathbf{H}_0^T,
\end{aligned}$$

where the minimum is over all semi-orthogonal matrices $\mathbf{G} \in \mathbb{R}^{v\times u}$. Let $\hat{L}_u^{(1)}$ and $\hat{L}_u^{(2)}$ denote the maximized values of $L_u^{(1)}$ and $L_u^{(2)}$:

$$\begin{aligned}
\hat{L}_u^{(1)} &= -(nv/2)\log(2\pi) - nv/2 - (n/2)\log|\mathbf{S}_{\mathbf{H}^T\mathbf{Y}}|\\
&\quad -(n/2)\log|\hat{\mathbf{h}}^T\mathbf{S}_{\mathbf{H}^T\mathbf{Y}|\mathbf{X}}\hat{\mathbf{h}}| - (n/2)\log|\hat{\mathbf{h}}^T\mathbf{S}_{\mathbf{H}^T\mathbf{Y}}^{-1}\hat{\mathbf{h}}|,\\
\hat{L}_u^{(2)} &= -(n(r-v)/2)\log(2\pi) - n(r-v)/2 - (n/2)\log|\mathbf{S}_{\mathbf{H}_0^T\mathbf{Y}}|,\\
\hat{L}_u(\mathcal{H}) &= \hat{L}_u^{(1)} + \hat{L}_u^{(2)}\\
&= -(nr/2)\log(2\pi) - (nr/2) - (n/2)\log|\mathbf{S}_{\mathbf{H}^T\mathbf{Y}}| - (n/2)\log|\mathbf{S}_{\mathbf{H}_0^T\mathbf{Y}}|\\
&\quad -(n/2)\log|\hat{\mathbf{h}}^T\mathbf{S}_{\mathbf{H}^T\mathbf{Y}|\mathbf{X}}\hat{\mathbf{h}}| - (n/2)\log|\hat{\mathbf{h}}^T\mathbf{S}_{\mathbf{H}^T\mathbf{Y}}^{-1}\hat{\mathbf{h}}|.
\end{aligned}$$

Combining these results with (1.27), we obtain

$$\begin{aligned}
\Lambda_u(\mathcal{H}) = n\Big\{&\log|\mathbf{S}_{\mathbf{H}^T\mathbf{Y}}| + \log|\mathbf{S}_{\mathbf{H}_0^T\mathbf{Y}}| - \log|\mathbf{S}_{\mathbf{Y}}| + \log|\hat{\mathbf{h}}^T\mathbf{S}_{\mathbf{H}^T\mathbf{Y}|\mathbf{X}}\hat{\mathbf{h}}|\\
&+ \log|\hat{\mathbf{h}}^T\mathbf{S}_{\mathbf{H}^T\mathbf{Y}}^{-1}\hat{\mathbf{h}}| - \log|\hat{\boldsymbol{\Gamma}}^T\mathbf{S}_{\mathbf{Y}|\mathbf{X}}\hat{\boldsymbol{\Gamma}}| - \log|\hat{\boldsymbol{\Gamma}}^T\mathbf{S}_{\mathbf{Y}}^{-1}\hat{\boldsymbol{\Gamma}}|\Big\}. \qquad (1.40)
\end{aligned}$$

Under the hypothesis, $\Lambda_u(\mathcal{H})$ is distributed asymptotically as a chi-square random variable with $v(r-v)$ degrees of freedom.

1.8.2 Testing Individual Responses

In this section, we discuss tests to determine if the distribution of a selected subset of the responses, $\mathbf{Y}_1 \in \mathbb{R}^s, s < r$, is unaffected by changing the predictors. Without loss of generality, assume that those responses are the first s components in $\mathbf{Y} = (\mathbf{Y}_1^T, \mathbf{Y}_2^T)^T$, so we can write a partitioned form of model (1.1) as

$$\begin{pmatrix}\mathbf{Y}_1\\ \mathbf{Y}_2\end{pmatrix} = \begin{pmatrix}\alpha_1\\ \alpha_2\end{pmatrix} + \begin{pmatrix}\boldsymbol{\beta}_1\mathbf{X}\\ \boldsymbol{\beta}_2\mathbf{X}\end{pmatrix} + \begin{pmatrix}\varepsilon_1\\ \varepsilon_2\end{pmatrix}, \qquad (1.41)$$

where $\mathbf{Y}_2 \in \mathbb{R}^{r-s}$, and the partitioning of $\boldsymbol{\beta}$ corresponds to the partitioning of $\mathbf{Y}$, with $\boldsymbol{\beta}_1 \in \mathbb{R}^{s \times p}$ and $\boldsymbol{\beta}_2 \in \mathbb{R}^{(r-s)\times p}$. The specific requirements on $\mathbf{Y}_1$ under this hypothesis are

$$\text{(i) } \mathbf{Y}_1 \mid (\mathbf{X} = \mathbf{x}_1) \sim \mathbf{Y}_1 \mid (\mathbf{X} = \mathbf{x}_2) \text{ and (ii) } \mathbf{Y}_1 \perp\!\!\!\perp \mathbf{Y}_2 \mid \mathbf{X}. \tag{1.42}$$

This hypothesis corresponds to (1.37) with $\mathbf{H} = (0, \mathbf{I}_{r-s})^T \in \mathbb{R}^{r\times(r-s)}$, $v = r - s$, and $\mathbf{H}_0 = (\mathbf{I}_s, 0)^T \in \mathbb{R}^{r\times s}$. Assuming that (1.42) holds, the responses in $\mathbf{Y}_1$ are X-invariant. Generally, the response subvector $\mathbf{Y}_2$ may hold X-invariant responses as well because they are not determined solely by the value of the corresponding regression coefficients $\boldsymbol{\beta}_1$. It follows from (1.42(i)) that $\boldsymbol{\beta}_1 = 0$ is a necessary condition for $\mathbf{Y}_1$ to be X-invariant, but it is not sufficient since we also require (1.42(ii)) to hold so information from $\mathbf{Y}_1$ does not contribute to the estimation of $\boldsymbol{\beta}_2$ via an association with $\mathbf{Y}_2$. Additionally, it follows from the discussion in Section 1.8.1 that under (1.42) semi-orthogonal bases $\boldsymbol{\Gamma}$ and $\boldsymbol{\Gamma}_0$ for $\mathcal{E}_{\boldsymbol{\Sigma}}(\mathcal{B})$ and $\mathcal{E}_{\boldsymbol{\Sigma}}^{\perp}(\mathcal{B})$ can be constructed as

$$\boldsymbol{\Gamma} = \begin{pmatrix} 0 \\ \boldsymbol{\Gamma}_2 \end{pmatrix}, \quad \boldsymbol{\Gamma}_0 = \begin{pmatrix} \mathbf{I}_s & 0 \\ 0 & \boldsymbol{\Gamma}_{2,0} \end{pmatrix},$$

where $\boldsymbol{\Gamma}_2 \in \mathbb{R}^{(r-s)\times u}$, $\boldsymbol{\Gamma}_{2,0} \in \mathbb{R}^{(r-s)\times(r-s-u)}$, and still $u = \dim(\mathcal{E}_{\boldsymbol{\Sigma}}(\mathcal{B}))$. We will return to this line of reasoning when considering sparse envelopes for response selection in Section 7.5. The estimator of $\mathbf{h}$ and the likelihood ratio statistic simplify a bit to

$$\begin{aligned}
\hat{\mathbf{h}} &= \arg\min_{\mathbf{h}}\{\log|\mathbf{h}^T\mathbf{S}_{\mathbf{Y}_2|\mathbf{X}}\mathbf{h}| + \log|\mathbf{h}^T\mathbf{S}_{\mathbf{Y}_2}^{-1}\mathbf{h}|\},\\
\Lambda_u(\mathcal{H}) &= n\{\log|\mathbf{S}_{\mathbf{Y}_2}| + \log|\mathbf{S}_{\mathbf{Y}_1}| - \log|\mathbf{S}_{\mathbf{Y}}| + \log|\hat{\mathbf{h}}^T\mathbf{S}_{\mathbf{Y}_2|\mathbf{X}}\hat{\mathbf{h}}|\\
&\quad + \log|\hat{\mathbf{h}}^T\mathbf{S}_{\mathbf{Y}_2}^{-1}\hat{\mathbf{h}}| - \log|\hat{\boldsymbol{\Gamma}}^T\mathbf{S}_{\mathbf{Y}|\mathbf{X}}\hat{\boldsymbol{\Gamma}}| - \log|\hat{\boldsymbol{\Gamma}}^T\mathbf{S}_{\mathbf{Y}}^{-1}\hat{\boldsymbol{\Gamma}}|\}.
\end{aligned}$$

A brief illustration on the use of this methodology is given in Section 2.8.

We conclude this section by giving additional discussion to emphasize an important difference between inference on responses and coefficients, using results from Su et al. (2016). Suppose that an oracle told us that in fact $\boldsymbol{\beta}_1 = 0$ in (1.41). Should we now estimate $\boldsymbol{\beta}_2$ from the constrained model

$$\begin{pmatrix} \mathbf{Y}_1 \\ \mathbf{Y}_2 \end{pmatrix} = \begin{pmatrix} \boldsymbol{\alpha}_1 \\ \boldsymbol{\alpha}_2 \end{pmatrix} + \begin{pmatrix} 0 \\ \boldsymbol{\beta}_2\mathbf{X} \end{pmatrix} + \begin{pmatrix} \boldsymbol{\varepsilon}_1 \\ \boldsymbol{\varepsilon}_2 \end{pmatrix} \tag{1.43}$$

or the reduced model

$$\mathbf{Y}_2 = \boldsymbol{\alpha}_2 + \boldsymbol{\beta}_2\mathbf{X} + \boldsymbol{\varepsilon}_2?$$

We address this question by comparing $\mathbf{B}_{2,\mathrm{C}}$, the maximum likelihood estimator of $\boldsymbol{\beta}_2$ from the constrained model, and $\mathbf{B}_{2,\mathrm{R}}$, the maximum likelihood estimator of $\boldsymbol{\beta}_2$ from the reduced model. We need some additional notation for this comparison. Let $\mathbf{B}_1$ and $\mathbf{B}_2$ denote the maximum likelihood estimators from the separate regressions of $\mathbf{Y}_1$ and $\mathbf{Y}_2$ on $\mathbf{X}$. Let $\mathbf{r}_1$ and $\mathbf{r}_2$ denote the residual vectors from the separate regressions of $\mathbf{Y}_1$ and $\mathbf{Y}_2$ on $\mathbf{X}$, and let $\mathbf{B}_{2|1}$ denote the coefficient matrix from the ordinary least squares regression of $\mathbf{r}_2$ on $\mathbf{r}_1$, which is the same as the estimate of $\boldsymbol{\beta}_2$ from the full model (1.41). Let $\boldsymbol{\Sigma}_{jj} = \text{var}(\mathbf{Y}_j \mid \mathbf{X})$, let $\boldsymbol{\Sigma}_{ij} = \text{cov}(\mathbf{Y}_i, \mathbf{Y}_j \mid \mathbf{X})$, $i, j = 1, 2$, and let $\boldsymbol{\Sigma}_{2|1} = \text{var}(\mathbf{Y}_2 \mid \mathbf{Y}_1, \mathbf{X}) = \boldsymbol{\Sigma}_{22} - \boldsymbol{\Sigma}_{21}\boldsymbol{\Sigma}_{11}^{-1}\boldsymbol{\Sigma}_{12}$. Then (Su et al. 2016),

$$\mathbf{B}_{2,\text{C}} = \mathbf{B}_2 - \mathbf{B}_{2|1}\mathbf{B}_1 \ \text{ with } \ \text{avar}(\sqrt{n}\mathbf{B}_{2,\text{C}}) = \boldsymbol{\Sigma}_{\mathbf{X}}^{-1} \otimes \boldsymbol{\Sigma}_{2|1},$$

$$\mathbf{B}_{2,\text{R}} = \mathbf{B}_2 \ \text{ with } \ \text{avar}(\sqrt{n}\mathbf{B}_{2,\text{R}}) = \boldsymbol{\Sigma}_{\mathbf{X}}^{-1} \otimes \boldsymbol{\Sigma}_{22}.$$

From this we see that $\mathbf{B}_{2,\text{C}} \neq \mathbf{B}_{2,\text{R}}$, with $\mathbf{B}_{2,\text{C}}$ being an adjusted version of $\mathbf{B}_{2,\text{R}}$ that accounts for the conditional correlation between $\mathbf{Y}_1$ and $\mathbf{Y}_2$. More importantly,

$$\text{avar}(\sqrt{n}\mathbf{B}_{2,\text{R}}) - \text{avar}(\sqrt{n}\mathbf{B}_{2,\text{C}}) = \boldsymbol{\Sigma}_{\mathbf{X}}^{-1} \otimes \boldsymbol{\Sigma}_{21}\boldsymbol{\Sigma}_{11}^{-1}\boldsymbol{\Sigma}_{12} \geq 0,$$

so the estimator $\mathbf{B}_{2,\text{C}}$ from the constrained model is always at least as good as the estimator $\mathbf{B}_{2,\text{R}}$ from the reduced model and may be substantially better depending on $\boldsymbol{\Sigma}_{12}$, the two estimators being asymptotically equivalent when $\boldsymbol{\Sigma}_{12} = 0$.

1.8.3 Testing Containment Only

In some regressions, we may wish to test the hypothesis $\mathcal{E}_{\boldsymbol{\Sigma}}(\mathcal{B}) \subseteq \mathcal{H}$ of containment only. This hypothesis satisfies (1.37(i)), but not necessarily (1.37(ii)). Under this hypothesis, we still can represent the basis $\boldsymbol{\Gamma}$ for $\mathcal{E}_{\boldsymbol{\Sigma}}(\mathcal{B})$ as $\boldsymbol{\Gamma} = \mathbf{H}\mathbf{h}$, but we no longer necessarily have $\mathbf{H}^T\mathbf{Y} \perp\!\!\!\perp \mathbf{H}_0^T\mathbf{Y} \mid \mathbf{X}$. As a consequence, the estimator of $\mathbf{h}$ is determined as

$$\hat{\mathbf{h}} = \arg\min_{\mathbf{G}}\{\log|\mathbf{G}^T\mathbf{S}_{\mathbf{H}^T\mathbf{Y}|\mathbf{X}}\mathbf{G}| + \log|\mathbf{G}^T\mathbf{S}_{\mathbf{H}^T\mathbf{Y}|\mathbf{H}_0\mathbf{Y}}^{-1}\mathbf{G}|\},$$

where the second term of the objective function is now the residual covariance matrix from the regression of $\mathbf{H}^T\mathbf{Y}$ on $\mathbf{H}_0^T\mathbf{Y}$, rather than the marginal covariance matrix of $\mathbf{H}^T\mathbf{Y}$. The remaining parameters and the likelihood ratio statistic can be determined following the steps in the previous development.

1.9 Nonnormal Errors

Again consider model (1.1), but now relax the condition that the errors are normally distributed. The structure of an envelope described in Definition 1.2 requires only $\boldsymbol{\beta}$ and $\boldsymbol{\Sigma}$; it does not require normality. This implies that the coordinate form of the envelope model (1.20) is still applicable with nonnormal errors, although now the condition $\boldsymbol{\Gamma}^T\mathbf{Y} \perp\!\!\!\perp \boldsymbol{\Gamma}_0^T\mathbf{Y} \mid \mathbf{X}$ is replaced with $\text{cov}(\boldsymbol{\Gamma}^T\mathbf{Y}, \boldsymbol{\Gamma}_0^T\mathbf{Y} \mid \mathbf{X}) = 0$. Nevertheless, the goal under model (1.20) remains the estimation of $\boldsymbol{\beta} = \boldsymbol{\Gamma}\boldsymbol{\eta}$ and $\boldsymbol{\Sigma} = \boldsymbol{\Gamma}\boldsymbol{\Omega}\boldsymbol{\Gamma}^T + \boldsymbol{\Gamma}_0\boldsymbol{\Omega}_0\boldsymbol{\Gamma}_0^T$.

Lacking knowledge of the distribution of the errors, we need to decide how to estimate $\boldsymbol{\beta}$ and $\boldsymbol{\Sigma}$. One natural route is to base estimation on the least squares estimators $\mathbf{B}$ and $\mathbf{S}_{\mathrm{Y}|\mathrm{X}}$ by selecting an objective function to fit the mean and variance structures of model (1.20). There are likely many ways to proceed, but one good way is to use the partially maximized log-likelihood $L_2(\mathcal{E}_{\boldsymbol{\Sigma}}(\mathcal{B}))$ (1.24) to fill this role for the purpose of estimating the envelope. It can be used straightforwardly since it is a function of only $\mathbf{B}$ and $\mathbf{S}_{\mathrm{Y}|\mathrm{X}}$. The remaining parameters are then estimated as described in Section 1.5. Since we are not assuming normality, these estimators no longer inherit optimality properties from general likelihood theory, so a different approach is needed to study them.

Lemma 1.1 *The sample matrices* $\mathbf{B}$, $\mathbf{S}_{\mathrm{Y}|\mathrm{X}}$ *and* $\mathbf{S}_{\mathrm{Y}}$ *are* $\sqrt{n}$ *consistent estimators of their population counterparts* $\boldsymbol{\beta}$, $\boldsymbol{\Sigma} = \boldsymbol{\Gamma}\boldsymbol{\Omega}\boldsymbol{\Gamma}^T + \boldsymbol{\Gamma}_0\boldsymbol{\Omega}_0\boldsymbol{\Gamma}_0^T$ *and* $\boldsymbol{\Sigma}_{\mathrm{Y}} = \boldsymbol{\Sigma} + \boldsymbol{\Gamma}\boldsymbol{\eta}\boldsymbol{\Sigma}_{\mathrm{X}}\boldsymbol{\eta}^T\boldsymbol{\Gamma}^T$.

Recall from (1.25) that $\widehat{\mathcal{E}}_{\boldsymbol{\Sigma}}(\mathcal{B}) = \mathrm{span}\{\arg\min_{\mathbf{G}}(\log|\mathbf{G}^T\mathbf{S}_{\mathrm{Y}|\mathrm{X}}\mathbf{G}| + \log|\mathbf{G}_0^T\mathbf{S}_{\mathrm{Y}}\mathbf{G}_0|)\}$. It follows from this lemma that $\log|\mathbf{G}^T\mathbf{S}_{\mathrm{Y}|\mathrm{X}}\mathbf{G}| + \log|\mathbf{G}_0^T\mathbf{S}_{\mathrm{Y}}\mathbf{G}_0|$ converges in probability to $\log|\mathbf{G}^T\boldsymbol{\Sigma}\mathbf{G}| + \log|\mathbf{G}_0^T(\boldsymbol{\Sigma} + \boldsymbol{\Gamma}\boldsymbol{\eta}\boldsymbol{\Sigma}_{\mathrm{X}}\boldsymbol{\eta}^T\boldsymbol{\Gamma}^T)\mathbf{G}_0|$. This population objective function is covered by Proposition 6.1, and consequently it follows that

$$\mathcal{E}_{\boldsymbol{\Sigma}}(\mathcal{B}) = \mathrm{span}\{\arg\min_{\mathbf{G}}\,(\log|\mathbf{G}^T\boldsymbol{\Sigma}\mathbf{G}| + \log|\mathbf{G}_0^T\boldsymbol{\Sigma}_{\mathrm{Y}}\mathbf{G}_0|)\},$$

and thus that the normal-theory objective function recovers $\mathcal{E}_{\boldsymbol{\Sigma}}(\mathcal{B})$ in the population without actually assuming normality.

Going further, assume that the errors have finite fourth moments and that as $n \to \infty$ the maximum diagonal element of $\mathbf{P}_{\mathbb{X}}$ converges to 0. Under these conditions,

$$\sqrt{n}\begin{pmatrix} \mathrm{vec}(\mathbf{B}) - \mathrm{vec}(\boldsymbol{\beta}) \\ \mathrm{vech}(\mathbf{S}_{\mathrm{Y}|\mathrm{X}}) - \mathrm{vech}(\boldsymbol{\Sigma}) \end{pmatrix}$$

converges to a normal random vector with mean 0 and nonsingular covariance matrix (Su and Cook 2012). It then follows from Shapiro (1986) that with known u

$$\sqrt{n}\begin{pmatrix} \mathrm{vec}(\widehat{\boldsymbol{\beta}}) - \mathrm{vec}(\boldsymbol{\beta}) \\ \mathrm{vech}(\widehat{\boldsymbol{\Sigma}}) - \mathrm{vech}(\boldsymbol{\Sigma}) \end{pmatrix}$$

also converges to a normal random vector with mean 0 and nonsingular covariance matrix. Consequently, using the normal likelihood for estimation under nonnormality still produces asymptotically normal $\sqrt{n}$-consistent estimators.

Efficiency gains, as illustrated in Figure 1.3, can still accrue without normality, but now they are judged relative to the least squares estimators $\mathbf{B}$ and $\mathbf{S}_{\mathrm{Y}|\mathrm{X}}$ rather than maximum likelihood estimators. However, the normal-theory asymptotic variances given in Section 1.6 are no longer applicable. While expressions for the asymptotic variances can be derived, it will likely be difficult to use them as the basis for estimated variances in practice. The residual bootstrap (see Section 1.11 and Freedman 1981) offers a practically useful alternative.

1.10 Selecting the Envelope Dimension, *u*

The dimension u is in effect a model-selection parameter, rather like the rank of $\boldsymbol{\beta}$ in reduced-rank regression (see Section 9.2.1), the power when transforming the response in linear regression, the degree in polynomial regression, or the number of components in a mixture regression. Our discussion so far has treated u as known, but this will typically not be so in applications.

1.10.1 Selection Methods

In this section, we discuss selecting u by using sequential likelihood ratio testing, an information criterion such as AIC or BIC, or cross-validation.

1.10.1.1 Likelihood Ratio Testing

As mentioned in Section 1.4, two envelope models with different values for u are not necessarily nested, but an envelope model is always nested within the standard model (1.1), which arises when $u = r$. The likelihood ratio for testing an envelope model against the standard model can be cast as a test of the hypothesis $u = u_0$ versus the alternative $u = r$. The likelihood ratio statistic for this hypothesis is $\Lambda(u_0) = 2(\widehat{L}_r - \widehat{L}_{u_0})$, where $\widehat{L}_{u_0}$ is the maximized envelope log-likelihood given in (1.27), and $\widehat{L}_r$ is the maximized log-likelihood under the standard model, $\widehat{L}_r = -(nr/2)\log(2\pi) - nr/2 - (n/2)\log|\mathbf{S}_{\mathrm{Y}|\mathrm{X}}|$, giving

$$\Lambda(u_0) = n\log|\mathbf{S}_{\mathrm{Y}}| + n\log|\widehat{\boldsymbol{\Gamma}}^T\mathbf{S}_{\mathrm{Y}|\mathrm{X}}\widehat{\boldsymbol{\Gamma}}| + n\log|\widehat{\boldsymbol{\Gamma}}^T\mathbf{S}_{\mathrm{Y}}^{-1}\widehat{\boldsymbol{\Gamma}}| - n\log|\mathbf{S}_{\mathrm{Y}|\mathrm{X}}|. \tag{1.44}$$

Under the null hypothesis, this statistic is distributed asymptotically as a chi-squared random variable with $p(r - u_0)$ degrees of freedom. Schott (2013) obtained improved approximations for the asymptotic null distribution of $\Lambda(u_0)$ by using a saddlepoint expansion and he demonstrated that his method can outperform the chi-squared approximation in some settings. He also demonstrated by simulation that the chi-squared approximation can produce very large significance levels when the sample size is small. Due to its simplicity, we will use the chi-squared approximation in illustrations.

The likelihood ratio test statistic $\Lambda(u_0)$ can be used sequentially to estimate u: Starting with $u_0 = 0$, test the hypothesis $u = u_0$ against $u = r$ at a selected level α. If the hypothesis is rejected, increment u_0 by 1 and test again.

The estimate $\hat{u}$ of u is the first hypothesized value that is not rejected. We indicate this estimator using the notation LRT(α). When testing $u = 0$ versus $u = r$, the envelope basis $\boldsymbol{\Gamma} = 0$ under the null hypothesis, the two terms in $\Lambda(0)$ involving $\widehat{\boldsymbol{\Gamma}}$ do not appear in (1.44) and the test statistic reduces to

$$\Lambda(0) = n \log \frac{|\mathbf{S}_{\mathbf{Y}}|}{|\mathbf{S}_{\mathbf{Y}|\mathbf{X}}|},$$

which is asymptotically distributed under the null hypothesis as a chi-square random variable with pr degrees of freedom. This is the same as the usual likelihood ratio statistic (1.13) for testing $\boldsymbol{\beta} = 0$ in model (1.1).

1.10.1.2 Information Criteria

The envelope dimension can also be selected by using an information criterion:

$$\hat{u} = \arg\min_{u}\{-2\hat{L}_u + h(n)N_u\}, \tag{1.45}$$

where N_u is the number of envelope parameters given in (1.21) and $h(n) = \log n$ for BIC and $h(n) = 2$ for AIC. Theoretical results (Su and Cook 2013, Proposition 4) supported by simulations indicate that AIC will tend to select a model that contains the true model and thus will tend to overestimate u. BIC will select the correct u with probability tending to 1 as $n \to \infty$ (Yang 2005), but it can be slow to respond in small samples. LRT(α) can perform well depending on the sample size, but asymptotically it makes an error with rate α. What constitutes a small or large sample in any particular application depends on other characteristics of the regression, including the strength of the signal. It may be useful to use all three methods in applications, giving a preference to BIC and LRT if there is disagreement, or using the largest estimate of u in cases where it is desirable to be conservative.

1.10.1.3 Cross-validation

m-Fold cross-validation is used to select the dimension of the envelope based on prediction performance. For each u, the data are randomly partitioned into m parts of approximately equal size and each part is used in turn for testing prediction performance while the remaining $m - 1$ parts are used for fitting. The dimension selected is the one that minimizes the average prediction errors. A positive definite inner product matrix $\mathbf{M}$ is necessary to map the fitted response vectors to $\mathbb{R}^1$, so the cross-validation criterion is

$$\mathrm{CV}(u) = n^{-1} \sum_{j=1}^{m} \sum_{k=1}^{n_j} (\mathbf{Y}_{jk} - \widehat{\mathbf{Y}}_{jk}^{(u)})^T \mathbf{M} (\mathbf{Y}_{jk} - \widehat{\mathbf{Y}}_{jk}^{(u)}),$$

where j indexes the part, k indexes observations within part j, and $\widehat{\mathbf{Y}}_{jk}^{(u)}$ indicates the fitted vector for a selected u. Then $\hat{u} = \arg\min_u \mathrm{CV}(u)$. For best results, this procedure should be repeated for several random partitions of the data and the conclusions based on the overall average. Cross-validation will tend to balance variance and bias in its selection of u and so may naturally lead to choices that are different from those indicated by LRT(α) or an information criterion. We use the identity inner product $\mathbf{M} = \mathbf{I}_r$ in illustrations, unless indicated otherwise.

1.10.2 Inferring About rank(β)

The dimension u of the envelope cannot be less than the rank, $k = \text{rank}(\boldsymbol{\beta}) \leq \min(r, p)$, of the population coefficient matrix. For this reason, it can be useful to have some knowledge of k before using the methods of Section 1.10.1 to select u. Bura and Cook (2003) developed a method for estimating k based on a series of chi-squared tests, similar to the LRT method for determining u as discussed in Section 1.10.1.

Let $\hat{\varphi}_1 \geq \hat{\varphi}_2 \geq \cdots \geq \hat{\varphi}_{\min(p,r)}$ denote the singular values of the standardized coefficient matrix

$$\mathbf{B}_{\text{std}} = ((n-p-1)/n)\tilde{\mathbf{B}} = ((n-p-1)/n)\mathbf{S}_{\mathbf{Y}|\mathbf{X}}^{-1/2}\mathbf{B}\mathbf{S}_{\mathbf{X}}^{1/2},$$

where $\tilde{\mathbf{B}}$ is as defined previously in (1.6). The statistic for the hypothesis $k = k_0$ is $\Lambda(k_0) = n \sum_{i=k_0+1}^{\min(p,r)} \hat{\varphi}_i^2$. Under the hypothesis $k = k_0$, Bura and Cook showed that $\Lambda(k_0)$ is distributed asymptotically as a chi-squared random variable with $(p-k_0)(r-k_0)$ degrees of freedom. This conclusion is based essentially on only the requirement that $\mathbf{B}$ be asymptotically normal. The statistic $\Lambda(k_0)$ can be used in a sequential manner to provide an estimator of k: beginning with $k_0 = 0$, test $k = k_0$ at a preselected level α. If the test is rejected, increment k_0 by 1 and test again, terminating the first time the hypothesis is not rejected, in which case the current value of k_0 is taken as the estimator of k.

1.10.3 Asymptotic Considerations

It has been a common practice in applied statistics to perform postselection inference, treating the selected model as if it had been known a priori. This practice can be problematic because it neglects the model selection process that can distort classical inference in the known-model context. Nevertheless, finding a general solution is challenging because model selection is often a complex process that defies characterization. In the context of this book, choosing u is the model selection step, although there could also be selection involved in the choice of the original multivariate model, prior to the introduction of envelopes.

The main point of this section is that in some settings it may be appropriate to treat $\hat{u}$ as if it had been selected a priori, provided that $\Pr(\hat{u} \neq u)$ is sufficiently small. This is achieved asymptotically if $\Pr(\hat{u} = u) \to 1$ as $n \to \infty$, as it does with BIC. To state this result in detail, let $\hat{\boldsymbol{\beta}}_u$ and $\hat{\boldsymbol{\beta}}_{\hat{u}}$ denote the envelope estimators of $\boldsymbol{\beta}$ at the true value and estimated values of u, and for any fixed vector $\mathbf{c} \in \mathbb{R}^{pr}$, let $\xi_n(u, \mathbf{c}) = \sqrt{n}\mathbf{c}^T \operatorname{vec}(\hat{\boldsymbol{\beta}}_u - \boldsymbol{\beta})$. Then we know from Proposition 1.1 that $\xi_n(u, \mathbf{c})$ is asymptotically normal. The essential idea for the following proposition, which characterizes the asymptotic distribution of $\hat{\boldsymbol{\beta}}_{\hat{u}}$, was provided by Zhang (2017).

Proposition 1.2 *Assume envelope model (1.20) holds. Then for any $\delta \in \mathbb{R}$ and any $\mathbf{c} \in \mathbb{R}^{pr}$,*

$$|\Pr(\xi_n(\hat{u}, \mathbf{c}) > \delta) - \Pr(\xi_n(u, \mathbf{c}) > \delta)| \leq \Pr(\hat{u} \neq u).$$

In addition, if the method of selecting u is consistent, $\Pr(\hat{u} = u) \to 1$ as $n \to \infty$, then as $n \to \infty$,

$$|\Pr(\xi_n(\hat{u}, \mathbf{c}) > \delta) - \Pr(\xi_n(u, \mathbf{c}) > \delta)| \to 0.$$

Proof: The proof of this proposition hinges on the fact that the parameter space for u is discrete.

$$\begin{aligned}
\Pr(\xi_n(\hat{u}, \mathbf{c}) > \delta) &= \Pr(\xi_n(\hat{u}, \mathbf{c}) > \delta, \hat{u} = u) + \Pr(\xi_n(\hat{u}, \mathbf{c}) > \delta, \hat{u} \neq u) \\
&= \Pr(\xi_n(u, \mathbf{c}) > \delta, \hat{u} = u) + \Pr(\xi_n(\hat{u}, \mathbf{c}) > \delta, \hat{u} \neq u) \\
&\leq \Pr(\xi_n(u, \mathbf{c}) > \delta) + \Pr(\hat{u} \neq u).
\end{aligned}$$

So

$$\Pr(\xi_n(\hat{u}, \mathbf{c}) > \delta) - \Pr(\xi_n(u, \mathbf{c}) > \delta) \leq \Pr(\hat{u} \neq u).$$

Similarly,

$$\begin{aligned}
\Pr(\xi_n(\hat{u}, \mathbf{c}) > \delta) &= 1 - \Pr(\xi_n(\hat{u}, \mathbf{c}) \leq \delta) \\
&= 1 - \Pr(\xi_n(\hat{u}, \mathbf{c}) \leq \delta, \hat{u} = u) - \Pr(\xi_n(\hat{u}, \mathbf{c}) \leq \delta, \hat{u} \neq u) \\
&\geq 1 - \Pr(\xi_n(u, \mathbf{c}) \leq \delta) - \Pr(\hat{u} \neq u) \\
&= \Pr(\xi_n(u, \mathbf{c}) > \delta) - \Pr(\hat{u} \neq u).
\end{aligned}$$

So

$$\Pr(\xi_n(u, \mathbf{c}) > \delta) - \Pr(\xi_n(\hat{u}, \mathbf{c}) > \delta) \leq \Pr(\hat{u} \neq u).$$

Consequently,

$$|\Pr(\xi_n(\hat{u}, \mathbf{c}) > \delta) - \Pr(\xi_n(u, \mathbf{c}) > \delta)| \leq \Pr(\hat{u} \neq u) \to 0,$$

which is the desired conclusion. The Cramer–Wold device can be used to extend this to the asymptotic distributions of $\sqrt{n}\operatorname{vec}(\hat{\boldsymbol{\beta}}_{\hat{u}} - \boldsymbol{\beta})$ and $\sqrt{n}\operatorname{vec}(\hat{\boldsymbol{\beta}}_u - \boldsymbol{\beta})$. □

To illustrate the implications of Proposition 1.2, the parameter estimates $\hat{\alpha}$, $\hat{\beta}$, and $\hat{\Sigma}$ from the fit of the cattle data with $u = 3$ were taken as population values. A new set of data was then generated as

$$\mathbf{Y}_i = \hat{\alpha} + \hat{\beta}X_i + \hat{\Sigma}^{1/2}\varepsilon_i, \quad i = 1, \dots, n^*,$$

where n^* denotes simulation sample size, and the errors ε_i are independent copies of a $N_{10}(0, \mathbf{I})$ random vector. Because the predictors are fixed, the simulation sample size $n^* = r \times n$ was increased in multiples r of the sample size $n = 60$ for the cattle data, repeating the 60 values of X for each multiple. The BIC estimates of u were then computed for 100 replicates of this scenario for each of the simulation sample sizes $n^* = 2^k n$, $k = 0, 1, \dots, 6$. The percentage of replicates in which BIC selected the true value $u = 3$ is shown in Figure 1.8 for each n^*.

The simulation results indicate that BIC nearly always selects $u = 3$ at $n^* = 16n$ and thereafter, with $\Pr(\hat{u} \neq u)$ estimated to be 0.01 at $n^* = 64n$. For regressions where the sample size may not be large enough to instill confidence that $\Pr(\hat{u} \neq u)$ is sufficiently small, additional intuition and bootstrap methods for gaining data-analytic guidance are discussed later in this chapter. An alternative weighted estimator that does not require choosing a value for u is presented in Section 1.11.4.

Proposition 1.2 may have implications in other model-selection problems as well. Consider, for instance, the common practice of using the Box–Cox method to estimate a power transformation $Y^{(\lambda)}$ of a univariate response Y to induce a linear regression model. Can we reasonably treat the estimated power $\hat{\lambda}$ as being nonstochastic in subsequent inference statements? Or are we obliged to take the variability in $\hat{\lambda}$ into account? This issue was the focus of considerable discussion in the early 1980s (e.g. Bickel and Doksum 1981; Box and Cox 1982;

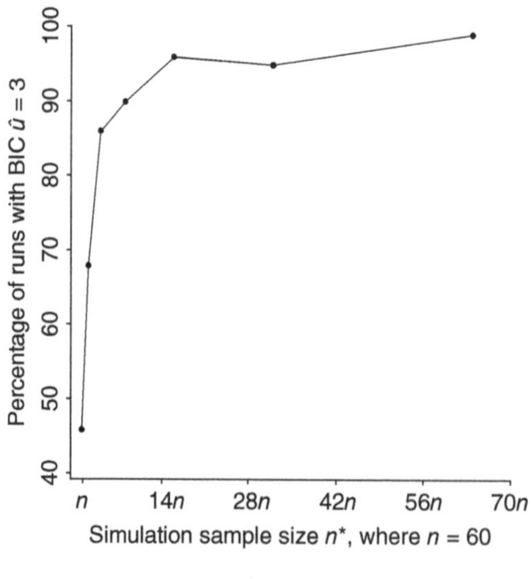

Figure 1.8 Cattle data: simulation results based on the fit of the cattle data with $u = 3$. The horizontal axis denotes the simulation sample size $n^* = 2^k n$, where $n = 60$ is the sample size for the original data.

Hinkley and Runger 1984, with discussion). If the transformation parameter λ is restricted to a small finite set of plausible values and $\Pr(\hat{\lambda} \neq \lambda)$ is sufficiently small, then it may well be reasonable to treat $\hat{\lambda}$ as nonstochastic.

1.10.4 Overestimation Versus Underestimation of *u*

For clarity in this section, we use u to denote the true dimension of the envelope, and let $u_{\rm f}$ denote the value used in fitting. In this discussion, we treat $u_{\rm f}$ as if it had been selected a priori. Overestimation occurs then $u_{\rm f} > u$, while underestimation occurs when $u_{\rm f} < u$. Overestimation is perhaps the less serious error. Fitting with $u_{\rm f} > u$ gives a $\sqrt{n}$-consistent estimator of $\boldsymbol{\beta}$, although the estimator will be more variable than that with $u_{\rm f} = u$. For instance, fitting with $u_{\rm f} = r$ reduces the regression to the standard model (1.1). Underestimation produces inconsistent estimators, and for this reason, it may be the more serious error.

We saw in Section 1.5.3 that the objective function serves to control a measure of bias. To gain intuition into the impact of underestimation of u, consider the population version of the representation of the objective function given in (1.29) allowing for fitting when $u_{\rm f} \neq u$. Let

$$(\tilde{\mathbf{G}}, \tilde{\mathbf{G}}_0) = \arg\min_{\mathcal{O}_{u_{\rm f}}} \Big\{ \log|\boldsymbol{\Sigma}_{\mathbf{G}^T\mathbf{Y}|\mathbf{X}}| + \log|\boldsymbol{\Sigma}_{\mathbf{G}_0^T\mathbf{Y}|\mathbf{X}}| \\ + \log|\mathbf{I}_{r-u_{\rm f}} + \tilde{\boldsymbol{\beta}}_{\mathbf{G}_0^T\mathbf{Y}|\mathbf{X}}\tilde{\boldsymbol{\beta}}^T_{\mathbf{G}_0^T\mathbf{Y}|\mathbf{X}}| \Big\},$$

where the minimum is taken over the set $\mathcal{O}_{u_{\rm f}}$ of all orthogonal $r \times r$ matrices $(\mathbf{G}, \mathbf{G}_0)$ with $\mathbf{G} \in \mathbb{R}^{r\times u_{\rm f}}$ and $\tilde{\boldsymbol{\beta}}_{\mathbf{G}_0^T\mathbf{Y}|\mathbf{X}} = (\mathbf{G}_0^T\boldsymbol{\Sigma}\mathbf{G}_0)^{-1/2}\mathbf{G}_0^T\boldsymbol{\beta}\boldsymbol{\Sigma}_{\mathbf{X}}^{1/2}$ is the population coefficient matrix for the regression of the standardized response $(\mathbf{G}_0^T\boldsymbol{\Sigma}\mathbf{G}_0)^{-1/2}$ $\mathbf{G}_0^T\mathbf{Y}$ on the standardized predictor $\boldsymbol{\Sigma}_{\mathbf{X}}^{-1/2}\mathbf{X}$. The population version of the envelope estimator at $\mathbf{G}$ is $\mathbf{P}_{\mathbf{G}}\boldsymbol{\beta}$, so the bias from underestimation is $\mathbf{Q}_{\mathbf{G}}\boldsymbol{\beta} = \mathbf{G}_0\mathbf{G}_0^T\boldsymbol{\beta}$. Thus, $\tilde{\boldsymbol{\beta}}_{\mathbf{G}_0^T\mathbf{Y}|\mathbf{X}}$ is a standardized version of the coordinates $\mathbf{G}_0^T\boldsymbol{\beta}$ of this bias.

To facilitate exposition, we consider the case $p = 1$ and for clarity use σ_X^2 instead of $\boldsymbol{\Sigma}_{\mathbf{X}}$. Then

$$B(\mathbf{G}_0) := \tilde{\boldsymbol{\beta}}^T_{\mathbf{G}_0^T\mathbf{Y}|\mathbf{X}}\tilde{\boldsymbol{\beta}}_{\mathbf{G}_0^T\mathbf{Y}|\mathbf{X}} = \boldsymbol{\beta}^T\mathbf{G}_0(\mathbf{G}_0^T\boldsymbol{\Sigma}\mathbf{G}_0)^{-1}\mathbf{G}_0^T\boldsymbol{\beta}\sigma_X^2$$

is a measure of the squared length of the bias at $\mathbf{G}_0$ for the standardized regression,

$$\log\left|\mathbf{I}_{r-u_{\rm f}} + \tilde{\boldsymbol{\beta}}_{\mathbf{G}_0^T\mathbf{Y}|X}\tilde{\boldsymbol{\beta}}^T_{\mathbf{G}_0^T\mathbf{Y}|X}\right| = \log\{1 + B(\mathbf{G}_0)\}$$

and

$$(\tilde{\mathbf{G}}, \tilde{\mathbf{G}}_0) = \arg\min_{\mathcal{O}_{u_{\rm f}}}\{\log|\boldsymbol{\Sigma}_{\mathbf{G}^T\mathbf{Y}|\mathbf{X}}| + \log|\boldsymbol{\Sigma}_{\mathbf{G}_0^T\mathbf{Y}|\mathbf{X}}| + \log\{1 + B(\mathbf{G}_0)\}\}. \tag{1.46}$$

Our goal now is to provide a bound on the bias $B(\tilde{\mathbf{G}}_0)$ that results from the minimization in (1.46). Assume for ease of exposition that the eigenvalues of

$\boldsymbol{\Sigma}$ are distinct and let $\mathcal{V}_m$ denote the collection of all subsets of m eigenvectors of $\boldsymbol{\Sigma}$. Then the columns of any $\mathbf{G} \in \mathcal{V}_{u_f}$ form a subset of u_f eigenvectors of $\boldsymbol{\Sigma}$, and $\mathbf{G}_0$ is the complementary subset of $r - u_f$ eigenvectors. A bound on $B(\tilde{\mathbf{G}}_0)$ can now be constructed by minimizing (1.46) over $\mathcal{V}_{u_f}$ instead of $\mathcal{O}_{u_f}$. We know from Lemma A.14 that the sum $\log|\boldsymbol{\Sigma}_{\mathbf{G}^T\mathbf{Y}|\mathbf{X}}| + \log|\boldsymbol{\Sigma}_{\mathbf{G}_0^T\mathbf{Y}|\mathbf{X}}|$ is minimized by any $\mathbf{G} \in \mathcal{V}_{u_f}$ and that its minimum value is $\log|\boldsymbol{\Sigma}|$. Let $\dot{\mathbf{G}}_0 = \arg\min_{\mathbf{G}_0 \in \mathcal{V}_{r-u_f}} B(\mathbf{G}_0)$. Then $B(\tilde{\mathbf{G}}_0) \le B(\dot{\mathbf{G}}_0)$ since, by construction,

$$\log|\boldsymbol{\Sigma}_{\tilde{\mathbf{G}}^T\mathbf{Y}|\mathbf{X}}| + \log|\boldsymbol{\Sigma}_{\tilde{\mathbf{G}}_0^T\mathbf{Y}|\mathbf{X}}| + \log\{1 + B(\tilde{\mathbf{G}}_0)\} \le \log|\boldsymbol{\Sigma}| + \log\{1 + B(\dot{\mathbf{G}}_0)\},$$

which implies that

$$\begin{aligned} 0 &\le \log|\boldsymbol{\Sigma}_{\tilde{\mathbf{G}}^T\mathbf{Y}|\mathbf{X}}| + \log|\boldsymbol{\Sigma}_{\tilde{\mathbf{G}}_0^T\mathbf{Y}|\mathbf{X}}| - \log|\boldsymbol{\Sigma}| \\ &\le \log\{1 + B(\dot{\mathbf{G}}_0)\} - \log\{1 + B(\tilde{\mathbf{G}}_0)\}. \end{aligned}$$

We can gain insights about the potential bias $B(\tilde{\mathbf{G}}_0)$ by studying its upper bound $B(\dot{\mathbf{G}}_0)$. Let $(\dot{\mathbf{G}}, \dot{\mathbf{G}}_0)$ be an orthogonal matrix. Since $\dot{\mathbf{G}} \in \mathcal{V}_{u_f}$ reduces $\boldsymbol{\Sigma}$, we have the decomposition $\boldsymbol{\Sigma} = \dot{\mathbf{G}}\boldsymbol{\Lambda}\dot{\mathbf{G}}^T + \dot{\mathbf{G}}_0\boldsymbol{\Lambda}_0\dot{\mathbf{G}}_0^T$, where the $\boldsymbol{\Lambda}$'s are diagonal matrices of eigenvalues, and thus

$$B(\dot{\mathbf{G}}_0) = \boldsymbol{\beta}^T\dot{\mathbf{G}}_0\boldsymbol{\Lambda}_0^{-1}\dot{\mathbf{G}}_0^T\boldsymbol{\beta}\sigma_X^2. \tag{1.47}$$

From this representation, we see that the bias bound depends on (i) the length of $\boldsymbol{\beta}$, (ii) the angles between $\boldsymbol{\beta}$ and the eigenvectors of $\boldsymbol{\Sigma}$ that comprise the columns of $\dot{\mathbf{G}}_0$, and (iii) the associated eigenvalues in $\boldsymbol{\Lambda}_0$. Since $\dot{\mathbf{G}}_0$ minimizes the bias, we expect that $\boldsymbol{\beta}$ will be orthogonal to multiple columns of $\dot{\mathbf{G}}_0$ when underestimating the dimension by one or two. Large eigenvalues $\boldsymbol{\Lambda}_0$ can also result in a small bias. As discussed in Section 1.4.2, envelopes result in substantial variance reduction when $\|\boldsymbol{\Omega}_0\| \gg \|\boldsymbol{\Omega}\|$. Consequently, we expect the bias due to underestimation to be small when the corresponding analysis shows substantial variance reduction. These comments are illustrated in Figure 1.9, which was constructed like the stylized illustration of Figure 1.3, except that $\mathcal{B} = \mathrm{span}(\boldsymbol{\beta})$ no longer aligns with the smallest eigenvector of $\boldsymbol{\Sigma}$ and thus the envelope $\mathcal{E}_{\boldsymbol{\Sigma}}(\mathcal{B}) = \mathbb{R}^2$. The figure illustrates the bias that can result when underestimating u with $u_f = 1$. In Figure 1.9a, the bias is small because the angle between $\mathcal{B}$ and $\mathbf{v}_2$ is small and $\boldsymbol{\Lambda}_0$, the eigenvalue associated with $\mathbf{v}_1$, is large. The bias is larger in Figure 1.9b because we increased the length of $\boldsymbol{\beta}$ and the angle between $\mathcal{B}$ and $\mathbf{v}_2$. In this case, the dimension selection methods discussed in Section 1.10.1 will likely indicate correctly that $u = 2$, depending on the sample size, so underestimation of u may not be a worrisome issue.

Another perhaps more relevant gauge of bias is to compare $\mathbf{B}(\dot{\mathbf{G}})$ to the squared length of $\tilde{\boldsymbol{\beta}} = \boldsymbol{\Sigma}^{-1/2}\boldsymbol{\beta}\sigma_X$, which is $\boldsymbol{\beta}$ in the standardized scale akin to $\boldsymbol{\beta}_{\mathbf{G}_0^T\mathbf{Y}|X}$:

$$\frac{B(\dot{\mathbf{G}})}{\|\tilde{\boldsymbol{\beta}}\|^2} = \frac{\boldsymbol{\beta}^T\dot{\mathbf{G}}_0\boldsymbol{\Lambda}_0^{-1}\dot{\mathbf{G}}_0^T\boldsymbol{\beta}\sigma_X^2}{\boldsymbol{\beta}^T\boldsymbol{\Sigma}^{-1}\boldsymbol{\beta}\sigma_X^2} = \frac{\tilde{\boldsymbol{\beta}}^T\dot{\mathbf{G}}_0\dot{\mathbf{G}}_0^T\tilde{\boldsymbol{\beta}}}{\tilde{\boldsymbol{\beta}}^T\tilde{\boldsymbol{\beta}}}.$$

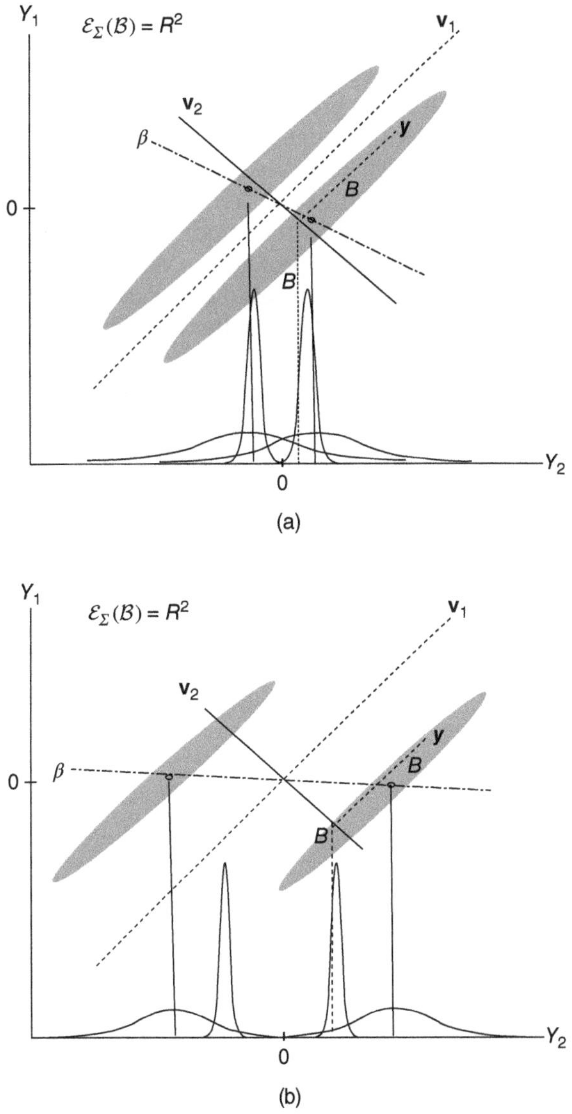

Figure 1.9 Illustrations of the potential bias that can result from underestimating u. The context is the same as that for Figure 1.3, $\mathbf{v}_1$ and $\mathbf{v}_2$ denote the eigenvectors of $\mathbf{\Sigma}$ and $\mathcal{B} = \text{span}(\boldsymbol{\beta})$. (a) Small bias; (b) large bias.

In this form, we see that the bias depends only on the angles between the standardized coefficients $\tilde{\boldsymbol{\beta}}$ and the eigenvectors in $\dot{\mathbf{G}}_0$.

1.10.5 Cattle Weights: Influence of *u*

A first step in the analysis of the cattle data in Section 1.3.4 was to determine the dimension $u = \dim(\mathcal{E}_{\mathbf{\Sigma}}(\mathcal{B}))$ of the envelope. Using the methods of this section, LRT(0.05) gave $u = 1$, which is the value used in the previous illustrations.

On the other hand, AIC and BIC gave $u = 3$, and fivefold cross-validation gave $u = 5$. In many applications, these methods agree on the value of u, but in some cases they disagree, as in this illustration. We tend to give preference to BIC and LRT since AIC has a propensity to overestimate u. Cross-validation is in effect a different criterion, and it is possible that $u = 5$ is best for prediction.

Nevertheless, when the selection methods disagree, we often find that the essential results are the same across the indicated values of u. To reinforce this point, Figure 1.10 shows the fitted profiles from envelope models with $u = 1$ and $u = 5$. These fitted profiles are very similar and lead to the same inference about their key features: the first detected differential treatment effects are around week 10 and persist thereafter.

Plots of the coordinates of $\mathbf{Q}_{\widehat{\Gamma}}\mathbf{Y}$ versus the predictors, which show the estimated X-invariant part of $\mathbf{Y}$, can also be informative diagnostics. In these plots, $\mathbf{Q}_{\widehat{\Gamma}}\mathbf{Y}$ should appear independent of the predictors if the model and choice of u are reasonable. Systematic variation with $\mathbf{X}$ may be an indication of model

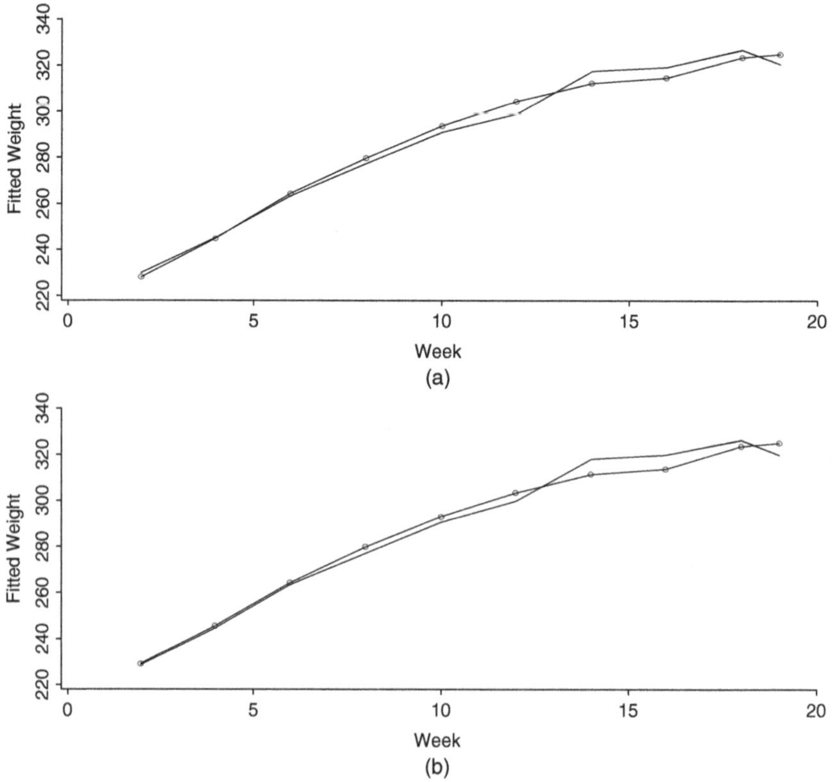

Figure 1.10 Cattle data: fitted profile plots from envelope models with (a) $u = 1$ and (b) $u = 5$.

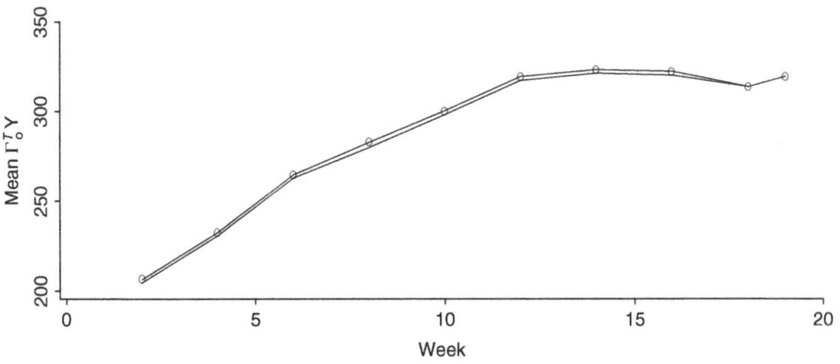

Figure 1.11 Cattle data: mean of immaterial variation $\mathbf{\Gamma}_0^T\mathbf{Y}$ for each treatment and time with $u = 3$.

failure. For instance, shown in Figure 1.11 is a plot of the 10 coordinates in the projection $\mathbf{Q}_{\widehat{\Gamma}}\bar{\mathbf{Y}}_j$ of the average response vector $\bar{\mathbf{Y}}_j$ by treatment $j = 1, 2$. The two curves, one for each treatment, are essentially identical and show no treatment clear effects. The variation in the curves themselves reflects the variation in $\mathbf{Y}$ that is common to the two treatments.

1.11 Bootstrap and Uncertainty in the Envelope Dimension

In the previous section, we discussed selection methods, when it might be reasonable to proceed as if $\widehat{u} = u$, overestimation versus underestimation, and the possibility that key aspects of inference are unaffected over a reasonable set of values for u. In this section, we turn to the bootstrap for standard errors and for assistance in addressing the uncertainty in $\widehat{u}$.

1.11.1 Bootstrap for Envelope Models

In this section, we describe how the residual bootstrap can be used to estimate the variance of $\widehat{\boldsymbol{\beta}}$ assuming that the envelope dimension u is known, perhaps relying on Proposition 1.2 to justify setting $u = \widehat{u}$. For emphasis, the envelope estimator at the true value is denoted $\widehat{\boldsymbol{\beta}}_u$ throughout our discussion of the bootstrap.

The variance of $\widehat{\boldsymbol{\beta}}_u$ can be obtained by using the asymptotic results of Section 1.6 or by using the residual bootstrap. Recall from the setup of Section 1.1 that $\mathbf{r}_i$ denotes the ith vector of residuals from the ordinary least squares fit of model (1.1). Let $\mathbf{R} \in \mathbb{R}^{n \times r}$ denote the matrix with residual vectors $\mathbf{r}_i^T$ as its rows, let $\mathbf{R}^*$ denote a resampled residual matrix constructed by sampling with replacement n rows of $\mathbf{R}$, and let $\mathbb{Y}^* = \mathbf{1}_n\widehat{\alpha}^T + \mathbb{X}\widehat{\beta}_u^T + \mathbf{R}^*$ denote the $n \times r$ matrix of bootstrapped responses. Then a single bootstrap envelope estimator $\widehat{\beta}_u^*$ of β is found from the envelope fit of model (1.20) to the data $(\mathbb{Y}^*, \mathbb{X})$. Repeating this operation B times gives the bootstrap estimators $\widehat{\beta}_{u,k}^*$, $k = 1, \ldots, B$. Then the sample variance of the $\text{vec}(\widehat{\beta}_{u,k}^*)$'s provides a bootstrap estimator of $\text{var}\{\text{vec}(\widehat{\beta}_u)\}$. The justification of this bootstrap follows from Andrews (2002, pp. 122–124 and Theorem 2) and from Eck (2017).

1.11.2 Wheat Protein: Bootstrap and Asymptotic Standard Errors, *u* Fixed

To illustrate application of the bootstrap, we consider the wheat protein data discussed in Section 1.3.3, but now with responses measured at six wavelengths instead of two. The dimension of the envelope model was estimated to be 1 by BIC, and for the purpose of this bootstrap it is assumed that $u = 1$. Two hundred (B) bootstrap samples were used throughout. The first part of Table 1.2 shows the estimated coefficients under the standard model (1.1) along with good agreement between their bootstrap and asymptotic standard errors. The second part of Table 1.2 shows the estimated envelope coefficients and the corresponding bootstrap and asymptotic standard errors. There is again a good agreement between the standard errors, which are much smaller than those for the standard model. The advantages of the envelope model in this fit are indicated roughly by the sizes of $\widehat{\Omega} = 7.88$ and $\|\widehat{\Omega}_0\| = 6517$. Thus, the envelope model has an apparent advantage because the variation $\widehat{\Omega}_0$ in the estimated X-invariant part of $\mathbf{Y}$ is considerably larger than the variation $\widehat{\Omega}$.

Table 1.2 Wheat protein data: bootstrap and asymptotic standard errors (SEs) of the six elements in $\widehat{\beta}$ under the standard (1.1) and envelope models (1.20) for the wheat protein data with six responses.

1. Standard model (1.1)						
B	3.27	8.03	7.52	−2.06	3.22	0.65
Bootstrap SE	9.87	8.12	8.70	9.65	13.90	5.48
SE	9.78	8.12	8.70	9.49	13.65	5.39
2. Envelope model with $u = 1$						
$\widehat{\beta}_u$	−1.06	4.47	3.68	−5.97	0.69	−1.60
Bootstrap SE	0.35	0.48	0.39	0.64	0.20	0.69
Asymptotic SE (1.33)	0.35	0.43	0.35	0.59	0.21	0.86

1.11.3 Cattle Weights: Bootstrapping *u*

In this section, we use the cattle data with BIC to illustrate how the bootstrap and other sampling scenarios might be used to gain data-analytic guidance on the choice of u. Part 1 of Table 1.3 gives the BIC bootstrap distribution of u based on 100 residual bootstrap datasets constructed as described in Section 1.11.1. We see from Part 1 of Table 1.3 that most of the mass is concentrated between $u = 1$ and $u = 4$. The question is how we might use these results to help with the choice of u.

Let $\widehat{\boldsymbol{\alpha}}_u$, $\widehat{\boldsymbol{\beta}}_u$, and $\widehat{\boldsymbol{\Sigma}}_u$ denote the estimates of $\boldsymbol{\alpha}$, $\boldsymbol{\beta}$, and $\boldsymbol{\Sigma}$ from the fit of the envelope model with the indicated value of u and consider generating reference data as

$$\widehat{\mathbf{Y}}_i^{(u)} = \widehat{\boldsymbol{\alpha}}_u + \widehat{\boldsymbol{\beta}}_u X_i + \widehat{\boldsymbol{\Sigma}}_u^{1/2} \boldsymbol{\varepsilon}_i, \quad i = 1, \ldots, 60, \tag{1.48}$$

where errors $\boldsymbol{\varepsilon}_i$ are independent copies of a $N_{10}(0, \mathbf{I})$ random vector. Part 2 of Table 1.3 shows the empirical distribution of the BIC estimate of u from 100 replications over the set of errors $\boldsymbol{\varepsilon}_i$, $i = 1, \ldots, 60$, with the value of u used in the fit shown in the first column. The distributions for $u = 3, 5, 7$ are similar and also similar to the distribution from the residual bootstrap in Part 1, while the distribution for $u = 1$ is notably different. One possible explanation for this

Table 1.3 Cattle data: distributions of $\hat{u}$ from BIC based on various sampling scenarios.

$\hat{u}$	0	1	2	3	4	5	6	Mean	sd
1. Residual bootstrap									
BIC	0	0.11	0.25	0.33	0.22	0.07	0.02	2.95	1.17
2. Normal errors									
$u = 1$	0	0.65	0.28	0.06	0.01	0	0	1.43	0.66
$u = 3$	0	0.11	0.22	0.43	0.20	0.03	0.01	2.85	1.02
$u = 5$	0	0.14	0.28	0.36	0.19	0.02	0.01	2.70	1.05
$u = 7$	0	0.13	0.21	0.36	0.25	0.03	0.02	2.90	1.13
3. Residual bootstrap from normal errors									
$u = 1$	0	0.43	0.27	0.22	0.04	0.04	0	1.99	1.09
$u = 3$	0	0.13	0.25	0.48	0.08	0.05	0.01	2.70	1.02
$u = 5$	0	0.16	0.23	0.33	0.15	0.10	0.03	2.89	1.29
$u = 7$	0	0.11	0.23	0.40	0.19	0.07	0	2.81	1.06

(1) 100 residual bootstrap datasets, (2) fit with the indicated value of u plus normal errors, (3) residual bootstrap from one dataset generated from a fit with the indicated value of u plus normal errors.

is as follows. If the true $u = 3$, then a population fit with $u > 3$ must give a subspace that contains the envelope. Hence, we are again led back to the model with $u = 3$. In other words, the results $u = 3, 5, 7$ are roughly as expected if in fact the true $u = 3$. Part 3 of Table 1.3 shows the BIC bootstrap distribution of $\hat{u}$ from one dataset generated from (1.48) based on 100 bootstrap samples. These distributions are similar to corresponding distributions shown in Part 2, so it again seems plausible that the true $u = 3$. These results give data-analytic support to the original BIC estimate $\hat{u} = 3$. The estimate $\hat{u} = 5$ might be used if we wanted to be conservative.

1.11.4 Bootstrap Smoothing

In this section, we expand our notation a bit and let $\hat{\boldsymbol{\beta}}_j$ denote the envelope estimator computed by assuming $u = j$. The envelope estimator at the true value is still denoted $\hat{\boldsymbol{\beta}}_u$. In the previous sections, we discussed how data-analytic methods might be used to help select the envelope dimension. In this section, we discussed one way to avoid choosing a particular dimension and instead use a weighted average of the estimators $\hat{\boldsymbol{\beta}}_j$, $j = 1, \dots, r$. By constraining $j > 0$, we are considering only regressions in which $\boldsymbol{\beta} \neq 0$, which will be reasonable in numerous applications. Extensions of the following to allow $j = 0$ are straightforward.

Let $b_j = -2\hat{L}_j + N_j \log(n)$ denote the BIC value (1.45) for the envelope model of dimension j, where $\hat{L}_j$ is the value of the maximized log-likelihood (1.27), and N_j is the number of parameters (1.21), both for the envelope model of dimension j. The weighted estimator we consider, which was proposed by Eck and Cook (2017), is of the form

$$\hat{\boldsymbol{\beta}}_w = \sum_{j=1}^{r} w_j \hat{\boldsymbol{\beta}}_j, \quad \text{where } w_j = \frac{e^{-b_j}}{\sum_{i=1}^{r} e^{-b_i}}. \tag{1.49}$$

Estimators of this form have been advocated in various contexts (Efron 2014; Nguefack-Tsague 2014; Hjort and Claeskens 2003; Burnham and Anderson 2004; Buckland et al. 1997). The estimator (1.49) may be advantageous because it bypasses the need to select a specific dimension and automatically incorporates model uncertainty in inference via BIC. It can be more variable than a specific estimator, however. For instance, if $j \geq u$, then $\hat{\boldsymbol{\beta}}_j$ corresponds to a true model, and the variability of $\hat{\boldsymbol{\beta}}_j$ can be noticeably less than that of $\hat{\boldsymbol{\beta}}_w$, depending on the sample size. The estimator $\hat{\boldsymbol{\beta}}_w$ is a $\sqrt{n}$-consistent estimator of $\boldsymbol{\beta}$, but analytic expressions of its asymptotic variance are unknown. However, the bootstrap can still be used to assess its variance.

Following the construction of $\hat{\boldsymbol{\beta}}_w$, generate n bootstrap samples $\hat{\boldsymbol{\beta}}^*_{w,k}$, $k = 1, \dots, B$, as described in Section 1.11.1, except replace $\hat{\boldsymbol{\beta}}_u$ with $\hat{\boldsymbol{\beta}}_w$ so the bootstrap responses are generated as $\mathbb{Y}^* = \mathbf{1}_n\widehat{\boldsymbol{\alpha}}^T + \mathbb{X}\widehat{\boldsymbol{\beta}}_w^T + \mathbf{R}^*$. The following proposition (Eck and Cook 2017) shows that the asymptotic distribution of $\widehat{\boldsymbol{\beta}}_w^*$ is the same as the asymptotic distribution of $\widehat{\boldsymbol{\beta}}_u^*$, the bootstrapped envelope estimator at the true dimension. In particular, the sample variance computed from the $\widehat{\boldsymbol{\beta}}_{w,k}^*$'s provides a $\sqrt{n}$-consistent estimator of the asymptotic variance of the envelope estimator $\widehat{\boldsymbol{\beta}}_u$.

Proposition 1.3 *Assume envelope model (1.20) for some $u = 1, \ldots, r$, and that $\mathbf{S}_\mathbf{X}$ converges to $\boldsymbol{\Sigma}_\mathbf{X} > 0$. Then as $n \to \infty$,*

$$\sqrt{n}\left\{\mathrm{vec}(\widehat{\boldsymbol{\beta}}_w^*) - \mathrm{vec}(\widehat{\boldsymbol{\beta}}_w)\right\} = \sqrt{n}\left\{\mathrm{vec}(\widehat{\boldsymbol{\beta}}_u^*) - \mathrm{vec}(\widehat{\boldsymbol{\beta}}_u)\right\} + O_p\left\{n^{(1/2-p)}\right\} + 2(u-1)O_p(1)\sqrt{n}e^{-n|O_p(1)|}. \tag{1.50}$$

To gain some intuition about the orders in (1.50) write

$$\begin{aligned}\sqrt{n}\{\mathrm{vec}(\widehat{\boldsymbol{\beta}}_w^*) - \mathrm{vec}(\widehat{\boldsymbol{\beta}}_w)\} &= \sqrt{n}\sum_{j=1}^{u-1}\{w_j^*\ \mathrm{vec}(\widehat{\boldsymbol{\beta}}_j^*) - w_j\mathrm{vec}(\widehat{\boldsymbol{\beta}}_j)\}\\ &+\sqrt{n}\{w_u^*\ \mathrm{vec}(\widehat{\boldsymbol{\beta}}_u^*) - w_u\ \mathrm{vec}(\widehat{\boldsymbol{\beta}}_u)\}\\ &+\sqrt{n}\sum_{j=u+1}^{r}\{w_j^*\ \mathrm{vec}(\widehat{\boldsymbol{\beta}}_j^*) - w_j\ \mathrm{vec}(\widehat{\boldsymbol{\beta}}_j)\}.\end{aligned}$$

Eck and Cook (2017) show that for $j \neq u$, $\sqrt{n}\{w_j^*\ \mathrm{vec}(\widehat{\boldsymbol{\beta}}_j^*) - w_j\ \mathrm{vec}(\widehat{\boldsymbol{\beta}}_j)\} \to 0$, so the first and third terms on the right-hand side vanish as $n \to 0$. The term $O_p\{n^{(1/2-p)}\}$ in (1.50) corresponds to the rate at which $\sqrt{n}w_j$ and $\sqrt{n}w_j^*$ converge to 0 for $j = u+1, \ldots, r$. This rate is a cost of overestimation of the envelope. It decreases quite fast, particularly when p is not small, because models with $j > u$ are true and thus have no systematic bias due to choosing the wrong dimension. The $2(u-1)\sqrt{n}e^{-n|O_p(1)|}$ term corresponds to the rate at which $\sqrt{n}w_j$ and $\sqrt{n}w_j^*$ vanish for $j = 1, \ldots, u-1$. This rate arises from under estimating the envelope space and it is affected by bias arising from choosing the wrong dimension. Because we consider only regressions in which $u \geq 1$, this term is 0 when $u = 1$. When $u = 1$, underestimation is not possible in our context and thus the term vanishes.

1.11.5 Cattle Data: Bootstrap Smoothing

Returning to the cattle data, Table 1.4 gives ratios of the standard errors of the elements of $\mathbf{B}$ to the bootstrap standard errors of the elements of $\widehat{\boldsymbol{\beta}}_w$ (second column) and to the bootstrap standard errors of the elements of $\widehat{\boldsymbol{\beta}}_j$ from

Table 1.4 Cattle data: ratios of standard errors of the elements of **B** to the bootstrap standard errors from the weighted envelope fit and the envelope fits for $u = 1, 2, 3, 4$.

		Envelope dimension			
Week	**Weighted**	**1**	**2**	**3**	**4**
2	1.48	2.38	1.83	1.34	1.19
4	1.61	3.77	1.99	1.44	1.22
6	1.81	3.30	2.48	1.57	1.29
8	2.01	4.07	3.30	1.58	1.31
10	2.09	5.30	3.80	1.59	1.34
12	1.88	3.70	2.59	1.56	1.31
14	2.02	3.94	3.19	1.59	1.33
16	2.07	4.01	3.15	1.58	1.3
18	2.14	5.08	3.38	1.51	1.28
19	1.96	5.26	2.69	1.48	1.23

the envelope fits with $j = 1, 2, 3, 4$ (columns 3–6). The table illustrates that the weighted estimator can be usefully less variable than the standard estimator. It can also be seen from the table that the standard errors for the weighted estimator all lie between those for envelope estimators with $u = 2$ and $u = 3$.

2

Illustrative Analyses Using Response Envelopes

In this chapter, we present several illustrations on the use of response envelopes. All illustrations show some advantages for envelopes, except for the example in Section 2.9, which was selected to illustrate cases where envelopes may be inappropriate. The goal here is to reinforce the theory and illustrate broadly how envelopes work and how they might be used effectively in data analyses. Several of the examples involve a bivariate response $\mathbf{Y} \in \mathbb{R}^2$ and a single binary predictor because comprehensive graphical representations are possible in this setting.

2.1 Wheat Protein: Full Data

The wheat protein data used for illustration in Section 1.3 contain two responses and a binary predictor, while the wheat protein data of Section 1.11 contain six responses and the same binary predictor. These datasets are subsets of a larger data set with $r = 6$ wavelengths as responses and actual wheat protein content as the $p = 1$ continuous predictor X. For these complete data, AIC, BIC, and LRT(0.05) each indicated that $u = 2$, and consequently changes in wheat protein are inferred to affect the distribution of only two linear combinations of $\mathbf{Y}$. Estimates of the 6×1 coefficient vector $\boldsymbol{\beta}$ are given in Table 2.1, with the standard error ratios for the coefficients shown in the last column of the table. The substantial gains shown in Table 2.1 are reflected by the relative magnitudes of $||\widehat{\boldsymbol{\Omega}}|| = 0.84$ and $||\widehat{\boldsymbol{\Omega}}_0|| = 6516.6$.

2.2 Berkeley Guidance Study

The Berkeley Guidance Study (Tuddenham and Snyder 1954) was designed to monitor the growth of children born in Berkeley, California, between 1928 and 1929. Data on $n = 93$ children, 39 boys and 54 girls, are available[1].

1 The data were obtained from http://rss.acs.unt.edu/Rdoc/library/fda/html/growth.html.

An Introduction to Envelopes: Dimension Reduction for Efficient Estimation in Multivariate Statistics,
First Edition. R. Dennis Cook.

Table 2.1 Wheat protein data: Coefficient estimates and their asymptotic standard errors from the envelope model with $u = 2$ and the standard model fitted to the complete data. $R = \text{se}(\mathbf{B})/\text{se}(\hat{\beta})$.

Y	$\hat{\beta}$	se($\hat{\beta}$)	B	se(B)	R
Y_1	−0.28	0.06	1.41	3.36	56.66
Y_1	1.86	0.07	3.21	2.76	37.71
Y_2	1.63	0.04	3.02	2.96	84.06
Y_3	−2.38	0.067	−0.94	3.26	48.85
Y_4	0.08	0.02	−0.40	4.70	191.49
Y_5	−0.89	0.09	0.04	1.85	20.45

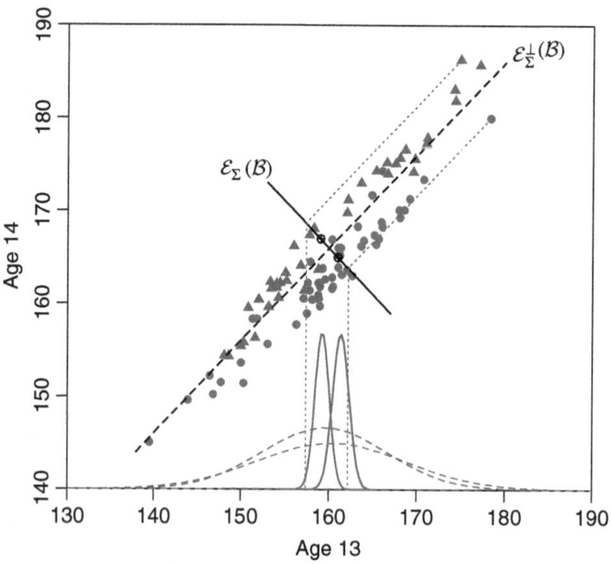

Figure 2.1 Berkeley guidance study: Envelope construction for the regression of height at ages 13 and 14, plotted on the horizontal and vertical axes, on a gender indicator $X \in \mathbb{R}^1$. Blue triangles represent males and red circles represent females. The lines marked by $\mathcal{E}_\Sigma(\mathcal{B})$ and $\mathcal{E}_\Sigma^\perp(\mathcal{B})$ denote the estimated envelope and its orthogonal complement.

We consider first the bivariate regression of the heights in centimeters at ages 13 and 14 on the single binary predictor indicating gender. Using BIC (see Section 1.10) in conjunction with a fit of the envelope model (1.20) led to the inference that $u = 1$. A plot of the data along with the estimated envelope and its orthogonal complement are shown in Figure 2.1. It can be seen from

Table 2.2 Berkeley guidance study: Bootstrap and estimated asymptotic standard errors of the coefficient estimates under the standard model (SM) and envelope model (EM) for two bivariate regressions from the Berkeley data. BSM and BEM designate the bootstrap standard errors for the standard and envelope models based on 200 bootstrap samples.

Response	SM	BSM	EM	BEM	SM/EM	BSM/BEM
Age 13	1.60	1.80	0.188	0.191	8.49	9.44
Age 14	1.61	1.81	0.187	0.190	8.61	9.64
Age 17	1.32	1.36	1.31	1.30	1.01	1.04
Age 18	1.33	1.37	1.34	1.37	0.99	1.01

this plot that the variation along $\widehat{\mathcal{E}}_{\Sigma}^{\perp}(\mathcal{B})$ is large relative to that along $\widehat{\mathcal{E}}_{\Sigma}(\mathcal{B})$. Consequently, we expect a substantial reduction in the standard errors relative to those from the usual bivariate linear model, as discussed broadly in Section 1.3. This is reflected also by the relative sizes of $\|\widehat{\mathbf{\Omega}}\| = 1.56$ and $\|\widehat{\mathbf{\Omega}}_0\| = 118.7$. The standard errors of the estimated elements of $\boldsymbol{\beta} \in \mathbb{R}^2$ under the standard and envelope models shown in Table 2.2 indicate that the envelope standard errors are about 11% of those from a fit of the standard model (SM).

We next consider the bivariate regression of the heights at ages 17 and 18 on gender. A plot of the data, the estimated envelope with $u = 1$, and its orthogonal complements are shown in Figure 2.2. In this case, the X-invariant (immaterial)

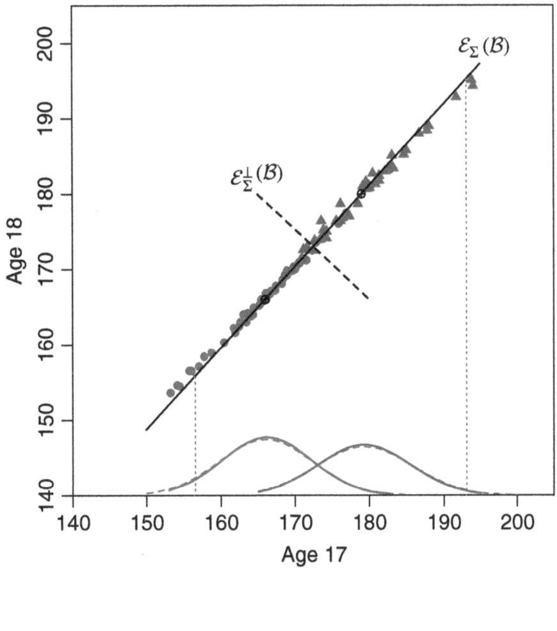

Figure 2.2 Berkeley guidance study: Envelope construction for the regression of height at ages 17 and 18, plotted on the horizontal and vertical axes, on a gender indicator $X \in \mathbb{R}^1$. Plot symbols are as in Figure 2.1.

variation, which lies in the direction of $\widehat{\mathcal{E}}_{\Sigma}^{\perp}(\mathcal{B})$, is small relative to the X-variant (material) variation, which lies in the direction of $\widehat{\mathcal{E}}_{\Sigma}(\mathcal{B})$. This is because the projections of the data directly onto the coordinate axes, as happens in an analysis based on the standard model, will be quite close to the projections of the data first onto $\widehat{\mathcal{E}}_{\Sigma}(\mathcal{B})$ and then onto the coordinate axes, as happens in an envelope analysis (Section 1.3). Consequently, we expect little reduction in the standard errors relative to those from the usual bivariate linear model. This expectation is supported by the relative sizes of $||\widehat{\Omega}|| = 79.5$ and $||\widehat{\Omega}_0|| = 0.156$. The standard errors of the estimated elements of $\boldsymbol{\beta} \in \mathbb{R}^2$ under the standard and envelope models are shown in Table 2.2.

2.3 Banknotes

Flury and Riedwyl (1988) reported six size measurements in millimeters on 100 counterfeit and 100 genuine Swiss bank notes. The six measurements, which were taken along the left, right, bottom, top, diagonal, and length of a note, constitute the six-dimensional response vector for our example. The single predictor indicates a note's status, genuine or counterfeit. The goal of the example is to understand how the distribution of the size measurements varies with note status.

BIC and LRT(0.01) both indicated that $u = 3$, while AIC indicated that $u = 4$. In view of AIC's propensity to overestimate and the fact that the AIC value for $u = 3$ was not much larger than that for $u = 4$, we chose to fit with $u = 3$. The ratios of the asymptotic standard errors from the ordinary least squares fit to those of the envelope fit varied between 0.991 and 1.019, so enveloping offers no estimative gain. This result is consistent with the relative magnitudes of $||\widehat{\Omega}|| = 1.07$ and $||\widehat{\Omega}_0|| = 0.24$. Additionally, the coefficient estimates are essentially the same, as shown in Table 2.3.

The conclusion then is that the ordinary least squares fit (equivalently the envelope fit with $u = 6$) and the envelope fit with $u = 3$ are essentially the

Table 2.3 Banknote data: Envelope $\widehat{\boldsymbol{\beta}}$ and ordinary least squares **B** coefficient estimates.

Responses	$\widehat{\boldsymbol{\beta}}$	**B**
Bottom	2.224	2.225
Diagonal	−2.067	−2.067
Left	0.360	0.357
Length	−0.148	−0.146
Right	0.478	0.473
Top	0.964	0.965

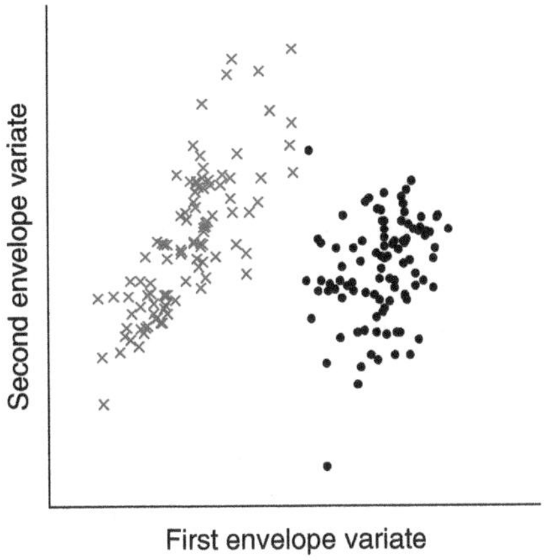

Figure 2.3 Banknote data: Plot of the first two envelope variates marked by status. Black • – genuine. red × – counterfeit.

same. However, the envelope fit still has advantages, since we are able to view a marked three-dimensional plot of $\widehat{\Gamma}^T\mathbf{Y}$ versus X to gain insights into how counterfeit and genuine banknotes differ. Shown in Figure 2.3 is a plot of the first two envelopes variates, formed using the first two coordinates of $\widehat{\Gamma}^T\mathbf{Y}$, marked by status. From this we see that, except for one genuine note lying at the edge of the counterfeit notes, the first two envelope variates nicely separate the distribution of $\mathbf{Y} \mid X$ by status. Additional intuition may be gained by viewing the full three-dimensional plot of $\widehat{\Gamma}^T\mathbf{Y}$ marked by status.

Figure 2.3 suggests that the conditional variances $\text{var}(\mathbf{Y} \mid X = 0)$ and $\text{var}(\mathbf{Y} \mid X = 1)$ might differ, which would violate a condition of the multivariate linear model (1.1). Envelope methods that allow for different covariance matrices are discussed in Section 5.2, and dimension reduction methods for comparing covariance matrices are discussed in Section 9.8.

2.4 Egyptian Skulls

The data for this illustration are a subset of measurements taken on Egyptian crania from five different epochs that cover a period of at least 4000 years. The crania were collected during excavations carried out between 1898 and 1901 in the region of Upper Egypt known as the Thebiad (Thomson and Randall-Maciver 1905). The dataset is unique because, according to Thomson and Randall-Maciver, it is the first instance of data capable of tracing the physical history of a people over a comparable period of 4000 years. The data[2] used here are a commonly available subset of the data reported by Thomson and Randall-Maciver (1905).

---

2 The data used here were obtained from http://www.dm.unibo.it/~simoncin/EgyptianSkulls.html. See also http://lib.stat.cmu.edu/DASL/Stories/EgyptianSkullDevelopment.html. The dataset can also be found in Hand et al. (1994) and Manly (1986).

The response vector $\mathbf{Y}$ consists of four measurements in millimeters on male crania: basibregmatic height (BH), basialveolar length (BL), maximal breadth (MB), and nasal height (NH). These responses were measured on 30 skulls in each of the five epochs indicated by the nominal dates 4000, 3300, 1850, and 200 B.C. and 150 A.D., giving a total sample size of $n = 150$. We regard epoch as categorical, leading to a regression of $\mathbf{Y}$ on four indicator variables $X_j, j = 1, \dots, 4$, plus a constant for the intercept:

$$\mathbf{Y} = \boldsymbol{\alpha} + \boldsymbol{\beta}_{3300}X_1 + \boldsymbol{\beta}_{1850}X_2 + \boldsymbol{\beta}_{200}X_3 + \boldsymbol{\beta}_{150}X_4 + \boldsymbol{\varepsilon}, \tag{2.1}$$

where $X_j = 1$ if $\mathbf{Y}$ was measured in the epoch indicated by the subscript on the corresponding coefficient vector and $X_j = 0$ otherwise. Let $\boldsymbol{\mu}_{(\cdot)} \in \mathbb{R}^5$ denote the epoch means. Model (2.1) is parameterized so that the coefficient vectors are the differences between the indicated epoch means and the mean for epoch 4000 B.C, $\boldsymbol{\beta}_{3300} = \boldsymbol{\mu}_{3300} - \boldsymbol{\mu}_{4000}$, $\boldsymbol{\beta}_{150} = \boldsymbol{\mu}_{150} - \boldsymbol{\mu}_{4000}$, and so on. In this way, an envelope structure is posited for the differences in epoch means and not the means themselves, which is similar to the parameterization used for the wheat protein data in Section 1.3.

As in other examples, AIC, BIC, and LRT(0.05) all indicated that $u = 1$, leading to the envelope model

$$\mathbf{Y} = \boldsymbol{\alpha} + \boldsymbol{\Gamma}(\eta_{3300}X_1 + \eta_{1850}X_2 + \eta_{200}X_3 + \eta_{150}X_4) + \boldsymbol{\varepsilon}, \tag{2.2}$$

with $\boldsymbol{\Sigma} = \boldsymbol{\Gamma}\boldsymbol{\Omega}\boldsymbol{\Gamma}^T + \boldsymbol{\Gamma}_0\boldsymbol{\Omega}_0\boldsymbol{\Gamma}_0^T$, where $\boldsymbol{\Gamma} \in \mathbb{R}^{4\times 1}$ is a basis vector for the one-dimensional envelope $\mathcal{E}_{\boldsymbol{\Sigma}}(\mathcal{B})$.

Fitting with $u = 1$, $\widehat{\boldsymbol{\Gamma}}$ is shown in Table 2.4 along with the ratios of the standard errors for the elements of $\mathbf{B}$ to those for $\widehat{\boldsymbol{\beta}}$ from the fit of the envelope model. The standard error ratios are all greater than one and some are large, indicating a strong overall improvement in efficiency. Table 2.5 gives the coefficients and Z-scores for the ordinary least squares fit of the standard model and the envelope fit with $u = 1$. One general impression is that the Z-scores for the envelope analysis are generally greater than those for the standard analysis. Additionally, the Z-scores for $\widehat{\boldsymbol{\beta}}_{3300}$ are all relatively small, indicating that there may be no detectable difference between $\boldsymbol{\mu}_{4000}$ and $\boldsymbol{\mu}_{3300}$ based on model

Table 2.4 Egyptian skull data: $\widehat{\boldsymbol{\Gamma}}$ and standard error ratios $\mathrm{se}((\mathbf{B})_{ij})/\mathrm{se}((\widehat{\boldsymbol{\beta}})_{ij})$.

Y	$\widehat{\boldsymbol{\Gamma}}$	$\widehat{\boldsymbol{\beta}}_{3300}$	$\widehat{\boldsymbol{\beta}}_{1850}$	$\widehat{\boldsymbol{\beta}}_{200}$	$\widehat{\boldsymbol{\beta}}_{150}$
BH	0.28	3.37	2.09	1.51	1.25
BL	0.72	1.36	1.29	1.19	1.12
MB	−0.62	1.48	1.33	1.17	1.06
NH	−0.11	5.23	2.67	1.84	1.50

Table 2.5 Egyptian skull data: Coefficients and absolute Z-scores in parentheses.

Y	$\widehat{\beta}_{3300}$	$\widehat{\beta}_{1850}$	$\widehat{\beta}_{200}$	$\widehat{\beta}_{150}$
		Envelope fit of (2.2)		
BH	−0.25 (0.68)	−1.15 (1.96)	−1.79 (2.20)	−2.24 (2.27)
BL	−0.66 (0.72)	−3.00 (3.09)	−4.67 (4.67)	−5.86 (5.26)
MB	0.56 (0.71)	2.57 (2.93)	4.00 (4.02)	5.01 (4.56)
NH	0.10 (0.65)	0.47 (1.55)	0.73 (1.66)	0.92 (1.83)
		Ordinary least squares fit of (2.1)		
BH	−0.90 (0.72)	0.20 (0.16)	−1.30 (1.04)	−3.27 (2.61)
BL	−0.10 (0.08)	−3.13 (2.47)	−4.63 (3.65)	−5.67 (4.46)
MB	1.00 (0.84)	3.10 (2.61)	4.13 (3.48)	4.80 (4.05)
NH	−0.30 (0.36)	0.03 (0.04)	1.43 (1.74)	0.83 (1.01)

(2.1). This difference could be tested more formally by using the likelihood ratio statistic to test the hypothesis that $\eta_{3300} = 0$ in model (2.2).

Table 2.6 shows the estimates of the residual covariance matrix $\mathbf{\Sigma}$ from the standard model and from the envelope model. The variance estimates for the two models are close, but some of the estimated covariances have different signs. Those covariances are not significantly different from 0 and so the change in sign is not necessarily noteworthy.

One particularly useful part of this analysis is the inference that $u = 1$ and thus that the X-variant information is embodied in the univariate response $\mathbf{\Gamma}^T\mathbf{Y}$. Boxplots of $\widehat{\mathbf{\Gamma}}^T\mathbf{Y}$ for each epoch are given in Figure 2.4. Although the envelope fit was based on epoch indicators, the box plots still show a linear

Table 2.6 Egyptian skull data: Estimated covariance matrices from fits of the multivariate linear model and envelope model with $u = 1$.

	Standard model, $\mathbf{S}_{\mathbf{Y}\mid\mathbf{X}}$				Envelope model, $\widehat{\mathbf{\Sigma}}$			
Y	BH	BL	MB	NH	BH	BL	MB	NH
BH	22.70	5.03[a)]	0.04	2.75[a)]	21.75	2.47	1.48	2.57
BL		23.37	0.08	1.10		22.17	−0.82	−0.24
MB			20.41	1.94			22.95	3.32
NH				9.81				10.42

a) indicates correlations significantly different from 0 at level 0.05.

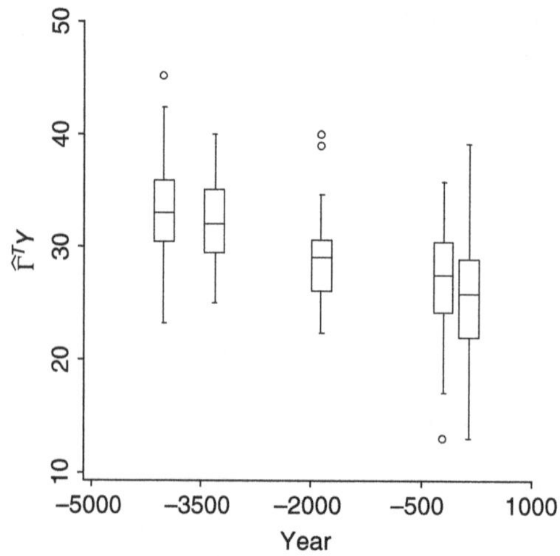

Figure 2.4 Egyptian skull data: Boxplots of $\widehat{\boldsymbol{\Gamma}}^T\mathbf{Y}$ versus year.

relationship between $\widehat{\boldsymbol{\Gamma}}^T\mathbf{Y}$ and the nominal epoch date. This analysis then may lead to useful findings, particularly if it were possible to reify the estimated material response $\widehat{\boldsymbol{\Gamma}}^T\mathbf{Y}$.

2.5 Australian Institute of Sport: Response Envelopes

Various allometric and hematological measurements were taken on 102 male and 100 female athletes at the Australian Institute of Sport.[3] We use a binary predictor to indicate gender : $X = 1$ for female and $X = 0$ for male. The response $\mathbf{Y} \in \mathbb{R}^2$ is bivariate consisting of the logarithms of white cell count (WCC) and hematocrit (Hc), leading to the multivariate linear model $\mathbf{Y} = \boldsymbol{\alpha} + \boldsymbol{\beta}X + \boldsymbol{\varepsilon}$.

A plot of log(Hc) versus log(WCC) with points marked by gender is shown in Figure 2.5. The mean difference, females minus males, for log(Hc) is −0.12 with a standard error of 0.0084, giving an absolute ratio of 14.4. The mean difference for log(WCC) is −0.027 with a standard error of 0.035, giving an absolute ratio of 0.77. Consequently, we see from the figure and the analysis that there is a clear gender difference in log(Hc), while the observed gender difference for log(WCC) is within estimated variation.

Since LRT(0.05), AIC and BIC all selected $u = 1$, we next fitted the envelope model $\mathbf{Y} = \boldsymbol{\alpha} + \boldsymbol{\Gamma}\eta X + \boldsymbol{\varepsilon}$, where $\boldsymbol{\Gamma} \in \mathbb{R}^{2\times 1}$ and η is a scalar. The estimated envelope, its orthogonal complement, and $\mathcal{B}$ are shown in Figure 2.5. The estimated basis $\widehat{\boldsymbol{\Gamma}}$ and the coefficient estimates under the standard model

3 The data are available from the website http://www.stat.umn.edu/arc/ for Cook and Weisberg (1998).

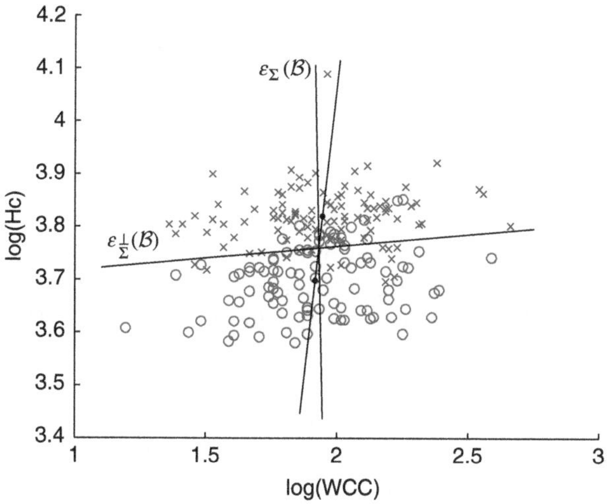

Figure 2.5 Australian Institute of Sport: Envelope construction for the regression of $(\log(\text{Hc}), \log(\text{WCC}))^T$ on gender X. Blue axes represent females, red circles represent males, and the two black dots represent the gender means. The lines marked by $\mathcal{E}_\Sigma(\mathcal{B})$ and $\mathcal{E}_\Sigma^\perp(\mathcal{B})$ denote the estimated envelope and its orthogonal complement.

Table 2.7 Australian Institute of Sport: $\widehat{\boldsymbol{\Gamma}}$ and coefficient estimates with standard errors in parentheses. The envelope fit is with $u = 1$.

Y	$\widehat{\Gamma}$	B	$\widehat{\beta}$
log(Hc)	−0.9990	−0.121 (0.008)	−0.119 (0.008)
log(WCC)	0.0448	−0.027 (0.035)	0.005 (0.002)

and the envelope model are shown in Table 2.7. We see from the table that the estimates and standard errors for log(Hc) are essentially the same, while the estimates for log(WCC) have different signs and the standard error from the fit of the standard model is about 17 times the standard error from the envelope model.

2.6 Air Pollution

For this illustration, we use data on air pollution obtained from Johnson and Wichern (2007, p. 39). The five responses are atmospheric concentrations of CO, NO, NO_2, O_3, and HC recorded at noon in the Los Angeles area on $n = 42$ different days. The two predictors are measurements on wind speed W and solar radiation S. We explore these data in more detail than other illustrations to give a feeling for additional aspects of an envelope analysis.

Consider first the standard multivariate linear regression of the vector $\mathbf{Y}$ of five concentrations on the two predictors:

$$\mathbf{Y} = \boldsymbol{\alpha} + \boldsymbol{\beta}\mathbf{X} + \boldsymbol{\varepsilon} = \boldsymbol{\mu} + \boldsymbol{\beta}_{\mathrm{W}} W + \boldsymbol{\beta}_{\mathrm{S}} S + \boldsymbol{\varepsilon}, \tag{2.3}$$

where $\boldsymbol{\beta} = (\boldsymbol{\beta}_{\mathrm{W}}, \boldsymbol{\beta}_{\mathrm{S}})$ and $\mathbf{X} = (W, S)^T$. The likelihood ratio test statistic for $\boldsymbol{\beta} = 0$ has the value $\Lambda = 19.14$ with 10 degrees of freedom, giving a nominal chi-square p-value of 0.038. This suggests that the mean of $\mathbf{Y}|\mathbf{X}$ might depend on $\mathbf{X}$ nontrivially, although that dependence is not likely to be crisp. The estimated coefficients and the coefficients relative to their standard errors are shown in the top half of Table 2.8. There seems to be some indication of an effect for solar radiation on O_3 and perhaps for wind speed on NO, but generally the individual significant characteristics of the regression seem unclear. The eigenvalues of $\widehat{\boldsymbol{\Sigma}}$ range between 26.2 and 0.21 so in view of (1.22) there is a possibility that an envelope will help clarify the analysis, particularly if the larger eigenvalues are associated with the X-invariant part of $\mathbf{Y}$.

Table 2.8 Ozone data: Estimated coefficients from fits of the standard multivariate linear model (1.1) and envelope model (1.20) with $u = 1$.

		Standard model (2.3)		
Y	$\mathbf{B}_W$	***Z*-score**	$\mathbf{B}_S$	***Z*-score**
CO	−0.138	−1.18	0.012	1.10
NO	−0.192	−1.88	−0.006	−0.69
NO_2	−0.211	−0.65	0.020	0.69
O_3	−0.787	−1.56	0.095	2.07
HC	0.071	1.07	0.003	0.45
		Envelope model, $u = 1$(2.4)		
Y	$\widehat{\boldsymbol{\beta}}_W$	***Z*-score**	$\widehat{\boldsymbol{\beta}}_S$	***Z*-score**
CO	0.071	2.76	0.0008	0.43
NO	−0.075	−2.89	−0.0009	−0.43
NO_2	−0.017	−2.62	−0.0002	−0.43
O_3	−0.011	−2.70	−0.0001	−0.43
HC	0.117	3.12	0.0013	0.44

S – solar radiation; W – wind speed. $\mathbf{B} = (\mathbf{B}_W, \mathbf{B}_S)$, and $\widehat{\boldsymbol{\beta}} = (\widehat{\boldsymbol{\beta}}_W, \widehat{\boldsymbol{\beta}}_S)$.

Table 2.9 Ozone data: Standard error ratios, standard errors from the standard model fit divided by the corresponding standard errors from the envelope model fit with $u = 1$.

Y	*W*	*S*
CO	4.5	5.7
NO	3.9	4.7
NO_2	51.8	67.9
O_3	128.1	163.2
HC	1.8	2.0

Turning to envelope regression, BIC and LRT(0.05) both indicate that $u = 1$ (see Section 1.10). Consequently, we base the envelope analysis on the model

$$\mathbf{Y} = \boldsymbol{\alpha} + \boldsymbol{\Gamma}(\eta_W W + \eta_S S) + \boldsymbol{\varepsilon}, \tag{2.4}$$

where the coefficient vectors in (2.3) are now modeled as $\boldsymbol{\beta}_W = \boldsymbol{\Gamma}\eta_W$ and $\boldsymbol{\beta}_S = \boldsymbol{\Gamma}\eta_S$, and $\boldsymbol{\Gamma} \in \mathbb{R}^5$. Fitting with $u = 1$, the relative magnitudes of $\widehat{\Omega} = 0.21$ and $\|\widehat{\boldsymbol{\Omega}}_0\| = 31.1$ indicate that the envelope model results in substantial reduction. Indeed, the standard error ratios shown in Table 2.9 range between 1.8 and 163.2.

The estimated envelope coefficients and the coefficients relative to their standard errors are shown in the bottom half of Table 2.8. Compared to the coefficient estimates from the standard model, we see all but one have been shrunk toward 0, and now wind speed seems relevant for all responses while solar radiation given wind speed seems irrelevant for all responses. The contributions of wind speed and solar radiation can be tested by comparing the envelope model (2.4) to the envelope models with only one of the predictors. The likelihood ratio statistic for $\eta_W = 0$ has the value 4.77, and the likelihood ratio statistic for $\eta_S = 0$ has the value 0.17, each on 1 degree of freedom. These statistics then support the indication in the lower half of Table 2.8: solar radiation is not contributing significantly to the regression given that wind speed is in the model. We can explore the implications of the envelope model further by using added variable plots (Cook and Weisberg 1982) from the univariate linear regression fit of $\widehat{\boldsymbol{\Gamma}}^T\mathbf{Y}$ on (W, S). The plots shown in Figure 2.6 seem to be in qualitative agreement with the envelope fit of Table 2.8.

For contrast, Table 2.10 gives the estimated coefficients and their standard error ratios for separate fits of $\mathbf{Y}$ on W and $\mathbf{Y}$ on S. The three dimension selection methods again all indicated that $u = 1$ for these two regressions. We see from Table 2.10 that the coefficients and standard errors for W are quite similar to those for the envelope fit with both variables in the lower half of Table 2.8. The corresponding quantities for S are very different, however.

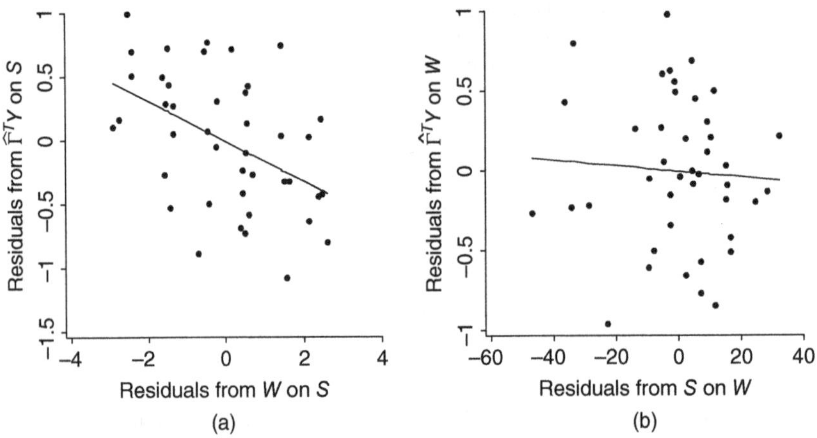

Figure 2.6 Ozone data: Added variable plots for W and S from the fit of $\widehat{\mathbf{\Gamma}}^T\mathbf{Y}$ on (W, S). The lines on the plots represent the linear ordinary least squares fits of the vertical axis variable on the horizontal axis variable. (a) AVP for W. (b) AVP for S.

Table 2.10 Ozone data: Estimated coefficients from separate envelope fits with $u = 1$ of $\mathbf{Y}$ on W and $\mathbf{Y}$ on S.

Y	$\widehat{\beta}_W$	*Z*-score	$\widehat{\beta}_S$	*Z*-score
CO	0.07	2.74	0.01	1.76
NO	0.07	−2.87	−0.002	−0.50
NO_2	−0.016	−2.62	−0.65	1.07
O_3	−0.011	−2.63	−1.56	2.22
HC	0.12	3.13	0.003	1.05

S – solar radiation; W – wind speed; Z-score denotes the estimate divided by its standard error.

The estimated X-invariant part of $\mathbf{Y}$ can be visualized by plotting linear combinations $\mathbf{b}^T\widehat{\mathbf{\Gamma}}_0^T\mathbf{Y}$ versus $\widehat{\eta}\mathbf{X}$ or other linear combinations of $\mathbf{X}$, where $\mathbf{b} \in \mathbb{R}^4$ and $\|\mathbf{b}\| = 1$. To be consistent with the envelope model, such plots should leave the impression that $\mathbf{b}^T\widehat{\mathbf{\Gamma}}_0^T\mathbf{Y} \perp\!\!\!\perp \widehat{\eta}\mathbf{X}$, and the variation of $\mathbf{b}^T\widehat{\mathbf{\Gamma}}_0^T\mathbf{Y}$ should be substantially greater than that of $\widehat{\mathbf{\Gamma}}^T\mathbf{Y}$. Figure 2.7 shows a plot of $\mathbf{b}^T\widehat{\mathbf{\Gamma}}_0^T\mathbf{Y}$ versus $\widehat{\eta}\mathbf{X}$ with $\mathbf{b}$ chosen to be the first eigenvector of the sample variance matrix of $\widehat{\mathbf{\Gamma}}_0^T\mathbf{Y}$, so $\mathbf{b}^T\widehat{\mathbf{\Gamma}}_0^T\mathbf{Y}$ is simply the first principal component of $\widehat{\mathbf{\Gamma}}_0^T\mathbf{Y}$. There is no clear dependence between $\mathbf{b}^T\widehat{\mathbf{\Gamma}}_0^T\mathbf{Y}$ and $\widehat{\eta}\mathbf{X}$ that is discernible from the plot, and the range of the vertical axis is much larger than that for the added variable plots in Figure 2.6.

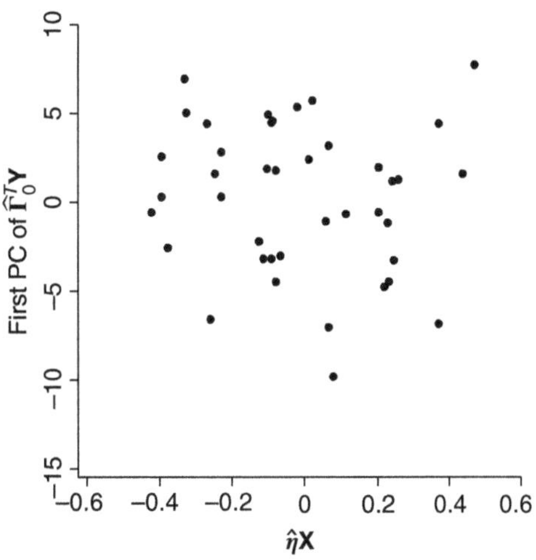

Figure 2.7 Ozone data: Scatterplot illustrating the immaterial variation.

Table 2.11 Ozone data: Estimated residual covariance matrices $\mathbf{S}_{\mathrm{Y|X}}$ and $\widehat{\boldsymbol{\Sigma}}$ from fits of the multivariate linear model and envelope model with $u = 1$.

	Standard model (2.3)					Envelope model, $u = 1$ (2.4)				
Y	CO	NO	NO_2	O_3	HC	CO	NO	NO_2	O_3	HC
CO	1.39	0.61	2.10	2.10	0.15	1.55	0.65	2.28	2.88	0.19
NO		1.05	0.99	−1.02	0.21		1.09	1.02	−0.93	0.22
NO_2			10.84	1.97	1.04			11.08	3.02	1.07
O_3				25.67	0.65				30.20	0.79
HC					0.45					0.46

Table 2.11 shows the estimated residual covariance matrices $\mathbf{S}_{\mathrm{Y|X}}$ and $\widehat{\boldsymbol{\Sigma}}$ from the fits of the standard model and the envelope model with $u = 1$. The two matrices seem to be in good agreement.

2.7 Multivariate Bioassay

In this illustration, we consider the results for the first two days of a bioassay of insulin by the rabbit blood sugar method (Vølund 1980). The test and standard preparations were each represented at two levels, coded −1 and +1. Each of the four treatment combinations was administered to nine rabbits whose blood sugar concentration (mg 100 ml^{-1}) was measured at 0, 1, 2, 3, 4, and 5 hours.

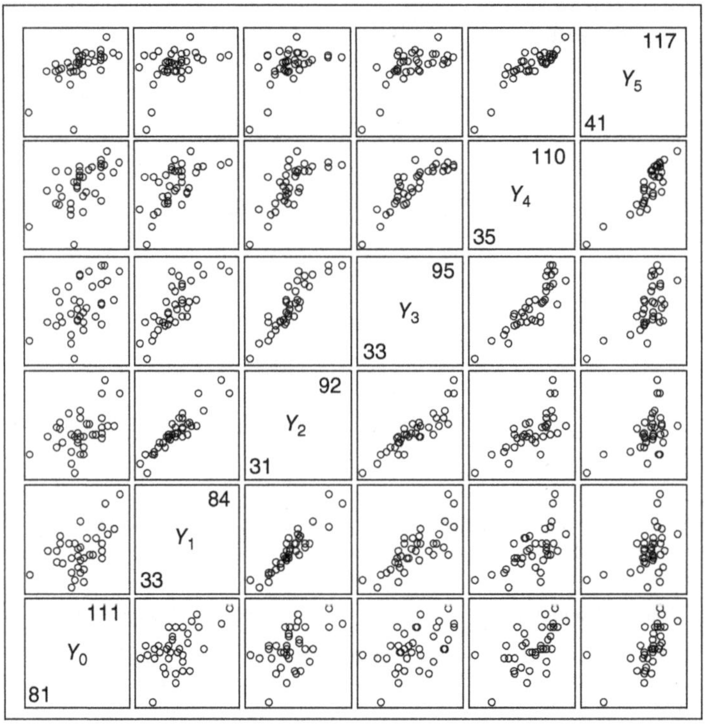

Figure 2.8 Multivariate bioassay: Scatterplot matrix of the six responses in the rabbit assay. The subscripts on Y_j indicate the hour at which the response was measured.

Let $\mathbf{Y} = (Y_0, Y_1, \dots, Y_5)^T \in \mathbb{R}^6$ denote the random vector of blood sugar concentrations, let $X_1 = \pm 1$ indicate the treatment and standard preparations, and let $X_2 = \pm 1$ indicate the dose level. The model relating the treatments to the concentrations has three predictors, the two treatment indicators plus their interaction,

$$\mathbf{Y} = \boldsymbol{\alpha} + \boldsymbol{\beta}_1 X_1 + \boldsymbol{\beta}_2 X_2 + \boldsymbol{\beta}_{12} X_1 X_2 + \boldsymbol{\varepsilon}. \tag{2.5}$$

A scatterplot matrix of the six responses is given in Figure 2.8 as background.

The presence of an interaction $\boldsymbol{\beta}_{12} \neq 0$ is often a general concern in models like (2.5). Twice the difference Λ of the log-likelihoods under the alternative $\boldsymbol{\beta}_{12} \neq 0$ and null $\boldsymbol{\beta}_{12} = 0$ models has the value $\Lambda = 5.8$ on 6 degrees of freedom, and thus there is little to suggest that $\boldsymbol{\beta}_{12} \neq 0$. Turning to the envelope models, BIC and LRT(0.05) both indicated that $u = 1$, so the envelope version of (2.5) becomes

$$\mathbf{Y} = \boldsymbol{\alpha} + \boldsymbol{\Gamma}(\eta_1 X_1 + \eta_2 X_2 + \eta_{12} X_1 X_2) + \boldsymbol{\varepsilon}, \tag{2.6}$$

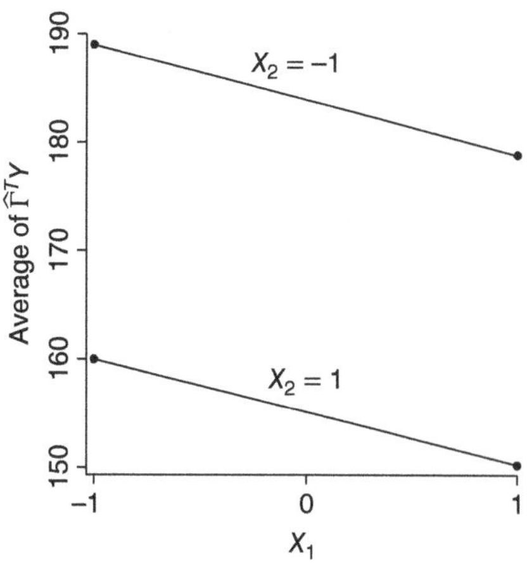

Figure 2.9 Multivariate bioassay: Scatterplot of $\widehat{\Gamma}^T\bar{\mathbf{Y}}$ versus X_1 and X_2 from an envelope analysis of the rabbit assay using model (2.5).

where $\boldsymbol{\Gamma} \in \mathbb{R}^6$, the η's are scalars, and $\boldsymbol{\Sigma}$ has the corresponding envelope structure. The presence of an interaction in this model can be checked by using the log-likelihood ratio to test $\eta_{12} = 0$. The test statistic has the value $\Lambda = 0.008$ on one degree of freedom, indicating that the effects in model (2.6) are additive. Shown in Figure 2.9 is a plot of $\widehat{\boldsymbol{\Gamma}}^T\bar{\mathbf{Y}}$ versus (X_1, X_2). The lines on the plot appear to be parallel, although they are not exactly so, which gives visual support for outcome of the test of $\eta_{12} = 0$. Depending on application-specific goals, the analysis could now be continued based on the additive model.

The scatterplot matrix of Figure 2.8 has a few observations that seem to be set apart from the trends exhibited by the remaining data. This leads to the possibility of influential observations having an impact on the conclusions. Although influence measures have not been developed specifically for envelope models, influence can still be studied by adapting the case deletion and local influence measures proposed by Cook (1977, 1986).

2.8 Brain Volumes

The response vector $\mathbf{Y}$ for this illustration consists of the log volumes of $r = 93$ regions of the brains of $n = 442$ male patients with Alzheimer's disease. It is known that overall brain volume decreases with age. One goal of this illustration is to see if the same phenomenon is manifested across different regions of the brain by using an envelope regression of $\mathbf{Y}$ on age. This is part of a larger dataset analyzed by Zhu et al. (2014) using a Bayesian generalized low rank regression model.

LRT(0.05) indicated that $u = 4$ while BIC indicated that $u = 5$, so it appears that the decrease in brain volume is concentrated in four or five linear combinations of the log volumes, while the remaining 88–89 linear combinations

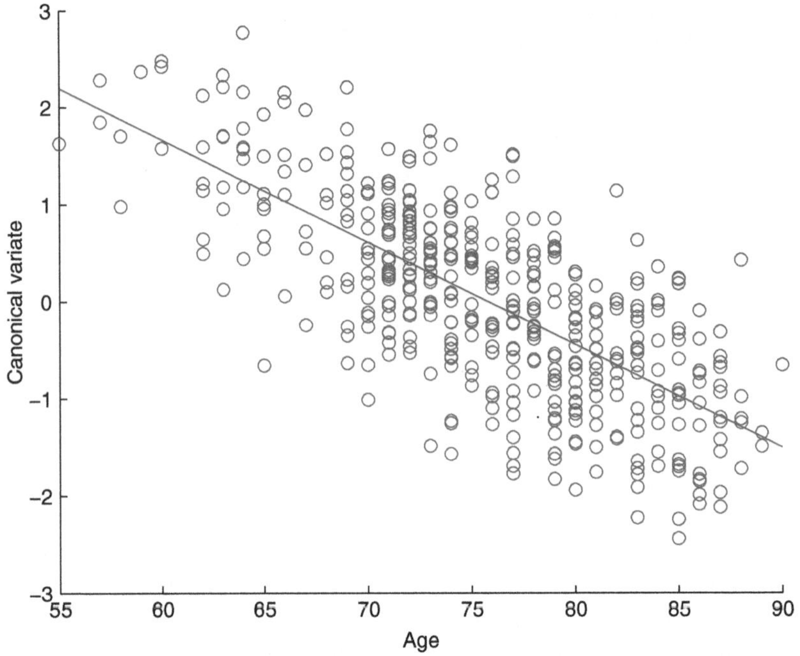

Figure 2.10 Brain volumes: Scatterplot of the canonical variate for $\widehat{\boldsymbol{\Gamma}}^T\bar{\mathbf{Y}}$ versus age from an envelope analysis with $u = 5$.

are age-invariant. Although u was estimated to be relatively small, the standard error ratios $\text{se}(\mathbf{B})/\text{se}(\widehat{\boldsymbol{\beta}})$ from the fit with $u = 5$ ranged between 1.01 and 1.24 so the envelope reduction in estimative variation is modest. We continue to use $u = 5$ in the rest of this illustration.

In cases where u is not small, a canonical correlation analysis (see Section 9.1 for background) of the linear relations between $\widehat{\boldsymbol{\Gamma}}^T\mathbf{Y}$ and $\mathbf{X}$ may aid in visualization of the results. The plot of the first canonical variate between $\widehat{\boldsymbol{\Gamma}}^T\mathbf{Y}$ and age given in Figure 2.10 shows that brain volume is clearly decreasing with age, as expected. The corresponding canonical correlation was about 0.71. The canonical correlation for the 88 age-invariants $\widehat{\boldsymbol{\Gamma}}_0^T\mathbf{Y}$ and age was about 0.21, which supports the notion that $\widehat{\boldsymbol{\Gamma}}^T\mathbf{Y}$ captures the part of $\mathbf{Y}$ that is most affected by age.

Using a one-sided test and Bonferroni's inequality to adjust for multiple testing with a nominal experiment-wise error rate of 0.05, 73 of the age coefficients were significantly negative based on their asymptotic envelope standard errors (1.33), leaving open the possibility that about 20 regions show no decrease with age. Using the procedure described in Section 1.8.2, we also tested the hypotheses that each of the individual responses is age-invariant. All of these 93 hypotheses were clearly rejected using the same Bonferroni adjustment to achieve a nominal experiment-wise rate of 0.05. In other words, each response is either marginally affected by age or correlated with other responses that are affected by age or both.

Allowing for sparsity in the rows of $\mathbf{\Gamma}$ is an alternative way of fitting that might also be used to study the change in the individual responses with age, as discussed in Section 7.5.

2.9 Reducing Lead Levels in Children

We conclude the illustrations with an example showing various issues that can impact the applicability of envelopes. The data arose from an experiment to assess the efficacy of a treatment to reduce the blood lead levels in children (Fitzmaurice et al. 2004). Two treatments, a placebo and a chelating agent, were randomly assigned to 100 children with blood lead levels between 20 and 44 μg dl^{-1}. The data consist of four repeated measurements of blood lead levels obtained at week 0 (baseline), 1, 4, and 6. Let Y_j denote the blood lead level at week $j = 0, 1, 4, 6$, $\mathbf{Y}_{01} = (Y_0, Y_1)^T$, and $\mathbf{Y}_{46} = (Y_4, Y_6)^T$. Also, let $X = 0, 1$ indicate the placebo and the treatment.

Figure 2.11a shows a scatterplot of the blood lead for week 0 versus week 1 marked by treatment. It seems clear from the plot that the lead levels in week 1 for the treated children are less than those for the placebo group, but also that its variation in the treated group is greater than that in the placebo group. This differential variation $\text{var}(\mathbf{Y}_{01}|X = 0) \neq \text{var}(\mathbf{Y}_{01}|X = 1)$ violates the

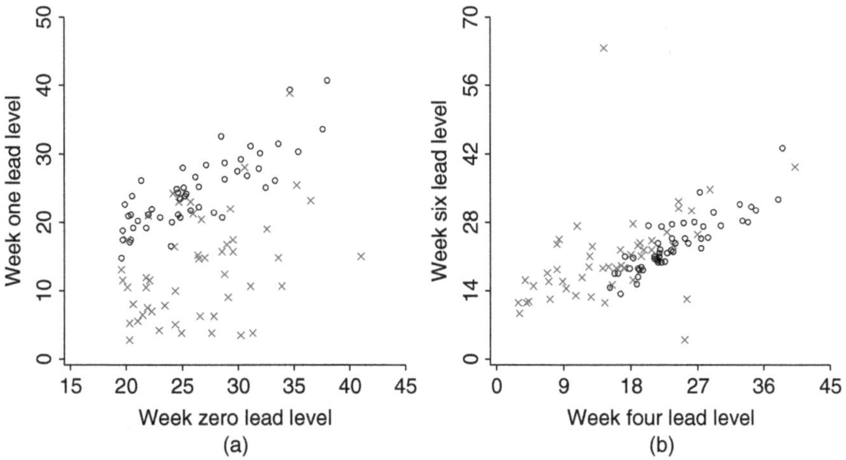

Figure 2.11 Blood lead levels in children: An ex represents the chelating agent, and an open circle represents the placebo. (a) Week zero versus week one lead levels. (b) Week four versus week six lead levels.

constant variance assumption of the multivariate linear model (1.1). Extensions of envelope methods to allow for heteroscedasticity across treatments require a more general definition of an envelope, leading to somewhat different methodology. Such extensions are considered in Section 5.2

Figure 2.11b shows a scatterplot of blood lead for week four versus week six marked by treatment. One and possibly three measurements outlie the main point cloud. Outliers can potentially have a noticeable impact on an envelope analysis, but robust methodology has not yet been developed. Graphical representations such as that in Figure 2.11b may often show extreme outliers. Neglecting the outliers, it seems clear from the figure that neither eigenvector of $\mathrm{var}(\mathbf{Y}_{46}|X)$ aligns with $\mathrm{span}\{\mathrm{E}(\mathbf{Y}_{46}|X = 0) - \mathrm{E}(\mathbf{Y}_{46}|X = 1)\}$, and thus the envelope for the regression of $\mathbf{Y}_{46}$ on X has dimension 2. As a consequence, envelope estimation offers no gains over standard methodology since all linear combinations of $\mathbf{Y}_{46}$ respond to changes in X.

3

Partial Response Envelopes

A subset of the predictors may often be of special interest in multivariate regression. For instance, some predictors may correspond to treatments while the remaining predictors are included to account for heterogeneity among the experimental units. In such cases, the columns of $\boldsymbol{\beta}$ that correspond to treatments will be of particular interest. Partial envelopes (Su and Cook 2011) were designed to focus consideration on the coefficients corresponding to the predictors of interest. The development of partial envelopes closely follows that for envelopes given in Chapter 1.

3.1 Partial Envelope Model

Partition $\mathbf{X}$ into two sets of predictors $\mathbf{X}_1 \in \mathbb{R}^{p_1}$ and $\mathbf{X}_2 \in \mathbb{R}^{p_1}$, $p_1 + p_2 = p$, and conformably partition the columns of $\boldsymbol{\beta}$ into $\boldsymbol{\beta}_1$ and $\boldsymbol{\beta}_2$. Then, as described in Section 1.1.1, model (1.1) can be rewritten as $\mathbf{Y} = \boldsymbol{\mu} + \boldsymbol{\beta}_1\mathbf{X}_1 + \boldsymbol{\beta}_2\mathbf{X}_2 + \boldsymbol{\varepsilon}$, where $\boldsymbol{\beta}_1$ corresponds to the coefficients of interest. We can now consider the $\boldsymbol{\Sigma}$-envelope of $\mathcal{B}_1 = \text{span}(\boldsymbol{\beta}_1)$, which is symbolized by $\mathcal{E}_{\boldsymbol{\Sigma}}(\mathcal{B}_1)$ and referred to as the partial envelope for $\mathcal{B}_1$, leaving $\boldsymbol{\beta}_2$ as an unconstrained parameter. This leads to the covariance structure $\boldsymbol{\Sigma} = \mathbf{P}_{\mathcal{E}_1}\boldsymbol{\Sigma}\mathbf{P}_{\mathcal{E}_1} + \mathbf{Q}_{\mathcal{E}_1}\boldsymbol{\Sigma}\mathbf{Q}_{\mathcal{E}_1}$, where $\mathbf{P}_{\mathcal{E}_1}$ denotes the projection onto $\mathcal{E}_{\boldsymbol{\Sigma}}(\mathcal{B}_1)$. This is the same as the envelope structure discussed in Chapter 1, except that the enveloping is on $\mathcal{B}_1$ instead of the larger space $\mathcal{B}$. For clarity, we will in this chapter refer to $\mathcal{E}_{\boldsymbol{\Sigma}}(\mathcal{B})$ as the full envelope. Because $\mathcal{B}_1 \subseteq \mathcal{B}$, the partial envelope is contained in the full envelope, $\mathcal{E}_{\boldsymbol{\Sigma}}(\mathcal{B}_1) \subseteq \mathcal{E}_{\boldsymbol{\Sigma}}(\mathcal{B})$, which allows the partial envelope to offer gains that may not be possible with the full envelope.

Let $u_1 = \dim(\mathcal{E}_{\boldsymbol{\Sigma}}(\mathcal{B}_1))$, let $\boldsymbol{\Gamma} \in \mathbb{R}^{r \times u_1}$ be a semi-orthogonal basis matrix for $\mathcal{E}_{\boldsymbol{\Sigma}}(\mathcal{B}_1)$, let $(\boldsymbol{\Gamma}, \boldsymbol{\Gamma}_0) \in \mathbb{R}^{r \times r}$ be an orthogonal matrix, and let $\boldsymbol{\eta} \in \mathbb{R}^{u_1 \times p_1}$ be the coordinates of $\boldsymbol{\beta}_1$ in terms of the basis matrix $\boldsymbol{\Gamma}$. Then the partial envelope model can be written as

$$\mathbf{Y} = \boldsymbol{\mu} + \boldsymbol{\Gamma}\boldsymbol{\eta}\mathbf{X}_1 + \boldsymbol{\beta}_2\mathbf{X}_2 + \boldsymbol{\varepsilon}, \quad \boldsymbol{\Sigma} = \boldsymbol{\Gamma}\boldsymbol{\Omega}\boldsymbol{\Gamma}^T + \boldsymbol{\Gamma}_0\boldsymbol{\Omega}_0\boldsymbol{\Gamma}_0^T, \tag{3.1}$$

An Introduction to Envelopes: Dimension Reduction for Efficient Estimation in Multivariate Statistics, First Edition. R. Dennis Cook.

where $\boldsymbol{\Omega} \in \mathbb{R}^{u_1 \times u_1}$ and $\boldsymbol{\Omega}_0 \in \mathbb{R}^{(r-u_1)\times(r-u_1)}$ are positive definite. From this we can see how partial envelopes describe properties of $\mathbf{Y}$ in the same way that (1.18) describes properties of $\mathbf{Y}$ for full envelopes:

$$\text{(i)}\quad \mathbf{Q}_{\mathcal{E}_1}\mathbf{Y} \mid (\mathbf{X}_1 = \mathbf{x}_{11}, \mathbf{X}_2 = \mathbf{x}_2) \sim \mathbf{Q}_{\mathcal{E}_1}\mathbf{Y} \mid (\mathbf{X}_1 = \mathbf{x}_{12}, \mathbf{X}_2 = \mathbf{x}_2)$$
$$\text{(ii)}\quad \mathbf{P}_{\mathcal{E}_1}\mathbf{Y} \perp\!\!\!\perp \mathbf{Q}_{\mathcal{E}_1}\mathbf{Y} \mid \mathbf{X}.$$

Condition (i) stipulates that, with $\mathbf{X}_2$ fixed, the marginal distribution of $\mathbf{Q}_{\mathcal{E}_1}\mathbf{Y}$ must be unaffected by changes in $\mathbf{X}_1$ and condition (ii) requires that $\mathbf{Q}_{\mathcal{E}_1}\mathbf{Y}$ and $\mathbf{P}_{\mathcal{E}_1}\mathbf{Y}$ be independent given $\mathbf{X}$. Thus, in any experiment with $\mathbf{X}_2$ fixed, $\mathcal{E}_{\boldsymbol{\Sigma}}(\mathcal{B}_1)$ can be interpreted as the full envelope for the regression of $\mathbf{Y}$ on $\mathbf{X}_1$. This implies that, with $\mathbf{X}_2$ fixed, changes in $\mathbf{X}_1$ affect $\mathbf{Y}$ only via $\mathbf{P}_{\mathcal{E}_1}\mathbf{Y}$. We can then see as a consequence of this interpretation that partial envelopes may be inappropriate when $\mathbf{X}_1$ and $\mathbf{X}_2$ contain functionally related components.

A second interpretation arises from linearly transforming (3.1) with $(\boldsymbol{\Gamma}, \boldsymbol{\Gamma}_0)$ to get

$$\boldsymbol{\Gamma}^T\mathbf{Y} = \boldsymbol{\Gamma}^T\boldsymbol{\mu} + \boldsymbol{\eta}\mathbf{X}_1 + \boldsymbol{\Gamma}^T\boldsymbol{\beta}_2\mathbf{X}_2 + \boldsymbol{\Gamma}^T\boldsymbol{\varepsilon},$$
$$\boldsymbol{\Gamma}_0^T\mathbf{Y} = \boldsymbol{\Gamma}_0^T\boldsymbol{\mu} + \boldsymbol{\Gamma}_0^T\boldsymbol{\beta}_2\mathbf{X}_2 + \boldsymbol{\Gamma}_0^T\boldsymbol{\varepsilon},$$

where $\boldsymbol{\Gamma}^T\boldsymbol{\varepsilon} \perp\!\!\!\perp \boldsymbol{\Gamma}_0^T\boldsymbol{\varepsilon}$. Consequently, the distribution of $\boldsymbol{\Gamma}^T\mathbf{Y}$ can be affected by changes in both $\mathbf{X}_1$ and $\mathbf{X}_2$, while the distribution of $\boldsymbol{\Gamma}_0^T\mathbf{Y}$ is affected only by changes in $\mathbf{X}_2$. Since the dimension of $\mathcal{E}_{\boldsymbol{\Sigma}}(\mathcal{B}_1)$ is minimal, the dimension of $\mathcal{E}_{\boldsymbol{\Sigma}}^{\perp}(\mathcal{B}_1)$ must be maximal and consequently $\boldsymbol{\Gamma}_0^T\mathbf{Y}$ can be interpreted as the maximal part of $\mathbf{Y}$ that is unaffected by changes in $\mathbf{X}_1$, but can be affected by changes in $\mathbf{X}_2$.

Possibilities for the partial envelope model (3.1) are represented schematically in Figure 3.1 that shows three alternatives for a $\boldsymbol{\beta}_1 \in \mathbb{R}^3$. The axes of the plot represent the eigenvectors ℓ_j, $j = 1, 2, 3$, of $\boldsymbol{\Sigma}$ with corresponding eigenvalues $\varphi_1 > \varphi_2 \geq \varphi_3$ for a regression having $r = 3$ responses and $p = 2$ predictors

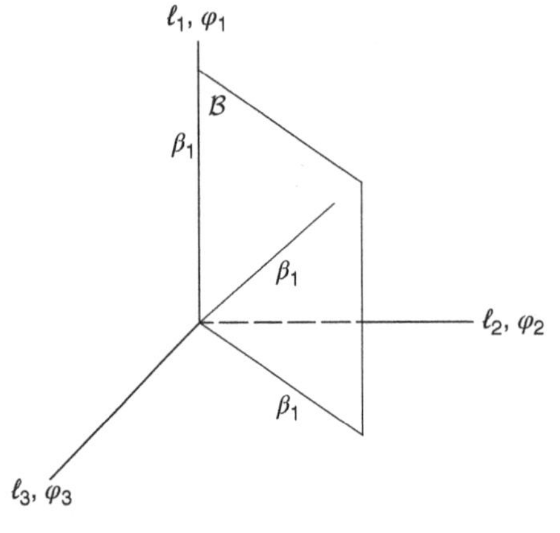

Figure 3.1 Schematic representation of a partial envelope with three possibilities for $\boldsymbol{\beta}_1$.

with coefficient vectors $\boldsymbol{\beta}_1$ and $\boldsymbol{\beta}_2$, so $\boldsymbol{\beta} = (\boldsymbol{\beta}_1, \boldsymbol{\beta}_2) \in \mathbb{R}^{3\times 2}$. The two-dimensional coefficient subspace $\mathcal{B} = \text{span}(\boldsymbol{\beta})$ is depicted as a plane in the plot. If the eigenvalues φ_j are distinct, then all three eigenvectors of $\boldsymbol{\Sigma}$ are needed to fully envelop $\mathcal{B}$ and $\mathcal{E}_{\boldsymbol{\Sigma}}(\mathcal{B}) = \mathbb{R}^3$ so there is no response reduction. On the other hand, if $\varphi_2 = \varphi_3$, then both edges of the plane representing $\mathcal{B}$ correspond to eigenvectors of $\boldsymbol{\Sigma}$ and $\mathcal{B} = \mathcal{E}_{\boldsymbol{\Sigma}}(\mathcal{B})$. The partial envelope for $\boldsymbol{\beta}_1$ depends on where it lies relative to the configuration in the plot. If $\boldsymbol{\beta}_1$ falls in $\text{span}(\boldsymbol{\ell}_1)$, then $\mathcal{B}_1 = \text{span}(\boldsymbol{\beta}_1) = \text{span}(\boldsymbol{\ell}_1) = \mathcal{E}_{\boldsymbol{\Sigma}}(\mathcal{B}_1)$ is one-dimensional. If $\mathcal{B}_1 \subset \text{span}(\boldsymbol{\ell}_2, \boldsymbol{\ell}_3)$, then $\mathcal{E}_{\boldsymbol{\Sigma}}(\mathcal{B}_1) = \mathcal{B}_1$ if $\varphi_2 = \varphi_3$ and $\mathcal{E}_{\boldsymbol{\Sigma}}(\mathcal{B}_1) = \text{span}(\boldsymbol{\ell}_2, \boldsymbol{\ell}_3)$ if $\varphi_2 \neq \varphi_3$. On the other hand, if $\boldsymbol{\beta}_1$ is the third possibility represented in Figure 3.1 and $\varphi_2 \neq \varphi_3$, then $\mathcal{E}_{\boldsymbol{\Sigma}}(\mathcal{B}_1) = \mathcal{E}_{\boldsymbol{\Sigma}}(\mathcal{B}) = \mathbb{R}^3$, and there is no partial reduction.

3.2 Estimation

In preparation for stating the normal-theory maximum likelihood estimators from model (3.1), let $\widehat{\mathbf{R}}_{1|2} \in \mathbb{R}^{p_1}$ and $\widehat{\mathbf{R}}_{\mathbf{Y}|2} \in \mathbb{R}^r$ denote residual vectors from the multivariate linear regressions of $\mathbf{X}_1$ on $\mathbf{X}_2$ and $\mathbf{Y}$ on $\mathbf{X}_2$, as defined previously in Section 1.1.1. Let $\mathbf{S}_{\mathbf{R}_{\mathbf{Y}|2}} \in \mathbb{R}^{r\times r}$ denote the sample covariance matrix of $\widehat{\mathbf{R}}_{\mathbf{Y}|2}$, and let $\mathbf{S}_{\mathbf{R}_{\mathbf{Y}|2}|\mathbf{R}_{1|2}} \in \mathbb{R}^{r\times r}$ denote the sample covariance matrix of the residuals from the linear regression of $\widehat{\mathbf{R}}_{\mathbf{Y}|2}$ on $\widehat{\mathbf{R}}_{1|2}$. This sample covariance matrix is the same as the sample covariance matrix from the fit of $\mathbf{Y}$ on $\mathbf{X}$, $\mathbf{S}_{\mathbf{R}_{\mathbf{Y}|2}|\mathbf{R}_{1|2}} = \mathbf{S}_{\mathbf{Y}|\mathbf{X}}$, but we use the $\mathbf{S}_{\mathbf{R}_{\mathbf{Y}|2}|\mathbf{R}_{1|2}}$ notation to emphasize the partitioned form of the envelope. We obtain the maximum likelihood estimator for $\mathcal{E}_{\boldsymbol{\Sigma}}(\mathcal{B}_1)$ given its dimension u_1 by following the derivation in Section 1.5 of the maximum likelihood estimators for the full envelope model:

$$\begin{aligned}\widehat{\mathcal{E}}_{\boldsymbol{\Sigma}}(\mathcal{B}_1) &= \arg\min_{\mathcal{S}\in\mathcal{G}(u_1,r)} \{\log|\mathbf{P}_{\mathcal{S}}\mathbf{S}_{\mathbf{R}_{\mathbf{Y}|2}|\mathbf{R}_{1|2}}\mathbf{P}_{\mathcal{S}}|_0 + \log|\mathbf{Q}_{\mathcal{S}}\mathbf{S}_{\mathbf{R}_{\mathbf{Y}|2}}\mathbf{Q}_{\mathcal{S}}|_0\} \qquad (3.2)\\ &= \text{span}\{\arg\min_{\mathbf{G}} (\log|\mathbf{G}^T\mathbf{S}_{\mathbf{R}_{\mathbf{Y}|2}|\mathbf{R}_{1|2}}\mathbf{G}| + \log|\mathbf{G}_0^T\mathbf{S}_{\mathbf{R}_{\mathbf{Y}|2}}\mathbf{G}_0|)\}\\ &= \text{span}\{\arg\min_{\mathbf{G}} (\log|\mathbf{G}^T\mathbf{S}_{\mathbf{Y}|\mathbf{X}}\mathbf{G}| + \log|\mathbf{G}^T\mathbf{S}^{-1}_{\mathbf{R}_{\mathbf{Y}|2}}\mathbf{G}|)\},\end{aligned}$$

where $\min_{\mathbf{G}}$ is over $r \times u_1$ semi-orthogonal matrices, $(\mathbf{G}, \mathbf{G}_0) \in \mathbb{R}^{r\times r}$ is an orthogonal matrix, and $\mathcal{G}(u_1, r)$ still denotes the Grassmannian of dimension u_1 in $\mathbb{R}^r$. The function to be optimized is of the general form given in (6.3), and consequently all of the algorithms discussed in Chapter 6 can be used in this context.

Following determination of $\widehat{\mathcal{E}}_{\boldsymbol{\Sigma}}(\mathcal{B}_1)$, the maximum likelihood estimators of the remaining parameters are as follows. The maximum likelihood estimator $\widehat{\boldsymbol{\beta}}_1$ of $\boldsymbol{\beta}_1$ is the projection onto $\widehat{\mathcal{E}}_{\boldsymbol{\Sigma}}(\mathcal{B}_1)$ of the estimator $\mathbf{B}_1$ of $\boldsymbol{\beta}_1$ from the standard model: $\widehat{\boldsymbol{\beta}}_1 = \mathbf{P}_{\widehat{\mathcal{E}}_1}\mathbf{B}_1$, where $\mathbf{P}_{\widehat{\mathcal{E}}_1}$ denotes the projection operator for $\widehat{\mathcal{E}}_{\boldsymbol{\Sigma}}(\mathcal{B}_1)$. The maximum likelihood estimator $\widehat{\boldsymbol{\beta}}_2$ of $\boldsymbol{\beta}_2$ is the coefficient matrix from the ordinary least squares fit of the residuals $\mathbf{Y} - \bar{\mathbf{Y}} - \widehat{\boldsymbol{\beta}}_1\mathbf{X}_1$ on $\mathbf{X}_2$. If $\mathbf{X}_1$ and $\mathbf{X}_2$ are orthogonal, then $\widehat{\boldsymbol{\beta}}_2$ reduces to the maximum likelihood estimator of $\boldsymbol{\beta}_2$ from the standard model. Let $\widehat{\boldsymbol{\Gamma}}$ be a semi-orthogonal basis matrix for $\widehat{\mathcal{E}}_{\boldsymbol{\Sigma}}(\mathcal{B}_1)$ and let $(\widehat{\boldsymbol{\Gamma}}, \widehat{\boldsymbol{\Gamma}}_0)$ be an orthogonal matrix. The maximum likelihood estimator $\widehat{\boldsymbol{\Sigma}}$ of $\boldsymbol{\Sigma}$ is then

$$\widehat{\boldsymbol{\Sigma}} = \mathbf{P}_{\widehat{\mathcal{E}}_1}\mathbf{S}_{\mathbf{R}_{\mathbf{Y}|2}|\mathbf{R}_{1|2}}\mathbf{P}_{\widehat{\mathcal{E}}_1} + \mathbf{Q}_{\widehat{\mathcal{E}}_1}\mathbf{S}_{\mathbf{R}_{\mathbf{Y}|2}}\mathbf{Q}_{\widehat{\mathcal{E}}_1} = \widehat{\boldsymbol{\Gamma}}\widehat{\boldsymbol{\Omega}}\widehat{\boldsymbol{\Gamma}}^T + \widehat{\boldsymbol{\Gamma}}_0\widehat{\boldsymbol{\Omega}}_0\widehat{\boldsymbol{\Gamma}}_0^T$$
$$\widehat{\boldsymbol{\Omega}} = \widehat{\boldsymbol{\Gamma}}^T\mathbf{S}_{\mathbf{R}_{\mathbf{Y}|2}|\mathbf{R}_{1|2}}\widehat{\boldsymbol{\Gamma}}$$
$$\widehat{\boldsymbol{\Omega}}_0 = \widehat{\boldsymbol{\Gamma}}_0^T\mathbf{S}_{\mathbf{R}_{\mathbf{Y}|2}}\widehat{\boldsymbol{\Gamma}}_0.$$

Comparing the maximum likelihood estimator of $\mathcal{E}_{\boldsymbol{\Sigma}}(\mathcal{B})$ in (1.25) with the maximum likelihood estimator of $\mathcal{E}_{\boldsymbol{\Sigma}}(\mathcal{B}_1)$ in (3.2) we see that the two estimators differ only by the inner product matrices: The maximum likelihood estimator of the full envelope $\mathcal{E}_{\boldsymbol{\Sigma}}(\mathcal{B})$ is constructed using $\mathbf{S}_{\mathbf{Y}|\mathbf{X}}$ and $\mathbf{S}_{\mathbf{Y}}$, while the maximum likelihood estimator of the partial envelope $\mathcal{E}_{\boldsymbol{\Sigma}}(\mathcal{B}_1)$ uses $\mathbf{S}_{\mathbf{R}_{\mathbf{Y}|2}|\mathbf{R}_{1|2}} = \mathbf{S}_{\mathbf{Y}|\mathbf{X}}$ and $\mathbf{S}_{\mathbf{R}_{\mathbf{Y}|2}}$. We see from this correspondence that $\widehat{\mathcal{E}}_{\boldsymbol{\Sigma}}(\mathcal{B}_1)$ is the same as the estimator of the full envelope applied in the context of the working model $\widehat{\mathbf{R}}_{\mathbf{Y}|2} = \boldsymbol{\beta}_1\widehat{\mathbf{R}}_{1|2} + \boldsymbol{\varepsilon}^*$. The X-variant and X-invariant parts of $\mathbf{Y}$ in the partial envelope model can be interpreted in terms of the model $\mathbf{R}_{\mathbf{Y}|2} = \boldsymbol{\beta}_1\mathbf{R}_{1|2} + \boldsymbol{\varepsilon}$, where $\mathbf{R}_{\mathbf{Y}|2}$ and $\mathbf{R}_{1|2}$ denote the population versions of $\widehat{\mathbf{R}}_{\mathbf{Y}|2}$ and $\widehat{\mathbf{R}}_{1|2}$. In particular, $\boldsymbol{\Gamma}_0^T\mathbf{R}_{\mathbf{Y}|2}$ can be interpreted as the $\mathbf{R}_{1|2}$-invariant part of $\mathbf{R}_{\mathbf{Y}|2}$.

3.2.1 Asymptotic Distribution of $\widehat{\boldsymbol{\beta}}_1$

The correspondence between partial and full envelopes carries over to asymptotic variances as well. Partition $\boldsymbol{\Sigma}_{\mathbf{X}} = (\boldsymbol{\Sigma}_{\mathbf{X}}^{(ij)})$ according to the partitioning of $\mathbf{X}$ $(i, j = 1, 2)$ and let $\boldsymbol{\Sigma}_{\mathbf{X}}^{(1|2)} = \boldsymbol{\Sigma}_{\mathbf{X}}^{(11)} - \boldsymbol{\Sigma}_{\mathbf{X}}^{(12)}\boldsymbol{\Sigma}_{\mathbf{X}}^{-(22)}\boldsymbol{\Sigma}_{\mathbf{X}}^{(21)}$. The matrix $\boldsymbol{\Sigma}_{\mathbf{X}}^{(1|2)}$ is constructed in the same way as the covariance matrix for the conditional distribution of $\mathbf{X}_1 \mid \mathbf{X}_2$ when $\mathbf{X}$ is normally distributed, although here $\mathbf{X}$ is fixed. Define

$$\mathbf{U}_{1|2} = \boldsymbol{\eta}\boldsymbol{\Sigma}_{\mathbf{X}}^{(1|2)}\boldsymbol{\eta}^T \otimes \boldsymbol{\Omega}_0^{-1} + \boldsymbol{\Omega} \otimes \boldsymbol{\Omega}_0^{-1} + \boldsymbol{\Omega}^{-1} \otimes \boldsymbol{\Omega}_0 - 2\mathbf{I}_{u_1(r-u_1)}.$$

The limiting distribution of $\widehat{\boldsymbol{\beta}}_1$ is stated in the following proposition. See Su and Cook (2011) for justification and for the limiting distribution of $\widehat{\boldsymbol{\beta}}_2$.

Proposition 3.1 *Under the partial envelope model (3.1)* $\sqrt{n}\{\mathrm{vec}(\widehat{\boldsymbol{\beta}}_1) - \mathrm{vec}(\boldsymbol{\beta}_1)\}$ *converge in distribution to a normal random vector with mean 0 and covariance matrix*

$$\begin{aligned}\mathrm{avar}\{\sqrt{n}\,\mathrm{vec}(\widehat{\boldsymbol{\beta}}_1)\} &= \boldsymbol{\Sigma}_{\mathbf{X}}^{-(1|2)} \otimes \boldsymbol{\Gamma}\boldsymbol{\Omega}\boldsymbol{\Gamma}^T + (\boldsymbol{\eta}^T \otimes \boldsymbol{\Gamma}_0)\mathbf{U}_{1|2}^{\dagger}(\boldsymbol{\eta} \otimes \boldsymbol{\Gamma}_0^T),\\ &= \mathrm{avar}\{\sqrt{n}\,\mathrm{vec}(\widehat{\boldsymbol{\beta}}_{1\boldsymbol{\Gamma}})\} + \mathrm{avar}\{\sqrt{n}\,\mathrm{vec}(\mathbf{Q}_{\boldsymbol{\Gamma}}\widehat{\boldsymbol{\beta}}_{1\boldsymbol{\eta}})\},\end{aligned}$$

where $\widehat{\boldsymbol{\beta}}_{1\eta}$ *and* $\widehat{\boldsymbol{\beta}}_{1\Gamma}$ *are the maximum likelihood estimators of* $\boldsymbol{\beta}_1$ *when* $\boldsymbol{\eta}$ *and* $\boldsymbol{\Gamma}$ *are known,* $\text{avar}[\sqrt{n}\,\text{vec}(\widehat{\boldsymbol{\beta}}_{1\Gamma})] = \boldsymbol{\Sigma}_{\text{X}}^{-(1|2)} \otimes \boldsymbol{\Gamma}\boldsymbol{\Omega}\boldsymbol{\Gamma}^T$ *and* $\text{avar}\{\sqrt{n}\,\text{vec}(\mathbf{Q}_{\Gamma}\widehat{\boldsymbol{\beta}}_{1\eta})\}$ *is defined implicitly.*

The form of the asymptotic variance of $\widehat{\boldsymbol{\beta}}_1$ given in this proposition is identical to the form of the asymptotic variance of $\widehat{\boldsymbol{\beta}}$ from the full envelope model given in Section 1.6, although the definitions of components are different. For instance, $\text{avar}\{\sqrt{n}\,\text{vec}(\widehat{\boldsymbol{\beta}})\}$ requires $\boldsymbol{\Sigma}_{\text{X}}$, while $\text{avar}\{\sqrt{n}\,\text{vec}(\widehat{\boldsymbol{\beta}}_1)\}$ uses $\boldsymbol{\Sigma}_{\text{X}}^{(1|2)}$ in its place. Several additional characteristics of this proposition may be of interest. First consider regressions in which $\boldsymbol{\Sigma}_{\text{X}}^{(12)} = 0$. This will arise when $\mathbf{X}_1$ and $\mathbf{X}_2$ have been chosen by design to be orthogonal. Because $\mathbf{X}$ is nonrandom, this condition can always be forced without an inferential cost by replacing $\mathbf{X}_1$ with $\widehat{\mathbf{R}}_{1|2}$. This replacement alters the definition of $\boldsymbol{\beta}_2$ but does not alter the parameter of interest $\boldsymbol{\beta}_1$. When $\boldsymbol{\Sigma}_{\text{X}}^{(12)} = 0$, $\text{avar}\{\sqrt{n}\,\text{vec}(\widehat{\boldsymbol{\beta}}_1)\}$ reduces to the asymptotic covariance matrix for the full envelope estimator of $\boldsymbol{\beta}_1$ in the working model $\widehat{\mathbf{R}}_{\text{Y}|2} = \boldsymbol{\beta}_1\mathbf{X}_1 + \boldsymbol{\varepsilon}^*$. No longer requiring that $\boldsymbol{\Sigma}_{\text{X}}^{(12)} = 0$, we can carry out asymptotic inference for $\boldsymbol{\beta}_1$ based on a partial envelope by using the full envelope for $\boldsymbol{\beta}_1$ in the working model $\widehat{\mathbf{R}}_{\text{Y}|2} = \boldsymbol{\beta}_1\widehat{\mathbf{R}}_{1|2} + \boldsymbol{\varepsilon}^*$. If we choose $\boldsymbol{\beta}_1 = \boldsymbol{\beta}$, so $\mathbf{X}_1 = \mathbf{X}$, $\boldsymbol{\beta}_2$ is nil and $\boldsymbol{\Sigma}_{\text{X}}^{(1|2)} = \boldsymbol{\Sigma}_{\text{X}}$, then $\text{avar}\{n^{1/2}\,\text{vec}(\widehat{\boldsymbol{\beta}}_1)\}$ reduces to the asymptotic covariance matrix for the full envelope estimator of $\boldsymbol{\beta}$ given in Section 1.6.

The asymptotic standard error for an element of $\widehat{\boldsymbol{\beta}}_1$ can be determined using the rational for (1.33):

$$\text{se}\{(\widehat{\boldsymbol{\beta}}_1)_{ij}\} = \frac{\{\widehat{\text{avar}}[\sqrt{n}(\widehat{\boldsymbol{\beta}}_1)_{ij}]\}^{1/2}}{\sqrt{n}},$$

with corresponding Z-score equal to $(\widehat{\boldsymbol{\beta}}_1)_{ij}/\text{se}\{(\widehat{\boldsymbol{\beta}}_1)_{ij}\}$. The bootstrap can also be used, as described in Section 1.11.

3.2.2 Selecting u_1

The methods discussed in Section 1.10 for selecting the dimension u of the full envelope can be adapted straightforwardly for the selection of the dimension u_1 of the partial envelope.

To use a sequence of likelihood ratio tests, the hypothesis, $u_1 = u_{10}$, $u_{10} < r$, can be tested by using the likelihood ratio statistic $\Lambda(u_{10}) = 2\{\widehat{L}_r - \widehat{L}_{u_{10}}\}$, where $\widehat{L}_r$ is the maximized log-likelihood for the standard model given in Section 1.10, and

$$\begin{aligned}\widehat{L}_{u_{10}} &= -(nr/2)\{1 + \log(2\pi)\} - (n/2)\log|\widehat{\boldsymbol{\Gamma}}^T\mathbf{S}_{\mathbf{R}_{\text{Y}|2}|\mathbf{R}_{1|2}}\widehat{\boldsymbol{\Gamma}}| \\ &\quad - (n/2)\log|\widehat{\boldsymbol{\Gamma}}^T\mathbf{S}_{\mathbf{R}_{\text{Y}|2}}^{-1}\widehat{\boldsymbol{\Gamma}}|\end{aligned}$$

denotes the maximum value of the likelihood for the partial envelope model with $u_1 = u_{10}$. When $u_1 = p_1$, $\hat{L}_{u_{10}} = \hat{L}_r$. Following standard likelihood theory, under the null hypothesis, $\Lambda(u_{10})$ is distributed asymptotically as a chi-squared random variable with $p_1(r - u_{10})$ degrees of freedom. This test statistic can be used in a sequential scheme to choose u_1 as described in Section 1.10.

The dimension of the partial envelope could also be selected by using an information criterion:

$$\hat{u}_1 = \arg\min_{u_1}\{-2\hat{L}_{u_1} + h(n)N_{u_1}\},$$

where $h(n)$ is equal to $\log(n)$ for Bayes information criterion and is equal to 2 for Akaike's information criterion, and N_{u_1} is the number of real parameters in the partial envelope model,

$$N_{u_1} = r + u_1(r - u_1) + u_1p_1 + rp_2 + \frac{u_1(u_1+1)}{2} + \frac{(r-u_1)(r-u_1+1)}{2}.$$

Subtracting N_{u_1} from the number of parameters $r + pr + r(r+1)/2$ for the standard model gives the degrees of freedom for $\Lambda(u_1)$ mentioned previously.

3.3 Illustrations

3.3.1 Cattle Weight: Incorporating Basal Weight

In this illustration, we return to the cattle weight data introduced in Section 1.3.4, this time incorporating the basal weight X_2 as a covariate:

$$\mathbf{Y} = \boldsymbol{\alpha} + \boldsymbol{\beta}_1X_1 + \boldsymbol{\beta}_2X_2 + \boldsymbol{\varepsilon},$$

where $\mathbf{Y}$ is the 10×1 vector of weights, and X_1 is the indicator for treatment. Interest centers on $\boldsymbol{\beta}_1$, the vector of treatment effects adjusted for the basal weights. From the results shown in Table 3.1 based on the BIC inferred dimension $u_1 = 2$, we see that adjusting by basal weight seems to strengthen the results based on the unadjusted weights. The estimate of $\boldsymbol{\beta}_2$ given in the last column of Table 3.1 suggests that the weights are being adjusted by essentially subtracting the basal weight.

3.3.2 Mens' Urine

This illustration makes use of data from a study of the relationship between the composition of a man's urine and his weight (Smith et al. 1962). The eight responses $\mathbf{Y}$ that we use are the concentrations of phosphate, calcium,

Table 3.1 Cattle data: Estimated coefficients for the partial envelope model with $u_1 = 2$.

Week	$\widehat{\beta}_1$	Z-score	R	$\widehat{\beta}_2$
2	0.03	0.05	1.45	0.94
4	1.48	1.34	1.76	0.97
6	2.09	1.97	2.56	0.90
8	2.74	3.00	3.21	0.96
10	2.58	3.72	4.62	0.99
12	3.77	3.32	3.03	0.92
14	−6.08	−5.95	3.42	0.95
16	−5.93	−5.25	3.35	1.01
18	−3.51	−3.57	5.00	0.96
19	5.17	4.99	5.08	1.01

The Z-score is the estimate divided by its asymptotic standard error, and $R = \text{se}(\mathbf{B}_1)/\text{se}(\widehat{\beta}_1)$.

phosphorus, creatinine, and chloride in mg ml^{-1} and boron, choline, and copper in µg ml^{-1}. These responses plus two covariates, volume (ml) and $(\text{specificgravity} - 1) \times 10^3$, were measured on 45 men in four weight categories. The multivariate analysis of covariance model that (Smith et al. 1962) used for these data is

$$\begin{aligned} \mathbf{Y} &= \boldsymbol{\alpha} + \boldsymbol{\beta}_1 \mathbf{X}_1 + \boldsymbol{\beta}_2 \mathbf{X}_2 + \boldsymbol{\varepsilon} \\ &= \boldsymbol{\alpha} + \boldsymbol{\beta}_{11} X_{11} + \boldsymbol{\beta}_{12} X_{12} + \boldsymbol{\beta}_{13} X_{13} + \boldsymbol{\beta}_{21} X_{21} + \boldsymbol{\beta}_{22} X_{22} + \boldsymbol{\varepsilon}, \end{aligned} \tag{3.3}$$

where $\mathbf{X}_1 = (X_{11}, X_{12}, X_{13})^T$, $X_{1k} = 1$ for category k and 0 otherwise, $k = 1, 2, 3$, and the covariates are represented by $\mathbf{X}_2 = (X_{21}, X_{22})^T$. The primary question of interest is whether the mean response varies with the weight category. In terms of the model, the question is whether or not there is evidence to indicate that elements of $\boldsymbol{\beta}_1 = (\boldsymbol{\beta}_{11}, \boldsymbol{\beta}_{12}, \boldsymbol{\beta}_{13})^T$ are nonzero. This is the kind of question for which partial envelopes are appropriate.

The assumption of normal errors tends to be most important with small data sets, as is the case here. Consequently, we first consider if coordinate-wise power transformations of the response vector might move it closer to normality. The likelihood ratio test (Cook and Weisberg 1998) that no transformation is required has a p-value close to 0, while the likelihood ratio test that all responses should be in the log scale has a p-value of 0.14. Consequently, we use log concentrations as the responses, a decision that could have been taken based on past experience without recourse to likelihood ratio testing.

The dimension selection criteria AIC, BIC, and LRT(0.05) all indicated that $u = 3$ for fitting the full envelope model to these data. Consequently, we would normally expect that $u_1 < 3$ when fitting the partial envelope model (3.1). As it turned out AIC, BIC, and LRT(0.05) as described in Section 3.2.2 all indicated that $u_1 = 1$, so in effect only a single linear combination of the responses responds to mens' weight given the two covariates. The corresponding partial response envelope model is

$$\mathbf{Y} = \boldsymbol{\alpha} + \boldsymbol{\Gamma}(\eta_{11}X_{11} + \eta_{12}X_{12} + \eta_{13}X_{13}) + \boldsymbol{\beta}_{21}X_{21} + \boldsymbol{\beta}_{22}X_{22} + \boldsymbol{\varepsilon}, \tag{3.4}$$

where $\boldsymbol{\Gamma} \in \mathbb{R}^8$ and the η's are scalars.

Table 3.2 shows the Z-scores, the ratios of the estimated elements of $\boldsymbol{\beta}_1$ to their standard errors, for the ordinary least squares and partial envelope (with $u_1 = 1$) fits of models (3.3) and (3.4). The estimated basis $\widehat{\boldsymbol{\Gamma}}$ for $\mathcal{E}_{\boldsymbol{\Sigma}}(\mathcal{B}_1)$ is shown in the last column. The results in this table suggest that in this illustration the partial envelope model has strengthened and clarified the conclusions indicated by ordinary least squares. In particular, except for boron and choline, the mean responses of the three indicated weight categories all differ from weight category 4, the basal category in the parameterization of model (3.3).

According to the discussion in Section 3.2, we may be able to gain insights into the regression by considering the working model $\widehat{\boldsymbol{\Gamma}}^T\widehat{\mathbf{R}}_{Y|2} = \boldsymbol{\eta}_1\widehat{\mathbf{R}}_{1|2} + \text{error}$, where $\widehat{\boldsymbol{\Gamma}}^T\widehat{\mathbf{R}}_{Y|2}$ is a scalar. For instance, the added variable plot in Figure 3.2 gives a visual representation of the contribution of the first category after accounting for the other categories and the two covariates. The line on the plot is the ordinary least squares fit of the vertical axis residuals on the horizontal axis residuals. The slope of the line 0.878 is equal to $\widehat{\eta}_{11}$ from the fit of the partial envelope model (3.4).

Table 3.2 Mens' urine: Ratios of coefficient estimates to their standard errors (Z-scores) for the ordinary least squares (OLS) fit and partial envelope fit with $u_1 = 1$.

	B_{11}	B_{12}	B_{13}	$\widehat{\beta}_{11}$	$\widehat{\beta}_{12}$	$\widehat{\beta}_{13}$	
Y		OLS Z-score			Partial envelope Z-score		$\widehat{\boldsymbol{\Gamma}}$
Phosphate	−3.86	−2.86	−2.44	−3.81	−3.32	−3.11	−0.39
Calcium	−1.35	−2.30	−3.31	−2.23	−2.16	−2.16	−0.45
Phosphorus	−3.93	−2.22	−1.78	−4.33	−3.67	−3.37	−0.39
Creatinine	−3.14	−2.41	−2.29	−3.72	−3.26	−3.06	−0.56
Chloride	4.08	3.26	1.76	2.85	2.63	2.52	0.29
Boron	1.04	−0.01	−0.35	0.26	0.26	0.26	0.04
Choline	−1.21	1.05	0.73	−0.42	−0.42	0.42	−0.04
Copper	−2.64	−1.48	−0.29	−3.44	−3.07	−2.90	−0.30

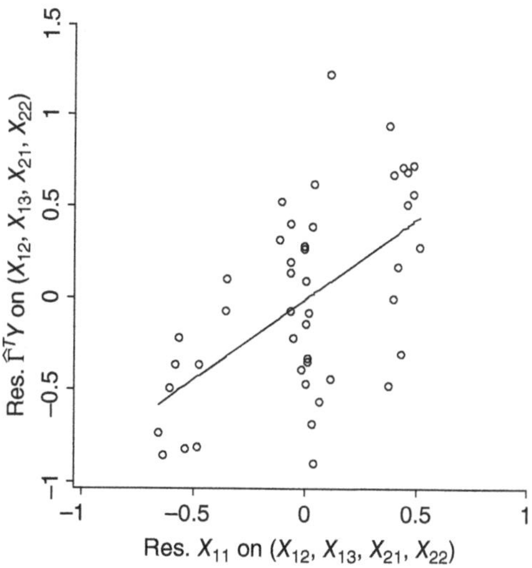

Figure 3.2 Mens' urine: Added variable plot for X_{11}. The line is the ordinary least squares fit of the plotted data.

3.4 Partial Envelopes for Prediction

3.4.1 Rationale

In Section 1.7, we described the prediction of the response vector at a new value $\mathbf{X}_{\text{new}}$ of $\mathbf{X}$ as $\widehat{\mathbf{Y}} = \widehat{\boldsymbol{\alpha}} + \widehat{\boldsymbol{\beta}}\mathbf{X}_{\text{new}}$, where $\widehat{\boldsymbol{\beta}}$ is the envelope estimator of $\boldsymbol{\beta}$ based on the full envelope $\mathcal{E}_{\boldsymbol{\Sigma}}(\mathcal{B})$. In this section, we use partial envelopes to envelop $\boldsymbol{\beta}\mathbf{X}_{\text{new}}$, which leads to a new method that has the potential to yield predicted and fitted values with smaller variation than the standard predictions based on model (1.1) or the full envelope predictions described in Section 1.7.

Select an $\mathbf{A}_0 \in \mathbb{R}^{p\times(p-1)}$ so that $\mathbf{A} = (\mathbf{X}_{\text{new}}, \mathbf{A}_0) \in \mathbb{R}^{p\times p}$ has full rank. It may be helpful computationally, although not necessary, to choose the columns of $\mathbf{A}_0$ to be orthogonal to $\mathbf{X}_{\text{new}}$. Let $\boldsymbol{\phi}_1 = \boldsymbol{\beta}\mathbf{X}_{\text{new}}$, $\boldsymbol{\phi}_2 = \boldsymbol{\beta}\mathbf{A}_0$, $\boldsymbol{\phi} = (\boldsymbol{\phi}_1, \boldsymbol{\phi}_2)$, and $\mathbf{Z} = \mathbf{A}^{-1}\mathbf{X} = (Z_1, \mathbf{Z}_2^T)^T$, where $\mathbf{Z}_2 \in \mathbb{R}^{p-1}$. Then we can parameterize model (1.1) as

$$\begin{aligned}\mathbf{Y} &= \boldsymbol{\alpha} + \boldsymbol{\beta}\mathbf{X} + \boldsymbol{\varepsilon}\\ &= \boldsymbol{\alpha} + \boldsymbol{\beta}\mathbf{A}\mathbf{A}^{-1}\mathbf{X} + \boldsymbol{\varepsilon}\\ &= \boldsymbol{\alpha} + (\boldsymbol{\phi}_1, \boldsymbol{\phi}_2)\mathbf{Z} + \boldsymbol{\varepsilon}\\ &= \boldsymbol{\alpha} + \boldsymbol{\phi}_1 Z_1 + \boldsymbol{\phi}_2\mathbf{Z}_2 + \boldsymbol{\varepsilon}.\end{aligned}$$

We can now parameterize the model in terms of a basis $\boldsymbol{\Gamma}$ for the $\boldsymbol{\Sigma}$-envelope of span($\boldsymbol{\phi}_1$), leading to a partial envelope representation for prediction at $\mathbf{X}_{\text{new}}$:

$$\mathbf{Y} = \boldsymbol{\alpha} + \boldsymbol{\Gamma}\boldsymbol{\eta}Z_1 + \boldsymbol{\phi}_2\mathbf{Z}_2 + \boldsymbol{\varepsilon}, \quad \boldsymbol{\Sigma} = \boldsymbol{\Gamma}\boldsymbol{\Omega}\boldsymbol{\Gamma}^T + \boldsymbol{\Gamma}_0\boldsymbol{\Omega}_0\boldsymbol{\Gamma}_0^T. \tag{3.5}$$

At this point, we can apply any of the methods discussed previously in this chapter.

The reparameterization $\boldsymbol{\beta} \mapsto \boldsymbol{\phi}$ is full rank by construction so $\mathcal{B} = \mathrm{span}(\boldsymbol{\phi})$ and thus it does not change the envelope, $\mathcal{E}_{\boldsymbol{\Sigma}}(\mathcal{B}) = \mathcal{E}_{\boldsymbol{\Sigma}}(\mathrm{span}(\boldsymbol{\phi}))$. However, it does change the coordinate system for moving inside $\mathcal{B}$ as represented in Figure 3.1. As $\mathbf{X}_{\mathrm{new}}$ changes, $\boldsymbol{\phi} = \boldsymbol{\beta}\mathbf{X}_{\mathrm{new}}$ will change locations within $\mathcal{B}$ and the discussion of Figure 3.1 in terms of $\boldsymbol{\beta}_1$ applies also to $\boldsymbol{\phi}_1$. For instance, if for some value of $\mathbf{X}_{\mathrm{new}}$, $\mathrm{span}(\boldsymbol{\phi}_1) = \mathrm{span}(\boldsymbol{\ell}_1)$, then we can expect the partial envelope for $\boldsymbol{\phi}_1$ to yield a better prediction at $\mathbf{X}_{\mathrm{new}}$ than that based on the standard model or on the full envelope $\mathcal{E}_{\boldsymbol{\Sigma}}(\mathcal{B})$. A practical implication of this discussion is that we may wish to do a partial envelope analysis for each prediction.

3.4.2 Pulp Fibers: Partial Envelopes and Prediction

The pulp fibers dataset (Johnson and Wichern 2007), which was used by Su and Cook (2011) to introduce partial envelopes, is on properties of pulp fibers and the paper made from them. There are 62 measurements on four paper properties that form the responses: breaking length, elastic modulus, stress at failure, and burst strength. Three aspects of the pulp fiber are the predictors: arithmetic fiber length, long fiber fraction, and fine fiber fraction. We first review the findings of Su and Cook (2011) and then turn to prediction envelopes.

Fitting the full envelope model to all three predictors, BIC and LRT(0.01) both indicated that $u = 2$. The ratios of the standard deviations of the element of $\mathbf{B}$ to those of the envelope estimator $\widehat{\boldsymbol{\beta}}$ with $u = 2$ range from 0.98 to 1.10, with an average of 1.03, so the envelope model does not bring notable reduction. The reason is evidently that $\|\boldsymbol{\Omega}_0\| \ll \|\boldsymbol{\Omega}\|$. Suppose that we are particularly interested in how the fine fiber fraction affects the paper quality. Fitting the partial envelope model to the column of $\boldsymbol{\beta}$ that corresponds to fine fiber fraction, BIC and LRT(0.05) both selected $u_1 = 1$. The standard deviation ratios between the full envelope model and the partial envelope model for the four elements in the column of $\boldsymbol{\beta}$ that corresponds to fine fiber fraction range between about 7 and 66. This reduction comes about because for the partial envelope model $\|\boldsymbol{\Omega}_0\| \gg \|\boldsymbol{\Omega}\|$.

Turning to prediction, consider predicting $\mathbf{Y}$ at $\mathbf{X}_{\mathrm{new}} = (-0.324, 17.639, 28.937)^T$, which corresponds to the last observation in the data file. Fitting a prediction envelope (3.5), both BIC and LRT(0.05) signaled that $u_1 = 1$. Table 3.3 gives the predictions of the elements of $\mathbf{Y}$ from the full envelope model and the prediction envelope (3.5) along with the standard errors of the fitted values and actual predictions. The standard errors from the full envelope model were determined by plugging in sample versions of the parameters in (1.36), using $n^{-1}\mathrm{avar}\{\sqrt{n}\mathrm{vec}(\widehat{\boldsymbol{\beta}}\mathbf{X})\}$ for the fitted values and $n^{-1}\mathrm{avar}\{\sqrt{n}\mathrm{vec}(\widehat{\boldsymbol{\beta}}\mathbf{X})\} + \widehat{\boldsymbol{\Sigma}}$ for the predictions. Standard errors for the fitted values and predictions from the partial model were determined similarly. It can be seen from the table that fitted values change little, while the standard errors change appreciably.

Table 3.3 Pulp fibers: Fitted and predicted values at $\mathbf{X}_{\text{new}}$.

	Envelope prediction			Partial envelope prediction		
Y	$\hat{Y}$	SE fit	SE predict	$\hat{Y}$	SE fit	SE predict
Breaking length	18.86	1.53	2.45	19.25	0.25	2.01
Elastic modulus	6.71	0.39	0.67	6.89	0.09	0.56
Stress at failure	4.04	0.71	1.16	4.30	0.14	0.96
Burst strength	0.25	0.33	0.54	0.33	0.06	0.44

Envelope prediction is from the full envelope model with $u = 2$. Partial envelope prediction is from model (3.5) with $u = 1$. "SE fit" is the standard error for a fitted value, and "SE predict" is the standard error for a predicted value of the response vector.

3.5 Reducing Part of the Response

The partial envelopes described in Section 3.1 were designed to reduce $\mathbf{Y}$ while focusing consideration on selected columns of $\boldsymbol{\beta}$. In this section, we sketch a different form of partial reduction that may be relevant in some applications.

Partition the response vector as $\mathbf{Y} = (\mathbf{Y}_1^T, \mathbf{Y}_2^T)^T$ with $\mathbf{Y}_1 \in \mathbb{R}^{r_1}$ and $\mathbf{Y}_2 \in \mathbb{R}^{r_2}$, and ask if there are linear combinations of $\mathbf{Y}_1$ whose distribution is invariant under changes in $\mathbf{X}$, leaving $\mathbf{Y}_2$ out of the reduction. This type of reduction might be useful when the elements of $\mathbf{Y}_1$ are longitudinal measures on an experimental unit, while the elements of $\mathbf{Y}_2$ are cross sectional measures. Partition $\boldsymbol{\beta} = (\boldsymbol{\beta}_1^T, \boldsymbol{\beta}_2^T)^T$, $\boldsymbol{\beta}_j \in \mathbb{R}^{r_j \times p}$ $(j = 1, 2)$, to conform to the partitioning of $\mathbf{Y}$. The reduction of $\mathbf{Y}_1$ can be addressed with the envelope $\mathcal{E}_{\boldsymbol{\Sigma}}(\text{span}(\boldsymbol{\beta}_1 \oplus \boldsymbol{\beta}_2))$, while restricting bases to be semi-orthogonal matrices of the form $\boldsymbol{\Gamma} = \boldsymbol{\Gamma}_1 \oplus \mathbf{I}_{r_2}$, where $\oplus$ denotes the direct sum and $\boldsymbol{\Gamma}_1 \in \mathbb{R}^{r_1 \times u_1}$ with $r_1 \geq u_1$. Since $\mathcal{E}_{\boldsymbol{\Sigma}}(\mathcal{B}) \subseteq \mathcal{E}_{\boldsymbol{\Sigma}}(\text{span}(\boldsymbol{\beta}_1 \oplus \boldsymbol{\beta}_2))$, we are asking for a particular type of upper bound on the usual envelope. Let $\boldsymbol{\Gamma}_0 = (\boldsymbol{\Gamma}_{10}^T, 0)^T$ denote a semi-orthogonal basis matrix for the orthogonal complement of $\text{span}(\boldsymbol{\Gamma}_1 \oplus \mathbf{I}_{r_2})$. Then following (1.18), we seek reductions that satisfy

$$\boldsymbol{\Gamma}_{10}^T \mathbf{Y}_1 \mid \mathbf{X} \sim \boldsymbol{\Gamma}_{10}^T \mathbf{Y}_1 \text{ and } (\boldsymbol{\Gamma}_1^T \mathbf{Y}, \mathbf{Y}_2) \perp\!\!\!\perp \boldsymbol{\Gamma}_{10}^T \mathbf{Y}_1 \mid \mathbf{X}.$$

The model statement is the same as (1.20), but with $\boldsymbol{\Gamma}$ and $\boldsymbol{\Gamma}_0$ as defined here.

Correspondingly, the derivation of the maximum likelihood estimators given in Section 1.5.1 is applicable here as well, except that the optimization required in (1.25) is over semi-orthogonal matrices $\boldsymbol{\Gamma}_1 \in \mathbb{R}^{r_1 \times u_1}$ as they occur in $\boldsymbol{\Gamma} = (\boldsymbol{\Gamma}_1 \oplus \mathbf{I}_{r_2})$. Beginning with (1.25) and using the forms of $\boldsymbol{\Gamma}$ and $\boldsymbol{\Gamma}_0 = (\boldsymbol{\Gamma}_{10}^T, 0^T)^T$ for this application, we have

$$\begin{aligned} \log |\boldsymbol{\Gamma}_0^T \mathbf{S}_{\mathbf{Y}} \boldsymbol{\Gamma}_0| &= \log |\boldsymbol{\Gamma}_{10}^T \mathbf{S}_{\mathbf{Y}_1} \boldsymbol{\Gamma}_{10}| \\ &= \log |\mathbf{S}_{\mathbf{Y}_1}| + \log |\boldsymbol{\Gamma}_1^T \mathbf{S}_{\mathbf{Y}_1}^{-1} \boldsymbol{\Gamma}_1|. \end{aligned}$$

For the other term in (1.25), partition $\mathbb{Y} = (\mathbb{Y}_1, \mathbb{Y}_2)$, $\mathbb{Y}_j \in \mathbb{R}^{n \times r_j}$ $(j = 1, 2)$, and write

$$\mathbf{S}_{\mathbf{Y}|\mathbf{X}} = n^{-1}\mathbb{Y}^T\mathbf{Q}_{\mathbb{X}}\mathbb{Y} = n^{-1}\begin{pmatrix} \mathbb{Y}_1^T\mathbf{Q}_{\mathbb{X}}\mathbb{Y}_1 & \mathbb{Y}_1^T\mathbf{Q}_{\mathbb{X}}\mathbb{Y}_2 \\ \mathbb{Y}_2^T\mathbf{Q}_{\mathbb{X}}\mathbb{Y}_1 & \mathbb{Y}_2^T\mathbf{Q}_{\mathbb{X}}\mathbb{Y}_2 \end{pmatrix}.$$

Then

$$\log|\boldsymbol{\Gamma}^T\mathbf{S}_{\mathbf{Y}|\mathbf{X}}\boldsymbol{\Gamma}| = \log|\mathbf{S}_{\mathbf{Y}_2}| + \log|\boldsymbol{\Gamma}_1^T\mathbf{S}_{\mathbf{R}_{\mathbf{Y}_1|\mathbf{X}}|\mathbf{R}_{\mathbf{Y}_2|\mathbf{X}}}\boldsymbol{\Gamma}_1|$$

where

$$\mathbf{S}_{\mathbf{R}_{\mathbf{Y}_1|\mathbf{X}}|\mathbf{R}_{\mathbf{Y}_2|\mathbf{X}}} = n^{-1}(\mathbb{Y}_1^T\mathbf{Q}_{\mathbb{X}}\mathbb{Y}_1 - \mathbb{Y}_1^T\mathbf{Q}_{\mathbb{X}}\mathbb{Y}_2(\mathbb{Y}_2^T\mathbf{Q}_{\mathbb{X}}\mathbb{Y}_2)^{-1}\mathbb{Y}_2^T\mathbf{Q}_{\mathbb{X}}\mathbb{Y}_2)$$

denotes the sample covariance matrix from the regression of $\mathbf{R}_{\mathbf{Y}_1|\mathbf{X}}$, the residual vector from the regression of $\mathbf{Y}_1$ on $\mathbf{X}$, on $\mathbf{R}_{\mathbf{Y}_2|\mathbf{X}}$, the residual vector from the regression of $\mathbf{Y}_2$ on $\mathbf{X}$. We then have the full log-likelihood maximized over all parameters except $\boldsymbol{\Gamma}_1$:

$$\begin{aligned} L_{u_1}(\boldsymbol{\Gamma}_1) = &-(nr/2)\log(2\pi) - nr/2 - (n/2)\log|\mathbf{S}_{\mathbf{Y}_1}| - (n/2)\log|\mathbf{S}_{\mathbf{Y}_2}| \\ &-(n/2)\log|\boldsymbol{\Gamma}_1^T\mathbf{S}_{\mathbf{R}_{\mathbf{Y}_1|\mathbf{X}}|\mathbf{R}_{\mathbf{Y}_2|\mathbf{X}}}\boldsymbol{\Gamma}_1| - (n/2)\log|\boldsymbol{\Gamma}_1^T\mathbf{S}_{\mathbf{Y}_1}^{-1}\boldsymbol{\Gamma}_1|. \end{aligned}$$

Partitioning $\mathbf{B} = (\mathbf{B}_1^T, \mathbf{B}_2^T)^T$ as we partitioned $\boldsymbol{\beta}$, we have in consequence

$$\widehat{\mathcal{E}}_{\boldsymbol{\Sigma}}(\mathrm{span}(\boldsymbol{\beta}_1 \oplus \boldsymbol{\beta}_2)) = \mathrm{span}\{\arg\min_{\mathbf{G}_1} L_{u_1}(\mathbf{G}_1)\}$$

$$\widehat{\boldsymbol{\beta}} = \begin{pmatrix} \mathbf{P}_{\widehat{\boldsymbol{\Gamma}}_1}\mathbf{B}_1 \\ \mathbf{B}_2 \end{pmatrix}$$

and

$$\widehat{\boldsymbol{\Sigma}} = \begin{pmatrix} \mathbf{P}_{\widehat{\boldsymbol{\Gamma}}_1}\mathbf{S}_{\mathbf{Y}_1}\mathbf{P}_{\widehat{\boldsymbol{\Gamma}}_1} + \mathbf{Q}_{\widehat{\boldsymbol{\Gamma}}_1}\mathbf{S}_{\mathbf{Y}_1}\mathbf{Q}_{\widehat{\boldsymbol{\Gamma}}_1} & \mathbf{P}_{\widehat{\boldsymbol{\Gamma}}_1}\mathbf{S}_{\mathbf{Y}_1,\mathbf{Y}_2} \\ \mathbf{S}_{\mathbf{Y}_2,\mathbf{Y}_1}\mathbf{P}_{\widehat{\boldsymbol{\Gamma}}_1} & \mathbf{S}_{\mathbf{Y}_2} \end{pmatrix},$$

where $\mathbf{S}_{\mathbf{Y}_1,\mathbf{Y}_2} = n^{-1}\mathbb{Y}_1^T\mathbf{Q}_{\mathbb{X}}\mathbb{Y}_2$ and $\min_{\mathbf{G}_1}$ is over all semi-orthogonal matrices $\mathbf{G}_1 \in \mathbb{R}^{r \times u_1}$.

Other aspects of this application can be developed by following the ideas of Chapter 1, and it can be extended to allow for simultaneous reduction of $\mathbf{Y}_1$ and $\mathbf{Y}_2$ using bases of the form $\boldsymbol{\Gamma} = \boldsymbol{\Gamma}_1 \oplus \boldsymbol{\Gamma}_2$.

4

Predictor Envelopes

In Chapter 1, we considered reduction of $\mathbf{Y}$, relying on the notion of material (X-variant) and immaterial (X-invariant) components for motivation and using $\mathcal{E}_{\boldsymbol{\Sigma}}(\mathcal{B})$ with $\mathcal{B} = \text{span}(\boldsymbol{\beta})$ as the reduction construct. In this chapter, we consider reducing the dimensionality of $\mathbf{X}$, again using envelopes as the essential device. In Section 4.1.1, we use a motivation based on direct predictor reduction to derive an envelope model. In Section 4.1.2, an envelope model is developed by using a formulation based on latent variables that is commonly associated with partial least squares (PLS) regression. Section 4.1.3 contains a brief discussion of the potential advantages of envelopes for predictor reduction. We turn to estimation in Sections 4.2 and 4.3, giving a few illustrations in Section 4.4. In Section 4.5, we expand our discussion by considering simultaneous predictor-response reduction motivated in part by the envelope model described in Section 4.1.2.

4.1 Model Formulations

4.1.1 Linear Predictor Reduction

Again consider the multivariate linear model (1.1), but now requiring the predictors to be stochastic. Restating it for ease of reference,

$$\mathbf{Y} = \boldsymbol{\alpha} + \boldsymbol{\beta}(\mathbf{X} - \boldsymbol{\mu}_{\mathbf{X}}) + \boldsymbol{\varepsilon}, \tag{4.1}$$

where the error vector $\boldsymbol{\varepsilon}$ has mean 0 and covariance matrix $\boldsymbol{\Sigma}$, the now random predictor vector $\mathbf{X}$ has mean $\boldsymbol{\mu}_{\mathbf{X}}$ and variance $\boldsymbol{\Sigma}_{\mathbf{X}}$, and $\boldsymbol{\varepsilon} \perp\!\!\!\perp \mathbf{X}$. Given n independent copies $(\mathbf{Y}_i, \mathbf{X}_i)$, $i = 1, \ldots, n$, of $(\mathbf{Y}, \mathbf{X})$, consider regressing $\mathbf{Y}$ on $\mathbf{X}$ in two steps. The first is the reduction step: reduce $\mathbf{X}$ linearly to $\boldsymbol{\Phi}^T\mathbf{X}$ using some methodology that produces $\boldsymbol{\Phi} \in \mathbb{R}^{p\times q}$, $q \leq p$. The matrix $\boldsymbol{\Phi}$ will typically depend on the data, although we temporarily assume that $\boldsymbol{\Phi}$ is known and nonstochastic to facilitate discussion. The second step consists of using ordinary least squares to estimate the coefficient matrix $\boldsymbol{\beta} \in \mathbb{R}^{r\times p}$ of $\mathbf{X}$.

An Introduction to Envelopes: Dimension Reduction for Efficient Estimation in Multivariate Statistics,
First Edition. R. Dennis Cook.

In preparation for describing the form of the resulting estimator $\widehat{\boldsymbol{\beta}}_{\boldsymbol{\Phi}}$ of $\boldsymbol{\beta}$, recall that $\mathbf{S}_{\mathbf{X}}$ denotes the sample version of $\boldsymbol{\Sigma}_{\mathbf{X}}$, $\mathbf{S}_{\mathbf{X},\mathbf{Y}}$ denotes the sample version of $\boldsymbol{\Sigma}_{\mathbf{X},\mathbf{Y}}$, $\mathbf{S}_{\mathbf{X},\mathbf{Y}}^T = \mathbf{S}_{\mathbf{Y},\mathbf{X}}$, and $\mathbf{B} = \mathbf{S}_{\mathbf{Y},\mathbf{X}}\mathbf{S}_{\mathbf{X}}^{-1}$ is the estimator of $\boldsymbol{\beta}$ from the ordinary least squares fit of $\mathbf{Y}$ on $\mathbf{X}$ when $\mathbf{S}_{\mathbf{X}} > 0$.

Following the reduction $\mathbf{X} \mapsto \boldsymbol{\Phi}^T\mathbf{X}$, we use ordinary least squares to fit the multivariate regression of $\mathbf{Y}$ on $\boldsymbol{\Phi}^T\mathbf{X}$, giving coefficient matrix $\mathbf{B}_{\boldsymbol{\Phi}} = \mathbf{S}_{\mathbf{Y},\mathbf{X}}\boldsymbol{\Phi}(\boldsymbol{\Phi}^T\mathbf{S}_{\mathbf{X}}\boldsymbol{\Phi})^{-1}$, assuming that $\boldsymbol{\Phi}^T\mathbf{S}_{\mathbf{X}}\boldsymbol{\Phi} > 0$. The estimator $\widehat{\boldsymbol{\beta}}_{\boldsymbol{\Phi}}$ of $\boldsymbol{\beta}$ is then

$$\widehat{\boldsymbol{\beta}}_{\boldsymbol{\Phi}} = \mathbf{B}_{\boldsymbol{\Phi}}\boldsymbol{\Phi}^T = \mathbf{S}_{\mathbf{Y},\mathbf{X}}\boldsymbol{\Phi}(\boldsymbol{\Phi}^T\mathbf{S}_{\mathbf{X}}\boldsymbol{\Phi})^{-1}\boldsymbol{\Phi}^T \tag{4.2}$$

$$= \mathbf{B}\mathbf{P}^T_{\mathcal{W}_q(\mathbf{S}_{\mathbf{X}})}, \tag{4.3}$$

$$\boldsymbol{\beta}_{\boldsymbol{\Phi}} = \boldsymbol{\beta}\mathbf{P}^T_{\mathcal{W}_q(\boldsymbol{\Sigma}_{\mathbf{X}})}, \tag{4.4}$$

where $\mathcal{W}_q = \text{span}(\boldsymbol{\Phi})$. The form (4.2) does not require computation of $\mathbf{S}_{\mathbf{X}}^{-1}$ and this could be useful when $n < p$, depending on the size of q. Form (4.3) describes $\widehat{\boldsymbol{\beta}}^T_{\boldsymbol{\Phi}}$ as the projection of $\mathbf{B}^T$ onto $\mathcal{W}_q$ in the $\mathbf{S}_{\mathbf{X}}$ inner product. The population version (4.4) is constructed by replacing the sample quantities with their population values. The relationships shown in (4.2)–(4.4) depend on $\boldsymbol{\Phi}$ only through its span $\mathcal{W}_q$, and thus we take $\boldsymbol{\Phi}$ to be a semi-orthogonal matrix without loss of generality. Let $(\boldsymbol{\Phi}, \boldsymbol{\Phi}_0) \in \mathbb{R}^{p\times p}$ be an orthogonal matrix. If we choose $q = p$, then $\boldsymbol{\Phi} = \mathbf{I}_q$ and $\widehat{\boldsymbol{\beta}}_{\boldsymbol{\Phi}} = \mathbf{B}$, which achieves nothing beyond $\mathbf{B}$. If we choose the columns of $\boldsymbol{\Phi}$ to be the first q eigenvectors of $\mathbf{S}_{\mathbf{X}}$, then $\boldsymbol{\Phi}^T\mathbf{X}$ consists of the first q principal components of $\mathbf{X}$ and $\widehat{\boldsymbol{\beta}}_{\boldsymbol{\Phi}}$ is the standard principal component regression estimator.

We use two conditions to guide the choice of $\boldsymbol{\Phi}$ and define the material and immaterial parts of $\mathbf{X}$. We first require that $\mathbf{Y}$ and $\boldsymbol{\Phi}_0^T\mathbf{X}$ be uncorrelated given $\boldsymbol{\Phi}^T\mathbf{X}$, $\text{cov}(\mathbf{Y}, \boldsymbol{\Phi}_0^T\mathbf{X} \mid \boldsymbol{\Phi}^T\mathbf{X}) = 0$. This condition insures that there is no linear relation between $\mathbf{Y}$ and $\boldsymbol{\Phi}_0^T\mathbf{X}$ when $\boldsymbol{\Phi}^T\mathbf{X}$ is known. The second condition is that $\boldsymbol{\Phi}^T\mathbf{X}$ and $\boldsymbol{\Phi}_0^T\mathbf{X}$ are uncorrelated, which insures that there is no marginal linear relationship between the reduced predictor and its complement. We think of $\boldsymbol{\Phi}_0^T\mathbf{X}$ as the (linearly) immaterial part of $\mathbf{X}$. These conditions play the same role in $\mathbf{X}$ reduction as conditions (1.18) did previously for $\mathbf{Y}$ reduction:

Lemma 4.1 *Under model (4.1), (i)* $\text{cov}(\mathbf{Y}, \boldsymbol{\Phi}_0^T\mathbf{X} \mid \boldsymbol{\Phi}^T\mathbf{X}) = 0$ *if and only if* $\text{span}(\boldsymbol{\beta}^T) \in \mathcal{W}_q$, *and (ii)* $\text{cov}(\boldsymbol{\Phi}^T\mathbf{X}, \boldsymbol{\Phi}_0^T\mathbf{X}) = 0$ *if and only if* $\mathcal{W}_q$ *reduces* $\boldsymbol{\Sigma}_{\mathbf{X}}$.

Proof: Replacing $\mathbf{Y}$ with its model (4.1), we have

$$\begin{aligned}\text{cov}(\mathbf{Y}, \boldsymbol{\Phi}_0^T\mathbf{X} \mid \boldsymbol{\Phi}^T\mathbf{X}) &= \text{cov}(\boldsymbol{\beta}\mathbf{X}, \boldsymbol{\Phi}_0^T\mathbf{X} \mid \boldsymbol{\Phi}^T\mathbf{X}) \\ &= \text{cov}(\boldsymbol{\beta}\mathbf{P}_{\boldsymbol{\Phi}}\mathbf{X} + \boldsymbol{\beta}\mathbf{Q}_{\boldsymbol{\Phi}}\mathbf{X}, \boldsymbol{\Phi}_0^T\mathbf{X} \mid \boldsymbol{\Phi}^T\mathbf{X}) \\ &= \text{cov}(\boldsymbol{\beta}\mathbf{Q}_{\boldsymbol{\Phi}}\mathbf{X}, \boldsymbol{\Phi}_0^T\mathbf{X} \mid \boldsymbol{\Phi}^T\mathbf{X}) \\ &= \boldsymbol{\beta}\mathbf{Q}_{\boldsymbol{\Phi}}\text{var}(\mathbf{X} \mid \boldsymbol{\Phi}^T\mathbf{X})\boldsymbol{\Phi}_0.\end{aligned}$$

The conclusion follows since $\text{span}\{\text{var}(\mathbf{X} \mid \boldsymbol{\Phi}^T\mathbf{X})\} = \text{span}(\boldsymbol{\Phi}_0)$ and thus $\text{cov}(\mathbf{Y}, \boldsymbol{\Phi}_0^T\mathbf{X} \mid \boldsymbol{\Phi}^T\mathbf{X}) = 0$ if and only if $\text{span}(\boldsymbol{\beta}^T) \in \mathcal{W}_q$.

For statement (ii), $\text{cov}(\boldsymbol{\Phi}^T\mathbf{X}, \boldsymbol{\Phi}_0^T\mathbf{X}) = 0$ if and only if $\mathcal{W}_q$ decomposes $\boldsymbol{\Sigma}_\mathbf{X} = \mathbf{P}_{\mathcal{W}_q}\boldsymbol{\Sigma}_\mathbf{X}\mathbf{P}_{\mathcal{W}_q} + \mathbf{Q}_{\mathcal{W}_q}\boldsymbol{\Sigma}_\mathbf{X}\mathbf{Q}_{\mathcal{W}_q}$. Consequently, $\mathcal{W}_q$ must be a reducing subspace of $\boldsymbol{\Sigma}_\mathbf{X}$ (Proposition A.1). □

If $\mathbf{X}$ and $\mathbf{Y}$ are jointly normally distributed then the two conditions of Lemma 4.1 can be restated using independence conditions that are similar to those (1.18) used for response envelopes: (i) $\mathbf{Y} \perp\!\!\!\perp \boldsymbol{\Phi}_0^T\mathbf{X} \mid \boldsymbol{\Phi}^T\mathbf{X}$ and (ii) $\boldsymbol{\Phi}^T\mathbf{X} \perp\!\!\!\perp \boldsymbol{\Phi}_0^T\mathbf{X}$. The objective function that we use later to develop an estimation method is based on joint normality.

4.1.1.1 Predictor Envelope Model

Let $\mathcal{B}' = \text{span}(\boldsymbol{\beta}^T)$. It then follows from Lemma 4.1 that its conditions (i) and (ii) hold if and only if $\mathcal{W}_q$ is a reducing subspace of $\boldsymbol{\Sigma}_\mathbf{X}$ that contains $\mathcal{B}'$. The intersection of all subspaces with these properties is the $\boldsymbol{\Sigma}_\mathbf{X}$ envelope of $\mathcal{B}'$, $\mathcal{E}_{\boldsymbol{\Sigma}_\mathbf{X}}(\mathcal{B}')$ (see Definition 1.2). From this point on we work in terms of the envelope parameter $\mathcal{E}_{\boldsymbol{\Sigma}_\mathbf{X}}(\mathcal{B}')$ with basis $\boldsymbol{\Phi}$ and dimension q. As before, we use $\mathcal{E}$ to denote this subspace when used as a subscript.

Since $\mathcal{B}' \subseteq \mathcal{E}_{\boldsymbol{\Sigma}_\mathbf{X}}(\mathcal{B}')$, we can write write $\boldsymbol{\beta}^T = \boldsymbol{\Phi}\boldsymbol{\eta}$ for some coordinate matrix $\boldsymbol{\eta} \in \mathbb{R}^{q\times r}$ and then rewrite model (4.1) as

$$\begin{aligned} \mathbf{Y} &= \boldsymbol{\alpha} + \boldsymbol{\eta}^T\boldsymbol{\Phi}^T(\mathbf{X} - \boldsymbol{\mu}_\mathbf{X}) + \boldsymbol{\varepsilon} \\ \boldsymbol{\Sigma}_\mathbf{X} &= \boldsymbol{\Phi}\boldsymbol{\Delta}\boldsymbol{\Phi}^T + \boldsymbol{\Phi}_0\boldsymbol{\Delta}_0\boldsymbol{\Phi}_0^T, \end{aligned} \tag{4.5}$$

where $\boldsymbol{\Delta} = \boldsymbol{\Phi}^T\boldsymbol{\Sigma}_\mathbf{X}\boldsymbol{\Phi}$ and $\boldsymbol{\Delta}_0 = \boldsymbol{\Phi}_0^T\boldsymbol{\Sigma}_\mathbf{X}\boldsymbol{\Phi}_0$. In contrast to the approach when reducing $\mathbf{Y}$, here there are no envelope constraints placed on $\boldsymbol{\Sigma}$. The number of free real parameters in this model is the sum of r for $\boldsymbol{\alpha}$, rq for $\boldsymbol{\eta}$, $q(p-q)$ for $\mathcal{E}_{\boldsymbol{\Sigma}_\mathbf{X}}(\mathcal{B}')$, p for $\boldsymbol{\mu}_\mathbf{X}$, $q(q+1)/2$ for $\boldsymbol{\Delta}$, $(p-q)(p-q+1)/2$ for $\boldsymbol{\Delta}_0$, and $r(r+1)/2$ for $\boldsymbol{\Sigma}$, giving a total of

$$N_q = r + p + rq + p(p+1)/2 + r(r+1)/2.$$

This amounts to a reduction of $N_p - N_q = r(p-q)$ parameters over the standard model.

4.1.1.2 Expository Example

The following example may help to fix basic ideas underlying model (4.5). Suppose that we have a univariate response Y and 10 predictors $\mathbf{X} \sim N_{10}(0, \boldsymbol{\Sigma}_\mathbf{X})$, where the first three predictors $\mathbf{X}_1$ are uncorrelated with the last seven predictors $\mathbf{X}_2$, $\mathbf{X}^T = (\mathbf{X}_1^T, \mathbf{X}_2^T)$. Let $\text{var}(\mathbf{X}_1) = \boldsymbol{\Sigma}_1$ and $\text{var}(\mathbf{X}_2) = \boldsymbol{\Sigma}_2$ so that $\boldsymbol{\Sigma}_\mathbf{X} = \text{bdiag}(\boldsymbol{\Sigma}_1, \boldsymbol{\Sigma}_2)$. Suppose also that the coefficient vector has the form

$\beta = (\beta_1, \beta_2, 0, \dots, 0)$ so Y depends on only the first two predictors. For this regression, we can take $\mathbf{\Delta} = \mathbf{\Sigma}_1$, $\mathbf{\Delta}_0 = \mathbf{\Sigma}_2$,

$$\mathbf{\Phi} = \begin{pmatrix} \mathbf{I}_3 \\ 0_{7\times 3} \end{pmatrix}, \quad \mathbf{\Phi}_0 = \begin{pmatrix} 0_{3\times 7} \\ \mathbf{I}_7 \end{pmatrix}, \quad \text{and} \quad \eta = \begin{pmatrix} \beta_1 \\ \beta_2 \\ 0 \end{pmatrix}.$$

The envelope basis in this case is three dimensional because the response depends only on the first two predictors, and the first three predictors are correlated and uncorrelated with the last seven predictors. Additionally, $(Y, \mathbf{\Phi}^T\mathbf{X}) \perp\!\!\!\perp \mathbf{\Phi}_0^T\mathbf{X}$ so the conditions of Lemma 4.1 are satisfied.

For clarity, this example was constructed so that the envelope aligns with a subset of the predictors. The envelope model (4.5) does not require nor take advantage of this relatively simple structure since it allows for a general linear reduction of $\mathbf{X}$. A numerical version of this illustration is given in Section 4.4.1.

4.1.2 Latent Variable Formulation of Partial Least Squares Regression

Originally developed for Chemometrics (Wold 1982), PLS regression has for decades been used as a method of reducing the predictor dimension in multivariate linear regression (Wold et al. 1983, 2001; de Jong 1993). It is now widely used across the applied sciences as a dimension reduction method that improves prediction performance over ordinary least squares (e.g. Boulesteix and Strimmer 2006; Krishnan et al. 2011; Nguyen and Rocke 2002, 2004; Rönkkö et al. 2016). Although studies have appeared in statistics literature from time to time (e.g. Cook et al. 2013; Delaigle and Hall 2012; Frank and Friedman 1993; Helland 1990,1992), the development of PLS regression has taken place mainly within the Chemometrics community where Martens and Næs (1992) is a classical reference. Perhaps the two most touted aspects of PLS regression are its ability to give useful predictions in rank-deficient regressions with $n < p$ and to provide insights through graphical interpretation of PLS loadings.

PLSs regression is sometimes associated with a model that describes the predictors $\mathbf{X} \in \mathbb{R}^p$ and response $\mathbf{Y} \in \mathbb{R}^r$ in terms of a pair of correlated latent vectors $\mathbf{u} \in \mathbb{R}^{d_u}$ and $\mathbf{v} \in \mathbb{R}^{d_v}$ with $d_u < r$ and $d_v < p$:

$$\mathbf{X} = \mu_{\mathbf{X}} + \mathbf{C}\mathbf{v} + \mathbf{e} \quad \text{and} \quad \mathbf{Y} = \mu_{\mathbf{Y}} + \mathbf{D}\mathbf{u} + \mathbf{f}, \tag{4.6}$$

where $\mathbf{C} \in \mathbb{R}^{p\times d_v}$ and $\mathbf{D} \in \mathbb{R}^{r\times d_u}$ are full-rank and nonstochastic, and $(\mathbf{u}^T, \mathbf{v}^T, \mathbf{e}^T, \mathbf{f}^T)^T$ follows a multivariate normal distribution with mean 0, $\mathbf{e} \perp\!\!\!\perp \mathbf{f}$, $(\mathbf{e}, \mathbf{f}) \perp\!\!\!\perp (\mathbf{u}, \mathbf{v})$, $\mathbf{u} \sim N(0, \mathbf{I}_{d_u})$ and $\mathbf{v} \sim N(0, \mathbf{I}_{d_v})$. The restrictions on the means and variances of $\mathbf{u}$ and $\mathbf{v}$ involve no loss of generality since otherwise these variables can always be centered and standardized, leading to an equivalent model. There are apparently no widely accepted constraints on $\text{var}(\mathbf{e}) = \boldsymbol{\Sigma}_{\mathbf{e}}$ and $\text{var}(\mathbf{f}) = \boldsymbol{\Sigma}_{\mathbf{f}}$. Some authors require

$$\boldsymbol{\Sigma}_{\mathbf{e}} = \sigma_{\mathbf{e}}^2 \mathbf{I}_p \quad \text{and} \quad \boldsymbol{\Sigma}_{\mathbf{f}} = \sigma_{\mathbf{f}}^2 \mathbf{I}_r, \tag{4.7}$$

other authors (e.g. Rao et al. 2008) state explicitly that PLS methodology does not require constraints on **e** and **f**, while still others describe (4.6) without explicitly mentioning properties of **e** and **f**. Nevertheless, the stochastic behavior of these errors can have an impact on properties of methodology based on (4.6). If (4.7) holds, then the individual models for **X** and **Y** are instances of the model for probabilistic principal components developed by Tipping and Bishop (1999) and discussed in Section 9.9, where the impact of changes to the stochastic properties of **e** and **f** is discussed. The rationale for requiring (4.7) is that the latent variables **u** and **v** must capture the covariance structures of **X** and **Y** up to the isotropic residual errors **e** and **f**.

In PLS path analysis (e.g. Tenenhaus et al. 2005; Haenlein and Kaplan 2004), the latent variables **u** and **v** may be taken as real, but unobservable characterizations of particular theoretical constructs and then **X** and **Y** are often called reflexive indicators. In this scenario, estimates of quantities related to **u** and **v**, like their covariance $\boldsymbol{\Sigma}_{\mathbf{u},\mathbf{v}}$ or mean functions $\text{E}(\mathbf{u} \mid \mathbf{Y})$ and $\text{E}(\mathbf{v} \mid \mathbf{X})$, are typically of interest. These considerations are outside the scope of this book since our central focus is the regression of **Y** on **X**. The latent variables may then be seen as imaginable constructs or simply as a convenient device to guide dimension reduction.

We continue by adopting an error structure that subsumes condition (4.7). Let $\boldsymbol{\Phi} \in \mathbb{R}^{p\times q}$ and $\boldsymbol{\Gamma} \in \mathbb{R}^{r\times u}$, where $q \geq d_v$ and $u \geq d_u$, be semi-orthogonal basis matrices for the $\boldsymbol{\Sigma}_{\mathbf{e}}$-envelope of $\mathcal{C} := \text{span}(\mathbf{C})$ and the $\boldsymbol{\Sigma}_{\mathbf{f}}$-envelope of $\mathcal{D} := \text{span}(\mathbf{D})$. Then $\mathbf{C} = \boldsymbol{\Phi}\mathbf{c}$ and $\mathbf{D} = \boldsymbol{\Gamma}\mathbf{d}$, where the coordinate matrices $\mathbf{c} \in \mathbb{R}^{q\times d_v}$ and $\mathbf{d} \in \mathbb{R}^{u\times d_u}$, and the error covariances $\boldsymbol{\Sigma}_{\mathbf{e}}$ and $\boldsymbol{\Sigma}_{\mathbf{f}}$ can be written in the usual envelope form:

$$\begin{aligned}
\boldsymbol{\Sigma}_{\mathbf{e}} &= \boldsymbol{\Phi}\mathbf{W}_{\mathbf{e}}\boldsymbol{\Phi}^T + \boldsymbol{\Phi}_0^T\mathbf{W}_{\mathbf{e},0}\boldsymbol{\Phi}_0^T \\
\boldsymbol{\Sigma}_{\mathbf{f}} &= \boldsymbol{\Gamma}\mathbf{W}_{\mathbf{f}}\boldsymbol{\Gamma}^T + \boldsymbol{\Gamma}_0\mathbf{W}_{\mathbf{f},0}\boldsymbol{\Gamma}_0^T.
\end{aligned}$$

In consequence, we have the following structure.

$$\begin{aligned}
\boldsymbol{\Sigma}_{\mathbf{X}} &= \boldsymbol{\Phi}(\mathbf{c}\mathbf{c}^T + \mathbf{W}_{\mathbf{e}})\boldsymbol{\Phi}^T + \boldsymbol{\Phi}_0\mathbf{W}_{\mathbf{e},0}\boldsymbol{\Phi}_0^T \\
\boldsymbol{\Sigma}_{\mathbf{Y},\mathbf{X}} &= \boldsymbol{\Gamma}\mathbf{d}\boldsymbol{\Sigma}_{\mathbf{u},\mathbf{v}}\mathbf{c}^T\boldsymbol{\Phi}^T \\
\boldsymbol{\beta} &= \boldsymbol{\Gamma}\mathbf{d}\boldsymbol{\Sigma}_{\mathbf{u},\mathbf{v}}\mathbf{c}^T(\mathbf{c}\mathbf{c}^T + \mathbf{W}_{\mathbf{e}})^{-1}\boldsymbol{\Phi}^T \\
\boldsymbol{\Sigma} &= \boldsymbol{\Gamma}\left\{\mathbf{d}\mathbf{d}^T + \mathbf{W}_{\mathbf{f}} - \mathbf{d}\boldsymbol{\Sigma}_{\mathbf{u},\mathbf{v}}\mathbf{c}^T(\mathbf{c}\mathbf{c}^T + \mathbf{W}_{\mathbf{e}})^{-1}\mathbf{c}\boldsymbol{\Sigma}_{\mathbf{v},\mathbf{u}}\mathbf{d}^T\right\}\boldsymbol{\Gamma}^T + \boldsymbol{\Gamma}_0\mathbf{W}_{\mathbf{f},0}\boldsymbol{\Gamma}_0^T,
\end{aligned}$$

where, as before, $\boldsymbol{\Sigma} = \text{var}(\mathbf{Y} \mid \mathbf{X})$, and $\boldsymbol{\beta}$ is the coefficient matrix for the regression of **Y** on **X**. From this we see that the latent variable model (4.6) implies the necessary structure for both response (1.20) and predictor (4.5) envelopes.

The independence conditions $\mathbf{e} \perp\!\!\!\perp \mathbf{f}$ and $(\mathbf{e}, \mathbf{f}) \perp\!\!\!\perp (\mathbf{u}, \mathbf{v})$ stated for (4.6) are important, but deviations from normality are less so since we can always rephrase the model in terms of correlations as we did when arriving at model (4.5). Condition (4.7) is not necessary.

Envelope methodology is serviceable in this context if either $\mathcal{E}_{\boldsymbol{\Sigma}_\mathbf{f}}(\mathcal{D})$ is a proper subset of $\mathbb{R}^r$ or $\mathcal{E}_{\boldsymbol{\Sigma}_\mathbf{e}}(\mathcal{C})$ is a proper subset of $\mathbb{R}^p$. If $\mathcal{E}_{\boldsymbol{\Sigma}_\mathbf{f}}(\mathcal{D}) \subset \mathbb{R}^r$ and $\mathcal{E}_{\boldsymbol{\Sigma}_\mathbf{e}}(\mathcal{C}) = \mathbb{R}^p$, then we can take $\boldsymbol{\Phi} = \mathbf{I}_p$ and $\mathbf{c} = \mathbf{C}$. This leads to the following connections between parameters in the response envelope model (1.20) and the PLS latent variable formulation,

$$\begin{aligned}
\boldsymbol{\Gamma} &= \boldsymbol{\Gamma} \\
\eta &= \mathbf{d}\boldsymbol{\Sigma}_{\mathbf{u},\mathbf{v}}\mathbf{C}^T(\mathbf{C}\mathbf{C}^T + \mathbf{W}_\mathbf{e})^{-1} \\
\boldsymbol{\Omega} &= \mathbf{d}\mathbf{d}^T + \mathbf{W}_\mathbf{f} - \mathbf{d}\boldsymbol{\Sigma}_{\mathbf{u},\mathbf{v}}\mathbf{C}^T(\mathbf{C}\mathbf{C}^T + \mathbf{W}_\mathbf{e})^{-1}\mathbf{C}\boldsymbol{\Sigma}_{\mathbf{v},\mathbf{u}}\mathbf{d}^T \\
\boldsymbol{\Omega}_0 &= \mathbf{W}_{\mathbf{f},0}.
\end{aligned}$$

If $\mathcal{E}_{\boldsymbol{\Sigma}_\mathbf{f}}(\mathcal{D}) = \mathbb{R}^r$ and $\mathcal{E}_{\boldsymbol{\Sigma}_\mathbf{e}}(\mathcal{C}) \subset \mathbb{R}^p$ then we can take $\boldsymbol{\Gamma} = \mathbf{I}_r$ and $\mathbf{d} = \mathbf{D}$, leading to the following connections between parameters in the model for predictor envelopes (4.5) and the PLS latent variable formulation,

$$\begin{aligned}
\boldsymbol{\Phi} &= \boldsymbol{\Phi} \\
\eta &= (\mathbf{c}\mathbf{c}^T + \mathbf{W}_\mathbf{e})^{-1}\mathbf{c}\boldsymbol{\Sigma}_{\mathbf{v},\mathbf{u}}\mathbf{D} \\
\boldsymbol{\Delta} &= \mathbf{c}\mathbf{c}^T + \mathbf{W}_\mathbf{e} \\
\boldsymbol{\Delta}_0 &= \mathbf{W}_{\mathbf{e},0}.
\end{aligned}$$

If $\mathcal{E}_{\boldsymbol{\Sigma}_\mathbf{f}}(\mathcal{D}) \subset \mathbb{R}^r$ and $\mathcal{E}_{\boldsymbol{\Sigma}_\mathbf{e}}(\mathcal{C}) \subset \mathbb{R}^p$, then response and predictor reduction may be pursued simultaneously, a possibility that is discussed in Section 4.5. Until then we focus on predictor envelopes using model (4.5).

4.1.3 Potential Advantages

Returning to model (4.5), the following proposition provides intuition about the potential advantages of the envelope estimator by contrasting its variance with that of the ordinary least squares estimator when a basis $\boldsymbol{\Phi}$ of $\mathcal{E}_{\boldsymbol{\Sigma}_\mathbf{X}}(\mathcal{B}')$ is known and the predictors are normally distributed.

Proposition 4.1 *Let $f_p = n - p - 2 > 0$ and $f_q = n - q - 2 > 0$. Assume that model (4.5) holds, that $\mathbf{X} \sim N_p(\boldsymbol{\mu}_\mathbf{X}, \boldsymbol{\Sigma}_\mathbf{X})$ and that a semi-orthogonal basis $\boldsymbol{\Phi}$ of $\mathcal{E}_{\boldsymbol{\Sigma}_\mathbf{X}}(\mathcal{B}')$ is known. Then*

$$\widehat{\boldsymbol{\beta}}_{\boldsymbol{\Phi}} = \mathbf{B}\mathbf{P}^T_{\mathcal{E}(\mathbf{S}_\mathbf{X})}, \tag{4.8}$$

$$\mathrm{var}\{\mathrm{vec}(\mathbf{B})\} = f_p^{-1}\boldsymbol{\Sigma}_\mathbf{X}^{-1} \otimes \boldsymbol{\Sigma}, \tag{4.9}$$

$$\mathrm{var}\{\mathrm{vec}(\widehat{\boldsymbol{\beta}}_{\boldsymbol{\Phi}})\} = f_q^{-1}\boldsymbol{\Phi}\boldsymbol{\Delta}^{-1}\boldsymbol{\Phi}^T \otimes \boldsymbol{\Sigma}. \tag{4.10}$$

where the variances are computed over both $\mathbf{X}$ *and* $\mathbf{Y}$.

Proof: Conclusion (4.8) follows immediately from (4.3) by replacing $\mathcal{W}_q$ with the basis $\mathbf{\Phi}$ for the envelope $\mathcal{E}_{\mathbf{\Sigma}_\mathbf{X}}(\mathcal{B}')$. To see (4.9) write

$$\text{var}\{\text{vec}(\mathbf{B})\} = \text{E}[\text{var}\{\text{vec}(\mathbf{B})|\mathbf{X}_1, \dots, \mathbf{X}_n\}] + \text{var}[\text{E}\{\text{vec}(\mathbf{B})|\mathbf{X}_1, \dots, \mathbf{X}_n\}].$$

Since $\text{E}\{\text{vec}(\mathbf{B})|\mathbf{X}_1, \dots, \mathbf{X}_n\} = \text{vec}(\boldsymbol{\beta})$, the second addend on the right side is 0. Then substituting $\text{var}\{\text{vec}(\mathbf{B})|\mathbf{X}_1, \dots, \mathbf{X}_n\} = n^{-1}\mathbf{S}_\mathbf{X}^{-1} \otimes \mathbf{\Sigma}$ from (1.8) we have

$$\begin{aligned}\text{var}\{\text{vec}(\mathbf{B})\} &= \text{E}(n^{-1}\mathbf{S}_\mathbf{X}^{-1} \otimes \mathbf{\Sigma})\\ &= \text{E}(n^{-1}\mathbf{S}_\mathbf{X}^{-1}) \otimes \mathbf{\Sigma}\\ &= f_p^{-1}\mathbf{\Sigma}_\mathbf{X}^{-1} \otimes \mathbf{\Sigma},\end{aligned}$$

where the last step follows because $n^{-1}\mathbf{S}_\mathbf{X}^{-1}$ follows an inverse Wishart distribution with mean $f_p^{-1}\mathbf{\Sigma}_\mathbf{X}^{-1}$ (see von Rosen (1988) for background on the inverse Wishart distribution).

Conclusion (4.10) follows from (4.9) after a little preparation. Write $\widehat{\boldsymbol{\beta}}_\mathbf{\Phi} = \widehat{\boldsymbol{\eta}}_\mathbf{\Phi}^T\mathbf{\Phi}^T$, where $\widehat{\boldsymbol{\eta}}_\mathbf{\Phi}^T$ is the coefficient matrix from the ordinary least squares fit of $\mathbf{Y}$ on $\mathbf{\Phi}^T\mathbf{X}$. Replacing $\mathbf{X} \in \mathbb{R}^p$ with $\mathbf{\Phi}^T\mathbf{X} \in \mathbb{R}^q$, it follows from part (ii) that

$$\begin{aligned}\text{var}\{\text{vec}(\widehat{\boldsymbol{\eta}}_\mathbf{\Phi}^T)\} &= f_q^{-1}\mathbf{\Sigma}_{\mathbf{\Phi}^T\mathbf{X}}^{-1} \otimes \mathbf{\Sigma}\\ &= f_q^{-1}(\mathbf{\Phi}^T\mathbf{\Sigma}_\mathbf{X}\mathbf{\Phi})^{-1} \otimes \mathbf{\Sigma}\\ &= f_q^{-1}\mathbf{\Delta}^{-1} \otimes \mathbf{\Sigma}.\end{aligned}$$

Then

$$\begin{aligned}\text{var}\{\text{vec}(\widehat{\boldsymbol{\beta}}_\mathbf{\Phi})\} &= (\mathbf{\Phi} \otimes \mathbf{I}_r)\text{var}\{\text{vec}(\widehat{\boldsymbol{\eta}}_\mathbf{\Phi}^T)\}(\mathbf{\Phi}^T \otimes \mathbf{I}_r)\\ &= f_q^{-1}\mathbf{\Phi}\mathbf{\Delta}^{-1}\mathbf{\Phi}^T \otimes \mathbf{\Sigma}.\end{aligned}$$

□

It follows from Proposition 4.1 that

Corollary 4.1 *Under the conditions of Proposition 4.1*

$$\text{var}\{\text{vec}(\widehat{\boldsymbol{\beta}}_\mathbf{\Phi})\} = f_p f_q^{-1}\text{var}\{\text{vec}(\mathbf{B})\} - f_q^{-1}\mathbf{\Phi}_0\mathbf{\Delta}_0^{-1}\mathbf{\Phi}_0^T.$$

Since $f_p f_q^{-1} \leq 1$, the first addend on the right-hand side reflects gain though reducing the dimension. The factor $f_p f_q^{-1}$ could be small if q is small relative to p. But experience has shown that the largest gains are associated with the second addend: If $\mathbf{\Sigma}_\mathbf{X}$ has some small eigenvalues and if some of the corresponding eigenvectors are associated with the immaterial variation, then $\mathbf{\Delta}_0^{-1}$ may be large. In short, if collinearity among the predictors is associated with immaterial variation then predictor envelopes could result in substantial gains over ordinary least squares. In practice, it will be necessary to estimate a basis $\mathbf{\Phi}$ for $\mathcal{E}_{\mathbf{\Sigma}_\mathbf{X}}(\mathcal{B}')$, and this will mitigate the gains suggested here. Nevertheless, experience has shown that these calculations give a useful qualitative indication for the kinds of regressions in which envelope reduction of the predictors will be useful.

In the discussion of (1.23), we found that substantial gains are possible using response envelopes when the immaterial variation in $\mathbf{Y}$ is associated with the *larger* eigenvalues of $\boldsymbol{\Sigma}$. In contrast, here we find that substantial gains are possible using predictor envelopes when the immaterial variation in $\mathbf{X}$ is associated with the *smaller* eigenvalues of $\boldsymbol{\Sigma}_{\mathbf{X}}$.

We next turn to methods for estimating the predictor envelope $\mathcal{E}_{\boldsymbol{\Sigma}_{\mathbf{X}}}(\mathcal{B}')$.

4.2 SIMPLS

Historically , PLS regression has been defined in terms of iterative algorithms, the two most common being NIPALS (Wold 1975) and SIMPLS (de Jong 1993). While PLS algorithms have existed for decades and much has been written about them, their asymptotic statistical properties have been largely unknown. We focus on SIMPLS in this section. SIMPLS is an instance of a more general algorithm studied in Section 6.5.1, where some attributes claimed in this section are justified. In particular, Cook et al. (2013) showed that the SIMPLS algorithm produces a $\sqrt{n}$-consistent estimator of the projection onto $\mathcal{E}_{\boldsymbol{\Sigma}_{\mathbf{X}}}(\mathcal{B}')$, and thus it provides a first method for estimating a basis $\boldsymbol{\Phi}$.

4.2.1 SIMPLS Algorithm

The population SIMPLS algorithm for predictor reduction produces a sequence of p-dimensional vectors $\mathbf{w}_0, \mathbf{w}_1, \dots, \mathbf{w}_k$, with initial $\mathbf{w}_0 = 0$, whose cumulative spans are strictly increasing and converge after q steps to the predictor envelope $\mathcal{E}_{\boldsymbol{\Sigma}_{\mathbf{X}}}(\mathcal{B}')$. Let $\mathbf{W}_k = (\mathbf{w}_0, \mathbf{w}_1, \dots, \mathbf{w}_k) \in \mathbb{R}^{p\times(k+1)}$ and let $\mathcal{W}_k = \text{span}(\mathbf{W}_k)$. Then the $\mathbf{w}_k$'s are constructed so that $\mathcal{W}_0 \subset \mathcal{W}_1 \subset \cdots \subset \mathcal{W}_{q-1} \subset \mathcal{W}_q = \mathcal{E}_{\boldsymbol{\Sigma}_{\mathbf{X}}}(\mathcal{B}')$. Given $\mathbf{W}_k$, $k < q$, the $k+1$st direction is constructed as

$$\begin{aligned} \mathbf{w}_{k+1} &= \arg\max_{\mathbf{w}} \mathbf{w}^T\boldsymbol{\Sigma}_{\mathbf{X},\mathbf{Y}}\boldsymbol{\Sigma}^T_{\mathbf{X},\mathbf{Y}}\mathbf{w}, \text{ subject to} \\ & \mathbf{w}^T\boldsymbol{\Sigma}_{\mathbf{X}}\mathbf{W}_k = 0 \text{ and } \mathbf{w}^T\mathbf{w} = 1. \end{aligned}$$

At termination, $\boldsymbol{\Phi}$ can be any semi-orthogonal basis matrix for $\mathcal{W}_q$. If the length constraint $\mathbf{w}^T\mathbf{w} = 1$ is modified to $\mathbf{w}^T\mathbf{Q}_{\mathcal{W}_k}\mathbf{w} = 1$, then we obtain the population version of the NIPALS algorithm, which also produces the desired envelope after q steps.

Since $\mathbf{w}^T\boldsymbol{\Sigma}_{\mathbf{X},\mathbf{Y}}\boldsymbol{\Sigma}^T_{\mathbf{X},\mathbf{Y}}\mathbf{w} = \boldsymbol{\Sigma}_{(\mathbf{w}^T\mathbf{X}),\mathbf{Y}}\boldsymbol{\Sigma}^T_{(\mathbf{w}^T\mathbf{X}),\mathbf{Y}}$, the SIMPLS algorithm is sometimes described as a sequential method for producing linear combinations of the predictors that have maximal covariances with the response. However, the constraints are more than just a convenience, but are key in producing an envelope solution. If the length constraint is changed to $\mathbf{w}^T\boldsymbol{\Sigma}_{\mathbf{X}}\mathbf{w} = 1$, which might be considered more natural in view of the orthogonality constraint, then $\mathcal{W}_q$ is equal to $\boldsymbol{\Sigma}_{\mathbf{X}}^{-1/2}$ times the span of the first q eigenvectors of $\mathbf{A} = \boldsymbol{\Sigma}_{\mathbf{X}}^{-1/2}\boldsymbol{\Sigma}_{\mathbf{X},\mathbf{Y}}\boldsymbol{\Sigma}_{\mathbf{X},\mathbf{Y}}^{T}\boldsymbol{\Sigma}_{\mathbf{X}}^{-1/2}$. But $q \geq \text{rank}(\boldsymbol{\Sigma}_{\mathbf{X},\mathbf{Y}}) = \text{rank}(\mathbf{A})$ and thus the last $q - \text{rank}(\boldsymbol{\Sigma}_{\mathbf{X},\mathbf{Y}})$ eigenvectors of $\mathbf{A}$ are not uniquely defined so the algorithm with length constraint $\mathbf{w}^T\boldsymbol{\Sigma}_{\mathbf{X}}\mathbf{w} = 1$ cannot return the envelope.

Let $\boldsymbol{\ell}_{\max}(\mathbf{A})$ be a normalized eigenvector associated with the largest eigenvalue of a symmetric matrix $\mathbf{A}$, $\boldsymbol{\ell}_{\max}(\mathbf{A}) = \arg\max_{\boldsymbol{\ell}^T\boldsymbol{\ell}=1} \boldsymbol{\ell}^T\mathbf{A}\boldsymbol{\ell}$. The SIMPLS algorithm can be stated equivalently without explicit constraints as follows. Again set $\mathbf{w}_0 = 0$ and $\mathbf{W}_0 = \mathbf{w}_0$. For $k = 0, \ldots, q-1$, set

$$
\begin{aligned}
\mathcal{S}_k &= \text{span}(\boldsymbol{\Sigma}_{\mathbf{X}}\mathbf{W}_k) \\
\mathbf{w}_{k+1} &= \boldsymbol{\ell}_{\max}(\mathbf{Q}_{\mathcal{S}_k}\boldsymbol{\Sigma}_{\mathbf{X},\mathbf{Y}}\boldsymbol{\Sigma}_{\mathbf{X},\mathbf{Y}}^{T}\mathbf{Q}_{\mathcal{S}_k}) \\
\mathbf{W}_{k+1} &= (\mathbf{w}_0, \ldots, \mathbf{w}_k, \mathbf{w}_{k+1}).
\end{aligned}
$$

At termination, $\mathcal{E}_{\boldsymbol{\Sigma}_{\mathbf{X}}}(\mathcal{B}') = \mathcal{W}_q = \text{span}(\mathbf{W}_q)$. Details behind this algorithm are available in Section 6.5.1 and Appendix B.2. For now we look at the first two steps to provide a little intuition. Expressing $\mathcal{S}_k$ and $\mathbf{w}_{k+1}$ in terms of the envelope basis $\boldsymbol{\Phi}$ we have

$$
\begin{aligned}
\mathcal{S}_k &= \text{span}\{(\boldsymbol{\Phi}\boldsymbol{\Delta}\boldsymbol{\Phi}^T + \boldsymbol{\Phi}_0\boldsymbol{\Delta}_0\boldsymbol{\Phi}_0^T)\mathbf{W}_k\} \\
\mathbf{w}_{k+1} &= \boldsymbol{\ell}_{\max}(\mathbf{Q}_{\mathcal{S}_k}\boldsymbol{\Phi}\boldsymbol{\gamma}\boldsymbol{\gamma}^T\boldsymbol{\Phi}^T\mathbf{Q}_{\mathcal{S}_k}),
\end{aligned}
$$

where by Proposition A.5 we have expressed $\boldsymbol{\Sigma}_{\mathbf{X},\mathbf{Y}} = \boldsymbol{\Phi}\boldsymbol{\gamma}$. For the first iteration, $\mathcal{S}_0 = \text{span}(0)$ and $\mathbf{w}_1 = \boldsymbol{\ell}_{\max}(\boldsymbol{\Phi}\boldsymbol{\gamma}\boldsymbol{\gamma}^T\boldsymbol{\Phi}^T) \in \mathcal{E}_{\boldsymbol{\Sigma}_{\mathbf{X}}}(\mathcal{B}')$. For the second iteration,

$$
\mathcal{S}_1 = \text{span}(\boldsymbol{\Phi}\boldsymbol{\Delta}\boldsymbol{\Phi}^T\mathbf{w}_1) := \text{span}(\boldsymbol{\Phi}\mathbf{a}),
$$

where $\mathbf{a} = \boldsymbol{\Delta}\boldsymbol{\Phi}^T\mathbf{w}_1$, and

$$
\mathbf{w}_2 = \boldsymbol{\ell}_{\max}(\mathbf{Q}_{\mathcal{S}_1}\boldsymbol{\Phi}\boldsymbol{\gamma}\boldsymbol{\gamma}^T\boldsymbol{\Phi}^T\mathbf{Q}_{\mathcal{S}_1}) = \boldsymbol{\ell}_{\max}(\boldsymbol{\Phi}\mathbf{Q}_{\mathbf{a}}\boldsymbol{\gamma}\boldsymbol{\gamma}^T\mathbf{Q}_{\mathbf{a}}\boldsymbol{\Phi}^T) \in \mathcal{E}_{\boldsymbol{\Sigma}_{\mathbf{X}}}(\mathcal{B}').
$$

The iterations continue to extract vectors from $\mathcal{E}_{\boldsymbol{\Sigma}_{\mathbf{X}}}(\mathcal{B}')$ until the required number q is reached.

The SIMPLS algorithm depends on only three population quantities, $\boldsymbol{\Sigma}_{\mathbf{X},\mathbf{Y}}$, $\boldsymbol{\Sigma}_{\mathbf{X}}$, and $q = \dim(\mathcal{E}_{\boldsymbol{\Sigma}_{\mathbf{X}}}(\mathcal{B}'))$. The sample version of SIMPLS is constructed straightforwardly by replacing $\boldsymbol{\Sigma}_{\mathbf{X},\mathbf{Y}}$ and $\boldsymbol{\Sigma}_{\mathbf{X}}$ by their sample counterparts, terminating after q steps with $\widehat{\boldsymbol{\Phi}}$ (cf. Eq. (4.3)) or generally any basis matrix for $\mathcal{W}_q$. In particular, SIMPLS does not make use of $\mathbf{S}_{\mathbf{X}}^{-1}$ and so does not require $\mathbf{S}_{\mathbf{X}}$ to be nonsingular, but it does require $q \leq \min(p, n-1)$. Of course, there is no sample counterpart to q, which must be inferred using one of the available rules. Five- and ten-fold cross-validations of predictive performance are commonly used methods for choosing a suitable q. Given $q = \dim(\mathcal{E}_{\boldsymbol{\Sigma}_{\mathbf{X}}}(\mathcal{B}'))$ then, with r and p fixed, this algorithm provides a $\sqrt{n}$-consistent estimator of the projection onto $\mathcal{E}_{\boldsymbol{\Sigma}_{\mathbf{X}}}(\mathcal{B}')$ (Cook et al. 2013).

Consider the case where $r = 1$, so Y is a scalar. Let

$$\begin{aligned} \mathbf{G} &= \left(\boldsymbol{\Sigma}_{\mathrm{X},Y}, \boldsymbol{\Sigma}_{\mathrm{X}}\boldsymbol{\Sigma}_{\mathrm{X},Y}, \ldots, \boldsymbol{\Sigma}_{\mathrm{X}}^{q-1}\boldsymbol{\Sigma}_{\mathrm{X},Y}\right) \\ \widehat{\mathbf{G}} &= \left(\widehat{\boldsymbol{\Sigma}}_{\mathrm{X},Y}, \widehat{\boldsymbol{\Sigma}}_{\mathrm{X}}\widehat{\boldsymbol{\Sigma}}_{\mathrm{X},Y}, \ldots, \widehat{\boldsymbol{\Sigma}}_{\mathrm{X}}^{q-1}\widehat{\boldsymbol{\Sigma}}_{\mathrm{X},Y}\right) \end{aligned} \tag{4.11}$$

denote population and sample Krylov matrices. Helland (1990) showed that $\mathrm{span}(\mathbf{G}) = \mathcal{W}_q$, giving a closed-form expression for a basis of the population PLS subspace, and that the sample version of the SIMPLS algorithm gives $\mathrm{span}(\widehat{\mathbf{G}})$ (see Section 6.5.2 for a sketch of the justification for this result).

Generally, $\mathrm{rank}(\boldsymbol{\Sigma}_{\mathrm{X,Y}}) \leq q \leq p$, where $\mathrm{rank}(\boldsymbol{\Sigma}_{\mathrm{X,Y}}) \leq \min(r, p)$. Consequently, if $\mathrm{rank}(\boldsymbol{\Sigma}_{\mathrm{X,Y}}) = p$ then $q = p$ and the SIMPLS estimator of $\boldsymbol{\beta}$ is the same as the ordinary least squares estimator. In particular, if $r \geq p$ and if $\boldsymbol{\Sigma}_{\mathrm{X,Y}}$ has full row rank then again SIMPLS and ordinary least squares are the same. The SIMPLS algorithm is most useful for reducing the dimension of $\mathbf{X}$ when $r < p$.

4.2.2 SIMPLS When $n < p$

4.2.2.1 Behavior of the SIMPLS Algorithm

One often cited practical advantage of PLS regression is that it is useable when $n < p$. In particular, the population SIMPLS algorithm is serviceable even when $q \leq \mathrm{rank}(\boldsymbol{\Sigma}_{\mathrm{X}}) < p$. The same is true of the sample algorithm when $\mathrm{rank}(\mathbf{S}_{\mathrm{X}}) < p$, provided that $q \leq \min\{\mathrm{rank}(\mathbf{S}_{\mathrm{X}}), n-1\}$. Let $\mathcal{S} = \mathrm{span}(\boldsymbol{\Sigma}_{\mathrm{X}})$, so that $\mathbf{P}_{\mathcal{S}}\boldsymbol{\Sigma}_{\mathrm{X}} = \mathbf{P}_{\mathcal{S}}\boldsymbol{\Sigma}_{\mathrm{X}}\mathbf{P}_{\mathcal{S}} = \boldsymbol{\Sigma}_{\mathrm{X}}$ and $\mathbf{P}_{\mathcal{S}}\boldsymbol{\Sigma}_{\mathrm{X,Y}} = \boldsymbol{\Sigma}_{\mathrm{X,Y}}$. To see how SIMPLS operates at the population level when $\boldsymbol{\Sigma}_{\mathrm{X}}$ is less than full rank, write the algorithm as

$$\begin{aligned} \mathbf{w}_{k+1} = \arg\max_{\mathbf{w}} \mathbf{w}^T\mathbf{P}_{\mathcal{S}}\boldsymbol{\Sigma}_{\mathrm{X,Y}}\boldsymbol{\Sigma}_{\mathrm{X,Y}}^T\mathbf{P}_{\mathcal{S}}\mathbf{w}, \text{ subject to} \\ \mathbf{w}^T\mathbf{P}_{\mathcal{S}}\boldsymbol{\Sigma}_{\mathrm{X}}\mathbf{P}_{\mathcal{S}}\mathbf{W}_k = 0 \text{ and } \mathbf{w}^T\mathbf{P}_{\mathcal{S}}\mathbf{w} + \mathbf{w}^T\mathbf{Q}_{\mathcal{S}}\mathbf{w} = 1. \end{aligned}$$

The objective function and the first constraint depend only on the part $\mathbf{P}_{\mathcal{S}}\mathbf{w}$ of $\mathbf{w}$ that lies in $\mathcal{S}$. It is the length constraint that forces a unique solution. Consider any vector $\mathbf{w}$ that satisfies SIMPLS and let $\mathbf{w}_{\mathrm{new}} = \mathbf{P}_{\mathcal{S}}\mathbf{w}/\|\mathbf{P}_{\mathcal{S}}\mathbf{w}\|$, so $\mathbf{w}_{\mathrm{new}}$ satisfies the two constraints in the SIMPLS algorithm. If $\mathbf{Q}_{\mathcal{S}}\mathbf{w} \neq 0$, then $\|\mathbf{w}\| > \|\mathbf{P}_{\mathcal{S}}\mathbf{w}\|$ and thus $\|\mathbf{P}_{\mathcal{S}}\mathbf{w}\| < 1$ since $\|\mathbf{w}\| = 1$. Then

$$\begin{aligned} \mathbf{w}^T\mathbf{P}_{\mathcal{S}}\boldsymbol{\Sigma}_{\mathrm{X,Y}}\boldsymbol{\Sigma}_{\mathrm{X,Y}}^T\mathbf{P}_{\mathcal{S}}\mathbf{w} &= \mathbf{w}_{\mathrm{new}}^T\mathbf{P}_{\mathcal{S}}\boldsymbol{\Sigma}_{\mathrm{X,Y}}\boldsymbol{\Sigma}_{\mathrm{X,Y}}^T\mathbf{P}_{\mathcal{S}}\mathbf{w}_{\mathrm{new}}\|\mathbf{P}_{\mathcal{S}}\mathbf{w}\|^2 \\ &< \mathbf{w}_{\mathrm{new}}^T\mathbf{P}_{\mathcal{S}}\boldsymbol{\Sigma}_{\mathrm{X,Y}}\boldsymbol{\Sigma}_{\mathrm{X,Y}}^T\mathbf{P}_{\mathcal{S}}\mathbf{w}_{\mathrm{new}}. \end{aligned}$$

Consequently, if $\mathbf{Q}_{\mathcal{S}}\mathbf{w} \neq 0$, we arrive at a contradiction: since the objective function at $\mathbf{w}$ is strictly smaller than that at $\mathbf{w}_{\mathrm{new}}$, $\mathbf{w}$ cannot satisfy SIMPLS, which implies that $\mathbf{Q}_{\mathcal{S}}\mathbf{w} = 0$ and thus that all SIMPLS vectors lie in $\mathcal{S}$. The same conclusion holds for the sample version of SIMPLS, replacing $\boldsymbol{\Sigma}_{\mathrm{X,Y}}$ and $\boldsymbol{\Sigma}_{\mathrm{X}}$ with their sample counterparts.

For use in practice when $\mathrm{rank}(\mathbf{S}_{\mathrm{X}}) < p$, the SIMPLS algorithm can be restated in terms of the eigenvectors of $\mathbf{S}_{\mathrm{X}}$. Let $\mathbf{V} \in \mathbb{R}^{p \times p_1}$ denote the matrix whose columns are the eigenvectors of $\mathbf{S}_\mathbf{X}$ corresponding to its nonzero eigenvalues, and let $\mathbf{D} \in \mathbb{R}^{p_1 \times p_1}$ denote the corresponding diagonal matrix of nonzero eigenvalues. As argued previously, any SIMPLS vector $\mathbf{w}$ must be contained in span($\mathbf{S}_\mathbf{X}$), so we can replace $\mathbf{w}$ with $\mathbf{Vc}$ in the algorithm and optimize over coordinates $\mathbf{c}$: Let $\mathbf{C}_k = (\mathbf{c}_0, \mathbf{c}_1, \dots, \mathbf{c}_k)$. Then the $k+1$-st coordinate vector is determined as

$$\mathbf{c}_{k+1} = \arg\max_{\mathbf{c}} \mathbf{c}^T\mathbf{V}^T\mathbf{S}_{\mathrm{X,Y}}\mathbf{S}_{\mathrm{X,Y}}^T\mathbf{Vc}, \text{ subject to}$$
$$\mathbf{c}^T\mathbf{DC}_k = 0 \text{ and } \mathbf{c}^T\mathbf{c} = 1.$$

At termination, $\mathbf{W}_q = \mathbf{VC}_q$.

4.2.2.2 Asymptotic Properties of SIMPLS

Chun and Keleş (2010) showed that, in a certain context, the SIMPLS estimator of the coefficient vector in the univariate linear regression of Y on $\mathbf{X}$ is inconsistent unless $p/n \to 0$ and this raises questions about the usefulness of the SIMPLS algorithm when $n < p$. They also developed a sparse version that has better asymptotic behavior and that can mitigate this concern in some regressions. On the other hand, SIMPLS is typically used for prediction rather than parameter estimation and, judging from its wide-spread use across the applied sciences, many have found it serviceable. A little dilemma arises when the Chun–Keleş result is pitted against the long history of using PLS in regressions with $n < p$: How can an inconsistent method apparently be so useful in practice?

Cook and Forzani (2017) studied the asymptotic properties of SIMPLS predictions as n and p approach infinity in various alignments, restricting consideration to single-component $q = 1$ univariate $r = 1$ linear regressions in which $(Y, \mathbf{X})$ follows a multivariate normal distribution. Although these constraints notably restrict the scope of their study, their conclusions still give useful clues about settings in which SIMPLS predictions might be useful. We next discuss some of their results.

Under model (4.1) with a single response, $Y = \mu + \boldsymbol{\beta}(\mathbf{X} - \boldsymbol{\mu}_\mathbf{X}) + \varepsilon$, where now $\boldsymbol{\beta}$ is a $1 \times p$ vector since $r = 1$. From this we obtain $\mathrm{var}(Y) = \boldsymbol{\beta}\boldsymbol{\Sigma}_\mathbf{X}\boldsymbol{\beta}^T + \mathrm{var}(\varepsilon)$. As we increase p, $\boldsymbol{\beta}\boldsymbol{\Sigma}_\mathbf{X}\boldsymbol{\beta}^T$ may increase as well, but since var(Y) is constant in p, var(ε) must correspondingly decrease. We assume throughout this discussion that var(ε) is bounded away from 0 as $p \to \infty$ so the predictors under consideration cannot explain all of the variation in Y asymptotically.

Let $Y_N = \alpha + \boldsymbol{\beta}(\mathbf{X}_N - \boldsymbol{\mu}_\mathbf{X}) + \varepsilon_N$ denote a new observation on Y at a new independent observation $\mathbf{X}_N$ of $\mathbf{X}$. It follows from (4.2) to (4.4) and the SIMPLS algorithm in Section 4.2.1 that $\boldsymbol{\beta}$ and its SIMPLS estimator $\widehat{\boldsymbol{\beta}}$ can be expressed as

$$\boldsymbol{\beta}^T = \boldsymbol{\Sigma}_{\mathrm{X},Y}\left(\boldsymbol{\Sigma}_{\mathrm{X},Y}^T\boldsymbol{\Sigma}_\mathbf{X}\boldsymbol{\Sigma}_{\mathrm{X},Y}\right)^{-1}\boldsymbol{\Sigma}_{\mathrm{X},Y}^T\boldsymbol{\Sigma}_{\mathrm{X},Y}$$
$$\widehat{\boldsymbol{\beta}}^T = \mathbf{S}_{\mathrm{X},Y}\left(\mathbf{S}_{\mathrm{X},Y}^T\mathbf{S}_\mathbf{X}\mathbf{S}_{\mathrm{X},Y}\right)^{-1}\mathbf{S}_{\mathrm{X},Y}^T\mathbf{S}_{\mathrm{X},Y}.$$

Expressing the predicted value of Y at $\mathbf{X}_N$ as $\widehat{Y}_N = \bar{Y} + \widehat{\boldsymbol{\beta}}(\mathbf{X}_N - \bar{\mathbf{X}})$ gives a difference of

$$\begin{aligned}\widehat{Y}_N - Y_N &= (\bar{Y} - \mu) + (\widehat{\boldsymbol{\beta}} - \boldsymbol{\beta})(\mathbf{X}_N - \boldsymbol{\mu}_{\mathbf{X}}) - (\widehat{\boldsymbol{\beta}} - \boldsymbol{\beta})(\bar{\mathbf{X}} - \boldsymbol{\mu}_{\mathbf{X}}) \\ &\quad -\boldsymbol{\beta}(\bar{\mathbf{X}} - \boldsymbol{\mu}_{\mathbf{X}}) + \varepsilon_N \\ &= O_p\{(\widehat{\boldsymbol{\beta}} - \boldsymbol{\beta})(\mathbf{X}_N - \boldsymbol{\mu}_{\mathbf{X}})\} + \varepsilon_N \text{ as } n, p \to \infty.\end{aligned}$$

Since ε_N is the intrinsic error in the new observation, the n, p-asymptotic behavior of the prediction $\widehat{Y}_N$ is governed by

$$D_N := (\widehat{\boldsymbol{\beta}} - \boldsymbol{\beta})\mathbf{e}_N, \tag{4.12}$$

where $\mathbf{e}_N = \mathbf{X}_N - \boldsymbol{\mu}_{\mathbf{X}} \sim N(0, \boldsymbol{\Sigma}_{\mathbf{X}})$. Cook and Forzani (2017) characterized SIMPLS predictions by determining the order of D_N as $n, p \to \infty$. Since $\text{var}(D_N \mid \widehat{\boldsymbol{\beta}}) = (\widehat{\boldsymbol{\beta}} - \boldsymbol{\beta})\boldsymbol{\Sigma}_{\mathbf{X}}(\widehat{\boldsymbol{\beta}} - \boldsymbol{\beta})^T$, their results for D_N also tell us about the large-sample behavior of $\widehat{\boldsymbol{\beta}}$ in the $\boldsymbol{\Sigma}_{\mathbf{X}}$ inner product. The following three propositions from Cook and Forzani (2017) are arranged according to the eigenvalues of $\boldsymbol{\Delta}_0$, as defined in (4.5). In each case, we assume that the univariate linear model holds with normal $(Y, \mathbf{X})$, $q = 1$, and $\text{var}(\varepsilon)$ is bounded away from 0.

Proposition 4.2 *Assume that the eigenvalues of $\boldsymbol{\Delta}_0$ are bounded as $p \to \infty$. Then*

I. Abundance: *If $\|\boldsymbol{\Sigma}_{\mathbf{X},Y}\|^2 \asymp p$ then $D_N = O_p\{(1/n)^{1/2}\}$.*
II. Sparsity: *If $\|\boldsymbol{\Sigma}_{\mathbf{X},Y}\|^2 \asymp 1$ then $D_N = O_p\{(p/n)^{1/2}\}$.*

According to conclusion I of this proposition, if most predictors are correlated with Y, so $\|\boldsymbol{\Sigma}_{\mathbf{X},Y}\|^2/p$ converges to a nonzero limit then SIMPLS predictions will converge at the usual root-n rate, even if $n < p$. Conclusion II says that if few predictors are correlated with the response, so $\|\boldsymbol{\Sigma}_{\mathbf{X},Y}\|^2$ converges, then for predictive consistency the sample size needs to be large relative to the number of predictors. The second case clearly suggests a sparse solution, while the first case does not. Sparse versions of PLS regression have been proposed by Chun and Keleş (2010) and Liland et al. (2013).

The next proposition is a refinement of Proposition 4.2 for regressions in which $p > n$.

Proposition 4.3 *Assume that the eigenvalues of $\boldsymbol{\Delta}_0$ are bounded as $p \to \infty$, that $p \asymp n^a$ for $a \geq 1$ and that $\|\boldsymbol{\Sigma}_{\mathbf{X},Y}\|^2 \asymp p^s$ for $0 \leq s \leq 1$ and $a(1-s) < 1$. Then*

I. $D_N = O_p(n^{-1/2})$ if $a(1-s) \leq 1/2$.
II. $D_N = O_p(n^{-1+a(1-s)})$ if $1/2 \leq a(1-s) < 1$.

Conclusion I of this proposition gives an idea about the relationship between n and p that is needed to achieve $\sqrt{n}$-consistency. For instance, if $a = 2$ so $p \asymp n^2$, then we need $s \geq 3/4$ for root-n convergence. Conclusion II gives a similar idea of what is needed to just get consistency. Again, if $a = 2$, then we need $s > 1/2$ for consistency and $s \geq 3/4$ for $\sqrt{n}$-consistency, although if s is close to $1/2$, the rate could be very slow. If $a = 2$ and $s = 9/16$, then according to conclusion II, $D_N = O_p(n^{-1/8})$.

The final proposition allows some of the eigenvalues $\varphi_j(\boldsymbol{\Delta}_0)$ of $\boldsymbol{\Delta}_0$ to be unbounded.

Proposition 4.4 *Assume that* $\varphi_j(\boldsymbol{\Delta}_0) \asymp p$ *for a finite collection of indices* j *while the other eigenvalues of* $\boldsymbol{\Delta}_0$ *are bounded as* $p \to \infty$. *Assume also that* $p \asymp n^a$ *for* $a \geq 1$, *that* $\|\boldsymbol{\Sigma}_{\mathrm{X},Y}\|^2 \asymp p^s$ *for* $0 \leq s \leq 1$ *and that* $a(1-s) < 1/2$. *Then*

$$D_N = O_p(n^{-1/2+a(1-s)}).$$

The conclusion of this proposition implies a near root-n convergence rate when $a(1-s)$ is small. It also indicates that there is a price to be paid when finitely many of the eigenvalues of $\boldsymbol{\Delta}_0$ are unbounded. For instance, if $a = 2$, $s = 13/16$ and the eigenvalues of $\boldsymbol{\Delta}_0$ are bounded then $a(1-s) = 3/8$ and from Proposition 4.3 we have $D_N = O_p(n^{-1/2})$. However, if finitely many of the eigenvalues of $\boldsymbol{\Delta}_0$ are unbounded, then from Proposition 4.4 we have $D_N = O_p(n^{-1/8})$.

This series of three propositions gives clues to settings in which SIMPLS predictions have good asymptotic properties. If many predictors are correlated with the response, so $\|\boldsymbol{\Sigma}_{\mathrm{X},Y}\|^2 \asymp p$, and the eigenvalues of $\boldsymbol{\Delta}_0$ are bounded then from Proposition 4.2 we get the usual root-n convergence rate, regardless of the relationship between n and p. If fewer predictors are correlated with the response, so $\|\boldsymbol{\Sigma}_{\mathrm{X},Y}\|^2 \asymp p^s$ with $s < 1$, then the behavior of SIMPLS predictions depends on the relationship between n and p and on the behavior of the eigenvalues of $\boldsymbol{\Delta}_0$, as described in Propositions 4.3 and 4.4. In reference to PLS regression with a univariate response and $n < p$, Helland (1990) commented prophetically that

> What one could hope eventually would be to find methods/formulations which made it transparent that the observation of more **x**-variables gives us more information, and hence should make prediction easier, not more difficult.

Propositions 4.2–4.4 indicate that PLS regressions may satisfy Helland's wish. For instance, adding predictors that are correlated with the response causes $\mathrm{var}(\epsilon)$ to decrease and, from Proposition 4.2, is sufficient for predictions to achieve $\sqrt{n}$-convergence.

As mentioned previously in this section, Chun and Keleş (2010) showed that, in a certain context, the SIMPLS estimator of $\boldsymbol{\beta}$ is inconsistent unless $p/n \to 0$. Their setup included the requirement that the eigenvalues of $\boldsymbol{\Sigma}_{\mathbf{X}}$ be bounded as $p \to \infty$. The implication of this condition in the present context is given in the next lemma.

Lemma 4.2 *If the eigenvalues of* $\boldsymbol{\Sigma}_{\mathbf{X}}$ *are bounded as* $p \to \infty$, *then* $\|\boldsymbol{\Sigma}_{\mathbf{X},Y}\|^2$ *is bounded.*

Proof: The proof makes use of the facts that $\boldsymbol{\beta}\boldsymbol{\Sigma}_{\mathbf{X}}\boldsymbol{\beta}^T$ is bounded and that $\boldsymbol{\Sigma}_{\mathbf{X},Y}$ is an eigenvector of $\boldsymbol{\Sigma}_{\mathbf{X}}$. Let φ_k denote the eigenvalue of $\boldsymbol{\Sigma}_{\mathbf{X}}$ corresponding to $\boldsymbol{\Sigma}_{\mathbf{X},Y}$. Then

$$\boldsymbol{\beta}\boldsymbol{\Sigma}_{\mathbf{X}}\boldsymbol{\beta}^T = \boldsymbol{\Sigma}_{\mathbf{X},Y}^T\boldsymbol{\Sigma}_{\mathbf{X}}^{-1}\boldsymbol{\Sigma}_{\mathbf{X},Y} = \varphi_k^{-1}\|\boldsymbol{\Sigma}_{\mathbf{X},Y}\|^2.$$

Consequently, if φ_k is bounded, then $\|\boldsymbol{\Sigma}_{\mathbf{X},Y}\|^2$ must be bounded since $\boldsymbol{\beta}\boldsymbol{\Sigma}_{\mathbf{X}}\boldsymbol{\beta}^T$ is bounded. □

From this lemma we see that bounding the eigenvalues of $\boldsymbol{\Sigma}_{\mathbf{X}}$ implies that $\|\boldsymbol{\Sigma}_{\mathbf{X},Y}\|^2 \asymp 1$ and so the signal is bounded. This is covered by conclusion II of Proposition 4.2, which agrees with the Chun–Keleş result. By requiring that the eigenvalues of $\boldsymbol{\Sigma}_{\mathbf{X}}$ be bounded, Chun and Keleş (2010) essentially assumed sparsity to motivate a sparse solution. Going further, Wold et al. (1996) argued against the notion of sparsity (although "sparsity" was not a common term at the time),

> In situations with many variables, more than say 50 or 100, there is therefore a strong temptation to drastically reduce the number of variables in the model. This temptation is further strengthened by the 'regression tradition' to reduce the variables as far as possible to get the X-matrix well conditioned. As discussed below, however, this reduction of variables often removes information, makes the interpretation misleading and increases the risk of spurious models.

They instead proposed a method called multiblock PLS based on dividing the predictors into conceptually meaningful sets and then performing dimension reduction within each set. From this and other authors, it seems that sparsity is not part of the PLS tradition.

4.3 Likelihood-Based Predictor Envelopes

It is traditional in regression to base estimation on the conditional likelihood from $\mathbf{Y} \mid \mathbf{X}$, treating the predictors as fixed even if they were randomly sampled. This practice arose because in many regressions the predictors provide only ancillary information and consequently estimation and inference should be conditioned on their observed values. (See Aldrich 2005 for a review and a historical perspective.) In contrast, PLS and the likelihood-based envelope method developed in this section both postulate a link – represented here by the envelope $\mathcal{E}_{\boldsymbol{\Sigma}_{\mathbf{X}}}(\mathcal{B}')$ – between $\boldsymbol{\beta}$, the parameter of interest, and $\boldsymbol{\Sigma}_{\mathbf{X}}$. As a consequence, $\mathbf{X}$ is not ancillary, and we pursue estimation through the joint distribution of $\mathbf{Y}$ and $\mathbf{X}$. The development here follows (Cook et al., 2013).

4.3.1 Estimation

Let $\mathbf{C} = (\mathbf{X}^T, \mathbf{Y}^T)^T$ denote the random vector constructed by concatenating $\mathbf{X}$ and $\mathbf{Y}$, and let $\mathbf{S}_{\mathbf{C}}$ denote the sample version of $\boldsymbol{\Sigma}_{\mathbf{C}} = \mathrm{var}(\mathbf{C})$. Given q, we base estimation on the objective function $F_q(\mathbf{S}_{\mathbf{C}}, \boldsymbol{\Sigma}_{\mathbf{C}}) = \log|\boldsymbol{\Sigma}_{\mathbf{C}}| + \mathrm{tr}(\mathbf{S}_{\mathbf{C}}\boldsymbol{\Sigma}_{\mathbf{C}}^{-1})$ that stems from the log likelihood of the multivariate normal family after replacing the population mean vector with the vector of sample means, although we do not require $\mathbf{C}$ to have a multivariate normal distribution. Rather we are using F_q as a multipurpose objective function in the same spirit as least squares objective functions are often used. The structure of the envelope $\mathcal{E}_{\boldsymbol{\Sigma}_{\mathbf{X}}}(\mathcal{B}')$ can be introduced into F_q by using the parameterizations $\boldsymbol{\Sigma}_{\mathbf{X}} = \boldsymbol{\Phi}\boldsymbol{\Delta}\boldsymbol{\Phi}^T + \boldsymbol{\Phi}_0\boldsymbol{\Delta}_0\boldsymbol{\Phi}_0^T$ and $\boldsymbol{\Sigma}_{\mathbf{X},\mathbf{Y}} = \boldsymbol{\Phi}\boldsymbol{\gamma}$, where $\boldsymbol{\Phi} \in \mathbb{R}^{p\times q}$ is a semi-orthogonal basis matrix for $\mathcal{E}_{\boldsymbol{\Sigma}_{\mathbf{X}}}(\mathcal{B}')$, $(\boldsymbol{\Phi}, \boldsymbol{\Phi}_0) \in \mathbb{R}^{p\times p}$ is an orthogonal matrix, and $\boldsymbol{\Delta} \in \mathbb{R}^{q\times q}$ and $\boldsymbol{\Delta}_0 \in \mathbb{R}^{(p-q)\times(p-q)}$ are symmetric positive definite matrices, as defined previously for model (4.5). From Corollary A.2, $\mathrm{span}(\boldsymbol{\Sigma}_{\mathbf{X},\mathbf{Y}}) \subseteq \mathcal{E}_{\boldsymbol{\Sigma}_{\mathbf{X}}}(\mathcal{B}')$, so we wrote $\boldsymbol{\Sigma}_{\mathbf{X},\mathbf{Y}}$ as linear combinations of the columns of $\boldsymbol{\Phi}$. The matrix $\boldsymbol{\gamma} \in \mathbb{R}^{q\times r}$ then gives the coordinates of $\boldsymbol{\Sigma}_{\mathbf{X},\mathbf{Y}}$ in terms of the basis $\boldsymbol{\Phi}$. With this we have

$$\begin{aligned}
\boldsymbol{\Sigma}_{\mathbf{C}} &= \begin{pmatrix} \boldsymbol{\Sigma}_{\mathbf{X}} & \boldsymbol{\Sigma}_{\mathbf{X},\mathbf{Y}} \\ \boldsymbol{\Sigma}_{\mathbf{X},\mathbf{Y}}^T & \boldsymbol{\Sigma}_{\mathbf{Y}} \end{pmatrix} = \begin{pmatrix} \boldsymbol{\Phi}\boldsymbol{\Delta}\boldsymbol{\Phi}^T + \boldsymbol{\Phi}_0\boldsymbol{\Delta}_0\boldsymbol{\Phi}_0^T & \boldsymbol{\Phi}\boldsymbol{\gamma} \\ \boldsymbol{\gamma}^T\boldsymbol{\Phi}^T & \boldsymbol{\Sigma}_{\mathbf{Y}} \end{pmatrix} \\
&= \begin{pmatrix} \boldsymbol{\Phi} & \boldsymbol{\Phi}_0 & 0 \\ 0 & 0 & \mathbf{I}_r \end{pmatrix} \begin{pmatrix} \boldsymbol{\Delta} & 0 & \boldsymbol{\gamma} \\ 0 & \boldsymbol{\Delta}_0 & 0 \\ \boldsymbol{\gamma}^T & 0 & \boldsymbol{\Sigma}_{\mathbf{Y}} \end{pmatrix} \begin{pmatrix} \boldsymbol{\Phi}^T & 0 \\ \boldsymbol{\Phi}_0^T & 0 \\ 0 & \mathbf{I}_r \end{pmatrix} \\
&= \mathbf{O}\boldsymbol{\Sigma}_{\mathbf{O}^T\mathbf{C}}\mathbf{O}^T, \qquad (4.13)
\end{aligned}$$

where

$$\mathbf{O} = \begin{pmatrix} \boldsymbol{\Phi} & \boldsymbol{\Phi}_0 & 0 \\ 0 & 0 & \mathbf{I}_r \end{pmatrix} \in \mathbb{R}^{(p+r)\times(p+r)}$$

is an orthogonal matrix and

$$\boldsymbol{\Sigma}_{\mathbf{O}^T\mathbf{C}} = \begin{pmatrix} \boldsymbol{\Delta} & 0 & \boldsymbol{\gamma} \\ 0 & \boldsymbol{\Delta}_0 & 0 \\ \boldsymbol{\gamma}^T & 0 & \boldsymbol{\Sigma}_{\mathbf{Y}} \end{pmatrix} \in \mathbb{R}^{(p+r)\times(p+r)}$$

is the covariance matrix of the transformed vector $\mathbf{O}^T\mathbf{C}$. The objective function $F_q(\mathbf{S}_\mathbf{C}, \mathbf{\Sigma}_\mathbf{C})$ can now be regarded as a function of the five constituent parameters – $\mathbf{\Phi}$, $\mathbf{\Delta}$, $\mathbf{\Delta}_0$, $\boldsymbol{\gamma}$, and $\mathbf{\Sigma}_\mathbf{Y}$ – that comprise $\mathbf{\Sigma}_\mathbf{C}$. The parameters $\boldsymbol{\beta}$ and $\boldsymbol{\eta}$ of model (4.5) can be written as $\boldsymbol{\eta} = \mathbf{\Delta}^{-1}\boldsymbol{\gamma}$ and $\boldsymbol{\beta}^T = \mathbf{\Phi}\boldsymbol{\eta} = \mathbf{\Phi}\mathbf{\Delta}^{-1}\boldsymbol{\gamma}$.

To minimize $F_q(\mathbf{S}_\mathbf{C}, \mathbf{\Sigma}_\mathbf{C})$, we first hold $\mathbf{\Phi}$ fixed and substitute (4.13) giving

$$\begin{aligned} F_q(\mathbf{S}_\mathbf{C}, \mathbf{\Sigma}_\mathbf{C}) &= \log|\mathbf{O}\mathbf{\Sigma}_{\mathbf{O}^T\mathbf{C}}\mathbf{O}^T| + \mathrm{tr}(\mathbf{O}^T\mathbf{S}_\mathbf{C}\mathbf{O}\mathbf{\Sigma}^{-1}_{\mathbf{O}^T\mathbf{C}}) \\ &= \log|\mathbf{\Sigma}_{\mathbf{O}^T\mathbf{C}}| + \mathrm{tr}(\mathbf{S}_{\mathbf{O}^T\mathbf{C}}\mathbf{\Sigma}^{-1}_{\mathbf{O}^T\mathbf{C}}). \end{aligned}$$

The form of $\mathbf{\Sigma}_{\mathbf{O}^T\mathbf{C}}$ allows us to factor this into a term that depends only on its covariance matrix $\mathbf{\Delta}_0$ and a term that depends only on $\mathbf{\Delta}$, $\boldsymbol{\gamma}$, and $\mathbf{\Sigma}_\mathbf{Y}$. The values of these parameters that minimize F_q for fixed $\mathbf{\Phi}$ are then $\mathbf{\Sigma}_\mathbf{Y} = \mathbf{S}_\mathbf{Y}$, $\mathbf{\Delta} = \mathbf{\Phi}^T\mathbf{S}_\mathbf{X}\mathbf{\Phi}$, $\mathbf{\Delta}_0 = \mathbf{\Phi}_0^T\mathbf{S}_\mathbf{X}\mathbf{\Phi}_0$, and $\boldsymbol{\gamma} = \mathbf{\Phi}^T\mathbf{S}_{\mathbf{X},\mathbf{Y}}$. Substituting these forms into F_q then leads to the following estimator of the envelope when q is assumed to be known:

$$\widehat{\mathcal{E}}_{\mathbf{\Sigma}_\mathbf{X}}(\mathcal{B}') = \mathrm{span}\{\arg\min_{\mathbf{G}} L_q(\mathbf{G})\}, \text{ where} \tag{4.14}$$

$$\begin{aligned} L_q(\mathbf{G}) &= \log|\mathbf{G}^T\mathbf{S}_{\mathbf{X}|\mathbf{Y}}\mathbf{G}| + \log|\mathbf{G}^T\mathbf{S}_\mathbf{X}^{-1}\mathbf{G}| \\ &= \log|\mathbf{G}^T(\mathbf{S}_\mathbf{X} - \mathbf{S}_{\mathbf{X},\mathbf{Z}}\mathbf{S}^T_{\mathbf{X},\mathbf{Z}})\mathbf{G}| + \log|\mathbf{G}^T\mathbf{S}_\mathbf{X}^{-1}\mathbf{G}|, \end{aligned} \tag{4.15}$$

$\mathbf{Z} = \mathbf{S}_\mathbf{Y}^{-1/2}\mathbf{Y}$ is the standardized response vector, and the minimization in (4.14) is taken over all semi-orthogonal matrices $\mathbf{G} \in \mathbb{R}^{p\times q}$. This optimization is the same as that encountered for $\mathbf{Y}$ reduction in Chapter 1 (see Eq. (1.25)), except that the roles of $\mathbf{Y}$ and $\mathbf{X}$ have been interchanged. Let $\widehat{\mathbf{\Phi}}$ be any semi-orthogonal basis of $\widehat{\mathcal{E}}_{\mathbf{\Sigma}_\mathbf{X}}(\mathcal{B}')$. The estimators of the constituent parameters are then

$$\begin{aligned} \widehat{\mathbf{\Sigma}}_\mathbf{Y} &= \mathbf{S}_\mathbf{Y}, \\ \widehat{\mathbf{\Delta}} &= \widehat{\mathbf{\Phi}}^T\mathbf{S}_\mathbf{X}\widehat{\mathbf{\Phi}}, \\ \widehat{\mathbf{\Delta}}_0 &= \widehat{\mathbf{\Phi}}_0^T\mathbf{S}_\mathbf{X}\widehat{\mathbf{\Phi}}_0, \\ \widehat{\boldsymbol{\gamma}} &= \widehat{\mathbf{\Phi}}^T\mathbf{S}_{\mathbf{X},\mathbf{Y}} \\ \widehat{\boldsymbol{\eta}} &= \widehat{\mathbf{\Delta}}^{-1}\widehat{\boldsymbol{\gamma}}. \end{aligned}$$

From these we construct the estimators of the parameters of interest:

$$\begin{aligned} \widehat{\mathbf{\Sigma}}_{\mathbf{X},\mathbf{Y}} &= \mathbf{P}_{\widehat{\mathbf{\Phi}}}\mathbf{S}_{\mathbf{X},\mathbf{Y}}, \\ \widehat{\mathbf{\Sigma}}_\mathbf{X} &= \widehat{\mathbf{\Phi}}\widehat{\mathbf{\Delta}}\widehat{\mathbf{\Phi}}^T + \widehat{\mathbf{\Phi}}_0\widehat{\mathbf{\Delta}}_0\widehat{\mathbf{\Phi}}_0^T = \mathbf{P}_{\widehat{\mathbf{\Phi}}}\mathbf{S}_\mathbf{X}\mathbf{P}_{\widehat{\mathbf{\Phi}}} + \mathbf{Q}_{\widehat{\mathbf{\Phi}}}\mathbf{S}_\mathbf{X}\mathbf{Q}_{\widehat{\mathbf{\Phi}}}, \\ \widehat{\boldsymbol{\beta}}^T &= \widehat{\mathbf{\Phi}}\widehat{\mathbf{\Delta}}^{-1}\widehat{\boldsymbol{\gamma}} = \widehat{\mathbf{\Phi}}(\widehat{\mathbf{\Phi}}^T\mathbf{S}_\mathbf{X}\widehat{\mathbf{\Phi}})^{-1}\widehat{\mathbf{\Phi}}^T\mathbf{S}_{\mathbf{X},\mathbf{Y}} = \mathbf{P}_{\widehat{\mathbf{\Phi}}(\mathbf{S}_\mathbf{X})}\mathbf{B}^T. \end{aligned} \tag{4.16}$$

The estimators $\widehat{\mathbf{\Delta}}$, $\widehat{\mathbf{\Delta}}_0$, and $\widehat{\boldsymbol{\gamma}}$ depend on the selected basis $\widehat{\mathbf{\Phi}}$. The parameters of interest – $\widehat{\mathbf{\Sigma}}_{\mathbf{X},\mathbf{Y}}$, $\widehat{\mathbf{\Sigma}}_\mathbf{X}$ and $\widehat{\boldsymbol{\beta}}$ – depend on $\widehat{\mathcal{E}}_{\mathbf{\Sigma}_\mathbf{X}}(\mathcal{B}')$ but do not depend on the particular basis selected.

4.3.2 Comparisions with SIMPLS and Principal Component Regression

There are consequential differences between the likelihood-based estimation method and SIMPLS. To see how these differences arise, we first describe some operating characteristics of $L_q(\mathbf{G})$ and then contrast those characteristics with the behavior of SIMPLS. Let $L_q^{(1)}(\mathbf{G}) = \log|\mathbf{G}^T\mathbf{S}_{\mathbf{X}}\mathbf{G}| + \log|\mathbf{G}^T\mathbf{S}_{\mathbf{X}}^{-1}\mathbf{G}|$ and $L_q^{(2)}(\mathbf{G}) = \log|\mathbf{S}_{\mathbf{Z}|\mathbf{G}^T\mathbf{X}}|$, where $\mathbf{S}_{\mathbf{Z}|\mathbf{G}^T\mathbf{X}}$ is the sample covariance matrix of the residual vectors from the ordinary least squares fit of $\mathbf{Z}$ on $\mathbf{G}^T\mathbf{X}$ and $\mathbf{Z}$ is the standardized response defined following (4.15). Then the objective function L_q can be represented as $L_q(\mathbf{G}) = L_q^{(1)}(\mathbf{G}) + L_q^{(2)}(\mathbf{G})$:

$$\begin{aligned} L_q(\mathbf{G}) &= \log|\mathbf{G}^T\mathbf{S}_{\mathbf{X}}\mathbf{G}| + \log|\mathbf{G}^T\mathbf{S}_{\mathbf{X}}^{-1}\mathbf{G}| + \log|\mathbf{I}_r - \mathbf{S}_{\mathbf{X},\mathbf{Z}}^T\mathbf{G}(\mathbf{G}^T\mathbf{S}_{\mathbf{X}}\mathbf{G})^{-1}\mathbf{G}^T\mathbf{S}_{\mathbf{X},\mathbf{Z}}| \\ &= \log|\mathbf{G}^T\mathbf{S}_{\mathbf{X}}\mathbf{G}| + \log|\mathbf{G}^T\mathbf{S}_{\mathbf{X}}^{-1}\mathbf{G}| + \log|\mathbf{I}_r - \mathbf{S}_{\mathbf{Z},\mathbf{G}^T\mathbf{X}}\mathbf{S}_{\mathbf{G}^T\mathbf{X}}^{-1}\mathbf{S}_{\mathbf{Z},\mathbf{G}^T\mathbf{X}}^T| \\ &= \log|\mathbf{G}^T\mathbf{S}_{\mathbf{X}}\mathbf{G}| + \log|\mathbf{G}^T\mathbf{S}_{\mathbf{X}}^{-1}\mathbf{G}| + \log|\mathbf{S}_{\mathbf{Z}|\mathbf{G}^T\mathbf{X}}|. \end{aligned}$$

The first addend $L_q^{(1)}(\mathbf{G}) \geq 0$ with $L_q^{(1)}(\mathbf{G}) = 0$ when the columns of $\mathbf{G}$ correspond to any subset of q eigenvectors of $\mathbf{S}_{\mathbf{X}}$ (Lemma A.15). Consequently, the role of $L_q^{(1)}$ is to pull the solution toward subsets of q eigenvectors of $\mathbf{S}_{\mathbf{X}}$. This in effect imposes a sample counterpart of the characterization in Proposition A.3, which states that in the population $\mathcal{E}_{\mathbf{\Sigma}_{\mathbf{X}}}(\mathcal{B}')$ is spanned by a subset of the eigenvectors of $\mathbf{\Sigma}_{\mathbf{X}}$. The second addend $L_q^{(2)}(\mathbf{G}) = \log|\mathbf{S}_{\mathbf{Z}|\mathbf{G}^T\mathbf{X}}|$ of $L_q(\mathbf{G})$ measures the goodness of fit of the regression of the standardized response $\mathbf{Z}$ on $\mathbf{G}^T\mathbf{X}$. As a consequence, $L_q(\mathbf{G})$ can be seen as balancing the closeness of span($\mathbf{G}$) to a reducing subspace of $\mathbf{S}_{\mathbf{X}}$ and the fit of $\mathbf{Z}$ on $\mathbf{G}^T\mathbf{X}$.

Let $\mathbf{V} = \mathbf{S}_{\mathbf{X}}^{-1/2}\mathbf{X}$ denote the sample standardized version of $\mathbf{X}$, let $\mathbf{S}_{\mathbf{V},\mathbf{Z}} = \mathbf{S}_{\mathbf{X}}^{-1/2}\mathbf{S}_{\mathbf{X},\mathbf{Z}}$ denote the matrix of sample covariances between $\mathbf{V}$ and $\mathbf{Z}$, and let $L_q^{(3)} = \log|\mathbf{I}_r - \mathbf{S}_{\mathbf{V},\mathbf{Z}}^T\mathbf{P}_{\mathbf{S}_{\mathbf{X}}^{1/2}\mathbf{G}}\mathbf{S}_{\mathbf{V},\mathbf{Z}}|$, then L_q can be expressed also as $L_q(\mathbf{G}) = L_q^{(1)}(\mathbf{G}) + L_q^{(3)}(\mathbf{G})$:

$$\begin{aligned} L_q(\mathbf{G}) &= \log|\mathbf{G}^T\mathbf{S}_{\mathbf{X}}\mathbf{G}| + \log|\mathbf{G}^T\mathbf{S}_{\mathbf{X}}^{-1}\mathbf{G}| + \log|\mathbf{I}_r - \mathbf{S}_{\mathbf{X},\mathbf{Z}}^T\mathbf{S}_{\mathbf{X}}^{-1/2}\mathbf{P}_{\mathbf{S}_{\mathbf{X}}^{1/2}\mathbf{G}}\mathbf{S}_{\mathbf{X}}^{-1/2}\mathbf{S}_{\mathbf{X},\mathbf{Z}}| \\ &= \log|\mathbf{G}^T\mathbf{S}_{\mathbf{X}}\mathbf{G}| + \log|\mathbf{G}^T\mathbf{S}_{\mathbf{X}}^{-1}\mathbf{G}| + \log|\mathbf{I}_r - \mathbf{S}_{\mathbf{V},\mathbf{Z}}^T\mathbf{P}_{\mathbf{S}_{\mathbf{X}}^{1/2}\mathbf{G}}\mathbf{S}_{\mathbf{V},\mathbf{Z}}| \\ &= L_q^{(1)}(\mathbf{G}) + L_q^{(3)}(\mathbf{G}). \end{aligned}$$

The addend $L_q^{(3)}(\mathbf{G})$ of $L_q(\mathbf{G})$ carries the covariance signal from $\mathbf{S}_{\mathbf{V},\mathbf{Z}}$ in terms of the standardized variables $\mathbf{V}$ and $\mathbf{Z}$. It is minimized alone by setting $\mathbf{G}$ to be $\mathbf{S}_{\mathbf{X}}^{-1/2}$ times the first q eigenvectors of $\mathbf{S}_{\mathbf{V},\mathbf{Z}}\mathbf{S}_{\mathbf{V},\mathbf{Z}}^T$. If $q > r$ only the first $q - r$ of these generalized eigenvectors are determined uniquely. The full objective function $L_q(\mathbf{G}) = L_q^{(1)}(\mathbf{G}) + L_q^{(3)}(\mathbf{G})$ can also be viewed as balancing the requirement that the optimal value should stay close to a subset of q eigenvectors of $\mathbf{S}_{\mathbf{X}}$ and to the generalized eigenvectors of $\mathbf{S}_{\mathbf{X},\mathbf{Z}}\mathbf{S}_{\mathbf{X},\mathbf{Z}}^T$ relative to $\mathbf{S}_{\mathbf{X}}$.

4.3.2.1 Principal Component Regression

The principal component regression estimator of $\boldsymbol{\beta}$ is obtained by setting the columns of $\widehat{\boldsymbol{\Phi}}$ to be the first q eigenvectors of $\mathbf{S}_{\mathbf{X}}$. While $L_q^{(1)}(\mathbf{G})$ pulls the solution toward the eigenspaces of $\mathbf{S}_{\mathbf{X}}$, there is no particular preference for the principal eigenspaces. In fact, $L_q^{(1)}(\mathbf{G})$ alone places no special a priori preference on any ordering of the eigenspaces. The function $L_q^{(2)}(\mathbf{G})$ can guide the solution toward any q-dimensional eigenspace. In short, while the eigenspaces of $\mathbf{S}_{\mathbf{X}}$ play a role in both envelopes and principal component regression, those roles are quite different.

4.3.2.2 SIMPLS

Turning to comparisons of the likelihood-based method with SIMPLS, we see first that $L_q(\mathbf{G})$ depends on the response only through its standardized version $\mathbf{Z} = \mathbf{S}_{\mathbf{Y}}^{-1/2}\mathbf{Y}$. On the other hand, SIMPLS depends on the scale of the response: when $q = 1$, the SIMPLS estimator of $\mathcal{E}_{\boldsymbol{\Sigma}_{\mathbf{X}}}(\mathcal{B}')$ is the span of the first eigenvector $\widehat{\mathbf{w}}_1$ of $\mathbf{S}_{\mathbf{X},\mathbf{Y}}\mathbf{S}_{\mathbf{X},\mathbf{Y}}^T$. After performing a full rank transformation of the response $\mathbf{Y} \mapsto \mathbf{A}\mathbf{Y}$, the SIMPLS estimator of $\mathcal{E}_{\boldsymbol{\Sigma}_{\mathbf{X}}}(\mathcal{B}')$ is the span of the first eigenvector $\widetilde{\mathbf{w}}_1$ of $\mathbf{S}_{\mathbf{X},\mathbf{Y}}\mathbf{A}^T\mathbf{A}\mathbf{S}_{\mathbf{X},\mathbf{Y}}^T$. Generally, $\text{span}(\widehat{\mathbf{w}}_1) \neq \text{span}(\widetilde{\mathbf{w}}_1)$, so the estimates of $\mathcal{E}_{\boldsymbol{\Sigma}_{\mathbf{X}}}(\mathcal{B}')$ differ, although $\boldsymbol{\Sigma}_{\mathbf{X},\mathbf{Y}}\boldsymbol{\Sigma}_{\mathbf{X},\mathbf{Y}}^T$ and $\boldsymbol{\Sigma}_{\mathbf{X},\mathbf{Y}}\mathbf{A}^T\mathbf{A}\boldsymbol{\Sigma}_{\mathbf{X},\mathbf{Y}}^T$ span the same subspace. It is customary to standardize the individual responses marginally $y_j \mapsto y_j/\{\widehat{\text{var}}(y_j)\}^{1/2}$, $j = 1, \ldots, r$, prior to application of SIMPLS, but it is evidently not customary to standardize the responses jointly $\mathbf{Y}_i \mapsto \mathbf{Z}_i = \mathbf{S}_{\mathbf{Y}}^{-1/2}\mathbf{Y}_i$. Of course, the SIMPLS algorithm could be applied after replacing $\mathbf{Y}$ with jointly standardized responses $\mathbf{Z}$.

The methods also differ on how they utilize information from $\mathbf{S}_{\mathbf{X}}$. In the likelihood-based objective function, $L_q^{(1)}(\mathbf{G})$ gauges how far $\text{span}(\mathbf{G})$ is from subsets of q eigenvectors of $\mathbf{S}_{\mathbf{X}}$, but there is no corresponding operation in the SIMPLS method. The first SIMPLS vector $\widehat{\mathbf{w}}_1$ does not incorporate direct information about $\mathbf{S}_{\mathbf{X}}$. The second SIMPLS vector incorporates $\mathbf{S}_{\mathbf{X}}$ by essentially removing the subspace $\text{span}(\mathbf{S}_{\mathbf{X}}\widehat{\mathbf{w}}_1)$ from consideration, but the choice of $\text{span}(\mathbf{S}_{\mathbf{X}}\widehat{\mathbf{w}}_1)$ is not guided by the relationship between $\widehat{\mathbf{w}}_1$ and the eigenvectors of $\mathbf{S}_{\mathbf{X}}$. Subsequent SIMPLS vectors operate similarly in successively smaller spaces. SIMPLS often requires more directions to match the performance of the likelihood-based method (Cook et al., 2013).

4.3.3 Asymptotic Properties

In this section, we describe asymptotic properties of the envelope estimator, starting with the case in which $\mathbf{C}$ is normally distributed (Cook et al., 2013).

Proposition 4.5 *Assume that q is known and that $\mathbf{C}$ is normally distributed with mean $\boldsymbol{\mu}_{\mathbf{C}}$ and covariance matrix $\boldsymbol{\Sigma}_{\mathbf{C}} > 0$. Then* $\sqrt{n}\{\text{vec}(\widehat{\boldsymbol{\beta}}) - \text{vec}(\boldsymbol{\beta})\}$

converges in distribution to a normal random vector with mean 0 and covariance matrix

$$\begin{aligned}\mathrm{avar}\{\sqrt{n}\mathrm{vec}(\hat{\boldsymbol{\beta}})\} &= \mathrm{avar}\{\sqrt{n}\mathrm{vec}(\hat{\boldsymbol{\beta}}_{\boldsymbol{\Phi}})\} + \mathrm{avar}\{\sqrt{n}\mathrm{vec}(\hat{\boldsymbol{\beta}}_{\eta}\mathbf{Q}_{\boldsymbol{\Phi}})\} \\ &= \boldsymbol{\Phi}\boldsymbol{\Delta}^{-1}\boldsymbol{\Phi}^T \otimes \boldsymbol{\Sigma} + (\boldsymbol{\Phi}_0 \otimes \boldsymbol{\eta}^T)\mathbf{M}^{\dagger}(\boldsymbol{\Phi}_0^T \otimes \boldsymbol{\eta}),\end{aligned}$$

where $\mathbf{M} = \boldsymbol{\Delta}_0 \otimes \boldsymbol{\eta}\boldsymbol{\Sigma}^{-1}\boldsymbol{\eta}^T + \boldsymbol{\Delta}_0^{-1} \otimes \boldsymbol{\Delta} + \boldsymbol{\Delta}_0 \otimes \boldsymbol{\Delta}^{-1} - 2\mathbf{I}_{p-q} \otimes \mathbf{I}_q$. *Additionally,* $T_q = n(F(\mathbf{S}_{\mathbf{C}}, \hat{\boldsymbol{\Sigma}}_{\mathbf{C}}) - F(\mathbf{S}_{\mathbf{C}}, \mathbf{S}_{\mathbf{C}}))$ *converges to a Chi-squared random variable with* $(p-q)r$ *degrees of freedom.*

The decomposition of $\mathrm{avar}\{\sqrt{n}\mathrm{vec}(\hat{\boldsymbol{\beta}})\}$ shown in Proposition 4.5 has the same algebraic form as the decomposition given in Proposition 1.1 with the roles of **X** and **Y** reversed. In particular, we also have that

$$\mathrm{avar}\{\sqrt{n}\mathrm{vec}(\hat{\boldsymbol{\beta}})\} \le \mathrm{avar}\{\sqrt{n}\mathrm{vec}(\mathbf{B})\},$$

so the envelope estimator never does worse asymptotically than the ordinary least squares estimator. The first term in the decomposition of $\mathrm{avar}\{\sqrt{n}\mathrm{vec}(\hat{\boldsymbol{\beta}})\}$ can also be represented as

$$\mathrm{avar}\{\sqrt{n}\mathrm{vec}(\hat{\boldsymbol{\beta}}_{\boldsymbol{\Phi}})\} = (\boldsymbol{\Phi}^T \otimes \mathbf{I}_r)\mathrm{avar}\{\sqrt{n}\mathrm{vec}(\hat{\boldsymbol{\eta}}_{\boldsymbol{\Phi}}^T)\}(\boldsymbol{\Phi} \otimes \mathbf{I}_r) = \boldsymbol{\Phi}\boldsymbol{\Delta}^{-1}\boldsymbol{\Phi}^T \otimes \boldsymbol{\Sigma},$$

the finite sample version of which was given in (4.10). The second term in the decomposition of $\mathrm{avar}\{\sqrt{n}\mathrm{vec}(\hat{\boldsymbol{\beta}})\}$ then represents the cost of estimating the envelope. We also see from these results that when performing a prediction at $\mathbf{X}_N - \boldsymbol{\mu}_{\mathbf{X}}$ the asymptotic covariance $\mathrm{avar}\{\sqrt{n}\hat{\boldsymbol{\beta}}(\mathbf{X}_N - \boldsymbol{\mu}_{\mathbf{X}})\}$ depends on the part $\boldsymbol{\Phi}^T(\mathbf{X}_N - \boldsymbol{\mu}_{\mathbf{X}})$ of $\mathbf{X}_N - \boldsymbol{\mu}_{\mathbf{X}}$ that lies in the envelope and on the part $\boldsymbol{\Phi}_0^T(\mathbf{X}_N - \boldsymbol{\mu}_{\mathbf{X}})$ that lies in the orthogonal complement, which is in contrast to the situation when $\boldsymbol{\Phi}$ is known as discussed previously.

The following corollary to Proposition 4.5 describes $\mathrm{avar}\{\sqrt{n}\mathrm{vec}(\hat{\boldsymbol{\beta}})\}$ when $\boldsymbol{\Sigma}_{\mathbf{X}} = \sigma_{\mathbf{X}}^2\mathbf{I}_p$ and provides a comparison with the ordinary least squares estimator.

Corollary 4.2 *Assume the conditions of Proposition 4.5 and additionally that* $\boldsymbol{\Sigma}_{\mathbf{X}} = \sigma_{\mathbf{X}}^2\mathbf{I}_p$ *and that the coefficient matrix* $\boldsymbol{\beta} \in \mathbb{R}^{p \times r}$ *has rank r. Then* $\mathrm{avar}\{\sqrt{n}\mathrm{vec}(\hat{\boldsymbol{\beta}})\} = \mathrm{avar}\{\sqrt{n}\mathrm{vec}(\mathbf{B})\}$.

This corollary says that if there is no collinearity among homoscedastic predictors, then the envelope and ordinary least square estimators are asymptotically equivalent. Since this conclusion is based on maximum likelihood estimation, the performance of SIMPLS or other PLS estimators will also be no better asymptotically than ordinary least squares, a conclusion that seems at odds with some popular impressions. However, envelope and PLS estimators could still have small-sample advantages over ordinary least squares.

The next proposition describes the asymptotic properties of the envelope estimator when **C** is not necessarily normal.

Proposition 4.6 *Assume that* $\mathbf{C}_1, \ldots, \mathbf{C}_n$ *are independent and identically distributed copies of* $\mathbf{C}$ *with finite fourth moments and assume that* q *is known. Then* $\sqrt{n}\{\text{vec}(\widehat{\boldsymbol{\beta}}) - \text{vec}(\boldsymbol{\beta})\}$ *converges in distribution to a normal random vector with mean 0.*

The justification of this proposition involves application of results by Shapiro (1986) on the asymptotic behavior of overparameterized structural models. The shifted objective function, $F_q^*(\mathbf{S_C}, \boldsymbol{\Sigma}_\mathbf{C}) = F_q(\mathbf{S_C}, \boldsymbol{\Sigma}_\mathbf{C}) - F_q(\mathbf{S_C}, \mathbf{S_C})$, is nonzero, twice continuously differentiable in $\mathbf{S_C}$ and $\boldsymbol{\Sigma}_\mathbf{C}$, and is equal to 0 if and only if $\boldsymbol{\Sigma}_\mathbf{C} = \mathbf{S_C}$. Additionally, $\sqrt{n}\{\text{vech}(\mathbf{S_C}) - \text{vech}(\boldsymbol{\Sigma}_\mathbf{C})\}$ is asymptotically normal. These conditions plus some minor technical restrictions enable us to apply Shapiro's Propositions 3.1 and 4.1, from which the conclusions can be shown to follow.

This proposition says that the envelope estimator $\widehat{\boldsymbol{\beta}}$ is $\sqrt{n}$-consistent and asymptotically normal when the original data are nonnormal. The asymptotic covariance matrix of $\widehat{\boldsymbol{\beta}}$ depends on fourth moments of $\mathbf{C}$, and its usefulness in providing an approximation to $\text{var}\{\text{vec}(\widehat{\boldsymbol{\beta}})\}$ is unknown. The bootstrap is a useful option in practice for estimating the covariance matrix of $\widehat{\boldsymbol{\beta}}$, as demonstrated in previous chapters.

4.3.4 Fitted Values and Prediction

Fitted values and predictions after predictor reduction follow the same structure as those described for response reduction in Section 1.7. In particular,

$$\sqrt{n}(\widehat{\mathbf{Y}} - \text{E}(\widehat{\mathbf{Y}})) \rightarrow N_r(0, \text{avar}\{\sqrt{n}\text{vec}(\bar{\mathbf{Y}} + \widehat{\boldsymbol{\beta}}\mathbf{X})\}),$$

where

$$\begin{aligned} &\text{avar}\{\sqrt{n}\text{vec}(\bar{\mathbf{Y}} + \widehat{\boldsymbol{\beta}}\mathbf{X})\} \\ &\quad = \boldsymbol{\Sigma} + \text{avar}\{\sqrt{n}\text{vec}(\widehat{\boldsymbol{\beta}}_{\boldsymbol{\Phi}}\mathbf{X})\} + \text{avar}\{\sqrt{n}\text{vec}(\widehat{\boldsymbol{\beta}}_\eta \mathbf{Q}_{\boldsymbol{\Phi}}\mathbf{X})\} \\ &\quad = \boldsymbol{\Sigma} + \mathbf{X}^T\boldsymbol{\Phi}\boldsymbol{\Delta}^{-1}\boldsymbol{\Phi}^T\mathbf{X} + \text{avar}\{\sqrt{n}\text{vec}(\widehat{\boldsymbol{\beta}}_\eta \mathbf{Q}_{\boldsymbol{\Phi}}\mathbf{X})\}. \end{aligned}$$

Consequently, the variance of a fitted value has the same essential decomposition as the variance of $\widehat{\boldsymbol{\beta}}$ discussed previously.

Turning to prediction, suppose that we wish to predict a future value $\mathbf{Y}_{\text{new}}$ as some value of $\mathbf{X}$. Then

$$\text{E}[(\widehat{\mathbf{Y}} - \mathbf{Y}_{\text{new}})(\widehat{\mathbf{Y}} - \mathbf{Y}_{\text{new}})^T] = n^{-1}\text{avar}\{\sqrt{n}\text{vec}(\widehat{\boldsymbol{\beta}}\mathbf{X})\} + (1 + n^{-1})\boldsymbol{\Sigma} + o(n^{-1}).$$

The operating characteristics of these predictions follow those of response envelopes discussed in Section 1.7.

4.3.5 Choice of Dimension

The statistic T_q described in Proposition 4.5 can be used in a sequential manner to estimate q: beginning with $q_0 = 0$ test the hypothesis $q = q_0$, terminating the first time it is not rejected. Otherwise, q_0 is incremented by one and then the hypothesis is tested again. Cross-validation, a hold-out sample, and an information criterion are also options that may be useful in practice. The relative advantages of these methods in the context of predictor reduction have not been studied in detail.

4.3.6 Relevant Components

Envelopes were first developed for response reduction by Cook et al. (2010) as described in Chapter 1, and later Cook et al. (2013) advanced a similar logic for predictor reduction. There are similarities in these developments to the evolution of PLS regression. In describing population PLS for regressions with a univariate response, Helland (1990) reasoned as follows. Let $\mathbf{\Sigma}_{\mathbf{X}} = \sum_{j=1}^{p} \varphi_j \boldsymbol{\ell}_j \boldsymbol{\ell}_j$ denote the spectral decomposition of $\mathbf{\Sigma}_{\mathbf{X}}$ so that we can express the population coefficient vector as

$$\boldsymbol{\beta}^T = \sum_{j=1}^{p} \varphi_j^{-1} \boldsymbol{\ell}_j \boldsymbol{\ell}_j^T \mathbf{\Sigma}_{\mathbf{X},Y}.$$

Suppose for convenience of discussion that $p - q$ eigenvectors of $\mathbf{\Sigma}_{\mathbf{X}}$ are orthogonal to $\mathbf{\Sigma}_{\mathbf{X},Y}$, while the remaining eigenvectors $\boldsymbol{\ell}_{j_i}, i = 1, \dots, q$, have $\boldsymbol{\ell}_{j_i}^T \mathbf{\Sigma}_{\mathbf{X},Y} \neq 0$. Then the coefficient vector can be written as

$$\boldsymbol{\beta}^T = \sum_{i=1}^{q} \varphi_{j_i}^{-1} \boldsymbol{\ell}_{j_i} \boldsymbol{\ell}_{j_i}^T \mathbf{\Sigma}_{\mathbf{X},Y}.$$

This representation of $\boldsymbol{\beta}$ in terms of subset of q eigenvectors of $\mathbf{\Sigma}_{\mathbf{X}}$ is one of the driving algebraic ideas behind PLS. Næs and Martens (1985) referred to the reduced predictors $\boldsymbol{\ell}_{j_i}^T \mathbf{X}$ as relevant factors, while Helland (1990) called them relevant components and referred to the corresponding eigenvectors $\boldsymbol{\ell}_{j_i}$ as relevant eigenvectors. If $\mathbf{\Sigma}_{\mathbf{X}} > 0$ with distinct eigenvalues, then it follows from Proposition A.3 that $\mathcal{E}_{\mathbf{\Sigma}_{\mathbf{X}}}(\mathcal{B}') = \text{span}(\boldsymbol{\ell}_{j_1}, \dots, \boldsymbol{\ell}_{j_q})$, so the envelope is the span of the relevant eigenvectors. The PLS algorithm described by Helland (1990) produces essentially a Gram–Schmidt orthogonalization of the Krylov sequence (4.11) , so it does not actually provide for estimation of the relevant eigenvectors, but in the population they span the same subspace. Næs and Helland (1993) refined the notion of relevant components to weakly and strongly relevant components, which are based on conditions similar to those of Lemma 4.1.

Still restricting consideration to regressions with a univariate response, Helland (1992) developed maximum likelihood estimation assuming that $(Y, \mathbf{X})$ follows a multivariate normal distribution. The likelihood objective function he derived is essentially the same as (4.14) reparametrized to remove the restriction to semi-orthogonal matrices. Envelope estimators based on (4.14) are not necessarily nested; that is, with $q_1 < q_2$, the envelope estimator with dimension q_1 is not necessarily contained in the envelope estimator with dimension q_2. Helland (1992) also proposed a modified maximum likelihood estimator that forces nesting as the number of relevant components increases. In the notation of (4.14) his modified estimators force for $r = 1$

$$\text{span}\{\arg\min_{\mathbf{G}} L_1(\mathbf{G})\} \subseteq \text{span}\{\arg\min_{\mathbf{G}} L_2(\mathbf{G})\} \subseteq \cdots \subseteq \text{span}\{\arg\min_{\mathbf{G}} L_q(\mathbf{G})\}.$$

The 1D likelihood-based algorithm discussed in Section 6.4 and the moment-based algorithm of Section 6.5 also produce nested envelope estimators.

4.4 Illustrations

In this section, we provide examples to illustrate the operating characteristics of predictor envelopes in relatively simple settings. More complicated examples involving predictive performance were given by Cook et al. (2013). One example from their work is described in Section 4.4.5.

4.4.1 Expository Example, Continued

In this section, we provide a numerical version of the expository example introduced at the end of Section 4.1.1. The parameters in that example were set as follows: The error variance, which is a scalar in this case, was set equal to 1, $\boldsymbol{\alpha} = 0$, $\beta_1 = \beta_2 = 3$, $\boldsymbol{\Sigma}_2 = 0.2\mathbf{I}_7$, and $\boldsymbol{\Sigma}_1 = \mathbf{N}\mathbf{N}^T$ where $\mathbf{N} \in \mathbb{R}^{3\times3}$ is a matrix of independent standard normal random variables. The sample size was taken to be $n = 100$.

Using the normal likelihood, Akaike's information criterion (AIC) overestimated the dimension at 5, but Bayes information criterion (BIC) and LRT (0.05) correctly estimated the dimension to be 3. Estimation with $q = 3$ was then carried out using the likelihood discussed in Section 4.3.1. The three columns of the semi-orthogonal matrix $\widehat{\boldsymbol{\Phi}}$ each contained a single dominant element that exceeded 0.999 in absolute value. These three dominant elements identified the first three predictors $\mathbf{X}_1$.

The second and third columns of Table 4.1 give the absolute Z-scores for the ordinary least squares coefficients and the envelope coefficients. Both methods clearly identified the active predictors, X_1 and X_2. However, ordinary least squares coefficients also indicate that two inactive predictors, X_5 and X_7, might

Table 4.1 Expository data: The absolute Z-scores, estimates divided by their asymptotic standard errors, for the coefficients estimated by using ordinary least squares (OLS) and the envelope method (EM). $R = \text{SE}(\mathbf{B})/\text{SE}(\widehat{\beta})$.

Predictor X_j	Z-score, OLS	Z-score, EM	R
1	64.4	63.8	1.04
2	55.4	53.9	1.02
3	0.11	0.32	1.03
4	0.52	0.18	1.96
5	2.27	0.12	2.43
6	1.36	0.42	3.10
7	2.06	0.17	2.51
8	0.33	0.14	2.33
9	0.77	0.18	2.10
10	0.61	0.09	2.85

be active. In contrast, the envelope estimates give no clear hints that any of the inactive predictors are active. Except for X_3, the Z-scores for the envelope estimates of the inactive predictors are all less that the corresponding scores for the ordinary least squares estimates. It seems then that using envelopes led to sharper inferences. The standard error ratios in the last column of Table 4.1 indicate that there is little difference between the ordinary least squares and envelope estimates for the active predictors, while the standard errors for the inactive predictors differ materially.

This example is sparse in the sense that the response depends on only two of the ten predictors. If sparsity is suspected a priori, then a sparse method may be appropriate. Sparse PLS methods for predictors were proposed by Chun and Keleş (2010) and Liland et al. (2013). A sparse predictor method is also available as a variant on the sparse response envelopes discussed in Section 7.5

4.4.2 Australian Institute of Sport: Predictor Envelopes

Consider the regression of red cell count on the $p = 2$ hematological measurements hematocrit and hemoglobin. Since there are only two predictors, there are only three options for the dimension q of an envelope. Either $q = 0$ and there is no immaterial variation in $\mathbf{X}$, or $q = 1$ and $\mathcal{E}_{\boldsymbol{\Sigma}_{\mathbf{X}}}(\mathcal{B}')$ aligns with the first or second eigenvector of $\boldsymbol{\Sigma}_{\mathbf{X}}$, or $q = 2$ and thus $\mathcal{E}_{\boldsymbol{\Sigma}_{\mathbf{X}}}(\mathcal{B}') = \mathbb{R}^2$. In higher dimensional examples, the envelope need not correspond to an eigenspace, but will be contained in a reducing subspace of $\boldsymbol{\Sigma}_{\mathbf{X}}$.

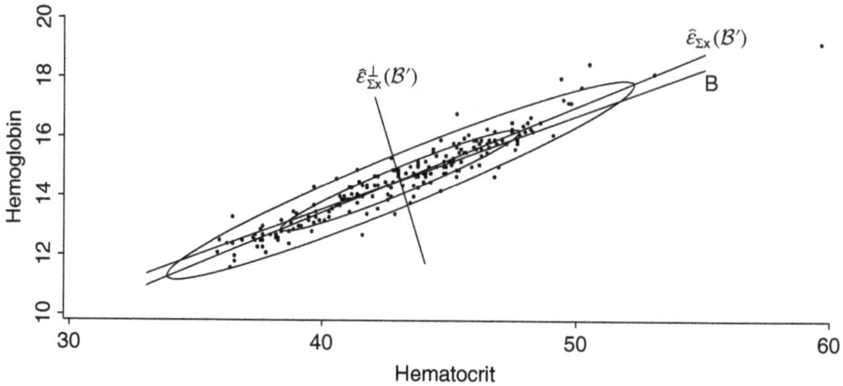

Figure 4.1 Australian Institute of Sport: scatterplot of the predictors hematocrit and hemoglobin, along with the estimated envelope and span($\mathbf{B}^T$) from the regression with red cell count as the response.

As in previous examples, a first step is to select a dimension for $\mathcal{E}_{\Sigma_X}(\mathcal{B}')$. In this example, AIC, BIC, and LRT(0.05) all indicated that $q = 1$ and thus that $\mathcal{E}_{\Sigma_X}(\mathcal{B}')$ aligns with either the first or second eigenvector of $\mathbf{\Sigma}_X$. Shown in Figure 4.1 is a plot of the $n = 202$ observations along with the contours of $\mathbf{S}_X$, the estimated envelope $\hat{\mathcal{E}}_{\Sigma_X}(\mathcal{B}')$ and its orthogonal complement $\hat{\mathcal{E}}^{\perp}_{\Sigma_X}(\mathcal{B}')$, and the subspace spanned by the ordinary least squares coefficient vector **B**. The plot indicates that the span of the first eigenvector of $\mathbf{S}_X$ and the envelope are quite close, and are not far from the ordinary least squares subspace span($\mathbf{B}^T$). Evidently, the response changes as we move along the envelope from left to right, which reflects the material variation in the data, but is relatively constant as we move along the orthogonal complement, which reflects the immaterial variation. The eigenvalues of $\mathbf{S}_X$ are 15.0 and 0.16, so according to the intuition provided by Propositions 4.1 and 4.5, we anticipate notable reduction in estimative variation.

Table 4.2 shows the estimated coefficients with their standard errors from the standard and envelope analyses. While the coefficients are quite close, as suggested by Figure 4.1, the standard errors from the standard analysis are about

Table 4.2 Australian Institute of Sport: coefficient estimates and their standard errors from the envelope model with $q = 1$ and the standard model.

X	$\hat{\beta}$	SE($\hat{\beta}$)	B	SE(B)	SE(B)/SE($\hat{\beta}$)
Hematocrit	0.103	0.005	0.104	0.011	2.25
Hemoglobin	0.037	0.010	0.033	0.029	2.75

two and a half times the standard errors from the envelope analysis. We would need a sample size about six times as large for the standard errors of $\mathbf{B}$ to match those of $\widehat{\boldsymbol{\beta}}$ based on the current sample size. Additionally, the sample $\mathbf{S}_{\mathbf{X}}$ version of $\boldsymbol{\Sigma}_{\mathbf{X}}$ is nearly identical to the envelope estimate $\widehat{\boldsymbol{\Sigma}}_{\mathbf{X}}$ in this example:

$$\widehat{\boldsymbol{\Sigma}}_{\mathbf{X}} = \begin{pmatrix} 13.3513 & 4.7211 \\ 4.7211 & 1.8468 \end{pmatrix}, \quad \mathbf{S}_{\mathbf{X}} = \begin{pmatrix} 13.3511 & 4.7214 \\ 4.7214 & 1.8471 \end{pmatrix}.$$

Since $q = 1$, the subspace estimated by SIMPLS is simply

$$\text{span}(\mathbf{S}_{\mathbf{X},Y}) = \text{span}((1.544, 0.552)^T).$$

The ratio of the first to the second element of $\mathbf{S}_{\mathbf{X},Y}$ is 2.798, while the same ratio from a basis $\widehat{\boldsymbol{\Phi}}$ for the estimated envelope is 2.795. Consequently, the estimated envelope and SIMPLS subspaces are nearly identical in this illustration, although that will not always be so, particularly when $q > 1$.

4.4.3 Wheat Protein: Predicting Protein Content

In Section 2.1, we used the wheat protein data by setting the response vector to be the logarithms of near infrared reflectance at six wavelengths and the univariate predictor to be the protein content. Here we interchange the roles of the variables, setting the univariate response to be protein content and the bivariate predictor to be measurements at the third and fourth wavelengths. Since there are only $p = 2$ predictors, we again have only three options for the envelope, and the three dimension selection methods again indicated that $q = 1$.

Figure 4.2 shows a scatterplot of the data along with $\text{span}(\mathbf{B}^T)$ from the regression with protein content as the response, and the estimated envelope and its orthogonal complement. In contrast to the previous illustration, the envelope is now estimated to coincide with the span of the second (rather than the first) eigenvector of $\boldsymbol{\Sigma}_{\mathbf{X}}$, while $\text{span}(\mathbf{B}^T)$ is still quite close to the envelope. The eigenvalues of $\mathbf{S}_{\mathbf{X}}$ are 2.051 and 0.018 so there might be little if any gain over ordinary least squares because the material variation in the direction of the second eigenvector is small relative to the immaterial variation in the direction of the first eigenvector, as discussed near the end of Section 4.1. The SIMPLS subspace $\text{span}(\mathbf{S}_{\mathbf{X},Y})$ is notably different from the estimated envelope. Table 4.3 gives the coefficient estimates and their standard errors. In this case, although there is a proper envelope, it essentially reproduced the ordinary least squares analysis, again because the immaterial variation in $\mathbf{X}$ is large relative to the material variation in $\mathbf{X}$.

Turning to the regression of protein content on all six wavelengths, we find a consistent dimension of $q = 4$ and the fitted envelope with $q = 4$ again reproduces the ordinary least squares analysis with unimportant differences, so the relationships illustrated in Figure 4.2 may be present in the larger data set.

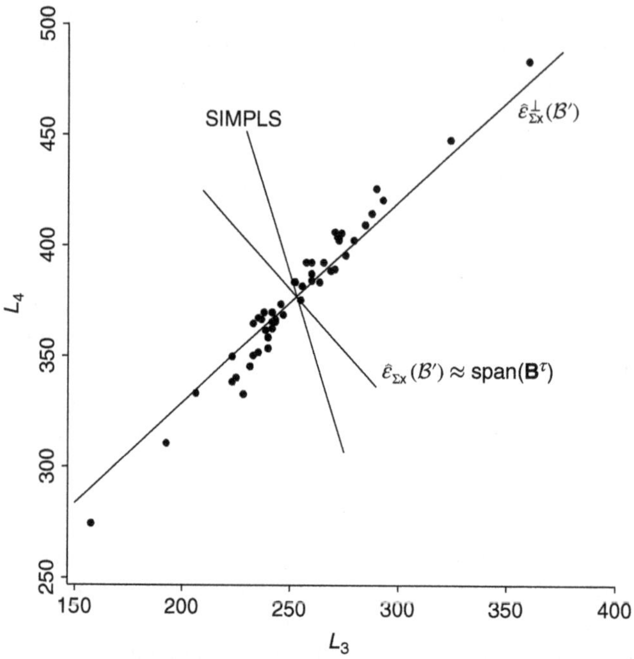

Figure 4.2 Wheat protein data: scatterplot of the predictors at the third and fourth wavelength, along with the estimated envelope and span($\mathbf{B}^T$) from the regression with protein content as the response.

Table 4.3 Wheat protein data: coefficient estimates and their standard errors from the envelope model with $q = 1$, the standard model and SIMPLS.

X	$\hat{\beta}$	SE($\hat{\beta}$)	B	SE(B)	SIMPLS
L_3	0.2470	0.0072	0.2476	0.0074	0.0165
L_4	−0.2249	0.0066	−0.2249	0.0068	−0.0052

4.4.4 Mussels' Muscles: Predictor Envelopes

The data[1] for this example came from an ecological study of horse mussels sampled from the Marlborough Sounds off the coast of New Zealand. The response variable is the logarithm of the mussel's muscle mass $M(g)$, the edible portion of the mussel. The four predictors are the logarithms of the mussel's shell height H, shell width W, shell length L, each in millimeters, and shell mass $S(g)$. The scatterplot matrix of the observations from $n = 82$ mussels given in Figure 4.3

1 Data are available from the website www.stat.umn.edu/RegGraph for Cook (1998).

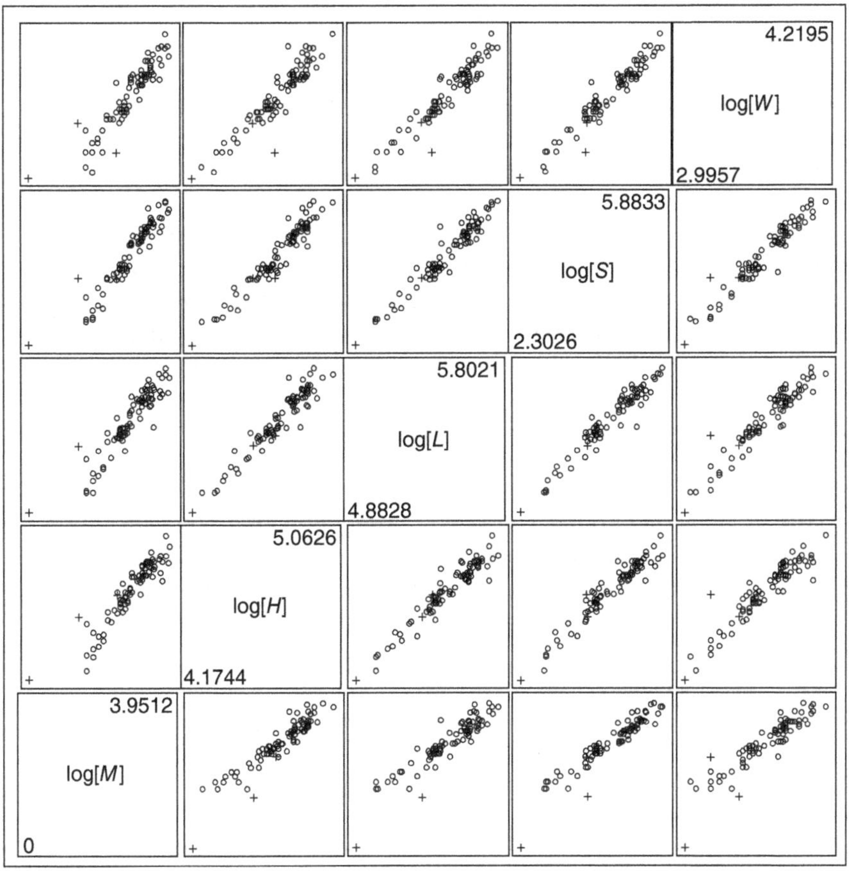

Figure 4.3 Mussels' muscles: scatterplot of the response log *M* and the four predictors. Three relatively outlying points are indicted with +.

shows that there are strong linear relationships between the five variables, the pairwise correlations ranging between 0.94 and 0.96. This is a clue that a predictor envelope could be effective. There are three points that seem to outlie trends in the scatterplot. These three points were deleted from the main analysis envelope analysis, leaving $n = 79$ observations; comments on the analysis with the full data are given at the end of this section.

Columns 2–4 of Table 4.4 give the estimated coefficients along with their standard errors and Z-scores from the ordinary least squares fit of the standard model. We see from those results that log L and log W have no detectable relationship with log M given log H and log S. This is a classic scenario in which it might be reasoned that log L and log W should be removed from the regression because they contribute little after log H and log S.

Turning to an envelope analysis, AIC, BIC, LRT (0.05), and 10-fold cross-validation all indicated that $q = 1$. The cross-validation results indicated that the prediction error from the envelope model with $q = 1$ is smaller than the prediction error for the standard model. The envelope prediction error from cross-validation is also smaller than that from the standard model for the regression of log M on log H and log S. Columns 5–7 of Table 4.4 give the estimated coefficients along with their standard errors and Z-scores from the fit of the envelope model with $q = 1$. In contrast to the standard analysis, all four predictors are now seen to contribute to the regression. The ratios of the standard errors $\text{SE}(\mathbf{B})/\text{SE}(\widehat{\boldsymbol{\beta}})$, which range between 5.9 and 76.3, indicate that the envelope results are a substantial improvement over the standard results. The bootstrap standard errors shown in Table 4.4 are in good agreement with the asymptotic standard errors; using them would not change any of the results materially. Table 4.5 shows the estimates of the predictor covariance matrix $\boldsymbol{\Sigma}_{\mathbf{X}}$ from the standard and envelope analyses. These estimates are essentially the same.

Finally, we discuss briefly the analysis based on the full data with all $n = 82$ observations. As in the analysis of the reduced data, AIC, BIC, LRT (0.05), and 10-fold cross-validation all indicated that $q = 1$. The results from fitting the full data with $q = 1$ are in qualitative agreement with those from the reduced data,

Table 4.4 Mussels' muscles: fits of the standard model and envelope model with $q = 1$.

	Standard model			Envelope model, $q = 1$		
X	**B**	**SE(B)**	***Z*-Score**	$\widehat{\boldsymbol{\beta}}$	**SE**($\widehat{\boldsymbol{\beta}}$)	***Z*-Score**
log H	0.742	0.397 (0.401)	1.87	0.141	0.0052 (0.0045)	27.0
log L	−0.113	0.387 (0.392)	−0.29	0.154	0.0056 (0.0050)	27.4
log S	0.567	0.115 (0.110)	4.93	0.625	0.0194 (0.0201)	32.2
log W	0.171	0.294 (0.282)	0.58	0.206	0.0073 (0.0067)	28.1

Bootstrap standard errors from 250 bootstrap samples are given in parentheses.

Table 4.5 Mussels' muscles: estimated covariance matrices from the standard and envelope ($q = 1$) fits.

	Envelope model, $\widehat{\boldsymbol{\Sigma}}_{\mathbf{X}}$				Standard model, $\mathbf{S}_{\mathbf{X}}$			
X	**log *H***	**log *L***	**log *S***	**log *W***	**log *H***	**log *L***	**log *S***	**log *W***
log H	0.0299	0.0310	0.1230	0.0411	0.0298	0.0310	0.1230	0.0411
log L		0.0354	0.1343	0.0452		0.0354	0.1343	0.0452
log S			0.5500	0.1801			0.5501	0.1802
log W				0.0632				0.0632

but are not as crisp. For instance, the agreement between $\mathbf{S}_{\mathbf{X}}$ and $\widehat{\boldsymbol{\Sigma}}_{\mathbf{X}}$ is not quite as good. Generally, the results from the full data parallel those from the reduced data, but are not quite as strong, as might be expected in view of the nature of the deleted observations.

4.4.5 Meat Properties

Sæbø et al. 2007 analyzed the absorbance spectra from infrared transmittance for fat, protein, and water in $n = 103$ pork or beef samples as an example with collinearity and multiple relevant components for soft-threshold-PLS. Cook et al. (2013) took the measurements at every fourth wavelength between 850 and 1050 nm as predictors, yielding $p = 50$. Predictions of protein were constructed as $\widehat{Y} = \bar{Y} + \widehat{\boldsymbol{\beta}}(\mathbf{X} - \bar{\mathbf{X}})$ with $\widehat{\boldsymbol{\beta}}$ obtained by using envelopes and SIMPLS.[2] Fivefold cross-validation was used to estimate the average prediction error. Their results indicate that envelopes perform much better than SIMPLS for a small q, while the two methods can have similar performance for q sufficiently large. For instance, at $q = 1$, the average SIMPLS prediction error was about four times the average envelope prediction error, and the minimum prediction error for envelopes occurred at a smaller value of q than that for SIMPLS.

4.5 Simultaneous Predictor–Response Envelopes

In this section, we sketch how envelopes can be used for simultaneous reduction of the predictors and the responses. The method was developed by Cook and Zhang (2015b), and additional details are available from their article.

4.5.1 Model Formulation

Again consider the multivariate linear model (1.1), allowing the predictors to be stochastic. Restating it for ease of reference,

$$\mathbf{Y} = \boldsymbol{\alpha} + \boldsymbol{\beta}(\mathbf{X} - \boldsymbol{\mu}_{\mathbf{X}}) + \boldsymbol{\varepsilon}, \tag{4.17}$$

where the error vector $\boldsymbol{\varepsilon}$ has mean 0 and covariance matrix $\boldsymbol{\Sigma}$, the random predictor vector $\mathbf{X}$ has mean $\boldsymbol{\mu}_{\mathbf{X}}$ and variance $\boldsymbol{\Sigma}_{\mathbf{X}}$, and $\boldsymbol{\varepsilon} \perp\!\!\!\perp \mathbf{X}$. Given n independent copies $(\mathbf{Y}_i, \mathbf{X}_i)$, $i = 1, \dots, n$, of $(\mathbf{Y}, \mathbf{X})$, we form the model for simultaneous reduction of $\mathbf{X}$ and $\mathbf{Y}$ by combining models (4.5) and (1.20) for predictor and response reduction:

$$\begin{aligned} \mathbf{Y} &= \boldsymbol{\alpha} + \boldsymbol{\Gamma}\boldsymbol{\eta}\boldsymbol{\Phi}^T(\mathbf{X} - \boldsymbol{\mu}_{\mathbf{X}}) + \boldsymbol{\varepsilon}, \\ \boldsymbol{\Sigma} &= \boldsymbol{\Gamma}\boldsymbol{\Omega}\boldsymbol{\Gamma}^T + \boldsymbol{\Gamma}_0\boldsymbol{\Omega}_0\boldsymbol{\Gamma}_0^T \\ \boldsymbol{\Sigma}_{\mathbf{X}} &= \boldsymbol{\Phi}\boldsymbol{\Delta}\boldsymbol{\Phi}^T + \boldsymbol{\Phi}_0\boldsymbol{\Delta}_0\boldsymbol{\Phi}_0^T, \end{aligned} \tag{4.18}$$

2 The SIMPLS estimator was obtained by using the MATLAB function *plsregress*.

where $\boldsymbol{\beta} = \boldsymbol{\Gamma}\boldsymbol{\eta}\boldsymbol{\Phi}^T$, $\boldsymbol{\Gamma} \in \mathbb{R}^{r\times u}$ and $\boldsymbol{\Phi} \in \mathbb{R}^{p\times q}$ are semi-orthogonal basis matrices for the response envelope $\mathcal{E}_{\boldsymbol{\Sigma}}(\mathcal{B})$ and the predictor envelope $\mathcal{E}_{\boldsymbol{\Sigma}_{\mathbf{X}}}(\mathcal{B}')$, and $\boldsymbol{\eta} \in \mathbb{R}^{u\times q}$ is an unconstrained matrix with rank $d = \text{rank}(\boldsymbol{\eta}) = \text{rank}(\boldsymbol{\beta})$. This model has the same essential structure as the envelope version of the PLS latent variable model (4.6) described in Section 4.1.2. The dimensions u and q of the response and predictor envelopes satisfy $d \le u \le r$ and $d \le q \le p$. From this we see that if $d = r = \min(r, p)$, then $u = r$, $\mathcal{E}_{\boldsymbol{\Sigma}}(\mathcal{B}) = \mathbb{R}^r$, only predictor reduction is possible and model (4.18) reduces to model (4.5) for predictor reduction. Similarly, if $d = p = \min(r, p)$, then $q = p$, $\mathcal{E}_{\boldsymbol{\Sigma}_{\mathbf{X}}}(\mathcal{B}') = \mathbb{R}^p$, only response reduction is possible and model (4.18) reduces to model (1.20) for response reduction. Consequently, if model (4.18) is to provide value beyond that for the individual response and predictor envelope models, then we must have $d < \min(r, p)$.

If $\mathcal{E}_{\boldsymbol{\Sigma}_{\mathbf{X}}}(\mathcal{B}') = \mathbb{R}^p$, then $\boldsymbol{\Phi} = \mathbf{I}_p$, and it follows from the discussion of Chapter 1 that $\text{cov}(\boldsymbol{\Gamma}^T\mathbf{Y}, \boldsymbol{\Gamma}_0^T\mathbf{Y} \mid \mathbf{X}) = 0$ and $\boldsymbol{\Gamma}_0^T\mathbf{Y} \mid \mathbf{X} \sim \boldsymbol{\Gamma}_0^T\mathbf{Y}$, which motivated the construction of response envelopes. If $\boldsymbol{\varepsilon}$ is normally distributed, then this pair of conditions is equivalent to $\boldsymbol{\Gamma}_0^T\mathbf{Y} \perp\!\!\!\perp \boldsymbol{\Gamma}^T\mathbf{Y} \mid \mathbf{X}$ and $\boldsymbol{\Gamma}_0^T\mathbf{Y} \mid \mathbf{X} \sim \boldsymbol{\Gamma}_0^T\mathbf{Y}$. If $(\mathbf{X}, \mathbf{Y})$ is jointly normal then this pair of conditions is equivalent to $\boldsymbol{\Gamma}_0^T\mathbf{Y} \perp\!\!\!\perp (\boldsymbol{\Gamma}^T\mathbf{Y}, \mathbf{X})$. If $\mathcal{E}_{\boldsymbol{\Sigma}}(\mathcal{B}) = \mathbb{R}^r$, then $\boldsymbol{\Gamma} = \mathbf{I}_r$, $\text{cov}(\mathbf{Y}, \boldsymbol{\Phi}_0^T\mathbf{X} \mid \boldsymbol{\Phi}^T\mathbf{X}) = 0$ and $\text{cov}(\boldsymbol{\Phi}^T\mathbf{X}, \boldsymbol{\Phi}_0^T\mathbf{X}) = 0$, which are the two conditions used in Chapter 4 for predictor reduction. If $(\mathbf{X}, \mathbf{Y})$ has a joint normal distribution, then this pair of conditions is equivalent to $\boldsymbol{\Phi}_0^T\mathbf{X} \perp\!\!\!\perp (\boldsymbol{\Phi}^T\mathbf{X}, \mathbf{Y})$.

The previous relationships follow from the response and predictor envelopes. The following lemma describes additional relationships between the material parts of $\mathbf{Y}$ and $\mathbf{X}$, and the immaterial part of $\mathbf{X}$ and $\mathbf{Y}$ that come with the simultaneous envelopes.

Lemma 4.3 *Assume the simultaneous envelope model (4.18). Then* $\text{cov}(\boldsymbol{\Gamma}^T\mathbf{Y}, \boldsymbol{\Phi}_0^T\mathbf{X}) = 0$ *and* $\text{cov}(\boldsymbol{\Gamma}_0^T\mathbf{Y}, \boldsymbol{\Phi}^T\mathbf{X}) = 0$.

This lemma, which does not require normality of $\mathbf{X}$ or $\mathbf{Y}$, is implied by the previous discussion if $\boldsymbol{\Phi} = \mathbf{I}_p$ or $\boldsymbol{\Gamma} = \mathbf{I}_r$.

4.5.2 Potential Gain

To gain intuition about the potential advantages of simultaneous envelopes, we next consider the case where $\mathcal{E}_{\boldsymbol{\Sigma}_{\mathbf{X}}}(\mathcal{B}')$ and $\mathcal{E}_{\boldsymbol{\Sigma}}(\mathcal{B})$ are known. Estimation of these envelopes in practice will mitigate the findings in this section, but we have nevertheless found them to be useful qualitative indicators of the benefits of simultaneous reduction.

The proposed envelope estimator for $\boldsymbol{\beta}$ based on model (4.18) with known semi-orthogonal basis matrices $\boldsymbol{\Phi}$ and $\boldsymbol{\Gamma}$, denoted by $\widehat{\boldsymbol{\beta}}_{\boldsymbol{\Gamma},\boldsymbol{\Phi}}$, can be written as

$$\widehat{\boldsymbol{\beta}}_{\boldsymbol{\Gamma},\boldsymbol{\Phi}} = \boldsymbol{\Gamma}\widehat{\boldsymbol{\eta}}_{\boldsymbol{\Gamma},\boldsymbol{\Phi}}\boldsymbol{\Phi}^T = \boldsymbol{\Gamma}\boldsymbol{\Gamma}^T\mathbf{S}_{\mathbf{Y},\mathbf{X}}\boldsymbol{\Phi}(\boldsymbol{\Phi}^T\mathbf{S}_{\mathbf{X}}\boldsymbol{\Phi})^{-1}\boldsymbol{\Phi}^T = \mathbf{P}_{\boldsymbol{\Gamma}}\mathbf{B}\mathbf{P}^T_{\boldsymbol{\Phi}(\mathbf{S}_{\mathbf{X}})}, \tag{4.19}$$

where $\mathbf{B} = \mathbf{S}_{\mathrm{Y,X}}\mathbf{S}_{\mathrm{X}}^{-1}$ is still the ordinary least squares estimator. The estimator $\hat{\beta}_{\Gamma,\Phi}$ is obtained by projecting $\mathbf{B}$ onto the reduced predictor space and onto the reduced response space, and so does not depend on the particular bases $\mathbf{\Phi}$ and $\mathbf{\Gamma}$ selected. The estimator

$$\hat{\eta}_{\Gamma,\Phi} = \mathbf{\Gamma}^T\mathbf{S}_{\mathrm{Y,X}}\mathbf{\Phi}(\mathbf{\Phi}^T\mathbf{S}_{\mathrm{X}}\mathbf{\Phi})^{-1} = \mathbf{S}_{\Gamma^T\mathrm{Y},\Phi^T\mathrm{X}}\mathbf{S}_{\Phi^T\mathrm{X}}^{-1}$$

is the ordinary least squares estimator of the coefficient matrix for the regression of $\mathbf{\Gamma}^T\mathbf{Y}$ on $\mathbf{\Phi}^T\mathbf{X}$.

The next proposition shows how var(vec($\mathbf{B}$)) can potentially be reduced by using simultaneous envelopes. Let $f_p = n - p - 2$ and $f_q = n - q - 2$.

Proposition 4.7 *Assume that* $\mathbf{X} \sim N_p(\boldsymbol{\mu}_{\mathrm{X}}, \mathbf{\Sigma}_{\mathrm{X}})$, $n > p + 2$ *and that semi-orthogonal basis matrices* $\mathbf{\Phi}$, $\mathbf{\Gamma}$ *for the predictor and response envelopes are known. Then* $\mathrm{var}\{\mathrm{vec}(\mathbf{B})\} = f_p^{-1}\mathbf{\Sigma}_{\mathrm{X}}^{-1} \otimes \mathbf{\Sigma}$ *and*

$$\begin{aligned}\mathrm{var}\{\mathrm{vec}(\hat{\beta}_{\Gamma,\Phi})\} &= f_q^{-1}(\mathbf{\Phi}\mathbf{\Delta}^{-1}\mathbf{\Phi}^T) \otimes (\mathbf{\Gamma}\mathbf{\Omega}\mathbf{\Gamma}^T)\\ &= f_p f_q^{-1}\mathrm{var}\{\mathrm{vec}(\mathbf{B})\} - f_q^{-1}\mathbf{\Phi}\mathbf{\Delta}^{-1}\mathbf{\Phi}^T \otimes \mathbf{\Gamma}_0\mathbf{\Omega}_0\mathbf{\Gamma}_0^T\\ &\quad - f_q^{-1}\mathbf{\Phi}_0\mathbf{\Delta}_0^{-1}\mathbf{\Phi}_0^T \otimes \mathbf{\Gamma}\mathbf{\Omega}\mathbf{\Gamma}^T - f_q^{-1}\mathbf{\Phi}_0\mathbf{\Delta}_0^{-1}\mathbf{\Phi}_0^T \otimes \mathbf{\Gamma}_0\mathbf{\Omega}_0\mathbf{\Gamma}_0^T,\end{aligned}$$

where $\mathbf{\Delta} = \mathbf{\Phi}^T\mathbf{\Sigma}_{\mathrm{X}}\mathbf{\Phi}$, $\mathbf{\Delta}_0 = \mathbf{\Phi}_0^T\mathbf{\Sigma}_{\mathrm{X}}\mathbf{\Phi}_0$, $\mathbf{\Omega} = \mathbf{\Gamma}^T\mathbf{\Sigma}\mathbf{\Gamma}$, $\mathbf{\Omega}_0 = \mathbf{\Gamma}_0^T\mathbf{\Sigma}\mathbf{\Gamma}_0$, *and the variances are computed over both* $\mathbf{Y}$ *and* $\mathbf{X}$.

Proof: Let $\mathbb{X} \in \mathbb{R}^{n\times p}$ and $\mathbb{Y} \in \mathbb{R}^{n\times r}$ be the centered data matrices. Then $\mathbf{B} = \mathbb{Y}^T\mathbb{X}(\mathbb{X}^T\mathbb{X})^{-1}$ and $\hat{\eta}_{\Gamma,\Phi} = \mathbf{\Gamma}\mathbf{\Gamma}^T\mathbb{Y}^T\mathbb{X}\mathbf{\Phi}(\mathbf{\Phi}^T\mathbb{X}^T\mathbb{X}\mathbf{\Phi})^{-1}\mathbf{\Phi}^T$. For the variance of vec($\mathbf{B}$), we decompose it into two terms, conditioning on $\mathbb{X}$.

$$\begin{aligned}\mathrm{var}\{\mathrm{vec}(\mathbf{B})\} &= \mathrm{var}\{\mathrm{vec}(\mathbb{Y}^T\mathbb{X}(\mathbb{X}^T\mathbb{X})^{-1})\}\\ &= \mathrm{var}\{((\mathbb{X}^T\mathbb{X})^{-1}\mathbb{X}^T \otimes \mathbf{I}_r)\mathrm{vec}(\mathbb{Y}^T)\}\\ &= \mathrm{var}\{\mathrm{E}\{((\mathbb{X}^T\mathbb{X})^{-1}\mathbb{X}^T \otimes \mathbf{I}_r)\mathrm{vec}(\mathbb{Y}^T) \mid \mathbb{X}\}\}\\ &\quad + \mathrm{E}\{\mathrm{var}\{((\mathbb{X}^T\mathbb{X})^{-1}\mathbb{X}^T \otimes \mathbf{I}_r)\mathrm{vec}(\mathbb{Y}^T) \mid \mathbb{X}\}\}\\ &:= VE + EV.\end{aligned}$$

The first term is

$$\begin{aligned}VE &= \mathrm{var}\{\mathrm{E}\{((\mathbb{X}^T\mathbb{X})^{-1}\mathbb{X}^T \otimes \mathbf{I}_r)\mathrm{vec}(\mathbb{Y}^T) \mid \mathbb{X}\}\}\\ &= \mathrm{var}\{((\mathbb{X}^T\mathbb{X})^{-1}\mathbb{X}^T \otimes \mathbf{I}_r)\mathrm{E}\{\mathrm{vec}(\mathbb{Y}^T) \mid \mathbb{X}\}\}\\ &= \mathrm{var}\{((\mathbb{X}^T\mathbb{X})^{-1}\mathbb{X}^T \otimes \mathbf{I}_r)\mathrm{vec}(\boldsymbol{\beta}\mathbb{X}^T)\}\\ &= \mathrm{var}\{\mathrm{vec}(\boldsymbol{\beta}\mathbb{X}^T\mathbb{X}(\mathbb{X}^T\mathbb{X})^{-1})\} = 0.\end{aligned}$$

Then

$$\begin{aligned}\operatorname{var}(\operatorname{vec}(\mathbf{B})) = EV &= \mathrm{E}\{\operatorname{var}\{((\mathbb{X}^T\mathbb{X})^{-1}\mathbb{X}^T \otimes \mathbf{I}_r)\operatorname{vec}(\mathbb{Y}^T) \mid \mathbb{X}\}\} \\ &= \mathrm{E}\{((\mathbb{X}^T\mathbb{X})^{-1}\mathbb{X}^T \otimes \mathbf{I}_r)\operatorname{var}(\operatorname{vec}(\mathbb{Y}^T) \mid \mathbb{X})((\mathbb{X}^T\mathbb{X})^{-1}\mathbb{X}^T \otimes \mathbf{I}_r)^T\} \\ &= \mathrm{E}\{((\mathbb{X}^T\mathbb{X})^{-1}\mathbb{X}^T \otimes \mathbf{I}_r)(\mathbf{I}_n \otimes \boldsymbol{\Sigma})((\mathbb{X}^T\mathbb{X})^{-1}\mathbb{X}^T \otimes \mathbf{I}_r)^T\} \\ &= \mathrm{E}\{(\mathbb{X}^T\mathbb{X})^{-1} \otimes \boldsymbol{\Sigma}\} = f_p^{-1}\boldsymbol{\Sigma}_{\mathbf{X}}^{-1} \otimes \boldsymbol{\Sigma},\end{aligned}$$

where the last equality comes from the fact that $\mathbb{X}^T\mathbb{X}$ follows a Wishart distribution, $\mathbb{X}^T\mathbb{X} \sim W_p(\boldsymbol{\Sigma}_{\mathbf{X}}, n-1)$. Since $n > p+2$, $(\mathbb{X}^T\mathbb{X})^{-1}$ follows an inverse Wishart distribution $W_p^{-1}(\boldsymbol{\Sigma}_{\mathbf{X}}^{-1}, n-1)$, with mean $\boldsymbol{\Sigma}_{\mathbf{X}}^{-1}/(n-p-2) = f_p^{-1}\boldsymbol{\Sigma}_{\mathbf{X}}^{-1}$.

Similarly for $\widehat{\boldsymbol{\beta}}_{\Gamma,\Phi}$, we have the following expressions by noticing that $\operatorname{vec}(\widehat{\boldsymbol{\beta}}_{\Gamma,\Phi}) = (\boldsymbol{\Phi} \otimes \boldsymbol{\Gamma})\operatorname{vec}(\widehat{\eta}_{\Gamma,\Phi})$ and $\widehat{\eta}_{\Gamma,\Phi}$ is the ordinary least squares estimator of $\boldsymbol{\Gamma}^T\mathbf{Y}$ on $\boldsymbol{\Phi}^T\mathbf{X}$.

$$\begin{aligned}\operatorname{var}(\operatorname{vec}(\widehat{\boldsymbol{\beta}}_{\Gamma,\Phi})) &= f_q^{-1}(\boldsymbol{\Phi} \otimes \boldsymbol{\Gamma})(\boldsymbol{\Sigma}_{\boldsymbol{\Phi}^T\mathbf{X}}^{-1} \otimes \boldsymbol{\Sigma}_{\boldsymbol{\Gamma}^T\mathbf{Y}|\boldsymbol{\Phi}^T\mathbf{X}})(\boldsymbol{\Phi}^T \otimes \boldsymbol{\Gamma}^T) \\ &= f_q^{-1}(\boldsymbol{\Phi} \otimes \boldsymbol{\Gamma})((\boldsymbol{\Phi}^T\boldsymbol{\Sigma}_{\mathbf{X}}\boldsymbol{\Phi})^{-1} \otimes (\boldsymbol{\Gamma}^T\boldsymbol{\Sigma}\boldsymbol{\Gamma}))(\boldsymbol{\Phi}^T \otimes \boldsymbol{\Gamma}^T) \\ &= f_q^{-1}(\boldsymbol{\Phi}(\boldsymbol{\Phi}^T\boldsymbol{\Sigma}_{\mathbf{X}}\boldsymbol{\Phi})^{-1}\boldsymbol{\Phi}^T) \otimes (\boldsymbol{\Gamma}\boldsymbol{\Gamma}^T\boldsymbol{\Sigma}\boldsymbol{\Gamma}\boldsymbol{\Gamma}^T) \\ &= f_q^{-1}(\boldsymbol{\Phi}\boldsymbol{\Delta}^{-1}\boldsymbol{\Phi}^T) \otimes (\boldsymbol{\Gamma}\boldsymbol{\Omega}\boldsymbol{\Gamma}^T).\end{aligned}$$

Substituting the decompositions $\boldsymbol{\Sigma}_{\mathbf{X}}^{-1} = \boldsymbol{\Phi}\boldsymbol{\Delta}^{-1}\boldsymbol{\Phi}^T + \boldsymbol{\Phi}_0\boldsymbol{\Delta}_0^{-1}\boldsymbol{\Phi}_0^T$ and $\boldsymbol{\Sigma} = \boldsymbol{\Gamma}\boldsymbol{\Omega}\boldsymbol{\Gamma}^T + \boldsymbol{\Gamma}_0\boldsymbol{\Omega}_0\boldsymbol{\Gamma}_0^T$ into $\operatorname{var}(\operatorname{vec}(\mathbf{B})) = f_p^{-1}\boldsymbol{\Sigma}_{\mathbf{X}}^{-1} \otimes \boldsymbol{\Sigma}$ and after a little algebra, we obtain the desired result:

$$\begin{aligned}\operatorname{var}(\operatorname{vec}(\mathbf{B})) = &f_q f_p^{-1}\operatorname{var}(\operatorname{vec}(\widehat{\boldsymbol{\beta}}_{\Gamma,\Phi}^T)) + f_p^{-1}\boldsymbol{\Phi}\boldsymbol{\Delta}^{-1}\boldsymbol{\Phi}^T \otimes \boldsymbol{\Gamma}_0\boldsymbol{\Omega}_0\boldsymbol{\Gamma}_0^T \\ &+ f_p^{-1}\boldsymbol{\Phi}_0\boldsymbol{\Delta}_0^{-1}\boldsymbol{\Phi}_0^T \otimes \boldsymbol{\Gamma}\boldsymbol{\Omega}\boldsymbol{\Gamma}^T + f_p^{-1}\boldsymbol{\Phi}_0\boldsymbol{\Delta}_0^{-1}\boldsymbol{\Phi}_0^T \otimes \boldsymbol{\Gamma}_0\boldsymbol{\Omega}_0\boldsymbol{\Gamma}_0^T.\end{aligned}$$

□

This proposition shows that the variation in $\widehat{\boldsymbol{\beta}}_{\Gamma,\Phi}$ can be seen in two parts: the first part is the variation in $\mathbf{B}$ times a constant $f_p f_q^{-1} \le 1$, and the second consists of terms that reduce this value depending on the variances associated with the immaterial information $\boldsymbol{\Phi}_0^T\mathbf{X}$ and $\boldsymbol{\Gamma}_0^T\mathbf{Y}$.

If $\boldsymbol{\Gamma} = \mathbf{I}_r$ then $\boldsymbol{\Omega} = \boldsymbol{\Sigma}$, and we recover Proposition 4.1:

$$\operatorname{var}(\operatorname{vec}(\widehat{\boldsymbol{\beta}}_{\Phi})) = f_p f_q^{-1}\operatorname{var}(\operatorname{vec}(\mathbf{B})) - f_q^{-1}\boldsymbol{\Phi}_0\boldsymbol{\Delta}_0^{-1}\boldsymbol{\Phi}_0^T \otimes \boldsymbol{\Sigma}. \tag{4.20}$$

When p is close to n, and the $\mathbf{X}$-envelope dimension q is small, the constant $f_p f_q^{-1}$ could be small and the gain by $\widehat{\boldsymbol{\beta}}_{\Gamma,\Phi}$ over $\mathbf{B}$ could be substantial. If there is substantial collinearity in the predictors, so $\boldsymbol{\Sigma}_{\mathbf{X}}$ has some small eigenvalues, and if the corresponding eigenvectors of $\boldsymbol{\Sigma}_{\mathbf{X}}$ fall in $\mathcal{E}_{\boldsymbol{\Sigma}_{\mathbf{X}}}^{\perp}(\mathcal{B}')$, then the variance of $\widehat{\boldsymbol{\beta}}_{\Phi}$ could be reduced considerably since $\boldsymbol{\Delta}_0^{-1}$ will be large. It is widely known that collinearity in $\mathbf{X}$ can increase the variance of $\mathbf{B}$. However, when the eigenvectors of $\boldsymbol{\Sigma}_{\mathbf{X}}$ corresponding to these small eigenvalues lie in $\mathcal{E}_{\boldsymbol{\Sigma}_{\mathbf{X}}}^{\perp}(\mathcal{B}')$, the variance of $\widehat{\boldsymbol{\beta}}_{\Gamma,\Phi}$ is not affected by collinearity.

Similarly, if $\mathbf{\Phi} = \mathbf{I}_p$, then $\mathbf{\Delta} = \mathbf{\Sigma}_{\mathbf{X}}$, and we get the following expression for $\mathbf{Y}$ reduction:

$$\text{var}(\text{vec}(\widehat{\boldsymbol{\beta}}_{\Gamma})) = f_p f_q^{-1} \text{var}(\text{vec}(\mathbf{B})) - f_q^{-1} \mathbf{\Sigma}_{\mathbf{X}}^{-1} \otimes \mathbf{\Gamma}_0 \mathbf{\Omega}_0 \mathbf{\Gamma}_0^T. \tag{4.21}$$

If the eigenvectors with larger eigenvalues of $\mathbf{\Sigma}$ lie in $\mathcal{E}_{\mathbf{\Sigma}}^{\perp}(\mathcal{B})$, then the variance of $\widehat{\boldsymbol{\beta}}_{\Gamma}$ may be reduced considerably since then $\mathbf{\Omega}_0$ will be large.

More importantly, the last term of the expansion in Proposition 4.7 represents a synergy between the $\mathbf{X}$ and $\mathbf{Y}$ reductions that is not present in the individual reductions. If the eigenvectors of $\mathbf{\Sigma}$ with large eigenvalues lie in $\mathcal{E}_{\mathbf{\Sigma}}(\mathcal{B})$ or if the eigenvectors of $\mathbf{\Sigma}_{\mathbf{X}}$ with small eigenvalues lie in $\mathcal{E}_{\mathbf{\Sigma}_{\mathbf{X}}}(\mathcal{B}')$, then the variance reductions in either (4.21) or (4.20) could be insignificant. However, the synergy in simultaneous $\mathbf{X}$ and $\mathbf{Y}$ reductions may still reduce the variance substantially because one of the factors in the Kronecker product $\mathbf{\Phi}_0 \mathbf{\Delta}_0^{-1} \mathbf{\Phi}_0^T \otimes \mathbf{\Gamma}_0 \mathbf{\Omega}_0 \mathbf{\Gamma}_0^T$ could still be large.

4.5.3 Estimation

Proceeding as in Section 4.3.1 recall that $\mathbf{C} = (\mathbf{X}^T, \mathbf{Y}^T)^T \in \mathbb{R}^{r+p}$ denotes the concatenation of $\mathbf{X}$ and $\mathbf{Y}$ into a single random vector with mean $\boldsymbol{\mu}_{\mathbf{C}}$ and covariance matrix $\mathbf{\Sigma}_{\mathbf{C}}$, and $\mathbf{S}_{\mathbf{C}}$ denotes the sample covariance matrix of $\mathbf{C}$. We estimate the parameters in the simultaneous envelope model (4.18) by minimizing the objective function

$$F(\mathbf{\Sigma}_{\mathbf{C}}) = \log|\mathbf{\Sigma}_{\mathbf{C}}| + \text{tr}(\mathbf{S}_{\mathbf{C}} \mathbf{\Sigma}_{\mathbf{C}}^{-1}). \tag{4.22}$$

Assuming that $\mathbf{C}$ is multivariate normal, this objective function arises from the negative log likelihood partially maximized over $\boldsymbol{\mu}_{\mathbf{C}}$. It is used here as it was in Section 4.3.1, as a multipurpose objective function that gives the maximum likelihood estimators when $\mathbf{C}$ is normal and $\sqrt{n}$-consistent estimators when $\mathbf{C}$ has finite fourth moments.

The first step in making (4.22) operational is to write $\mathbf{\Sigma}_{\mathbf{C}}$ explicitly as a function of the parameters in (4.18):

$$\begin{aligned}\mathbf{\Sigma}_{\mathbf{C}} &= \begin{pmatrix} \mathbf{\Sigma}_{\mathbf{X}} & \mathbf{\Sigma}_{\mathbf{X},\mathbf{Y}} \\ \mathbf{\Sigma}_{\mathbf{Y},\mathbf{X}} & \mathbf{\Sigma}_{\mathbf{Y}} \end{pmatrix} \\ &= \begin{pmatrix} \mathbf{\Phi}\mathbf{\Delta}\mathbf{\Phi}^T + \mathbf{\Phi}_0\mathbf{\Delta}_0\mathbf{\Phi}_0^T & \mathbf{\Phi}\mathbf{\Delta}\boldsymbol{\eta}^T\mathbf{\Gamma}^T \\ \mathbf{\Gamma}\boldsymbol{\eta}\mathbf{\Delta}\mathbf{\Phi}^T & \mathbf{\Gamma}(\mathbf{\Omega} + \boldsymbol{\eta}\mathbf{\Delta}\boldsymbol{\eta}^T)\mathbf{\Gamma}^T + \mathbf{\Gamma}_0\mathbf{\Omega}_0\mathbf{\Gamma}_0^T \end{pmatrix},\end{aligned}$$

where the form for $\mathbf{\Sigma}_{\mathbf{Y}}$ follows by substituting for $\mathbf{\Sigma}_{\mathbf{X}}$, $\mathbf{\Sigma}$, and $\mathbf{\Sigma}_{\mathbf{Y},\mathbf{X}}$ the forms from (4.18) and simplifying. It follows from this representation that

$$\begin{aligned}\mathbf{\Sigma}_{\mathbf{C}} &= \mathbf{P}_{\mathbf{\Phi}\oplus\mathbf{\Gamma}}\mathbf{\Sigma}_{\mathbf{C}}\mathbf{P}_{\mathbf{\Phi}\oplus\mathbf{\Gamma}} + \mathbf{Q}_{\mathbf{\Phi}\oplus\mathbf{\Gamma}}\mathbf{\Sigma}_{\mathbf{C}}\mathbf{Q}_{\mathbf{\Phi}\oplus\mathbf{\Gamma}} \\ &= \mathbf{P}_{\mathbf{\Phi}\oplus\mathbf{\Gamma}}\mathbf{\Sigma}_{\mathbf{C}}\mathbf{P}_{\mathbf{\Phi}\oplus\mathbf{\Gamma}} + \mathbf{Q}_{\mathbf{\Phi}\oplus\mathbf{\Gamma}}\mathbf{\Sigma}_{\mathbf{D}}\mathbf{Q}_{\mathbf{\Phi}\oplus\mathbf{\Gamma}},\end{aligned} \tag{4.23}$$

where $\mathbf{\Sigma_D} = \mathbf{\Sigma_X} \oplus \mathbf{\Sigma_Y}$ and $\mathbf{P}_{\mathbf{\Phi}\oplus\mathbf{\Gamma}} = \mathbf{P_\Phi} \oplus \mathbf{P_\Gamma}$ are the projection onto the direct sum of the predictor and response envelopes, $\mathcal{E}_{\mathbf{\Sigma_X}}(\mathcal{B}') \oplus \mathcal{E}_{\mathbf{\Sigma}}(\mathcal{B})$. It follows from (A.8) and Lemma A.6 that

$$\mathcal{E}_{\mathbf{\Sigma_X}}(\mathcal{B}') \oplus \mathcal{E}_{\mathbf{\Sigma}}(\mathcal{B}) = \mathcal{E}_{\mathbf{\Sigma_X}}(\mathcal{B}') \oplus \mathcal{E}_{\mathbf{\Sigma_Y}}(\mathcal{B}) = \mathcal{E}_{\mathbf{\Sigma_X}\oplus\mathbf{\Sigma_Y}}(\mathcal{B}' \oplus \mathcal{B}). \tag{4.24}$$

We refer to $\mathcal{E}_{\mathbf{\Sigma_X}\oplus\mathbf{\Sigma_Y}}(\mathcal{B}' \oplus \mathcal{B})$ as the simultaneous envelope for $\boldsymbol{\beta}$, which is the envelope estimated by using (4.22).

The objective function obtained after substituting the coordinate form for $\mathbf{\Sigma_C}$ into (4.22) can be minimized explicitly over $\mathbf{\Omega}$, $\mathbf{\Omega}_0$, $\mathbf{\Delta}$, $\mathbf{\Delta}_0$, and $\boldsymbol{\eta}$ with $\mathbf{\Phi}$ and $\mathbf{\Gamma}$ held fixed. The details of the minimization are omitted since the steps are similar to the minimizations described in other chapters. The resulting partially maximized function can be expressed as

$$F(\mathbf{\Phi} \otimes \mathbf{\Gamma}) = \log|(\mathbf{\Phi}^T \oplus \mathbf{\Gamma}^T)\mathbf{S_C}(\mathbf{\Phi} \oplus \mathbf{\Gamma})| + \log|(\mathbf{\Phi}^T \oplus \mathbf{\Gamma}^T)\mathbf{S_D}^{-1}(\mathbf{\Phi} \oplus \mathbf{\Gamma})|, \tag{4.25}$$

where $\mathbf{S_D} = \mathbf{S_X} \oplus \mathbf{S_Y}$ is the sample version of $\mathbf{\Sigma_D}$. This objective function is different from those encountered previously since it requires minimization over a direct sum of subspaces. However, alternating between minimization over $\mathbf{\Phi}$ and $\mathbf{\Gamma}$ yields an algorithm that can rely on previous objective functions. Minimizing (4.22) over $\mathbf{\Gamma}$ with $\mathbf{\Phi}$ fixed reduces the problem to minimizing

$$F(\mathbf{\Gamma} \mid \mathbf{\Phi}) = \log|\mathbf{\Gamma}^T\mathbf{S}_{\mathbf{Y}|\mathbf{\Phi}^T\mathbf{X}}\mathbf{\Gamma}| + \log|\mathbf{\Gamma}^T\mathbf{S_Y}^{-1}\mathbf{\Gamma}|. \tag{4.26}$$

This is the objective function for response reduction in the multivariate linear regression of $\mathbf{Y}$ on $\mathbf{\Phi}^T\mathbf{X}$ (cf. (1.27)). Similarly, minimizing (4.22) over $\mathbf{\Phi}$ with $\mathbf{\Gamma}$ fixed, reduces the problem to minimizing

$$F(\mathbf{\Phi} \mid \mathbf{\Gamma}) = \log|\mathbf{\Phi}^T\mathbf{S}_{\mathbf{X}|\mathbf{\Gamma}^T\mathbf{Y}}\mathbf{\Phi}| + \log|\mathbf{\Phi}^T\mathbf{S_X}^{-1}\mathbf{\Phi}|, \tag{4.27}$$

which is the objective function for predictor reduction in the multivariate linear regression of $\mathbf{\Gamma}^T\mathbf{Y}$ on $\mathbf{X}$ (cf. (4.14)).

Having obtained $\widehat{\mathbf{\Gamma}}$ and $\widehat{\mathbf{\Phi}}$ that minimize (4.22), the estimates of the remaining parameters are

$$\begin{aligned}
\widehat{\mathbf{\Delta}} &= \widehat{\mathbf{\Phi}}^T\mathbf{S_X}\widehat{\mathbf{\Phi}} \\
\widehat{\mathbf{\Delta}}_0 &= \widehat{\mathbf{\Phi}}_0^T\mathbf{S_X}\widehat{\mathbf{\Phi}}_0 \\
\widehat{\boldsymbol{\eta}} &= (\widehat{\mathbf{\Gamma}}^T\mathbf{S_{Y,X}}\widehat{\mathbf{\Phi}})(\widehat{\mathbf{\Phi}}^T\mathbf{S_X}\widehat{\mathbf{\Phi}})^{-1} \\
\widehat{\mathbf{\Omega}}_0 &= \widehat{\mathbf{\Gamma}}^T\mathbf{S_Y}\widehat{\mathbf{\Gamma}} \\
\widehat{\mathbf{\Omega}} &= \widehat{\mathbf{\Gamma}}^T(\mathbf{S}_Y - \mathbf{S_{Y,X}}\widehat{\mathbf{\Phi}}(\widehat{\mathbf{\Phi}}^T\mathbf{S_X}\widehat{\mathbf{\Phi}})^{-1}\widehat{\mathbf{\Phi}}^T\mathbf{S_{X,Y}})\widehat{\mathbf{\Gamma}} \\
&= \widehat{\mathbf{\Gamma}}^T\mathbf{S}_{\mathbf{Y}|\widehat{\mathbf{\Phi}}^T\mathbf{X}} \\
\widehat{\boldsymbol{\beta}} &= \widehat{\mathbf{\Gamma}}\widehat{\boldsymbol{\eta}}\widehat{\mathbf{\Phi}}^T \\
&= \mathbf{P}_{\widehat{\mathbf{\Gamma}}}\mathbf{B}\mathbf{P}^T_{\widehat{\mathbf{\Phi}}(\mathbf{S_X})}.
\end{aligned} \tag{4.28}$$

From (4.28) we see that the estimator of $\boldsymbol{\beta}$ is a direct combination of the individual estimators for $\mathbf{Y}$ reduction and $\mathbf{X}$ reduction (cf. (1.26) and (4.16)).

The asymptotic variance of $\widehat{\boldsymbol{\beta}}$ behaves in a manner similar to that for the estimators of $\boldsymbol{\beta}$ under $\mathbf{X}$ and $\mathbf{Y}$ reduction alone. Under normality of $\mathbf{C}$ and assuming that u and q are known, $\text{avar}(\sqrt{n}\widehat{\boldsymbol{\beta}}) \leq \text{avar}(\sqrt{n}\mathbf{B})$, so the asymptotic variance of $\boldsymbol{\beta}$ never exceeds that of the usual maximum likelihood estimator. Similarly, $\text{avar}(\sqrt{n}\widehat{\boldsymbol{\beta}})$ never exceeds that for the corresponding estimator under $\mathbf{X}$ or $\mathbf{Y}$ reduction alone. If $\mathbf{C}$ is not normal but still has finite fourth moments, then $\widehat{\boldsymbol{\beta}}$ is an asymptotically normal, $\sqrt{n}$-consistent estimator of $\boldsymbol{\beta}$.

Estimation of the envelope dimensions u and q is more difficult in this application than in $\mathbf{X}$ or $\mathbf{Y}$ reduction alone, because there are two dimensions to select and we must have $d < \min(r, p)$ for the method to be effective. The method described in Section 1.10.2 can help determine if the condition $d < \min(r, p)$ is met and may also provide a lower bound on u and q since $\min(u, q) \geq d$. Likelihood ratio testing, an information criterion or cross-validation, can be used to select the dimensions in a manner analogous to that described for $\mathbf{X}$ and $\mathbf{Y}$ reduction alone (Cook and Zhang 2015b).

5

Enveloping Multivariate Means

In this chapter, we discuss the possibility of using envelopes for estimation of a multivariate mean and for comparison of several multivariate means from heteroscedastic populations. For instance, the response vector $\mathbf{Y}$ might consist of longitudinal measurements at r time points that are the same across units. While longitudinal data are often modeled as a function of time, it could be useful in the initial stage of an analysis to view an efficient estimate of the mean as an aid to subsequent modeling as a function of time. The core ideas are the same as those introduced in Chapter 1, but the details and applications are sufficiently different to merit a separate treatment.

5.1 Enveloping a Single Mean

Let $\mathbf{Y}_1, \mathbf{Y}_2, \dots, \mathbf{Y}_n$ be independent copies of the normal random vector $\mathbf{Y} \in \mathbb{R}^r$ with mean $\boldsymbol{\mu}$ and covariance matrix $\boldsymbol{\Sigma} > 0$. The usual estimator of $\boldsymbol{\mu}$ is simply the sample mean $\bar{\mathbf{Y}} = n^{-1} \sum_{i=1}^{n} \mathbf{Y}_i$ with variance $\mathrm{var}(\bar{\mathbf{Y}}) = n^{-1}\boldsymbol{\Sigma}$. In this section, we discuss the envelope estimator of $\boldsymbol{\mu}$, which has the potential that have substantially smaller variance than $\bar{\mathbf{Y}}$.

5.1.1 Envelope Structure

To describe the rationale behind the envelope estimator of $\boldsymbol{\mu}$, let $\mathcal{S} \subseteq \mathbb{R}^r$ denote the smallest subspace with the properties

$$\text{(a)}\ \ \boldsymbol{\mu} \in \mathcal{S} \ \text{ and }\ \text{(b)}\ \ \mathbf{P}_{\mathcal{S}}\mathbf{Y} \perp\!\!\!\perp \mathbf{Q}_{\mathcal{S}}\mathbf{Y}. \tag{5.1}$$

Condition (a) implies that $\mathbf{P}_{\mathcal{S}}\mathbf{Y} \sim N(\boldsymbol{\mu}, \mathbf{P}_{\mathcal{S}}\boldsymbol{\Sigma}\mathbf{P}_{\mathcal{S}})$ and that $\mathbf{Q}_{\mathcal{S}}\mathbf{Y} \sim N(0, \mathbf{Q}_{\mathcal{S}}\boldsymbol{\Sigma}\mathbf{Q}_{\mathcal{S}})$. Marginal information on $\boldsymbol{\mu}$ is available from $\mathbf{P}_{\mathcal{S}}\mathbf{Y}$, while $\mathbf{Q}_{\mathcal{S}}\mathbf{Y}$ supplies no marginal information about $\boldsymbol{\mu}$. However, under condition (a) alone, $\mathbf{Q}_{\mathcal{S}}\mathbf{Y}$ could still carry information about $\boldsymbol{\mu}$ through an association with $\mathbf{P}_{\mathcal{S}}\mathbf{Y}$. This possibility is ruled out by condition (b). Since $\mathbf{Y}$ is multivariate normal, condition (b) holds if and only if $\mathbf{P}_{\mathcal{S}}\boldsymbol{\Sigma}\mathbf{Q}_{\mathcal{S}} = 0$, which is equivalent to requiring

An Introduction to Envelopes: Dimension Reduction for Efficient Estimation in Multivariate Statistics,
First Edition. R. Dennis Cook.

that $\mathcal{S}$ be a reducing subspace of $\boldsymbol{\Sigma}$. In short, we are led to the $\boldsymbol{\Sigma}$-envelope of $\mathcal{M} := \text{span}(\boldsymbol{\mu})$, $\mathcal{E}_{\boldsymbol{\Sigma}}(\mathcal{M})$. Then $\mathbf{P}_{\mathcal{E}}\mathbf{Y}$ contains all of the material information on $\boldsymbol{\mu}$ with material variation $\mathbf{P}_{\mathcal{E}}\boldsymbol{\Sigma}\mathbf{P}_{\mathcal{E}}$, and $\mathbf{Q}_{\mathcal{E}}\mathbf{Y}$ contains the immaterial information with variation $\mathbf{Q}_{\mathcal{E}}\boldsymbol{\Sigma}\mathbf{Q}_{\mathcal{E}}$.

As in the regression setting, we can gain intuition about the potential gain from an envelope analysis by supposing that the envelope $\mathcal{E}_{\boldsymbol{\Sigma}}(\mathcal{M})$ is known. Then the maximum likelihood estimator of $\boldsymbol{\mu}$ is just $\widehat{\boldsymbol{\mu}} = \mathbf{P}_{\mathcal{E}}\bar{\mathbf{Y}}$, which has variance $n^{-1}\mathbf{P}_{\mathcal{E}}\boldsymbol{\Sigma}\mathbf{P}_{\mathcal{E}}$. Since $\boldsymbol{\Sigma} = \mathbf{P}_{\mathcal{E}}\boldsymbol{\Sigma}\mathbf{P}_{\mathcal{E}} + \mathbf{Q}_{\mathcal{E}}\boldsymbol{\Sigma}\mathbf{Q}_{\mathcal{E}}$, we have straightforwardly

$$\text{var}(\bar{\mathbf{Y}}) - \text{var}(\widehat{\boldsymbol{\mu}}) = n^{-1}\mathbf{Q}_{\mathcal{E}}\boldsymbol{\Sigma}\mathbf{Q}_{\mathcal{E}},$$

so the difference between the variance of the standard and envelope estimators of $\boldsymbol{\mu}$ depends on the sample size and the immaterial variation. If $\mathbf{Q}_{\mathcal{E}}\boldsymbol{\Sigma}\mathbf{Q}_{\mathcal{E}}$ has eigenvalues that are large relative to the eigenvalues of the material variation $\mathbf{P}_{\mathcal{S}}\boldsymbol{\Sigma}\mathbf{P}_{\mathcal{S}}$, then the envelope estimator will have substantially smaller variation than the standard estimator $\bar{\mathbf{Y}}$. On the other hand, if $\mathbb{R}^r$ is the smallest reducing subspace of $\boldsymbol{\Sigma}$ that contains $\boldsymbol{\mu}$, $\mathcal{E}_{\boldsymbol{\Sigma}}(\mathcal{M}) = \mathbb{R}^r$, then $\mathbf{Q}_{\mathcal{E}} = 0$, and the envelope estimator reduces to $\bar{\mathbf{Y}}$.

The estimation process with a known envelope is illustrated in Figure 5.1 for two responses. The ellipses on the plot are centered at the origin and represent the contours of $\boldsymbol{\Sigma}$, and the ellipse axes represent the two eigenspaces of $\boldsymbol{\Sigma}$. The population mean lies in the second eigenspace that equals the envelope $\mathcal{E}_{\boldsymbol{\Sigma}}(\mathcal{M})$ in this illustration. The envelope estimator of $\boldsymbol{\mu}$ is obtained by projecting $\bar{\mathbf{Y}}$ onto the envelope, as represented by the dashed line. The gain in precision comes about because the material variation in the envelope is substantially smaller than the immaterial variation along the orthogonal complement of the envelope, which corresponds to the first eigenspace.

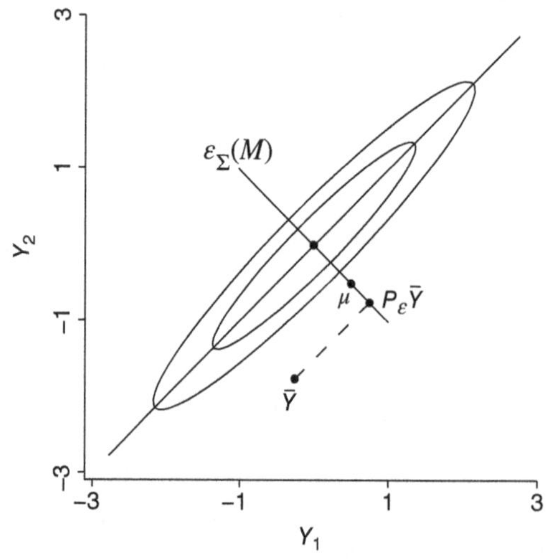

Figure 5.1 Graphical illustration of enveloping a population mean.

If $\boldsymbol{\mu}$ fell in the first eigenspace of $\boldsymbol{\Sigma}$, then the envelope would become the major axis of the ellipses in Figure 5.1. The envelope would still provide estimative gain in this case, although it would not be as great because the nonzero eigenvalue of $\mathbf{Q}_{\mathcal{E}}\boldsymbol{\Sigma}\mathbf{Q}_{\mathcal{E}}$ would now be the smaller eigenvalue of $\boldsymbol{\Sigma}$.

If $\boldsymbol{\mu}$ fell in neither eigenspace of Figure 5.1, then the envelope would be $\mathcal{E}_{\boldsymbol{\Sigma}}(\mathcal{M}) = \mathbb{R}^2$ and, strictly speaking, there would be no immaterial information. However, gains in mean squared error might still be realized. Still in the context of Figure 5.1, temporarily think of $\mathcal{E}_{\boldsymbol{\Sigma}}(\mathcal{M})$ as the eigenspace of $\boldsymbol{\Sigma}$ that is closest to $\boldsymbol{\mu}$, which does not necessarily fall into the eigenspace:

$$\begin{aligned} \mathrm{E}(\mathbf{P}_{\mathcal{E}}\bar{\mathbf{Y}} - \boldsymbol{\mu})(\mathbf{P}_{\mathcal{E}}\bar{\mathbf{Y}} - \boldsymbol{\mu})^T &= \mathrm{E}\{\mathbf{P}_{\mathcal{E}}(\bar{\mathbf{Y}} - \boldsymbol{\mu}) - \mathbf{Q}_{\mathcal{E}}\boldsymbol{\mu}\}\{\mathbf{P}_{\mathcal{E}}(\bar{\mathbf{Y}} - \boldsymbol{\mu}) - \mathbf{Q}_{\mathcal{E}}\boldsymbol{\mu}\}^T \\ &= n^{-1}\mathbf{P}_{\mathcal{E}}\boldsymbol{\Sigma}\mathbf{P}_{\mathcal{E}} + \mathbf{Q}_{\mathcal{E}}\boldsymbol{\mu}\boldsymbol{\mu}^T\mathbf{Q}_{\mathcal{E}}. \end{aligned}$$

Comparing this mean squared error with $\mathrm{var}(\bar{\mathbf{Y}}) = n^{-1}\mathbf{P}_{\mathcal{E}}\boldsymbol{\Sigma}\mathbf{P}_{\mathcal{E}} + n^{-1}\mathbf{Q}_{\mathcal{E}}\boldsymbol{\Sigma}\mathbf{Q}_{\mathcal{E}}$, we see that there could still be substantial gains provided $\boldsymbol{\mu}$ is not too far from the closest eigenspace.

While the low-dimensional representation of Figure 5.1 might leave the impression that the applicability of envelopes is limited because $\boldsymbol{\mu}$ might rarely fall in an eigenspace of $\boldsymbol{\Sigma}$, conditions (5.1) describe a plausible statistical context in which this can happen. Envelopes are generally more serviceable when $r > 2$ because empirically the propensity for $\mathbf{Y}$ to contain immaterial information tends to increase with r. To illustrate one way in which this might arise, suppose that $\mathbf{Y}$ is a linear combination of $q < r$ latent variables $\mathbf{Z}$ plus an isotropic error, $\mathbf{Y} = \mathbf{A}\mathbf{Z} + \boldsymbol{\delta}$, where $\mathbf{Z} \sim N(\boldsymbol{\mu}_{\mathbf{Z}}, \boldsymbol{\Sigma}_{\mathbf{Z}})$, $\boldsymbol{\delta} \sim N(0, \sigma^2\mathbf{I}_r)$, and $\mathbf{A} \in \mathbb{R}^{r\times q}$ has full column rank. Without loss of generality, we take $\mathbf{A}$ to be a semi-orthogonal matrix. Then $\boldsymbol{\mu} = \mathbf{A}\boldsymbol{\mu}_{\mathbf{Z}} \in \mathrm{span}(\mathbf{A})$ and

$$\begin{aligned} \boldsymbol{\Sigma} &= \mathbf{A}\boldsymbol{\Sigma}_{\mathbf{Z}}\mathbf{A}^T + \sigma^2\mathbf{I}_q \\ &= \mathbf{A}(\boldsymbol{\Sigma}_{\mathbf{Z}} + \sigma^2\mathbf{I}_q)\mathbf{A}^T + \sigma^2\mathbf{A}_0\mathbf{A}_0^T, \end{aligned}$$

where $(\mathbf{A}, \mathbf{A}_0)$ is an orthogonal matrix. Clearly, $\mathrm{span}(\mathbf{A})$ is a reducing subspace of $\boldsymbol{\Sigma}$ that contains $\boldsymbol{\mu}$, although it may not be the smallest. In this illustration, the dimension of $\mathcal{E}_{\boldsymbol{\Sigma}}(\mathcal{M})$ is at most $q < r$.

5.1.2 Envelope Model

The first step in estimating $\boldsymbol{\mu}$ under the envelope model is to formally incorporate a basis for the envelope $\mathcal{E}_{\boldsymbol{\Sigma}}(\mathcal{M})$. Let $u = \dim(\mathcal{E}_{\boldsymbol{\Sigma}}(\mathcal{M}))$, let $\boldsymbol{\Gamma} \in \mathbb{R}^{r\times u}$ be a semi-orthogonal basis matrix for $\mathcal{E}_{\boldsymbol{\Sigma}}(\mathcal{M})$, and let $(\boldsymbol{\Gamma}, \boldsymbol{\Gamma}_0)$ be an orthogonal matrix so that $\boldsymbol{\Gamma}_0$ is a basis matrix for the orthogonal complement of $\mathcal{E}_{\boldsymbol{\Sigma}}(\mathcal{M})$. Since $\boldsymbol{\mu} \in \mathcal{E}_{\boldsymbol{\Sigma}}(\mathcal{M})$ by construction, we can write $\boldsymbol{\mu} = \boldsymbol{\Gamma}\boldsymbol{\eta}$ for some vector $\boldsymbol{\eta} \in \mathbb{R}^{u\times 1}$ that contains the coordinates of $\boldsymbol{\mu}$ relative to $\boldsymbol{\Gamma}$. Let $\boldsymbol{\Omega} = \boldsymbol{\Gamma}^T\boldsymbol{\Sigma}\boldsymbol{\Gamma} > 0$ and $\boldsymbol{\Omega}_0 = \boldsymbol{\Gamma}_0^T\boldsymbol{\Sigma}\boldsymbol{\Gamma}_0 > 0$. Then the envelope model can now be summarized as

$$\mathbf{Y} \sim N(\boldsymbol{\Gamma}\boldsymbol{\eta}, \boldsymbol{\Gamma}\boldsymbol{\Omega}\boldsymbol{\Gamma}^T + \boldsymbol{\Gamma}_0\boldsymbol{\Omega}_0\boldsymbol{\Gamma}_0^T). \tag{5.2}$$

As in previous envelope models, the basis $\mathbf{\Gamma}$ of $\mathcal{E}_{\mathbf{\Sigma}}(\mathcal{M})$ is not identifiable in this model since mapping $\mathbf{\Gamma} \mapsto \mathbf{\Gamma O}$ with an orthogonal matrix $\mathbf{O}$ leads to an equivalent model, but the envelope itself is identifiable. The number of real parameters in model (5.2) is

$$\begin{aligned} N_u &= u(r-u) + u + u(u+1)/2 + (r-u)(r-u+1)/2 \\ &= r(r+1)/2 + u. \end{aligned} \tag{5.3}$$

The first count $u(r-u)$ is the number of reals needed to determine $\mathcal{E}_{\mathbf{\Sigma}}(\mathcal{M})$, the second u is the dimension of $\boldsymbol{\eta} \in \mathbb{R}^u$, and the final two counts are for the positive definite matrices $\mathbf{\Omega} \in \mathbb{S}^{u\times u}$ and $\mathbf{\Omega}_0 \in \mathbb{S}^{(r-u)\times(r-u)}$.

5.1.3 Estimation

In this section, we consider the maximum likelihood estimator of $\boldsymbol{\mu}$, followed by its asymptotic variance and methods for selecting the envelope dimension.

5.1.3.1 Maximum Likelihood Estimation

Using the parameterization given in (5.2) and assuming temporarily that u is known, the log-likelihood L_u for the multivariate normal can be represented, apart from the constant $-(nr/2)\log 2\pi$, as

$$\begin{aligned} L_u(\mathbf{\Gamma}, \boldsymbol{\eta}, \mathbf{\Omega}, \mathbf{\Omega}_0) &= -(n/2)\log|\mathbf{\Sigma}| - (1/2)\sum_{i=1}^{n}(\mathbf{Y}_i - \boldsymbol{\mu})^T\mathbf{\Sigma}^{-1}(\mathbf{Y}_i - \boldsymbol{\mu}) \\ &= -(n/2)\log|\mathbf{\Omega}| - (n/2)\log|\mathbf{\Omega}_0| \\ &\quad -(1/2)\sum_{i=1}^{n}(\mathbf{Y}_i - \mathbf{\Gamma}\boldsymbol{\eta})^T(\mathbf{\Gamma\Omega}^{-1}\mathbf{\Gamma}^T + \mathbf{\Gamma}_0\mathbf{\Omega}_0^{-1}\mathbf{\Gamma}_0^T)(\mathbf{Y}_i - \mathbf{\Gamma}\boldsymbol{\eta}), \end{aligned}$$

where the second inequality comes from the relationships given in Corollary A.1. Since $\mathbf{\Gamma}^T\mathbf{Y} \perp\!\!\!\perp \mathbf{\Gamma}_0^T\mathbf{Y}$, the likelihood will factor accordingly. This can be done algebraically by replacing $\mathbf{Y}$ with $\mathbf{P}_{\mathcal{E}}\mathbf{Y} + \mathbf{Q}_{\mathcal{E}}\mathbf{Y}$ and simplifying L_u:

$$\begin{aligned} L_u(\mathbf{\Gamma}, \boldsymbol{\eta}, \mathbf{\Omega}, \mathbf{\Omega}_0) &= -(n/2)\log|\mathbf{\Omega}| - (n/2)\log|\mathbf{\Omega}_0| \\ &\quad -(1/2)\sum_{i=1}^{n}(\mathbf{P}_{\mathcal{E}}\mathbf{Y}_i + \mathbf{Q}_{\mathcal{E}}\mathbf{Y}_i - \mathbf{\Gamma}\boldsymbol{\eta})^T(\mathbf{\Gamma\Omega}^{-1}\mathbf{\Gamma}^T + \mathbf{\Gamma}_0\mathbf{\Omega}_0^{-1}\mathbf{\Gamma}_0^T) \\ &\quad \times(\mathbf{P}_{\mathcal{E}}\mathbf{Y}_i + \mathbf{Q}_{\mathcal{E}}\mathbf{Y}_i - \mathbf{\Gamma}\boldsymbol{\eta}) \\ &= -(n/2)\log|\mathbf{\Omega}| - (1/2)\sum_{i=1}^{n}(\mathbf{\Gamma}^T\mathbf{Y}_i - \boldsymbol{\eta})^T\mathbf{\Omega}^{-1}(\mathbf{\Gamma}^T\mathbf{Y}_i - \boldsymbol{\eta}) \\ &\quad -(n/2)\log|\mathbf{\Omega}_0| - (1/2)\sum_{i=1}^{n}\mathbf{Y}_i^T\mathbf{\Gamma}_0\mathbf{\Omega}_0^{-1}\mathbf{\Gamma}_0^T\mathbf{Y}_i. \end{aligned}$$

Recall that $\mathbf{S}_{\mathbf{Y}}$ denotes the sample covariance matrix of $\mathbf{Y}$, and let $\mathbf{T}_{\mathbf{Y}} = n^{-1}\sum_{i=1}^{n}\mathbf{Y}_i\mathbf{Y}_i^T$ denote the matrix of raw second moments of $\mathbf{Y}$.

Then $L_u(\mathbf{\Gamma}, \boldsymbol{\eta}, \mathbf{\Omega}, \mathbf{\Omega}_0)$ is maximized for fixed $\mathbf{\Gamma}$ at $\boldsymbol{\eta} = \mathbf{\Gamma}^T\bar{\mathbf{Y}}$, $\mathbf{\Omega} = \mathbf{\Gamma}^T\mathbf{S}_\mathbf{Y}\mathbf{\Gamma}$ and $\mathbf{\Omega}_0 = \mathbf{\Gamma}_0^T\mathbf{T}_\mathbf{Y}\mathbf{\Gamma}_0$. Substituting these relationships into the log-likelihood and simplifying leads to the partially maximized log-likelihood

$$\begin{aligned} L_u(\mathbf{\Gamma}) &= -(n/2)\log|\mathbf{\Gamma}^T\mathbf{S}_\mathbf{Y}\mathbf{\Gamma}| - (n/2)\log|\mathbf{\Gamma}_0^T\mathbf{T}_\mathbf{Y}\mathbf{\Gamma}_0| - nr/2 \\ &= -(n/2)\log|\mathbf{\Gamma}^T\mathbf{S}_\mathbf{Y}\mathbf{\Gamma}| - (n/2)\log|\mathbf{\Gamma}^T\mathbf{T}_\mathbf{Y}^{-1}\mathbf{\Gamma}| + c, \end{aligned} \tag{5.4}$$

where $c = -(n/2)\log|\mathbf{T}_\mathbf{Y}| - nr/2$ and the final step comes from Lemma A.13. For a given dimension u, the maximum likelihood estimators can now be summarized as

$$\begin{aligned} \widehat{\mathcal{E}}_\mathbf{\Sigma}(\mathcal{M}) &= \text{span}\{\arg\max L_u(\mathbf{\Gamma})\} \\ \widehat{\boldsymbol{\eta}} &= \widehat{\mathbf{\Gamma}}^T\bar{\mathbf{Y}} \\ \widehat{\mathbf{\Omega}} &= \widehat{\mathbf{\Gamma}}^T\mathbf{S}_\mathbf{Y}\widehat{\mathbf{\Gamma}} \\ \widehat{\mathbf{\Omega}}_0 &= \widehat{\mathbf{\Gamma}}_0^T\mathbf{T}_\mathbf{Y}\widehat{\mathbf{\Gamma}}_0 \\ \widehat{\boldsymbol{\mu}} &= \mathbf{P}_{\widehat{\mathbf{\Gamma}}}\bar{\mathbf{Y}} \\ \widehat{\mathbf{\Sigma}} &= \widehat{\mathbf{\Gamma}}\widehat{\mathbf{\Omega}}\widehat{\mathbf{\Gamma}}^T + \widehat{\mathbf{\Gamma}}_0\widehat{\mathbf{\Omega}}_0\widehat{\mathbf{\Gamma}}_0^T, \end{aligned}$$

where the maximum is over the set of all semi-orthogonal matrices $\mathbf{\Gamma} \in \mathbb{R}^{r\times u}$, and $\widehat{\mathbf{\Gamma}}$ can then be taken as any semi-orthogonal basis matrix for $\widehat{\mathcal{E}}_\mathbf{\Sigma}(\mathcal{M})$.

Characteristics of these estimators are similar to those encountered in the regression case. The partially maximized log-likelihood $L_u(\mathbf{\Gamma})$ has the property that $L_u(\mathbf{\Gamma}) = L_u(\mathbf{\Gamma}\mathbf{O})$ for all orthogonal matrices $\mathbf{O} \in \mathbb{R}^{u\times u}$. Consequently, a value $\widehat{\mathbf{\Gamma}}$ of the semi-orthogonal matrix $\mathbf{\Gamma}$ that maximizes $L_u(\mathbf{\Gamma})$ is not unique, but span($\widehat{\mathbf{\Gamma}}$) is unique. Since $\widehat{\boldsymbol{\mu}} = \mathbf{P}_{\widehat{\mathbf{\Gamma}}}\bar{\mathbf{Y}}$ any value of $\mathbf{\Gamma}$ that maximizes $L_u(\mathbf{\Gamma})$ produces the same estimator of $\boldsymbol{\mu}$. Similarly, any value of $\mathbf{\Gamma}$ that maximizes $L_u(\mathbf{\Gamma})$ also gives the same estimate $\widehat{\mathbf{\Sigma}}$. On the other hand, the estimators $\widehat{\boldsymbol{\eta}}$, $\widehat{\mathbf{\Omega}}$, and $\widehat{\mathbf{\Omega}}_0$ depend on the basis $\widehat{\mathbf{\Gamma}}$ selected. This may be of little consequence, because $\boldsymbol{\eta}$, $\mathbf{\Omega}$, and $\mathbf{\Omega}_0$ will typically be nuisance parameters, of minor interest in an analysis.

5.1.3.2 Asymptotic Variance of $\widehat{\mu}$

From standard likelihood theory, $\sqrt{n}(\widehat{\boldsymbol{\mu}} - \boldsymbol{\mu})$ is asymptotically normal with mean 0 and variance avar($\sqrt{n}\,\widehat{\boldsymbol{\mu}}$), which can be constructed as the inverse of the Fisher information matrix. Following the derivations of Cook et al. (2010),

$$\begin{aligned} \text{avar}(\sqrt{n}\,\widehat{\boldsymbol{\mu}}) &= \mathbf{\Gamma}\mathbf{\Omega}\mathbf{\Gamma}^T + (\boldsymbol{\eta}^T \otimes \mathbf{\Gamma}_0^T)\mathbf{V}^\dagger(\boldsymbol{\eta} \otimes \mathbf{\Gamma}_0) & (5.5) \\ &\le \text{avar}(\sqrt{n}\,\bar{\mathbf{Y}}) = \mathbf{\Sigma}, & (5.6) \end{aligned}$$

where $\mathbf{V} = \boldsymbol{\eta}\boldsymbol{\eta}^T \otimes \mathbf{\Omega}_0^{-1} + \mathbf{\Omega} \otimes \mathbf{\Omega}_0^{-1} + \mathbf{\Omega}^{-1} \otimes \mathbf{\Omega}_0 - 2\mathbf{I}_u \otimes \mathbf{I}_{r-u}$. The interpretation of this result is similar to that in regression. Equation (5.5) gives the algebraic form of the asymptotic variance, while (5.6) says that it can never be greater than the asymptotic variance of the standard estimator $\bar{\mathbf{Y}}$. To see how (5.6) arises from (5.5), the last three addends of $\mathbf{V}$ have the same eigenvectors and thus a typical eigenvalue of their sum is $(\varphi/\varphi_0) + (\varphi_0/\varphi) - 2 \geq 0$, where φ and φ_0 are typical eigenvalues of $\mathbf{\Omega}$ and $\mathbf{\Omega}_0$. Consequently, $\mathbf{V} \geq \boldsymbol{\eta}\boldsymbol{\eta}^T \otimes \mathbf{\Omega}_0^{-1}$ and the conclusion then follows algebraically.

If $\mathbf{\Sigma} = \sigma^2\mathbf{I}_r$ for $\sigma^2 > 0$, then $\mathbf{\Omega} = \sigma^2\mathbf{I}_u$, $\mathbf{\Omega}_0 = \sigma^2\mathbf{I}_{r-u}$, $\mathbf{\Gamma} = \boldsymbol{\mu}/\|\boldsymbol{\mu}\|$ and $\boldsymbol{\eta} = \|\boldsymbol{\mu}\| \in \mathbb{R}^1$. Then $\mathbf{V} = \sigma^{-2}\|\boldsymbol{\mu}\|^2\mathbf{I}_{r-u}$, $\mathbf{V}^{-1} = \sigma^2\|\boldsymbol{\mu}\|^{-2}\mathbf{I}_{r-u}$ and $\text{avar}(\sqrt{n}\,\widehat{\boldsymbol{\mu}}) = \sigma^2\mathbf{I}_r = \text{avar}(\sqrt{n}\,\bar{\mathbf{Y}})$. Consequently, if $\mathbf{\Sigma} = \sigma^2\mathbf{I}$, then the standard estimator is asymptotically equivalent to the envelope estimator, and enveloping offers no gains. This result is similar to that presented in Corollary 1.1.

The first term on the right-hand side of (5.5) corresponds to the asymptotic variance of the envelope estimator when the envelope is known, $\text{avar}(\sqrt{n}\mathbf{P}_{\mathcal{E}}\bar{\mathbf{Y}}) = \mathbf{\Gamma}\mathbf{\Omega}\mathbf{\Gamma}^T$, as discussed previously in Section 5.1.1. Consequently, we can think of the second term as the cost of estimating $\mathcal{E}_{\mathbf{\Sigma}}(\mathcal{M})$. The previous description of $\text{avar}(\sqrt{n}\,\widehat{\boldsymbol{\mu}})$ when $\mathbf{\Sigma} = \sigma^2\mathbf{I}_r$ describes an extreme case with maximal cost, so the asymptotic variance of $\widehat{\boldsymbol{\mu}}$ equals the that of $\bar{\mathbf{Y}}$.

5.1.3.3 Selecting $u = \dim(\mathcal{E}_{\mathbf{\Sigma}}(\mathcal{M}))$

The dimension u of the envelope can again be selected by using an information criterion such as AIC or BIC or likelihood ratio testing. Using an information criterion, the envelope dimension u is selected as

$$\hat{u} = \arg\min_u\{-2\widehat{L}_u + h(n)N_u\},$$

where the minimum is over the set $\{0, 1, \ldots, r\}$, $\widehat{L}_u$ is the value of the fully maximized log-likelihood function, N_u is the number of parameters in the model as given in (5.3), and $h(n) = 2$ for AIC and $h(n) = \log n$ for BIC.

Sequential likelihood ratio for testing an envelope model against the standard model, $u = u_0$ versus the alternative $u = r$, can be carried out by using the log-likelihood ratio statistic $\Lambda(u_0) = 2(\widehat{L}_r - \widehat{L}_{u_0})$, where $\widehat{L}_{u_0} = L_{u_0}(\widehat{\mathbf{\Gamma}})$ is the fully maximized envelope log-likelihood given by (5.4) evaluated at $\mathbf{\Gamma} = \widehat{\mathbf{\Gamma}}$ and $\widehat{L}_r = \widehat{L}_r(\mathbf{I}_r)$ is the maximized log-likelihood under the standard model, $\widehat{L}_r = -(nr/2)\log(2\pi) - nr/2 - (n/2)\log|\mathbf{S}_{\mathbf{Y}}|$, giving

$$\begin{aligned}\Lambda(u_0) &= n\log|\widehat{\mathbf{\Gamma}}^T\mathbf{S}_{\mathbf{Y}}\widehat{\mathbf{\Gamma}}| + n\log|\widehat{\mathbf{\Gamma}}^T\mathbf{T}_{\mathbf{Y}}^{-1}\widehat{\mathbf{\Gamma}}| + n\log|\mathbf{T}_{\mathbf{Y}}| - n\log|\mathbf{S}_{\mathbf{Y}}| \\ &= n\log|\widehat{\mathbf{\Gamma}}_0^T\mathbf{S}_{\mathbf{Y}}^{-1}\widehat{\mathbf{\Gamma}}_0| + n\log|\widehat{\mathbf{\Gamma}}_0^T\mathbf{T}_{\mathbf{Y}}\widehat{\mathbf{\Gamma}}_0|.\end{aligned}$$

Under the null hypothesis this statistic is distributed asymptotically as a chi-squared random variable with $N_r - N_{u_0} = r - u_0$ degrees of freedom.

5.1.4 Minneapolis Schools

The Minneapolis school data[1] consist of observations on various characteristics of $n = 63$ elementary schools in Minneapolis, including the percentage of

1 Available from the web page http://www.stat.umn.edu/RegGraph/ for Cook (1998).

fourth and sixth grade students scoring above and below the national average on standardized reading tests. We use these data to illustrate various aspects of an envelope analysis. The percentages for each grade do not add to 100% because there is also an average performance classification that is not included in the analysis.

5.1.4.1 Two Transformed Responses

We begin by considering $r = 2$ responses, the square root of the percentage of sixth grade students scoring above Y_1 and below Y_2 average, the square root transformation being used to bring the data closer to bivariate normality. Figure 5.2 gives an indication of the potential for an envelope to improve efficiency. The ellipses, which are centered at the origin, are contours of $\mathbf{S}_\mathbf{Y}$, and the lines extending from the origin mark the two eigenspaces of $\mathbf{S}_\mathbf{Y}$. The mean of the data, which are plotted as circles, is marked by an ex in the center of the point cloud. From this representation, we can see that the mean $\bar{\mathbf{Y}}$ falls quite close to the second eigenspace of $\mathbf{S}_\mathbf{Y}$. Consequently, we can expect to find that $u = 1$, that the estimated variance of $\hat{\boldsymbol{\mu}}$ from an envelope analysis is noticeably less than the estimated variance of $\bar{\mathbf{Y}}$, and that $\hat{\boldsymbol{\mu}}$ is close to $\bar{\mathbf{Y}}$.

As expected, AIC and BIC agreed that $u = 1$. The results from an envelope analysis with $u = 1$ are summarized in Table 5.1. The estimates $\bar{\mathbf{Y}}$ and $\hat{\boldsymbol{\mu}}$ are quite close, as anticipated from Figure 5.2. The standard error for $\hat{\boldsymbol{\mu}}$ is $\widehat{\text{avar}}^{1/2}(\sqrt{n}\,\hat{\boldsymbol{\mu}})/\sqrt{n}$, where the asymptotic variance was computed from (5.5) by plugging in the estimates of the unknown parameters. Table 5.1 shows that the envelope analysis reduced the standard error of $\bar{\mathbf{Y}}$ by about 30%. Around

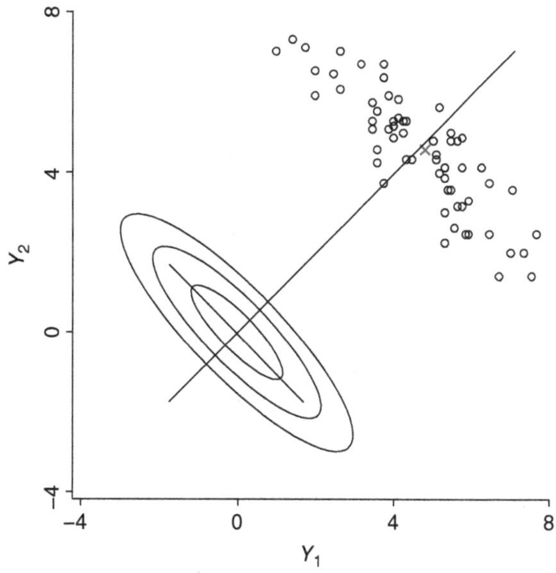

Figure 5.2 Minneapolis schools: Y_1 is the square root of the percentage of sixth graders scoring above average and Y_2 is the square root of the percentage of sixth graders scoring below average. The ellipses are contours of $\mathbf{S}_\mathbf{Y}$ with its eigenvectors shown, and the red ex marks the mean of the data.

Table 5.1 Minneapolis schools: Envelope analysis of the square root of the percent of sixth grade students scoring above Y_1 and below Y_2 average.

Variable	$\bar{\mathrm{Y}}$	se of $\bar{\mathrm{Y}}$	$\hat{\mu}$	se of $\hat{\mu}$	boot se of $\hat{\mu}$	se ratio	$\hat{\Gamma}$
Y_1	4.67	0.190	4.62	0.134	0.135	1.42	0.711
Y_2	4.52	0.189	4.57	0.135	0.138	1.40	0.703

"se" denotes a standard error, "boot se" indicates a bootstrap standard error based on 100 bootstrap samples, and "se ratio" is the ratio of the standard error for $\bar{\mathbf{Y}}$ to the standard error for $\hat{\boldsymbol{\mu}}$.

120 observations would be needed in a standard analysis using $\bar{\mathbf{Y}}$ to achieve the standard error of $\hat{\boldsymbol{\mu}}$ shown in Table 5.1, so the gain from using $\hat{\boldsymbol{\mu}}$ is roughly equivalent to doubling the sample size for $\bar{\mathbf{Y}}$. The bootstrap standard errors shown in Table 5.1 are in good agreement with the standard errors obtained by using the plugin estimate of the asymptotic variance of $\hat{\boldsymbol{\mu}}$.

The gains shown in Table 5.1 are indicated qualitatively by the relative magnitudes of the material variation summarized by $\hat{\boldsymbol{\Omega}} = 0.31$ and the immaterial variation summarized by $\hat{\boldsymbol{\Omega}}_0 = 4.21$. Since the material variation is evidently notably less than the immaterial variation, we can expect gains from an envelope analysis. Of course, these summary statistics agree qualitatively with the representation in Figure 5.2.

The final column of Table 5.1 gives the estimated basis $\hat{\boldsymbol{\Gamma}}$ for the envelope $\mathcal{E}_{\boldsymbol{\Sigma}}(\mathcal{M})$. Since the elements of $\boldsymbol{\Gamma}$ are nearly the same, the material information is essentially contained in the average response, while the immaterial information is captured by the difference between the responses.

5.1.4.2 Four Untransformed Responses

The square root transformations were used for two responses to move the data closer to bivariate normality. While normalizing transformations can be useful depending on the application, normality is not strictly required for an envelope analysis to produce gains over a standard analysis based on $\bar{\mathbf{Y}}$. The definition of an envelope requires first and second moments, but does not otherwise impose constraints on the distribution of $\mathbf{Y}$. If $\mathbf{Y}$ is not normal but has finite fourth moments, then the normal-theory estimators of $\boldsymbol{\mu}$ and $\boldsymbol{\Sigma}$ constructed in Section 5.1.3 are still $\sqrt{n}$-consistent envelope estimators. In particular, $\sqrt{n}(\hat{\boldsymbol{\mu}} - \boldsymbol{\mu})$ is asymptotically normal with mean 0 and finite variance. However, without normality $\mathrm{avar}(\sqrt{n}\,\hat{\boldsymbol{\mu}})$ could differ materially from the normal theory variance shown in (5.5), but the bootstrap can still be used.

To illustrate these comments, we next apply the envelope method to the Minneapolis School data using four untransformed responses, the percentages of fourth and sixth grade students scoring above and below average, which we

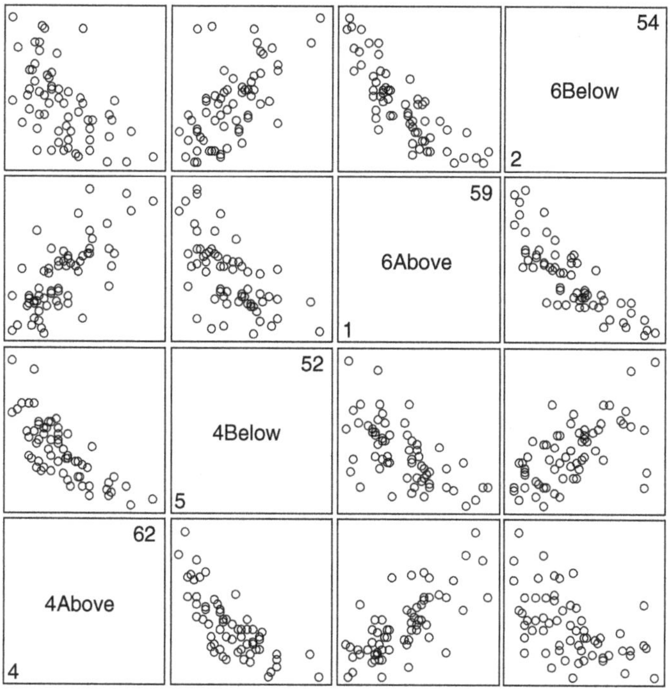

Figure 5.3 Minneapolis school data.

designate as "4 Above," "4 Below," "6 Above," "6 Below." A scatterplot matrix of these four variables is shown in Figure 5.3. Deviations from normality seem clear although they do not appear to be dramatic.

In contrast to the illustration with two responses, there is now disagreement between AIC, which yields the estimate $\hat{u} = 3$, and BIC, which gives $\hat{u} = 1$. For these data, the results of an envelope analysis are sensitive to the choice of u. With $\hat{u} = 1$, there are clear estimated gains in efficiency, while with $\hat{u} = 3$, there are negligible gains, the results in this case being essentially identical to those based on $\bar{\mathrm{Y}}$. One option in this situation is to adapt the weighted envelope estimator discussed in Section 1.11.4. Here we adopt a data-analytic approach, turning to likelihood ratio tests to help decide the issue. Table 5.2 gives

Table 5.2 Minneapolis schools: Estimates of u determined by using likelihood ratio tests sequentially for various test levels α on the Minneapolis school data with four untransformed responses.

α	0.4	0.3	0.25	0.1	0.05	0.005
$\hat{u}$	3	3	2	2	1	1

Table 5.3 Minneapolis schools: Summary of an envelope analysis with $u = 2$ of the percentage of fourth and sixth grade students scoring above and below average. Column headings are as in Table 5.1.

Y	$\bar{Y}$	se of $\bar{Y}$	$\hat{\mu}$	se[a] of $\hat{\mu}$	boot se of $\hat{\mu}$	se ratio
4 Above	24.60	1.58	22.91	1.12	1.16	1.41
4 below	22.19	1.26	23.60	0.96	0.93	1.32
6 Above	24.05	1.74	21.86	1.10	1.15	1.60
6 Below	22.71	1.69	24.85	1.19	1.30	1.42

a) Computed under normality.

estimates of u determined by using the sequential test procedure described in Section 5.1.3 for various test levels α. The results show that $\hat{u} = 2$ or 3 for unusually large $\alpha \geq 0.25$, while $\hat{u} = 1$ or 2 for the usual levels. This seems to support the notion that AIC overestimated the dimension. In some analyses, a compromise value, here $\hat{u} = 2$, might be reasonable, even if it is larger than the true dimension.

The results of an envelope analysis with $u = 2$ are shown in Table 5.3. The standard error ratios are similar to those shown in Table 5.1 and, although the data deviate noticeably from normality, the asymptotic standard errors are still reasonably accurate as judged against the results from 500 bootstrap samples.

5.1.5 Functional Data

Mallik (2017) adapted mean envelopes for modeling the regression of a univariate response Y on a functional predictor $X(t)$ observed at T equally spaced points, $t_1, \ldots, t_T$ so that the observed predictor data $\{X_1(t), \ldots, X_n(t)\}$ is a sample from a normal distribution with mean $\boldsymbol{\mu}$ and variance $\boldsymbol{\Sigma}_X$. The regression function of interest is of the form $Y = \boldsymbol{\mu}(X(t)) + \epsilon$, where the unknown function $\boldsymbol{\mu}$ maps the functional predictor to the real line and the error is independent of $X(t)$. Mallik combined a semi-metric distance function with kernel estimation and dimension reduction via envelopes to estimate $\boldsymbol{\mu}$ from the model with reduced predictors. In effect he developed an estimator for $\boldsymbol{\mu}$ after projecting the functional predictor onto $\mathcal{E}_{\Sigma}(\mathcal{M})$, and reported good results for envelopes from simulations and an analysis of Arctic oscillation.

5.2 Enveloping Multiple Means with Heteroscedastic Errors

5.2.1 Heteroscedastic Envelopes

Our standard multivariate model (1.1) could be adapted for contrasting multivariate means from h heteroscedastic normal populations, $N_r(\boldsymbol{\mu}_{(i)}, \boldsymbol{\Sigma}_{(i)})$,

$i = 1, \dots, h$, by using predictors to indicate the populations. But instead we use a one-way ANOVA model to guide the construction of envelopes. The subscript (i) will be used to indicate the ith population, and subscripts without parentheses will be used to index observations within a population. Let $\mathbf{Y}_{(i)j} \in \mathbb{R}^r$ denote the jth observation vector from the ith population, $j = 1, \dots, n_{(i)}$, let $n = \sum_i n_{(i)}$ denote the total sample size, let $\boldsymbol{\mu} = n^{-1} \sum_{i=1}^{h} n_{(i)}\boldsymbol{\mu}_{(i)}$ denote the grand mean over all observations, and let $\boldsymbol{\beta}_{(i)} = \boldsymbol{\mu}_{(i)} - \boldsymbol{\mu}$. Then we can model $\mathbf{Y}_{(i)j}$ as

$$\mathbf{Y}_{(i)j} = \boldsymbol{\mu} + \boldsymbol{\beta}_{(i)} + \boldsymbol{\varepsilon}_{(i)j},\ i = 1, \dots, h,\ j = 1, \dots, n_{(i)}, \tag{5.7}$$

where the error vector $\boldsymbol{\varepsilon}_{(i)j} \in \mathbb{R}^r$ follows the normal distribution with mean $\mathbf{0}$ and covariance matrix $\boldsymbol{\Sigma}_{(i)} > 0$. Enveloping in this context was studied by Su and Cook (2013) and we largely follow their approach.

Model (5.7) is different from others considered so far in this book because we now allow different dispersions $\boldsymbol{\Sigma}_{(i)}$. Nevertheless, the fundamental rationale for envelopes is the same. Let $\mathcal{B} = \text{span}(\boldsymbol{\beta}_{(1)}, \dots, \boldsymbol{\beta}_{(h)})$. Then the envelope is defined as the intersection of all subspaces $\mathcal{S} \subseteq \mathbb{R}^r$ so that the distribution of $\mathbf{Q}_{\mathcal{S}}\mathbf{Y}_{(i)}$ is constant across populations and $\mathbf{P}_{\mathcal{S}}\mathbf{Y}_{(i)} \perp\!\!\!\perp \mathbf{Q}_{\mathcal{S}}\mathbf{Y}_{(i)}$. These conditions hold if and only if $\mathcal{S}$ contains $\mathcal{B}$ and reduces $\boldsymbol{\Sigma}_{(i)}$ for each population:

$$\mathcal{B} \subseteq \mathcal{S} \text{ and } \boldsymbol{\Sigma}_{(i)} = \mathbf{P}_{\mathcal{S}}\boldsymbol{\Sigma}_{(i)}\mathbf{P}_{\mathcal{S}} + \mathbf{Q}_{\mathcal{S}}\boldsymbol{\Sigma}_{(i)}\mathbf{Q}_{\mathcal{S}},\ i = 1, \dots, h. \tag{5.8}$$

This structure is sufficiently different from the constant variance case to prompt a new definition:

Definition 5.1 Let $\mathcal{M}$ be a collection of real $p \times p$ symmetric matrices and let $\mathcal{V} \subseteq \text{span}(\mathbf{M})$ for all $\mathbf{M} \in \mathcal{M}$. The $\mathcal{M}$-envelope of $\mathcal{V}$, indicated with $\mathcal{E}_{\mathcal{M}}(\mathcal{V})$, is the intersection of all subspaces that contain $\mathcal{V}$ and that reduce each member of $\mathcal{M}$.

In our setup, $\mathcal{M} = \{\boldsymbol{\Sigma}_{(i)} : i = 1, \dots, h\}$, and $\mathcal{V} = \mathcal{B}$. As the $\boldsymbol{\Sigma}_{(i)}$ are all positive definite, $\text{span}(\boldsymbol{\Sigma}_{(i)}) = \mathbb{R}^r$ for all i. The condition $\mathcal{V} \subseteq \text{span}(\mathbf{M})$ is then satisfied by any subspace $\mathcal{V}$ of $\mathbb{R}^r$. The envelope $\mathcal{E}_{\mathcal{M}}(\mathcal{B})$ is the intersection of all subspaces that contain $\mathcal{B}$ and reduce $\mathcal{M}$, so it is the smallest subspace that satisfies (5.8). Any common eigenspace of the $\boldsymbol{\Sigma}_{(i)}$'s or the direct sum of the eigenspaces of the $\boldsymbol{\Sigma}_{(i)}$'s that contains $\mathcal{B}$ satisfies (5.8).

The coordinate form of the *heteroscedastic envelope model* can now be represented as

$$\begin{aligned} \mathbf{Y}_{(i)j} &= \boldsymbol{\mu} + \boldsymbol{\Gamma}\boldsymbol{\eta}_{(i)} + \boldsymbol{\varepsilon}_{(i)j}, \\ \boldsymbol{\Sigma}_{(i)} &= \boldsymbol{\Gamma}\boldsymbol{\Omega}_{(i)}\boldsymbol{\Gamma}^T + \boldsymbol{\Gamma}_0\boldsymbol{\Omega}_0\boldsymbol{\Gamma}_0^T, \end{aligned} \tag{5.9}$$

where $\boldsymbol{\Gamma} \in \mathbb{R}^{r \times u}$ is an orthogonal basis for $\mathcal{E}_{\mathcal{M}}(\mathcal{B})$, and $(\boldsymbol{\Gamma}, \boldsymbol{\Gamma}_0) \in \mathbb{R}^{r \times r}$ is an orthogonal matrix. For $i = 1, \dots, h$, $\boldsymbol{\eta}_{(i)} \in \mathbb{R}^{u \times 1}$ carries the coordinates of $\boldsymbol{\beta}_{(i)}$ with respect to basis $\boldsymbol{\Gamma}$, so $\boldsymbol{\beta}_{(i)} = \boldsymbol{\Gamma}\boldsymbol{\eta}_{(i)}$ and $\sum_{i=1}^{h} n_{(i)}\boldsymbol{\eta}_{(i)} = 0$, and $\boldsymbol{\Omega}_{(i)} \in \mathbb{R}^{u \times u}$ and $\boldsymbol{\Omega}_0 \in \mathbb{R}^{(r-u)\times(r-u)}$ are positive definite matrices. From this model, we see that the distribution of $\boldsymbol{\Gamma}_0^T\mathbf{Y}_{(i)}$ is constant across populations and that within each population $\boldsymbol{\Gamma}^T\mathbf{Y}_{(i)} \perp\!\!\!\perp \boldsymbol{\Gamma}_0^T\mathbf{Y}_{(i)}$.

5.2.2 Estimation

The maximum likelihood estimators based on model (5.9) are derived using the same general methods for other envelope models. Details are available from Su and Cook (2013). Here we emphasize the resulting estimators.

Let $\bar{\mathbf{Y}} = \sum_{i,j}\mathbf{Y}_{(i)j}/n$ be the sample mean, and let $\bar{\mathbf{Y}}_{(i)} = \sum_j \mathbf{Y}_{(i)j}/n_{(i)}$ be the sample mean for the ith population. We use $\mathbf{S}_{\mathbf{Y}} = \sum_{i,j}(\mathbf{Y}_{(i)j} - \bar{\mathbf{Y}})(\mathbf{Y}_{(i)j} - \bar{\mathbf{Y}})^T/n$ for the sample covariance matrix of $\mathbf{Y}$, and $\mathbf{S}_{\mathbf{Y}_{(i)}} = \sum_j(\mathbf{Y}_{(i)j} - \bar{\mathbf{Y}}_{(i)})(\mathbf{Y}_{(i)j} - \bar{\mathbf{Y}}_{(i)})^T/n_{(i)}$ for the sample covariance matrix of $\mathbf{Y}$ restricted to the ith population, $i = 1, \ldots, h$. Then, with $u = \dim(\mathcal{E}_{\mathcal{M}}(\mathcal{B}))$ specified, an orthogonal basis $\widehat{\boldsymbol{\Gamma}}$ of the maximum likelihood estimator of $\mathcal{E}_{\mathcal{M}}(\mathcal{B})$ can be obtained as

$$\widehat{\boldsymbol{\Gamma}} = \arg\min_{\mathbf{G}} n \log|\mathbf{G}^T\mathbf{S}_{\mathbf{Y}}^{-1}\mathbf{G}| + \sum_{i=1}^{h} n_{(i)} \log|\mathbf{G}^T\mathbf{S}_{\mathbf{Y}_{(i)}}\mathbf{G}|, \tag{5.10}$$

where the minimum is taken over the set of semi-orthogonal matrices $\mathbf{G} \in \mathbb{R}^{r\times u}$. Then $\widehat{\mathcal{E}}_{\mathcal{M}}(\mathcal{B}) = \operatorname{span}(\widehat{\boldsymbol{\Gamma}})$, $\widehat{\mathbf{P}}_{\mathcal{E}} = \widehat{\boldsymbol{\Gamma}}\widehat{\boldsymbol{\Gamma}}^T$, and the maximum likelihood estimators for the other parameters are for $i = 1, \ldots, h$

$$\begin{aligned}
\widehat{\boldsymbol{\mu}} &= \bar{\mathbf{Y}},\\
\widehat{\boldsymbol{\beta}}_{(i)} &= \widehat{\mathbf{P}}_{\mathcal{E}}(\bar{\mathbf{Y}}_{(i)} - \bar{\mathbf{Y}}),\\
\widehat{\boldsymbol{\mu}}_{(i)} &= \widehat{\boldsymbol{\mu}} + \widehat{\boldsymbol{\beta}}_{(i)} = \widehat{\mathbf{Q}}_{\mathcal{E}}\bar{\mathbf{Y}} + \widehat{\mathbf{P}}_{\mathcal{E}}\bar{\mathbf{Y}}_{(i)},\\
\widehat{\boldsymbol{\eta}}_{(i)} &= \widehat{\boldsymbol{\Gamma}}^T\widehat{\boldsymbol{\beta}}_{(i)},\\
\widehat{\boldsymbol{\Omega}}_{(i)} &= \widehat{\boldsymbol{\Gamma}}^T\mathbf{S}_{\mathbf{Y}_{(i)}}\widehat{\boldsymbol{\Gamma}},\\
\widehat{\boldsymbol{\Omega}}_0 &= \widehat{\boldsymbol{\Gamma}}_0^T\mathbf{S}_{\mathbf{Y}}\widehat{\boldsymbol{\Gamma}}_0,
\end{aligned}$$

and $\widehat{\boldsymbol{\Gamma}}_0$ is any orthogonal basis for $\widehat{\mathcal{E}}_{\mathcal{M}}^{\perp}(\mathcal{B})$. From this we see that, while the objective function is different, the pattern of estimation is the same as in other envelope models. For instance, the envelope estimator of $\boldsymbol{\beta}_{(i)}$ is just the projection of the usual estimator $\bar{\mathbf{Y}}_{(i)} - \bar{\mathbf{Y}}$ onto the envelope. Estimates of mean differences

$$\widehat{\boldsymbol{\mu}}_{(i)} - \widehat{\boldsymbol{\mu}}_{(k)} = \widehat{\boldsymbol{\beta}}_{(i)} - \widehat{\boldsymbol{\beta}}_{(k)} = \widehat{\mathbf{P}}_{\mathcal{E}}(\bar{\mathbf{Y}}_{(i)} - \bar{\mathbf{Y}}_{(k)})$$

and mean contrasts can be constructed straightforwardly from the parameter estimates. The estimation process is represented schematically in Figure 5.4 for two populations with two responses Y_1 and Y_2. The interpretation here is like that for Figure 1.3 except now the populations variances are unequal, with different eigenvalues but the same eigenvectors. Standard estimation for the first coordinate of the mean difference is represented by projecting data directly onto the horizontal axis (path A), while envelope estimation is represented by projecting data first onto the envelope and then onto the horizontal axis (path B).

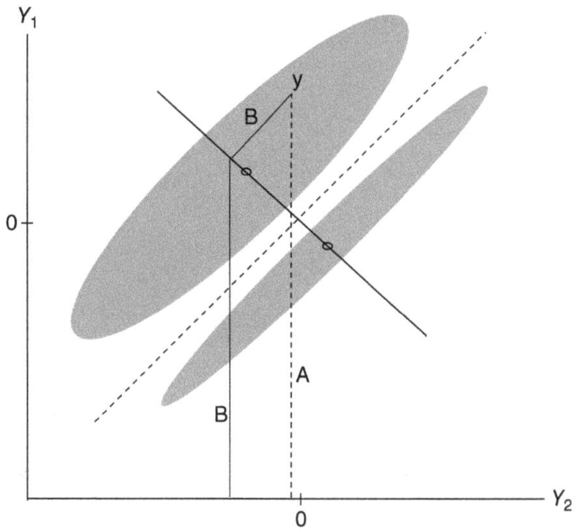

Figure 5.4 Schematic representation of heteroscedastic enveloping with projection paths A and B. The envelope and its orthogonal complement are represented by solid and dashed diagonal lines.

The dimension u of the envelope can again be selected by cross-validation, AIC, BIC, or likelihood ratio testing. To use the AIC criterion, we select the u that minimizes $2N_u - 2\hat{L}_u$, where

$$\hat{L}_u = -\frac{nr}{2}(1 + \log 2\pi) - \frac{n}{2}\log|\hat{\Gamma}^T \mathbf{S}_{\mathbf{Y}}^{-1}\hat{\Gamma}|$$
$$-\frac{n}{2}\log|\mathbf{S}_{\mathbf{Y}}| - \frac{1}{2}\sum_{i=1}^{h} n_{(i)} \log|\hat{\Gamma}^T \mathbf{S}_{\mathbf{Y}(i)}\hat{\Gamma}|$$

is the maximized log-likelihood and

$$N_u = r + u(r-u) + u(p-1) + pu(u+1)/2 + (r-u)(r-u+1)/2$$

is the number of free real parameters in model (5.9).

An expression for the asymptotic variance of $(\hat{\boldsymbol{\beta}}_{(1)}, \ldots, \boldsymbol{\beta}_{(p)})$ was given by Su and Cook (2013). That expression is not repeated here since it does not lend itself to useful interpretation. In practice, results based on the asymptotic variance would normally be accompanied by corresponding bootstrap results obtained by conditioning on the $n_{(i)}$s.

5.2.3 Cattle Weights: Heteroscedastic Envelope Fit

We return to the cattle data introduced in Section 1.3.4 for a brief illustration of fitting a heteroscedastic envelope. With a p-value about 0.005, Box's M test indicates that the 10×10 covariance matrices for the two treatments are unequal, and consequently it may be worthwhile to contrast previous results with those obtained by fitting a heteroscedastic envelope.

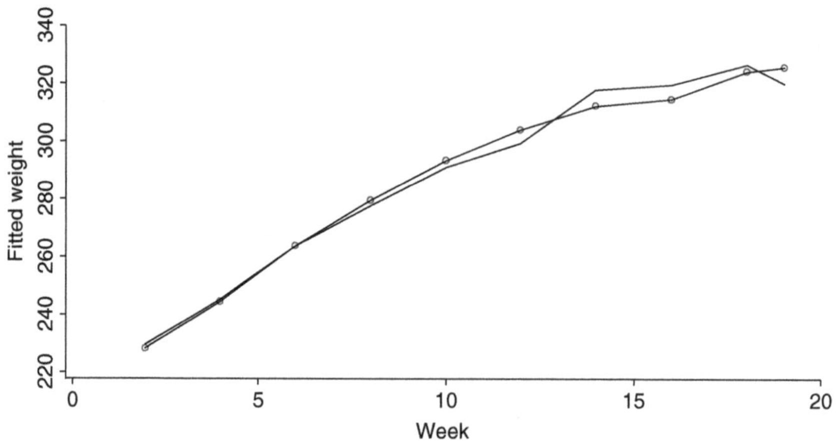

Figure 5.5 Cattle data: Profile plot of the fitted mean weights from a heteroscedastic envelope with $u = 5$.

The dimensions indicated by AIC, BIC, and LRT(0.05) varied between 1 and 5, as they did for the homoscedastic envelope (see Section 1.10.5). Using $u = 5$ to be conservative, the standard error ratios for the elements in $\hat{\boldsymbol{\beta}}_{(i)}$ $(i = 1, 2)$ varied between 3.3 and 5.9, so envelopes again provided substantial gains, as suggested by the eigenvalues $\|\hat{\boldsymbol{\Omega}}_1\| = 62.2$, $\|\hat{\boldsymbol{\Omega}}_2\| = 24.4$, and $\|\hat{\boldsymbol{\Omega}}_0\| = 2{,}350$. Figure 5.5 gives the profile plot of the fitted means from model (5.9). Comparing this with Figure 1.10b, it appears that the heteroscedastic and homoscedastic envelopes provided essentially the same fits to the data. Of course, this will not always be the case.

5.3 Extension to Heteroscedastic Regressions

Park et al. (2017) extended the envelope model (5.9) for estimations of means across heteroscedastic populations to heteroscedastic regressions. Their general goal was to develop efficient methods for studying associations between genetic variants and brain imaging phenotypes. In particular, their envelope model for groupwise heteroscedastic regressions provided substantial gains over standard methods in a study of how brain volumes for groups of male and female Alzheimer's patients relate to a variety of covariates. A small subset of these data was used for the illustration in Section 2.8.

To extend envelope model (5.9) to regression, we begin with the groupwise regression model

$$\mathbf{Y}_{(i)j} = \boldsymbol{\mu}_{(i)} + \boldsymbol{\beta}_{(i)}\mathbf{X}_{(i)j} + \boldsymbol{\varepsilon}_{(i)j}, \quad i = 1, \dots, h, \ j = 1, \dots, n_{(i)}, \tag{5.11}$$

where $\mathbf{Y}_{(i)j}$ and $\mathbf{X}_{(i)j}$ represent the jth observations on the response $\mathbf{Y}$ and predictor $\mathbf{X}$ within group i, the independent errors have $\mathrm{E}(\boldsymbol{\varepsilon}_{(i)j}) = 0$ and

$\mathrm{var}(\boldsymbol{\varepsilon}_{(i)j}) = \boldsymbol{\Sigma}_{(i)}$, and $\boldsymbol{\beta}_{(i)} \in \mathbb{R}^{r \times p}$ is an unconstrained matrix of regression coefficients for group i. The predictor vectors are assumed to be centered within each group so that $\sum_{j=1}^{n_i} \mathbf{X}_{(i)j} = 0$. Aside from the modifications necessary to accommodate a general predictor vector, this is a direct extension of model (5.7). In the case of the Alzheimer's study mentioned previously, Park et al. (2017) used $h = 2$ groups, males and females.

Let $\mathcal{B} = \mathrm{span}(\boldsymbol{\beta}_{(1)}, \dots, \boldsymbol{\beta}_{(h)})$, let $\mathcal{M} = \{\boldsymbol{\Sigma}_{(1)}, \dots, \boldsymbol{\Sigma}_{(h)}\}$ and reparameterize (5.11) in terms of $\mathcal{E}_{\mathcal{M}}(\boldsymbol{\beta})$, which by Definition 5.1 is the smallest subspace that contains $\mathcal{B}$ and reduced every member of $\mathcal{M}$. This leads to the following envelope version of (5.11)

$$\begin{aligned} \mathbf{Y}_{(i)j} &= \boldsymbol{\mu} + \boldsymbol{\Gamma}\boldsymbol{\eta}_{(i)}\mathbf{X}_{(i)j} + \boldsymbol{\varepsilon}_{(i)j}, \\ \boldsymbol{\Sigma}_{(i)} &= \boldsymbol{\Gamma}\boldsymbol{\Omega}_{(i)}\boldsymbol{\Gamma}^T + \boldsymbol{\Gamma}_0\boldsymbol{\Omega}_0\boldsymbol{\Gamma}_0^T, \end{aligned} \tag{5.12}$$

where $\boldsymbol{\Gamma} \in \mathbb{R}^{r \times u}$ is a semi-orthogonal basis matrix for $\mathcal{E}_{\mathcal{M}}(\boldsymbol{\beta})$ with $u = \dim(\mathcal{E}_{\mathcal{M}}(\boldsymbol{\beta}))$, $(\boldsymbol{\Gamma}, \boldsymbol{\Gamma}_0) \in \mathbb{R}^{r \times r}$ is orthogonal, and $\boldsymbol{\Omega}_{(i)} \in \mathbb{R}^{u \times u}$ and $\boldsymbol{\Omega}_0 \in \mathbb{R}^{(r-u)\times(r-u)}$ are positive definite. The derivation of the maximum likelihood estimators under normality for model (5.12) follows the same steps as used previously. In particular, the estimator of $\mathcal{E}_{\mathcal{M}}(\boldsymbol{\beta})$ uses objective function (5.10) with $\mathbf{S}_{\mathbf{Y}_{(i)}}$ replaced with the residuals from usual fit of the data from group i:

$$\hat{\mathcal{E}}_{\mathcal{M}}(\boldsymbol{\beta}) = \mathrm{span}\left\{ \arg\min_{\mathbf{G}} n \log|\mathbf{G}^T\mathbf{S}_{\mathbf{Y}}^{-1}\mathbf{G}| + \sum_{i=1}^{h} n_{(i)} \log|\mathbf{G}^T\mathbf{S}_{\mathbf{Y}_{(i)}|\mathbf{X}_{(i)}}\mathbf{G}| \right\}. \tag{5.13}$$

To write the maximum likelihood estimators of the remaining parameters, let $\mathbf{B}_{(i)}$ denote the ordinary least squares estimator of $\boldsymbol{\beta}_{(i)}$ from group i, let $\widehat{\boldsymbol{\Gamma}}$ be a semi-orthogonal basis matrix for $\hat{\mathcal{E}}_{\mathcal{M}}(\boldsymbol{\beta})$ and let $(\widehat{\boldsymbol{\Gamma}}, \widehat{\boldsymbol{\Gamma}}_0)$ be an orthogonal matrix. Then the maximum likelihood estimators for the other parameters are for $i = 1, \dots, h$

$$\begin{aligned} \widehat{\boldsymbol{\mu}}_{(i)} &= \bar{\mathbf{Y}}_{(i)}, \\ \widehat{\boldsymbol{\beta}}_{(i)} &= \widehat{\mathbf{P}}_{\mathcal{E}}\mathbf{B}_{(i)}, \\ \widehat{\boldsymbol{\eta}}_{(i)} &= \widehat{\boldsymbol{\Gamma}}^T\widehat{\boldsymbol{\beta}}_{(i)}, \\ \widehat{\boldsymbol{\Omega}}_{(i)} &= \widehat{\boldsymbol{\Gamma}}^T\mathbf{S}_{\mathbf{Y}_{(i)}|\mathbf{X}_{(i)}}\widehat{\boldsymbol{\Gamma}}, \\ \widehat{\boldsymbol{\Omega}}_0 &= \widehat{\boldsymbol{\Gamma}}_0^T\mathbf{S}_{\mathbf{Y}}\widehat{\boldsymbol{\Gamma}}_0. \end{aligned}$$

The dimension u of the envelope can be selected by adapting one of the methods discussed in Section 1.10. Park et al. (2017) show that these estimators of $\boldsymbol{\beta}_{(i)}$ and $\boldsymbol{\Sigma}_{(i)}$ are $\sqrt{n}$-consistent if the errors are not normal but have finite fourth moments. They also provide asymptotic covariances under normality and discuss other asymptotic properties of the estimators.

6

Envelope Algorithms

In this chapter, we describe algorithms for estimating a general envelope $\mathcal{E}_{\mathbf{M}}(\mathcal{U})$ with specified dimension $u = \dim(\mathcal{E}_{\mathbf{M}}(\mathcal{U}))$. Some of the algorithms are based on objective functions that stem from normal likelihoods, but normality itself is not required. The methods are all based on an estimator $\widehat{\mathbf{M}}$ of $\mathbf{M} \in \mathbb{S}^{r \times r}$ and an estimator $\widehat{\mathbf{U}}$ of a basis matrix $\mathbf{U} \in \mathbb{S}^{r \times r}$ for $\mathcal{U}$, and they all provide an estimated basis $\widehat{\mathbf{\Gamma}} \in \mathbb{R}^{r \times u}$ for $\mathcal{E}_{\mathbf{M}}(\mathcal{U})$. If the estimators $\widehat{\mathbf{M}}$ and $\widehat{\mathbf{U}}$ are $\sqrt{n}$-consistent estimators of $\mathbf{M}$ and $\mathbf{U}$, then, given the dimension u of $\mathcal{E}_{\mathbf{M}}(\mathcal{U})$, $\mathbf{P}_{\widehat{\mathbf{\Gamma}}}$ is a $\sqrt{n}$-consistent estimators of the projection $\mathbf{P}_{\mathcal{E}}$ onto $\mathcal{E}_{\mathbf{M}}(\mathcal{U})$. Subsequent use of the estimated envelope depends on context. While the choice of $\mathbf{M}$ and $\mathbf{U}$ may be clear from the application, one general way of choosing these quantities is described in Chapter 7. All of the envelope methods discussed so far are covered by the general algorithms in this chapter.

The estimator discussed in Section 6.1 was inspired by objective function (1.25) for estimation of a response envelope under normality. The estimator of Section 6.4 is a sequential version of the estimator discussed in Section 6.1, and a moment-based sequential estimator is discussed in Section 6.5.

6.1 Likelihood-Based Envelope Estimation

The sample and population versions of objective function (1.25) for estimating the response envelope $\mathcal{E}_{\mathbf{\Sigma}}(\mathcal{B})$ under normality of the errors in model (1.20) can be written as, recalling that $\mathcal{B} = \text{span}\,(\boldsymbol{\beta})$,

$$\widehat{\mathcal{E}}_{\mathbf{\Sigma}}(\mathcal{B}) = \text{span}\,\{\arg\min_{\mathbf{G}}\,(\log|\mathbf{G}^T\mathbf{S}_{\mathbf{Y}|\mathbf{X}}\mathbf{G}| + \log|\mathbf{G}^T(\mathbf{S}_{\mathbf{Y}|\mathbf{X}} + \mathbf{B}\mathbf{S}_{\mathbf{X}}\mathbf{B}^T)^{-1}\mathbf{G}|)\}, \tag{6.1}$$

$$\mathcal{E}_{\mathbf{\Sigma}}(\mathcal{B}) = \text{span}\,\{\arg\min_{\mathbf{G}}\,(\log|\mathbf{G}^T\mathbf{\Sigma}\mathbf{G}| + \log|\mathbf{G}^T(\mathbf{\Sigma} + \boldsymbol{\beta}\mathbf{\Sigma}_{\mathbf{X}}\boldsymbol{\beta}^T)^{-1}\mathbf{G}|)\}. \tag{6.2}$$

An Introduction to Envelopes: Dimension Reduction for Efficient Estimation in Multivariate Statistics,
First Edition. R. Dennis Cook.

Although (6.1) derives from the envelope model (1.20) with normal errors, we argued in Section 1.9 that it still results in an asymptotically normal $\sqrt{n}$-consistent estimator of $\boldsymbol{\beta}$ when the errors are nonnormal with finite fourth moments. The objective function in (6.1) can be adapted to estimate a general envelope $\mathcal{E}_{\mathbf{M}}(\mathcal{V})$. In (6.2), $\mathcal{B}$ occurs only via $\mathcal{B} = \text{span}\,(\boldsymbol{\beta}\boldsymbol{\Sigma}_{\mathbf{X}}\boldsymbol{\beta}^T)$; thus, to estimate $\mathcal{E}_{\mathbf{M}}(\mathcal{V})$, we associate $\boldsymbol{\Sigma}$ with $\mathbf{M}$, $\mathbf{U}$ with $\boldsymbol{\beta}\boldsymbol{\Sigma}_{\mathbf{X}}\boldsymbol{\beta}^T$, and $\mathcal{B}$ with $\mathcal{V}$, resulting in the general estimator

$$\widehat{\mathcal{E}}_{\mathbf{M}}(\mathcal{V}) = \text{span}\,\{\arg\min_{\mathbf{G}} (\log|\mathbf{G}^T\widehat{\mathbf{M}}\mathbf{G}| + \log|\mathbf{G}^T(\widehat{\mathbf{M}} + \widehat{\mathbf{U}})^{-1}\mathbf{G}|)\}. \quad (6.3)$$

The following proposition provides a population justification for $\widehat{\mathcal{E}}_{\mathbf{M}}(\mathcal{V})$, as well as providing alternative forms for objective function (6.3).

Proposition 6.1 *Let* $\mathbf{M} > 0$ *and* $\mathbf{U} \geq 0$ *be symmetric* $r \times r$ *matrices. Then the* $\mathbf{M}$*-envelope of* $\mathcal{V} = \text{span}\,(\mathbf{U})$ *can be constructed as*

$$\mathcal{E}_{\mathbf{M}}(\mathcal{V}) = \arg\min_{\mathcal{T}\in\mathcal{G}(r,u)} \{\log|\mathbf{P}_{\mathcal{T}}\mathbf{M}\mathbf{P}_{\mathcal{T}}|_0 + \log|\mathbf{Q}_{\mathcal{T}}(\mathbf{M}+\mathbf{U})\mathbf{Q}_{\mathcal{T}}|_0\},$$

where $u = \dim(\mathcal{E}_{\mathbf{M}}(\mathcal{V}))$. *A semi-orthogonal basis matrix* $\boldsymbol{\Gamma} \in \mathbb{R}^{r\times u}$ *for* $\mathcal{E}_{\mathbf{M}}(\mathcal{V})$ *can be obtained as*

$$\begin{aligned}\boldsymbol{\Gamma} &= \arg\min_{\mathbf{G}}\{\log|\mathbf{G}^T\mathbf{M}\mathbf{G}| + \log|\mathbf{G}_0^T(\mathbf{M}+\mathbf{U})\mathbf{G}_0|\}\\ &= \arg\min_{\mathbf{G}}\{\log|\mathbf{G}^T\mathbf{M}\mathbf{G}| + \log|\mathbf{G}^T(\mathbf{M}+\mathbf{U})^{-1}\mathbf{G}|\},\end{aligned}$$

where $\min_{\mathbf{G}}$ *is taken over all semi-orthogonal matrices* $\mathbf{G} \in \mathbb{R}^{r\times u}$ *and* $(\mathbf{G}, \mathbf{G}_0)$ *is an orthogonal matrix.*

Proof: Let $\mathbf{G}$ be a semi-orthogonal basis matrix for $\mathcal{T}$ and let $(\mathbf{G}, \mathbf{G}_0)$ be orthogonal. Then

$$|\mathbf{P}_{\mathcal{T}}\mathbf{M}\mathbf{P}_{\mathcal{T}}|_0 = \left|(\mathbf{G},\mathbf{G}_0)\begin{pmatrix}\mathbf{G}^T\mathbf{M}\mathbf{G} & 0\\ 0 & 0\end{pmatrix}(\mathbf{G},\mathbf{G}_0)^T\right|_0 = |\mathbf{G}^T\mathbf{M}\mathbf{G}|.$$

Consequently, we can work in terms of bases without loss of generality. Now,

$$\begin{aligned}\log|\mathbf{G}^T\mathbf{M}\mathbf{G}| + \log|\mathbf{G}_0^T(\mathbf{M}+\mathbf{U})\mathbf{G}_0| &= \log|\mathbf{G}^T\mathbf{M}\mathbf{G}| + \log|\mathbf{G}_0^T\mathbf{M}\mathbf{G}_0 + \mathbf{G}_0^T\mathbf{U}\mathbf{G}_0|\\ &\geq \log|\mathbf{G}^T\mathbf{M}\mathbf{G}| + \log|\mathbf{G}_0^T\mathbf{M}\mathbf{G}_0|\\ &\geq |\mathbf{M}|,\end{aligned}$$

where the second inequality follows from Lemma A.14. To achieve the lower bound, the second inequality requires that span ($\mathbf{G}$) reduce $\mathbf{M}$, while the first inequality requires that $\mathcal{V} \subseteq \text{span}\,(\mathbf{G})$. The first representation for $\boldsymbol{\Gamma}$ follows since u is the dimension of the smallest subspace that satisfies these two properties. The second representation for $\boldsymbol{\Gamma}$ follows immediately from Lemma A.13. □

The usual requirement that $\mathcal{U} \subseteq \text{span}(\mathbf{M})$ holds automatically in Proposition 6.1 because $\mathbf{M} > 0$. If $\mathbf{M} \geq 0$ and $\mathcal{U} \subseteq \text{span}(\mathbf{M})$, this proposition can still be used as a basis for the construction of $\mathcal{E}_{\mathbf{M}}(\mathcal{U})$ by using Proposition A.6 to first reduce the coordinates. Being based on a normal likelihood, some interpretations of the objective functions given in Proposition 6.1 parallel those discussed in Section 1.5.3 for objective function (6.1).

As mentioned in the preamble to this chapter, if $\widehat{\mathbf{M}}$ and $\widehat{\mathbf{U}}$ are $\sqrt{n}$-consistent estimators of $\mathbf{M}$ and $\mathbf{U}$, then the projection $\mathbf{P}_{\widehat{\mathcal{E}}}$ onto the corresponding solution $\widehat{\mathcal{E}}_{\mathbf{M}}(\mathcal{U})$ given by (6.3) provides a $\sqrt{n}$-consistent estimator of the population projection onto $\mathcal{E}_{\mathbf{M}}(\mathcal{U})$. A justification of this claim can be constructed in a manner similar to the proof of Proposition B.2.

Depending on the application, it might not be clear how to choose $\widehat{\mathbf{M}}$ and $\widehat{\mathbf{U}}$, particularly since there may be many choices that lead to consistent estimators of $\mathcal{E}_{\mathbf{M}}(\mathcal{U})$, as implied by the relationships described in Section A.3. For instance, without knowledge of the likelihood for response envelopes, we might choose $\widehat{\mathbf{U}} = \mathbf{B}\mathbf{B}^T$, leading to an estimator that is inferior to the likelihood estimator with $\widehat{\mathbf{U}} = \mathbf{B}\mathbf{S}_{\mathbf{X}}\mathbf{B}^T$. A general way to select $\widehat{\mathbf{M}}$ and $\widehat{\mathbf{U}}$ is indicated in Definition 7.1.

Objective function (6.3) is invariant under orthogonal transformations $\mathbf{O} \in \mathbb{R}^{u \times u}$ of $\mathbf{G}$, $\mathbf{G} \mapsto \mathbf{G}\mathbf{O}$, so the solution depends only on span $(\mathbf{G})$ and not on $\mathbf{G}$ itself. As a consequence, the optimization is again over the Grassmannian $\mathcal{G}(u, r)$. Since it takes $u(r-u)$ real numbers to uniquely specify a subspace in $\mathcal{G}(u, r)$, direct Grassmann optimization is usually computationally straightforward when $u(r-u)$ is not too large, but it can be slow and unreliable when $u(r-u)$ is large. Since our objective function is nonconvex, the solution returned may correspond to a local rather than global minimum, particularly when the signal is small relative to the noise. Good starting values are important for mitigating these issues.

6.2 Starting Values

In this section, we describe how to compute good starting values as proposed by Cook et al. (2016), and then in Section 6.3, we describe a non-Grassmann algorithm that is relatively straightforward to implement in standard statistics software such as R.

6.2.1 Choosing the Starting Value from the Eigenvectors of $\widehat{\mathbf{M}}$

Let $\widehat{\mathbf{u}}\widehat{\mathbf{u}}^T = \widehat{\mathbf{U}}$ be a factorization of the symmetric matrix $\widehat{\mathbf{U}}$, let $(\mathbf{G}, \mathbf{G}_0) \in \mathbb{R}^{r \times r}$ be an orthogonal matrix with $\mathbf{G} \in \mathbb{R}^{r \times u}$, and let $\mathbf{H}_0 = \widehat{\mathbf{M}}^{1/2}\mathbf{G}_0(\mathbf{G}_0^T\widehat{\mathbf{M}}\mathbf{G}_0)^{-1/2}$, so $\mathbf{H}_0^T\mathbf{H}_0 = \mathbf{I}_{r-u}$. The arguments that minimize the general objective function

$$L_u(\mathbf{G}) = \log|\mathbf{G}^T\widehat{\mathbf{M}}\mathbf{G}| + \log|\mathbf{G}^T(\widehat{\mathbf{M}} + \widehat{\mathbf{U}})^{-1}\mathbf{G}| \tag{6.4}$$

are the same as those that minimize the equivalent objective function

$$\begin{aligned} L_u^{(1)}(\mathbf{G}) &= \log|\mathbf{G}^T\widehat{\mathbf{M}}\mathbf{G}| + \log|\mathbf{G}_0^T(\widehat{\mathbf{M}}+\widehat{\mathbf{U}})\mathbf{G}_0| \\ &= \log|\mathbf{G}^T\widehat{\mathbf{M}}\mathbf{G}| + \log|\mathbf{G}_0^T(\widehat{\mathbf{M}}+\widehat{\mathbf{u}}\widehat{\mathbf{u}}^T)\mathbf{G}_0| \\ &= \log|\mathbf{G}^T\widehat{\mathbf{M}}\mathbf{G}| + \log|\mathbf{G}_0^T\widehat{\mathbf{M}}\mathbf{G}_0| + \log|\mathbf{I}_{r-u} + \widehat{\mathbf{u}}^T\widehat{\mathbf{M}}^{-1/2}\mathbf{H}_0\mathbf{H}_0^T\widehat{\mathbf{M}}^{-1/2}\widehat{\mathbf{u}}| \\ &= \log|\mathbf{G}^T\widehat{\mathbf{M}}\mathbf{G}| + \log|\mathbf{G}_0^T\widehat{\mathbf{M}}\mathbf{G}_0| + \log|\mathbf{I}_{r-u} + \mathbf{H}_0^T\widehat{\mathbf{M}}^{-1/2}\widehat{\mathbf{u}}\widehat{\mathbf{u}}^T\widehat{\mathbf{M}}^{-1/2}\mathbf{H}_0|. \end{aligned}$$

The sum of the first two terms is minimized globally when the columns of $\mathbf{G}$ are any u eigenvectors of $\widehat{\mathbf{M}}$. Restricting the columns of $\mathbf{G}$ to be eigenvectors of $\widehat{\mathbf{M}}$, we next choose the subset of u eigenvectors to minimize

$$\begin{aligned} L_u^{(2)}(\mathbf{G}) &= \log|\mathbf{I}_{r-u} + \mathbf{H}_0^T\widehat{\mathbf{M}}^{-1/2}\widehat{\mathbf{U}}\widehat{\mathbf{M}}^{-1/2}\mathbf{H}_0| \\ &= \log|\mathbf{I}_{r-u} + \mathbf{G}_0^T(\widehat{\mathbf{M}}^{-1/2}\widehat{\mathbf{U}}\widehat{\mathbf{M}}^{-1/2})\mathbf{G}_0| \\ &= \log|\mathbf{I}_{r-u} + \mathbf{G}_0^T\widehat{\mathbf{U}}_{\mathbf{M}}\mathbf{G}_0|, \end{aligned}$$

where $\widehat{\mathbf{U}}_{\mathbf{M}} = \widehat{\mathbf{M}}^{-1/2}\widehat{\mathbf{U}}\widehat{\mathbf{M}}^{-1/2}$ is a standardized version of $\widehat{\mathbf{U}}$ and the second equality follows because $\mathbf{G}_0$ reduces $\widehat{\mathbf{M}}$ and thus $\mathbf{H}_0 = \mathbf{G}_0$. There is evidently no closed-form expression for the minimizer of $L_u^{(2)}(\mathbf{G})$ over the eigenvectors of $\widehat{\mathbf{M}}$, and exact computation evaluating at all r-choose-u possibilities for $\mathbf{G}$ will be effectively impossible when r is large. For these reasons, we replace the log-determinant in $L_u^{(2)}(\mathbf{G})$ with the trace and minimize $\mathrm{tr}(\mathbf{I}_{r-u} + \mathbf{G}_0^T\widehat{\mathbf{U}}_{\mathbf{M}}\mathbf{G}_0)$, which is equivalent to minimizing

$$\mathrm{tr}(\mathbf{G}_0^T\widehat{\mathbf{U}}_{\mathbf{M}}\mathbf{G}_0) = \mathrm{tr}(\widehat{\mathbf{U}}_{\mathbf{M}}) - \mathrm{tr}(\mathbf{G}^T\widehat{\mathbf{U}}_{\mathbf{M}}\mathbf{G}).$$

Consequently, we need to only maximize

$$J_{\mathbf{M}}(\mathbf{G}) := \mathrm{tr}(\mathbf{G}^T\widehat{\mathbf{U}}_{\mathbf{M}}\mathbf{G}) = \sum_{i=1}^{u} \mathbf{g}_i^T\widehat{\mathbf{U}}_{\mathbf{M}}\mathbf{g}_i,$$

where $\mathbf{g}_i$ is the ith selected eigenvector of $\widehat{\mathbf{M}}$ (the ith column of $\mathbf{G}$). Computation is now easy, since we just select the u eigenvectors $\mathbf{g}_i$ of $\widehat{\mathbf{M}}$ that maximize $\mathbf{g}_i^T\widehat{\mathbf{U}}_{\mathbf{M}}\mathbf{g}_i$.

Applying this in response envelopes, $\widehat{\mathbf{M}} = \mathbf{S}_{\mathrm{Y|X}}$, $\widehat{\mathbf{U}} = \mathbf{B}\mathbf{S}_{\mathrm{X}}\mathbf{B}^T$, $\widehat{\mathbf{u}} = \mathbf{B}\mathbf{S}_{\mathrm{X}}^{1/2}$, and $\widehat{\mathbf{U}}_{\mathbf{M}} = \mathbf{S}_{\mathrm{Y|X}}^{-1/2}\mathbf{B}\mathbf{S}_{\mathrm{X}}\mathbf{B}^T\mathbf{S}_{\mathrm{Y|X}}^{-1/2}$, where $\mathbf{S}_{\mathrm{Y|X}}^{-1/2}\mathbf{B}\mathbf{S}_{\mathrm{X}}^{1/2}$ is just the standardized matrix of ordinary least squares regression coefficients (1.6).

This method for choosing starting values produces the envelope $\mathcal{E}_{\mathbf{M}}(\mathcal{U})$ in the population. To see this, it follows from Corollary A.2 and Proposition A.3 that

$$\mathcal{E}_{\mathbf{M}}(\mathcal{U}) = \mathcal{E}_{\mathbf{M}}(\mathbf{M}^{-1/2}\mathcal{U}) = \sum_{i=1}^{q} \mathbf{P}_i\mathbf{M}^{-1/2}\mathcal{U},$$

where $\mathbf{P}_i$ is the projection onto the ith of the q, say, eigenspaces of $\mathbf{M}$. In other words, $\mathcal{E}_{\mathbf{M}}(\mathcal{U})$ can be constructed as the sum of the reducing subspaces of $\mathbf{M}$ that are not orthogonal to $\mathbf{M}^{-1/2}\mathcal{U}$. The population version of $J_{\mathbf{M}}(\mathbf{G})$ gives exactly this set since $\mathbf{g}^T\mathbf{U}_{\mathbf{M}}\mathbf{g} = 0$ for any $\mathbf{g}$ that is orthogonal to $\mathbf{U}_{\mathbf{M}} = \mathbf{M}^{-1/2}\mathbf{U}\mathbf{M}^{-1/2}$.

The matrices $\widehat{\mathbf{M}}$ and $\widehat{\mathbf{M}} + \widehat{\mathbf{U}}$ both occur in $L_u(\mathbf{G})$, which balances their contribution to the solution. We could have equally begun by selecting from the eigenvectors of $\widehat{\mathbf{M}} + \widehat{\mathbf{U}}$. In fact, in some cases, using $\widehat{\mathbf{M}} + \widehat{\mathbf{U}}$ will produce better starting values than $\widehat{\mathbf{M}}$. To gain intuition into this conclusion, we turn to the population and represent

$$\begin{aligned}
\mathbf{U} &= \boldsymbol{\Gamma}\mathbf{U}_1\boldsymbol{\Gamma}^T,\\
\mathbf{M} &= \boldsymbol{\Gamma}\boldsymbol{\Omega}\boldsymbol{\Gamma}^T + \boldsymbol{\Gamma}_0\boldsymbol{\Omega}_0\boldsymbol{\Gamma}_0^T,\\
(\mathbf{M}+\mathbf{U})^{-1} &= \boldsymbol{\Gamma}(\boldsymbol{\Omega}+\mathbf{U}_1)^{-1}\boldsymbol{\Gamma}^T + \boldsymbol{\Gamma}_0\boldsymbol{\Omega}_0^{-1}\boldsymbol{\Gamma}_0^T,
\end{aligned}$$

where $\boldsymbol{\Gamma}$ is a population basis for $\mathcal{E}_{\mathbf{M}}(\mathcal{U})$ as in Proposition 6.1. For the starting values based on $\widehat{\mathbf{M}}$ to work well, the eigenvalues of $\boldsymbol{\Omega}$ need to be well distinguished from those of $\boldsymbol{\Omega}_0$. If some of the eigenvalues of $\boldsymbol{\Omega}$ are close to a subset of the eigenvalues of $\boldsymbol{\Omega}_0$, then in samples the corresponding eigenspaces may be confused when attempting to maximize $J_{\mathbf{M}}(\mathbf{G})$. In other words, we may well pick vectors near span $(\boldsymbol{\Gamma}_0)$ instead of eigenvectors near span $(\boldsymbol{\Gamma}) = \mathcal{E}_{\mathbf{M}}(\mathcal{U})$. In such cases, better starting values might be obtained from the eigenvectors of $\widehat{\mathbf{M}} + \widehat{\mathbf{U}}$ rather than the eigenvectors of $\widehat{\mathbf{M}}$.

The same argument applies to choosing the starting values from the eigenvectors of $\widehat{\mathbf{M}} + \widehat{\mathbf{U}}$: the eigenvalues of $\boldsymbol{\Omega} + \mathbf{U}_1$ need to be well distinguished from those of $\boldsymbol{\Omega}_0$. If some of the eigenvalues of $\boldsymbol{\Omega} + \mathbf{U}_1$ are close to a subset of the eigenvalues of $\boldsymbol{\Omega}_0$, then in samples the corresponding eigenspaces will again likely be confused. In such cases, we may obtain better starting values by starting with the eigenvectors of $\widehat{\mathbf{M}}$ rather than the eigenvectors of $\widehat{\mathbf{M}} + \widehat{\mathbf{U}}$.

The general conclusion from this discussion is that for effective starting values, we will need to consider both $\widehat{\mathbf{M}}$ and $\widehat{\mathbf{M}} + \widehat{\mathbf{U}}$. In Section 6.2.2, we describe how to select starting values from the eigenvectors of $\widehat{\mathbf{M}} + \widehat{\mathbf{U}}$.

6.2.2 Choosing the Starting Value from the Eigenvectors of $\widehat{\mathbf{M}} + \widehat{\mathbf{U}}$

Let $\widehat{\mathbf{W}} = \widehat{\mathbf{M}} + \widehat{\mathbf{U}}$ for notational convenience and express the objective function as follows:

$$\begin{aligned}
L_u(\mathbf{G}) &= \log|\mathbf{G}^T\widehat{\mathbf{M}}\mathbf{G}| + \log|\mathbf{G}^T\widehat{\mathbf{W}}^{-1}\mathbf{G}|\\
&= \log|\mathbf{G}^T(\widehat{\mathbf{M}}+\widehat{\mathbf{U}})\mathbf{G} - \mathbf{G}^T\widehat{\mathbf{U}}\mathbf{G}| + \log|\mathbf{G}^T\widehat{\mathbf{W}}^{-1}\mathbf{G}|\\
&= \log|\mathbf{G}^T\widehat{\mathbf{W}}\mathbf{G} - \mathbf{G}^T\hat{\mathbf{u}}\hat{\mathbf{u}}^T\mathbf{G}| + \log|\mathbf{G}^T\widehat{\mathbf{W}}^{-1}\mathbf{G}|\\
&= \log|\mathbf{G}^T\widehat{\mathbf{W}}\mathbf{G}| + \log|\mathbf{I} - \hat{\mathbf{u}}^T\mathbf{G}(\mathbf{G}^T\widehat{\mathbf{W}}\mathbf{G})^{-1}\mathbf{G}^T\hat{\mathbf{u}}| + \log|\mathbf{G}^T\widehat{\mathbf{W}}^{-1}\mathbf{G}|.
\end{aligned}$$

The sum of the first and last addends on the right side of this representation is always nonnegative and equals 0, its minimum value, when the columns of $\mathbf{G}$ span any reducing subspace of $\widehat{\mathbf{W}} = \widehat{\mathbf{M}} + \widehat{\mathbf{U}}$ (Lemma A.15). Restricting $\mathbf{G}$ in this way,

$$\begin{aligned}\hat{\mathbf{u}}^T\mathbf{G}(\mathbf{G}^T\widehat{\mathbf{W}}\mathbf{G})^{-1}\mathbf{G}^T\hat{\mathbf{u}} &= \hat{\mathbf{u}}^T\widehat{\mathbf{W}}^{-1/2}\widehat{\mathbf{W}}^{1/2}\mathbf{G}(\mathbf{G}^T\widehat{\mathbf{W}}\mathbf{G})^{-1}\mathbf{G}^T\widehat{\mathbf{W}}^{1/2}\widehat{\mathbf{W}}^{-1/2}\hat{\mathbf{u}}\\ &= \hat{\mathbf{u}}^T\widehat{\mathbf{W}}^{-1/2}\mathbf{G}\mathbf{G}^T\widehat{\mathbf{W}}^{-1/2}\hat{\mathbf{u}},\end{aligned}$$

and the middle term of $L_u(\mathbf{G})$ reduces to

$$\begin{aligned}\log|\mathbf{I} - \hat{\mathbf{u}}^T\mathbf{G}(\mathbf{G}^T(\widehat{\mathbf{M}}+\widehat{\mathbf{U}})\mathbf{G})^{-1}\mathbf{G}^T\hat{\mathbf{u}}| &= \log|\mathbf{I} - \hat{\mathbf{u}}^T\widehat{\mathbf{W}}^{-1/2}\mathbf{G}\mathbf{G}^T\widehat{\mathbf{W}}^{-1/2}\hat{\mathbf{u}}|\\ &= \log|\mathbf{I} - \mathbf{G}^T\widehat{\mathbf{W}}^{-1/2}\hat{\mathbf{u}}\hat{\mathbf{u}}^T\widehat{\mathbf{W}}^{-1/2}\mathbf{G}|\\ &= \log|\mathbf{I} - \mathbf{G}^T\widehat{\mathbf{U}}_{\mathbf{M+U}}\mathbf{G}|,\end{aligned}$$

where $\widehat{\mathbf{U}}_{\mathbf{M+U}} = \widehat{\mathbf{W}}^{-1/2}\hat{\mathbf{u}}\hat{\mathbf{u}}^T\widehat{\mathbf{W}}^{-1/2} = \widehat{\mathbf{W}}^{-1/2}\widehat{\mathbf{U}}\widehat{\mathbf{W}}^{-1/2}$ is $\widehat{\mathbf{U}}$ standardized by $\widehat{\mathbf{W}}^{-1/2} = (\widehat{\mathbf{M}}+\widehat{\mathbf{U}})^{-1/2}$. When we chose the starting values from the eigenvectors of $\widehat{\mathbf{M}}$, as described in the last section, $\widehat{\mathbf{U}}$ was standardized by $\widehat{\mathbf{M}}^{-1/2}$, not $(\widehat{\mathbf{M}}+\widehat{\mathbf{U}})^{-1/2}$ as is the case here. We now need to choose the starting values, i.e. the columns of $\mathbf{G}$, from the eigenvectors of $\widehat{\mathbf{M}}+\widehat{\mathbf{U}}$ to minimize $\log|\mathbf{I} - \mathbf{G}^T\widehat{\mathbf{U}}_{\mathbf{M+U}}\mathbf{G}|$. As in the previous case, this seems hard computationally, so we again replace the log-determinant with the trace and maximize

$$J_{\mathbf{M+U}}(\mathbf{G}) = \operatorname{tr}(\mathbf{G}^T\widehat{\mathbf{U}}_{\mathbf{M+U}}\mathbf{G}) = \sum_{i=1}^{u}\mathbf{g}_i^T\widehat{\mathbf{U}}_{\mathbf{M+U}}\mathbf{g}_i.$$

This is exactly the same as the previous case, except that the standardization of $\widehat{\mathbf{U}}$ is with $(\widehat{\mathbf{M}}+\widehat{\mathbf{U}})^{-1/2}$ instead of $\widehat{\mathbf{M}}^{-1/2}$.

Applying this in response envelopes, $\widehat{\mathbf{M}} = \mathbf{S}_{\mathbf{Y}|\mathbf{X}}$, $\widehat{\mathbf{U}} = \mathbf{B}\mathbf{S}_{\mathbf{X}}\mathbf{B}^T$, $\widehat{\mathbf{M}}+\widehat{\mathbf{U}} = \mathbf{S}_{\mathbf{Y}}$, $\hat{\mathbf{u}} = \mathbf{B}\mathbf{S}_{\mathbf{X}}^{1/2}$, and $\widehat{\mathbf{U}}_{\mathbf{M+U}} = \mathbf{S}_{\mathbf{Y}}^{-1/2}\mathbf{B}\mathbf{S}_{\mathbf{X}}\mathbf{B}^T\mathbf{S}_{\mathbf{Y}}^{-1/2}$. The population version of $J_{\mathbf{M+U}}(\mathbf{G})$ also reproduces the envelope $\mathcal{E}_{\mathbf{M}}(\mathcal{U})$ in the population for reasons similar to those stated at the end of Section 6.2.1

6.2.3 Summary

To summarize, recall that $\widehat{\mathbf{U}}_{\mathbf{M}} = \widehat{\mathbf{M}}^{-1/2}\widehat{\mathbf{U}}\widehat{\mathbf{M}}^{-1/2}$ and $\widehat{\mathbf{U}}_{\mathbf{M+U}} = (\widehat{\mathbf{M}}+\widehat{\mathbf{U}})^{-1/2}\widehat{\mathbf{U}}(\widehat{\mathbf{M}}+\widehat{\mathbf{U}})^{-1/2}$ denote the two different standardized versions of $\widehat{\mathbf{U}}$. Then to choose the starting value of $\mathbf{G}$ from the eigenvectors of $\widehat{\mathbf{M}}$, we maximize

$$J_{\mathbf{M}}(\mathbf{G}) = \sum_{i=1}^{u}\mathbf{g}_i^T\widehat{\mathbf{U}}_{\mathbf{M}}\mathbf{g}_i$$

and to choose from the eigenvectors of $\widehat{\mathbf{M}}+\widehat{\mathbf{U}}$, we maximize

$$J_{\mathbf{M+U}}(\mathbf{G}) = \sum_{i=1}^{u}\mathbf{g}_i^T\widehat{\mathbf{U}}_{\mathbf{M+U}}\mathbf{g}_i.$$

The standardized forms $\widehat{\mathbf{U}}_{\mathbf{M}}$ and $\widehat{\mathbf{U}}_{\mathbf{M+U}}$ are important when the scales involved in $\widehat{\mathbf{M}}$ and $\widehat{\mathbf{M}} + \widehat{\mathbf{U}}$ are very different. This can perhaps be appreciated readily in the context of response envelopes, where $\widehat{\mathbf{M}} = \mathbf{S}_{\mathbf{Y}|\mathbf{X}}$ and $\widehat{\mathbf{M}} + \widehat{\mathbf{U}} = \mathbf{S}_{\mathbf{Y}}$. In this case, the standardization may be important and effective if the scales of the elements of $\mathbf{Y}$ are very different. However, the standardization will be effectively unnecessary when the scales are similar. In the case of response envelopes, this means that the scales of the elements of $\mathbf{Y}$ are the same or similar. Depending on the scales involved, standardization can also be counterproductive when the sample size is not large enough to give sufficiently accurate estimates of $\mathbf{M}$ and $\mathbf{U}$. In such cases, we abandon the standardization and use either

$$J^*_{\mathbf{M}}(\mathbf{G}) = \sum_{i=1}^{u} \mathbf{g}_i^T \widehat{\mathbf{U}} \mathbf{g}_i$$

or

$$J^*_{\mathbf{M+U}}(\mathbf{G}) = \sum_{i=1}^{u} \mathbf{g}_i^T \widehat{\mathbf{U}} \mathbf{g}_i$$

as the objective function. The only difference between these is that $J^*_{\mathbf{M}}(\mathbf{G})$ confines $\mathbf{G}$ to the eigenvectors of $\widehat{\mathbf{M}}$, while $J^*_{\mathbf{M+U}}(\mathbf{G})$ confines $\mathbf{G}$ to the eigenvectors of $\widehat{\mathbf{M}} + \widehat{\mathbf{U}}$.

We now have four methods for choosing starting values: Choose from the eigenvectors of $\widehat{\mathbf{M}}$ or $\widehat{\mathbf{M}} + \widehat{\mathbf{U}}$ with standardized or unstandardized versions of $\widehat{\mathbf{U}}$ in the objective function. The actual starting value for $\mathbf{G}$, which is called $\widehat{\mathbf{G}}_{\text{start}}$, is the one that minimizes $L_u(\mathbf{G})$. This starting value is typically quite good and might also be used as a standalone estimator since Cook et al. (2016, Proposition 2.4) showed that the projection onto span $(\widehat{\mathbf{G}}_{\text{start}})$ is a $\sqrt{n}$-consistent estimator of $\mathcal{E}_{\mathbf{M}}(\mathcal{U})$. The asymptotic standard errors given in Section 1.6 may no longer apply, but the bootstrap can still be used.

6.3 A Non-Grassmann Algorithm for Estimating $\mathcal{E}_{\mathbf{M}}(\mathcal{U})$

In this section, we describe a reparameterized version of $L_u(\mathbf{G})$ that does not require optimization over a Grassmannian (Cook et al. 2016). The new parameterization requires that we first identify a nonsingular submatrix of u rows of a semi-orthogonal basis matrix $\mathbf{\Gamma}$ for $\mathcal{E}_{\mathbf{M}}(\mathcal{U})$. Arranging the responses so this submatrix corresponds to the first u rows of $\mathbf{\Gamma}$, we have the structure $\mathbf{\Gamma} = (\mathbf{\Gamma}_1^T, \mathbf{\Gamma}_2^T)^T$, where $\mathbf{\Gamma}_1 \in \mathbb{R}^{u \times u}$ is nonsingular. Imposing this structure on the argument of the objective function $L_u(\mathbf{G})$, we have

$$\mathbf{G} = \begin{pmatrix} \mathbf{G}_1 \\ \mathbf{G}_2 \end{pmatrix} = \begin{pmatrix} \mathbf{I}_u \\ \mathbf{A} \end{pmatrix} \mathbf{G}_1 = \mathbf{C}_{\mathbf{A}} \mathbf{G}_1,$$

where $\mathbf{G}_1 \in \mathbb{R}^{u\times u}$ is nonsingular, $\mathbf{A} = \mathbf{G}_2\mathbf{G}_1^{-1} \in \mathbb{R}^{(r-u)\times u}$ is an unconstrained matrix, and $\mathbf{C}_\mathbf{A} = (\mathbf{I}_u, \mathbf{A}^T)^T$. Since $\mathbf{G}^T\mathbf{G} = \mathbf{I}_u$, $\mathbf{G}_1\mathbf{G}_1^T = (\mathbf{C}_\mathbf{A}^T\mathbf{C}_\mathbf{A})^{-1}$. Using these relationships, $L_u(\mathbf{G})$ can be reparameterized as a function of only $\mathbf{A}$:

$$L_u(\mathbf{A}) = -2\log|\mathbf{C}_\mathbf{A}^T\mathbf{C}_\mathbf{A}| + \log|\mathbf{C}_\mathbf{A}^T\widehat{\mathbf{M}}\mathbf{C}_\mathbf{A}| + \log|\mathbf{C}_\mathbf{A}^T(\widehat{\mathbf{M}}+\widehat{\mathbf{U}})^{-1}\mathbf{C}_\mathbf{A}|.$$

With this objective function, minimization over $\mathbf{A}$ is unconstrained. The number of real parameters $u(r-u)$ comprising $\mathbf{A}$ is the same as the number of reals needed to specify a single element in the Grassmannian $\mathcal{G}(u, r)$.

If $u(r-u)$ is not too large, $L_u(\mathbf{A})$ might be minimized directly by using standard optimization software and the starting values described in Section 6.2.3. In other cases, minimization can be carried out iteratively over the rows of $\mathbf{A}$. Suppose that we wish to minimize over the last row $\mathbf{a}^T$ of $\mathbf{A}$. Partition

$$\mathbf{A} = \begin{pmatrix} \mathbf{A}_1 \\ \mathbf{a}^T \end{pmatrix},\quad \mathbf{C}_\mathbf{A} = \begin{pmatrix} \mathbf{C}_{\mathbf{A}_1} \\ \mathbf{a}^T \end{pmatrix},\quad \widehat{\mathbf{M}} = \begin{pmatrix} \widehat{\mathbf{M}}_{11} & \widehat{\mathbf{M}}_{12} \\ \widehat{\mathbf{M}}_{21} & \widehat{M}_{22} \end{pmatrix},\quad (\widehat{\mathbf{M}}+\widehat{\mathbf{U}})^{-1} = \begin{pmatrix} \widehat{\mathbf{V}}_{11} & \widehat{\mathbf{V}}_{12} \\ \widehat{\mathbf{V}}_{21} & \widehat{V}_{22} \end{pmatrix}.$$

Then after a little algebra, the objective function for minimizing over $\mathbf{a}$ with $\mathbf{A}_1$ held fixed can be written as

$$\begin{aligned} L_u(\mathbf{a} \mid \mathbf{A}_1) = {} & -2\log\{1 + \mathbf{a}^T(\mathbf{C}_{\mathbf{A}_1}^T\mathbf{C}_{\mathbf{A}_1})^{-1}\mathbf{a}\} \\ & + \log\{1 + \widehat{M}_{22}(\mathbf{a} + \widehat{M}_{22}^{-1}\mathbf{C}_{\mathbf{A}_1}^T\widehat{\mathbf{M}}_{12})^T\mathbf{W}_1^{-1}(\mathbf{a} + \widehat{M}_{22}^{-1}\mathbf{C}_{\mathbf{A}_1}^T\widehat{\mathbf{M}}_{12})\} \\ & + \log\{1 + \widehat{V}_{22}(\mathbf{a} + \widehat{V}_{22}^{-1}\mathbf{C}_{\mathbf{A}_1}^T\widehat{\mathbf{V}}_{12})^T\mathbf{W}_2^{-1}(\mathbf{a} + \widehat{V}_{22}^{-1}\mathbf{C}_{\mathbf{A}_1}^T\widehat{\mathbf{V}}_{12})\}, \end{aligned}$$

where

$$\begin{aligned} \mathbf{W}_1 &= \mathbf{C}_{\mathbf{A}_1}^T(\widehat{\mathbf{M}}_{11} - \widehat{M}_{22}^{-1}\widehat{\mathbf{M}}_{12}\widehat{\mathbf{M}}_{21})\mathbf{C}_{\mathbf{A}_1}, \\ \mathbf{W}_2 &= \mathbf{C}_{\mathbf{A}_1}^T(\widehat{\mathbf{V}}_{11} - \widehat{V}_{22}^{-1}\widehat{\mathbf{V}}_{12}\widehat{\mathbf{V}}_{21})\mathbf{C}_{\mathbf{A}_1}. \end{aligned}$$

The objective function $L_u(\mathbf{a} \mid \mathbf{A}_1)$ can now be minimized using an off-the-shelf algorithm, and iteration continues over rows of $\mathbf{A}$ until a convergence criterion is met.

The method described here uses the starting value $\widehat{\mathbf{G}}_{\text{start}}$ described in Section 6.2.3. Prior to application of the algorithm, we must identify u rows of $\widehat{\mathbf{G}}_{\text{start}}$ and then constrain the matrix $\widehat{\mathbf{G}}_{\text{start},1} \in \mathbb{R}^{u\times u}$ formed from those rows to be nonsingular. This of course implies that the matrix $\mathbf{\Gamma}_1$ formed from the corresponding rows of a basis matrix $\mathbf{\Gamma}$ for $\mathcal{E}_\mathbf{M}(\mathcal{U})$ should also be nonsingular. This can be achieved asymptotically by applying Gaussian elimination with partial pivoting to $\widehat{\mathbf{G}}_{\text{start}}$. The u rows identified during this process then form $\widehat{\mathbf{G}}_{\text{start},1}$. Cook et al. (2016, Prop. 3.1) showed that, as $n \to \infty$, this $\widehat{\mathbf{G}}_{\text{start},1}$ converges to a nonsingular matrix at the $\sqrt{n}$ rate, so asymptotically Gaussian elimination produces a nonsingular $u \times u$ matrix.

6.4 Sequential Likelihood-Based Envelope Estimation

In this section, we describe the sequential likelihood-based algorithm proposed by Cook and Zhang (2016) for estimating a general envelope $\mathcal{E}_{\mathbf{M}}(\mathcal{U})$. Called the *1D algorithm*, it requires optimization in r dimensions in the first step and reduces the optimization dimension by 1 in each subsequent step until an estimate $\widehat{\boldsymbol{\Gamma}}$ of a basis for $\mathcal{E}_{\mathbf{M}}(\mathcal{U})$ is obtained. Compared to (6.3), it is straightforward to implement and has the potential to reduce the computational burden with little loss of effectiveness. It could be used to produce a stand-alone estimator, or as an alternative method for getting starting values for a Grassmann optimization of (6.3). In the latter case, one Newton–Raphson iteration from $\widehat{\boldsymbol{\Gamma}}$ produces an estimator that is asymptotically equivalent to the maximum likelihood estimator (e.g. Small et al. 2000).

Relative to the algorithm discussed in Section 6.3, the 1D algorithm tends to be faster and just as precise for small values of u, but for large values of u, it can be relatively slow and imprecise. The envelope estimators from the algorithm of Section 6.3 are not necessarily nested, whereas the 1D algorithm produces nested envelope estimators: an envelope estimator of dimension u contains all the estimators of dimension less than u. Selecting u with AIC, BIC, and likelihood ratio testing is generally problematic under the 1D algorithm since it does not maximize a proper likelihood under normality. For that reason, it may be best used when prediction is a primary concern and cross-validation is employed to determine a suitable u. On the other hand, under mild regularity conditions without assuming a model, Zhang and Mai (2017) developed BIC-like consistent dimension selection methods for the 1D algorithm and the full objective function in (6.3). This means that consistent dimension selection is possible for those model-free algorithms without using cross-validation.

6.4.1 The 1D Algorithm

This algorithm requires $\mathbf{M} > 0$. It is stated here in the population; for the sample version $\mathbf{M}$ and $\mathbf{U}$ are replaced by $\widehat{\mathbf{M}}$ and $\widehat{\mathbf{U}}$. Let $\mathbf{g}_k \in \mathbb{R}^r$, $k = 1, \dots, u$, be the orthogonal stepwise directions constructed by the algorithm. Let $\mathbf{G}_k = (\mathbf{g}_0, \dots, \mathbf{g}_k)$, let $(\mathbf{G}_k, \mathbf{G}_{0k})$ be an orthogonal basis for $\mathbb{R}^r$ and set the initial value $\mathbf{g}_0 = \mathbf{G}_0 = 0$. Then for $k = 0, \dots, u-1$ get

$$\mathbf{w}_{k+1} = \arg \min_{\mathbf{w} \in \mathbb{R}^{r-k}} \phi_k(\mathbf{w}), \text{ subject to } \mathbf{w}^T\mathbf{w} = 1,$$

$$\mathbf{g}_{k+1} = \mathbf{G}_{0k}\mathbf{w}_{k+1},$$

where $\phi_k(\mathbf{w}) = \log(\mathbf{w}^T\mathbf{G}_{0k}^T\mathbf{M}\mathbf{G}_{0k}\mathbf{w}) + \log(\mathbf{w}^T\{\mathbf{G}_{0k}^T(\mathbf{M}+\mathbf{U})\mathbf{G}_{0k}\}^{-1}\mathbf{w})$.

On the first pass through the algorithm, $k = 0$ and we take $\mathbf{G}_{00} = \mathbf{I}_r$. With $\mathbf{M} = \boldsymbol{\Sigma}$ and $\mathbf{U} = \boldsymbol{\beta}\boldsymbol{\Sigma}_{\mathbf{X}}\boldsymbol{\beta}^T$, the result is the same as (6.2) with $u = 1$. The first iteration gives $\mathbf{g}_1 = \mathbf{w}_1 \in \mathcal{E}_{\mathbf{M}}(\mathcal{U})$, even if $u > 1$, as shown in Proposition B.1. For the second iteration, we construct $\mathbf{G}_{01}$ orthogonal to $\mathbf{w}_1$ and then use Proposition A.7 to reduce the coordinates, $\mathbf{M} \mapsto \mathbf{G}_{01}^T\mathbf{M}\mathbf{G}_{01}$ and $\mathbf{U} \mapsto \mathbf{G}_{01}^T\mathbf{U}\mathbf{G}_{01}$, and obtain $\mathbf{w}_2 \in \mathcal{E}_{\mathbf{G}_{01}^T\mathbf{M}\mathbf{G}_{01}}(\mathbf{G}_{01}^T\mathbf{U})$. It then follows from Proposition A.7 that $\mathbf{g}_2 = \mathbf{G}_{01}\mathbf{w}_2 \in \mathcal{E}_{\mathbf{M}}(\mathcal{U})$. The conclusion for subsequent directions is obtained similarly, and the algorithm terminates with span $(\mathbf{G}_u) = \mathcal{E}_{\mathbf{M}}(\mathcal{U})$. Because of the constraint $\mathbf{w}^T\mathbf{w} = 1$, the minimization at the kth step of the algorithm is over the Grassmannian $\mathcal{G}(1, r-k)$. This constraint could be handled by using an algorithm for Grassmann optimization, adapting the method of Section 6.3, or by reparameterizing $\mathbf{w} = \mathbf{v}/\|\mathbf{v}\|_2$ in terms of the unconstrained vector $\mathbf{v}$ to get

$$\phi_k(\mathbf{v}) = \log(\mathbf{v}^T\mathbf{G}_{0k}^T\mathbf{M}\mathbf{G}_{0k}\mathbf{v}) + \log(\mathbf{v}^T\{\mathbf{G}_{0k}^T(\mathbf{M}+\mathbf{U})\mathbf{G}_{0k}\}^{-1}\mathbf{v}) - 4\log(\|\mathbf{v}\|_2),$$

where $\|\mathbf{v}\|_2 = (\mathbf{v}^T\mathbf{v})^{1/2}$. Optimization of this unconstrained function gives the same essential solution as the constrained optimization:

$$\text{span}\{\arg\min_{\mathbf{w}\in\mathbb{R}^{r-k}} \phi_k(\mathbf{w}) \mid \mathbf{w}^T\mathbf{w} = 1\} = \text{span}\{\arg\min_{\mathbf{v}\in\mathbb{R}^{r-k}} \phi_k(\mathbf{v})\}.$$

The hardest part of implementing the 1D algorithm is minimizing $\phi_k(\mathbf{v})$ at each iteration. Since ϕ_k is nonconvex and has local minima, it may be hard to find an off-the-shelf algorithm that stably minimizes it at each iteration, particularly when r is large. The starting values summarized in Section 6.2.3 could be used to facilitate minimization of ϕ_k. Alternatively, Cook and Zhang (2017) developed an envelope coordinate decent (ECD) algorithm for minimizing ϕ_k. The ECD algorithm is built on two essential ideas. The first is to transform to canonical coordinates at each iteration by expressing $\mathbf{v}$ in terms of the eigenvectors of $\mathbf{G}_{0k}^T\mathbf{M}\mathbf{G}_{0k}$ and the second is to approximate the solution to $\partial\phi_k(\mathbf{v})/\partial\mathbf{v} = 0$. These two steps result in an algorithm that seems faster and more stable than standard nonlinear optimization algorithms.

Let $\widehat{\mathbf{M}}$ and $\widehat{\mathbf{U}}$ be $\sqrt{n}$-consistent estimators of $\mathbf{M}$ and $\mathbf{U}$ and let $\widehat{\boldsymbol{\Gamma}}$ be the basis from the sample version the 1D algorithm. It is shown in Section B.1 that then $\mathbf{P}_{\widehat{\boldsymbol{\Gamma}}}$ is a $\sqrt{n}$-consistent estimator of $\mathbf{P}_{\mathcal{E}_{\mathbf{M}}(\mathcal{U})}$ (Cook and Zhang 2016).

6.4.2 Envelope Component Screening

Cook and Zhang (2017) proposed a screening algorithm called envelope component Screening (ECS) that reduces the original dimension r to a more manageable dimension $d \leq n$, without losing notable structural information on the envelope. The ECS algorithm, which is discussed in this section, can be a useful tool for reducing the dimension prior to application of a full optimization algorithm.

The algorithms that we have discussed so far are all based on *minimizing* versions of the objective function $L_u(\mathbf{G})$ defined at (6.4). For instance, we know from the discussion of the 1D algorithm that span $\{\arg\min_{\mathbf{G}\in\mathbb{R}^r} L_u(\mathbf{G}) \mid \mathbf{G}^T\mathbf{G} = 1\}$ gives a $\sqrt{n}$-consistent estimator of a one-dimensional subspace of

$\mathcal{E}_{\mathbf{M}}(\mathcal{U})$. The basic idea behind the ECS algorithms is to eliminate directions by *maximizing* $L_u(\mathbf{G})$. The motivation for this comes from the following population proposition (Cook and Zhang 2017). In preparation, let $L_{\text{pop}}(\mathbf{G}) = \log|\mathbf{G}^T\mathbf{M}\mathbf{G}| + \log|\mathbf{G}^T(\mathbf{M}+\mathbf{U})^{-1}\mathbf{G}|$ denote the population version of $L_u(\mathbf{G})$, and let $L^0_{\text{pop}}(\mathbf{G}) = \log|\mathbf{G}^T\mathbf{M}\mathbf{G}| + \log|\mathbf{G}_0^T(\mathbf{M}+\mathbf{U})\mathbf{G}_0|$ denote the equivalent objective function that does not require computation of $(\mathbf{M}+\mathbf{U})^{-1}$.

Proposition 6.2 *Let* $\mathbf{A}$ *be a semi-orthogonal basis matrix for a reducing subspace of* $\mathbf{M}$ *and let* $(\mathbf{A},\mathbf{A}_0) \in \mathbb{R}^{r\times r}$ *be an orthogonal matrix. Then* $L_{\text{pop}}(\mathbf{A}_0) \le 0$ *and* $L_{\text{pop}}(\mathbf{A}_0) = 0$ *if and only if* $\mathcal{U} \subseteq \text{span}(\mathbf{A})$. *If* $L_{\text{pop}}(\mathbf{A}_0) = 0$ *then* $\mathcal{E}_{\mathbf{M}}(\mathcal{U}) = \mathbf{A}\mathcal{E}_{\mathbf{A}^T\mathbf{M}\mathbf{A}}(\text{span}(\mathbf{A}^T\mathbf{U}\mathbf{A}))$.

Proof: It follows from Lemma A.15 that

$$\begin{aligned} L_{\text{pop}}(\mathbf{A}_0) &= \log|\mathbf{A}_0^T\mathbf{M}\mathbf{A}_0| + \log|\mathbf{A}_0^T(\mathbf{M}+\mathbf{U})^{-1}\mathbf{A}_0| \\ &\le \log|\mathbf{A}_0^T\mathbf{M}\mathbf{A}_0| + \log|\mathbf{A}_0^T\mathbf{M}^{-1}\mathbf{A}_0| = 0. \end{aligned}$$

To see the second conclusion factor $\mathbf{U} = \mathbf{u}\mathbf{u}^T$, let $(\mathbf{I} + \mathbf{u}^T\mathbf{M}^{-1}\mathbf{u})^{-1} = \mathbf{C}^T$ and write

$$\begin{aligned} \log|\mathbf{A}_0^T(\mathbf{M}+\mathbf{U})^{-1}\mathbf{A}_0| &= \log|\mathbf{A}_0^T(\mathbf{M}+\mathbf{u}\mathbf{u}^T)^{-1}\mathbf{A}_0| \\ &= \log|\mathbf{A}_0^T\mathbf{M}^{-1}\mathbf{A}_0 - \mathbf{A}_0^T\mathbf{M}^{-1}\mathbf{u}\mathbf{C}\mathbf{u}^T\mathbf{M}^{-1}\mathbf{A}_0|. \end{aligned}$$

From this we see that $L_{\text{pop}}(\mathbf{A}_0) = 0$ if and only if $\mathbf{A}_0^T\mathbf{M}^{-1}\mathbf{u} = 0$ because then

$$L_{\text{pop}}(\mathbf{A}_0) = \log|\mathbf{A}_0^T\mathbf{M}\mathbf{A}_0| + \log|\mathbf{A}_0^T\mathbf{M}^{-1}\mathbf{A}_0| = 0,$$

the final equality holding because $\text{span}(\mathbf{A}_0)$ reduces $\mathbf{M}$. Now, $\mathbf{A}_0^T\mathbf{M}^{-1}\mathbf{u} = 0$ if and only if $\text{span}(\mathbf{M}^{-1}\mathbf{u}) \subseteq \text{span}(\mathbf{A})$. Since $\text{span}(\mathbf{A})$ reduces $\mathbf{M}$ this holds if and only if $\text{span}(\mathbf{u}) \subseteq \text{span}(\mathbf{A})$. We apply Proposition A.6 to reach the final conclusion that $\mathcal{E}_{\mathbf{M}}(\mathcal{U}) = \mathbf{A}\mathcal{E}_{\mathbf{A}^T\mathbf{M}\mathbf{A}}(\text{span}(\mathbf{A}^T\mathbf{U}\mathbf{A}))$. This is similar to the use of Proposition A.6 when we developed procedures for testing responses in Section 1.8. □

Proposition 6.2 says that if we have a reducing subspace of $\mathbf{M}$ with semi-orthogonal basis matrix $\mathbf{A}$ so that $L_u(\mathbf{A}_0) = 0$, then $\mathcal{U} \subseteq \text{span}(\mathbf{A})$, which implies that $\mathcal{E}_{\mathbf{M}}(\mathcal{U}) \subseteq \mathcal{E}_{\mathbf{M}}(\text{span}(\mathbf{A})) = \text{span}(\mathbf{A})$. This then leads to the following population version of the ECS algorithm. The sample version is constructed by replacing $\mathbf{M}$ and $\mathbf{U}$ by their sample counterparts. For simplicity, we assume that in the remainder of this section the eigenvalues of $\mathbf{M}$ are distinct. See Cook and Zhang (2017) for a treatment of multiplicities.

6.4.2.1 ECS Algorithm

Let $\mathbf{M} = \sum_{i=1}^r \varphi_i \boldsymbol{\ell}_i \boldsymbol{\ell}_i^T$ be the spectral decomposition of $\mathbf{M}$, and let

$$l_i = L_{\text{pop}}(\boldsymbol{\ell}_i) = \log(\varphi_i) + \log(\boldsymbol{\ell}_i^T(\mathbf{M}+\mathbf{U})^{-1}\boldsymbol{\ell}_i), \quad i = 1,\ldots,r.$$

Order the evaluations $l_{(r)} \leq \cdots \leq l_{(1)}$ and let $\boldsymbol{\ell}_{(r)}, \ldots, \boldsymbol{\ell}_{(1)}$ be the corresponding eigenvectors. Then $\mathbf{A}_0 = (\boldsymbol{\ell}_{(1)}, \ldots, \boldsymbol{\ell}_{r-d})$ and $\mathbf{A} = (\boldsymbol{\ell}_{r-d+1}, \ldots, \boldsymbol{\ell}_r)$ for a prespecified integer $r > d \geq u$.

Since we are assuming distinct eigenvalues for $\mathbf{M}$, the envelope $\mathcal{E}_{\mathbf{M}}(\mathcal{U})$ will be spanned by exactly u eigenvectors of $\mathbf{M}$. Accordingly, we have $l_{(1)} = \cdots = l_{(r-u)} = 0$ and $l_j < 0$ for $j = r - u + 1, \ldots, r$. So the population algorithm will in effect return $\mathcal{E}_{\mathbf{M}}(\mathcal{U})$. Of course this will not be so in the sample, necessitating the choice of d.

The following proposition describes the asymptotic behavior of the sample version of the ECS algorithm. Let $\widehat{\mathbf{A}}$ and $\widehat{\mathbf{A}}_0$ denote the estimators corresponding to $\mathbf{A}$ and $\mathbf{A}_0$ in the population ECS algorithm.

Proposition 6.3 *Let $\widehat{\mathbf{M}}$ and $\widehat{\mathbf{U}}$ be $\sqrt{n}$-consistent estimators of $\mathbf{M} > 0$ and $\mathbf{U} \geq 0$. Assume that the eigenvalues of $\mathbf{M}$ are distinct and that $d \geq u$. Then $L_{\text{pop}}(\widehat{\mathbf{A}}_0) = O_p(n^{-1/2})$ and $L_u(\widehat{\mathbf{A}}_0) = L_{\text{pop}}(\widehat{\mathbf{A}}_0) + O_p(n^{-1/2})$.*

The final ingredient needed for the ECS algorithm is a method to choose d that serves as an upper bound for u but otherwise does not have to be accurately specified. Cook and Zhang (2017) suggested a data-driven approach base on the observation that $L_u(\widehat{\mathbf{A}}_0)$ is monotonically increasing in d. Since d serves to establish a boundary between the large and small values of $L_u(\widehat{\mathbf{A}}_0)$, it seems reasonable to choose the largest d so that $L_u(\widehat{\mathbf{A}}_0) > C_0(n)$ for some cutoff value $C_0(n) < 0$ that has smaller order than the rate at which $L_u(\widehat{\mathbf{A}}_0)$ converges to 0 from Proposition 6.3. Cook and Zhang (2017) found that $C_0 = -1/n$ works well.

Recall that in Sections 6.2.1 and 6.2.2, we chose potential starting values from the eigenvectors of $\widehat{\mathbf{M}}$ and $\widehat{\mathbf{M}} + \widehat{\mathbf{U}}$. A similar rational applies here. Since Proposition 6.2 holds with $\mathbf{M}$ replaced by $\mathbf{M} + \mathbf{U}$, there is an alternative ECS algorithm based on the eigenvectors of $\widehat{\mathbf{M}} + \widehat{\mathbf{U}}$.

6.4.2.2 Alternative ECS Algorithm

Let $\mathbf{M} + \mathbf{U} = \sum_{i=1}^{r} \varphi_i \boldsymbol{\ell}_i \boldsymbol{\ell}_i^T$ be the spectral decomposition of $\mathbf{M} + \mathbf{U}$, and let

$$l_i = L_{\text{pop}}(\boldsymbol{\ell}_i) = \log(\boldsymbol{\ell}_i^T \mathbf{M} \boldsymbol{\ell}_i) - \log(\varphi_i), \quad i = 1, \ldots, r.$$

Order the evaluations $l_{(r)} \leq \cdots \leq l_{(1)}$ and let $\boldsymbol{\ell}_{(r)}, \ldots, \boldsymbol{\ell}_{(1)}$ be the corresponding eigenvectors. Then $\mathbf{A}_0 = (\boldsymbol{\ell}_{(1)}, \ldots, \boldsymbol{\ell}_{r-d})$ and $\mathbf{A} = (\boldsymbol{\ell}_{r-d+1}, \ldots, \boldsymbol{\ell}_r)$ for a prespecified integer $r > d \geq u$.

With the same d, the ECS algorithm and its alternative can produce different $\widehat{\mathbf{A}}$ and $\widehat{\mathbf{A}}_0$. These different solutions can be distinguished by selecting the one that achieves the smaller value of $L_u(\mathbf{A})$.

In regression with a relatively small sample size, $\widehat{\mathbf{M}}$ and $\widehat{\mathbf{M}} + \widehat{\mathbf{U}}$ may be singular, which can cause problems for the algorithms in this chapter. Definition 1.2 of the envelope $\mathcal{E}_{\mathbf{M}}(\mathcal{U})$ requires that $\mathcal{U} \subseteq \text{span}\,(\mathbf{M})$. If $\mathbf{M}$ is singular then that containment may not hold and there may be a tendency for methodology to return only an estimator of $\mathcal{U} \cap \text{span}(\mathbf{M})$, resulting in a loss of structural information. One way to approach regressions in which $\widehat{\mathbf{M}}$ is singular is to follow Proposition A.6 and reduce the size of the problem by downsizing via $\mathcal{E}_{\mathbf{M}}(\mathcal{U}) = \mathbf{A}\mathcal{E}_{\mathbf{A}^T\mathbf{MA}}(\text{span}(\mathbf{A}^T\mathbf{UA}))$, where the columns of $\mathbf{A}$ are the eigenvectors of $\widehat{\mathbf{M}}$ with nonzero eigenvalues. This is the same as the technique used by SIMPLS, as discussed in Section 4.2.2. The envelope $\mathcal{E}_{\mathbf{A}^T\mathbf{MA}}(\text{span}(\mathbf{A}^T\mathbf{UA}))$ can be estimated using any of the algorithms discussed in this chapter since $\mathbf{A}^T\widehat{\mathbf{M}}\mathbf{A} > 0$. Following the discussion at the end of Section 4.1, this method will likely be reasonable when reducing the predictors because the most desirable signals are associated with larger eigenvalues of $\mathbf{M}$. This method may not work well when reducing the response because then the most desirable signals are associated with smaller eigenvalues of $\mathbf{M}$. The sparse envelopes of Section 7.5.3 are applicable for response reduction, however.

6.5 Sequential Moment-Based Envelope Estimation

In this section, we describe a sequential moment-based algorithm. Like the 1D algorithm, it requires an r-dimensional optimization in the first step and reduces the optimization dimension by 1 in each subsequent step until an estimate $\widehat{\boldsymbol{\Gamma}}$ of a basis for $\mathcal{E}_{\mathbf{M}}(\mathcal{U})$ is obtained. It may be not only the easiest to implement but also tends to be the least accurate. Also like the 1D algorithm, it produces nested envelope estimators, but does not optimize a proper likelihood so u is best determined by predictive cross-validation.

6.5.1 Basic Algorithm

Let $\mathbf{u} \in \mathbb{R}^{r\times p}$ with $\text{rank}(\mathbf{u}) \le p$, let $\mathcal{U} = \text{span}(\mathbf{u}) \subseteq \text{span}(\mathbf{M})$, where $\mathbf{M} \in \mathbb{S}^{r\times r}$ is a positive semi-definite matrix. In this section, we describe a sequential method for finding $\mathcal{E}_{\mathbf{M}}(\mathcal{U})$ based only on knowledge of $\mathbf{u}$ and $\mathbf{M}$. Let $u = \dim(\mathcal{E}_{\mathbf{M}}(\mathcal{U}))$. Essentially a generalization of the SIMPLS method for PLS discussed in Chapter 4, the algorithm proceeds by finding a sequence of r-dimensional vectors $\mathbf{w}_0, \mathbf{w}_1, \dots, \mathbf{w}_u$, with $\mathbf{w}_0 = 0$, whose cumulative spans are strictly increasing and converge to $\mathcal{E}_{\mathbf{M}}(\mathcal{U})$ after u steps. Let $\mathbf{W}_k = (\mathbf{w}_0, \mathbf{w}_1, \dots, \mathbf{w}_k) \in \mathbb{R}^{r\times k}$ and let $\mathcal{W}_k = \text{span}(\mathbf{W}_k)$. Then the $\mathbf{w}_k$'s are constructed so that

$$\mathcal{W}_0 \subset \mathcal{W}_1 \subset \cdots \subset \mathcal{W}_{u-1} \subset \mathcal{W}_u = \mathcal{E}_{\mathbf{M}}(\mathcal{U}). \tag{6.5}$$

Let $\mathbf{U} = \mathbf{u}\mathbf{u}^T$, so $\text{span}(\mathbf{U}) = \text{span}(\mathbf{u})$. The algorithm is based on the following sequence of constrained optimizations, starting with $\mathbf{w}_1$ since $\mathbf{w}_0 = 0$ is a known starting vector. For $k = 0, 1, \dots, u-1$ find

$$\mathbf{w}_{k+1} = \arg\max_{\mathbf{w}} \mathbf{w}^T\mathbf{U}\mathbf{w}, \text{ subject to} \tag{6.6}$$

$$\mathbf{w}^T\mathbf{M}\mathbf{W}_k = 0, \tag{6.7}$$

$$\mathbf{w}^T\mathbf{w} = 1. \tag{6.8}$$

It follows from (6.7) that when solving for $\mathbf{w}_{k+1}$, the optimization variable $\mathbf{w}$ is constrained to lie in the orthogonal complement of $\mathcal{E}_k := \text{span}\,(\mathbf{M}\mathbf{W}_k)$, so we can write the algorithm step without explicit reference to (6.7) as

$$\begin{aligned}\mathbf{w}_{k+1} &= \arg\max_{\mathbf{w}} \mathbf{w}^T\mathbf{Q}_{\mathcal{E}_k}\mathbf{U}\mathbf{Q}_{\mathcal{E}_k}\mathbf{w}, \text{ subject to } \mathbf{w}^T\mathbf{w} = 1\\ &= \ell_{\max}(\mathbf{Q}_{\mathcal{E}_k}\mathbf{U}\mathbf{Q}_{\mathcal{E}_k}),\end{aligned}$$

where $\ell_{\max}(\mathbf{A})$ the normalized eigenvector associated with the largest eigenvalue of a symmetric matrix $\mathbf{A}$, as defined in Section 4.2.1. In consequence, the algorithm can be stated without explicit reference to the constraints as follows. Set $\mathbf{w}_0 = 0$ and $\mathbf{W}_0 = \mathbf{w}_0$. For $k = 0, 1, \ldots, u-1$, set

$$\begin{aligned}\mathcal{E}_k &= \text{span}\,(\mathbf{M}\mathbf{W}_k),\\ \mathbf{w}_{k+1} &= \ell_{\max}(\mathbf{Q}_{\mathcal{E}_k}\mathbf{U}\mathbf{Q}_{\mathcal{E}_k}),\\ \mathbf{W}_{k+1} &= (\mathbf{w}_0, \ldots, \mathbf{w}_k, \mathbf{w}_{k+1}).\end{aligned}$$

The algorithm continues until $\mathbf{Q}_{\mathcal{E}_k}\mathbf{U} = 0$, at which point $k = u$ and $\mathcal{E}_{\mathbf{M}}(\mathcal{U}) = \mathcal{W}_u = \text{span}\,(\mathbf{W}_u)$. Details underlying this algorithm, including a proof that it has property (6.5), are available in Section B.2

Viewing $\mathbf{U}$ and $\mathbf{M}$ as population parameters from a statistical perspective, this algorithm is Fisher consistent for the envelope $\mathcal{E}_{\mathbf{M}}(\mathcal{U})$. Substituting $\sqrt{n}$-consistent estimators for $\mathbf{U}$ and $\mathbf{M}$, the algorithm provides also a $\sqrt{n}$-consistent estimator for $\mathbf{P}_{\mathcal{E}}$. Even so, application of the algorithm could still seem elusive in practice because there are several ways in which to choose $\mathbf{U}$ and $\mathbf{M}$ that yield the same envelope in the population. For instance, choosing $\mathbf{U} = \mathbf{u}\mathbf{A}\mathbf{u}^T$ for any nonsingular $\mathbf{A} \in \mathbb{S}^{p\times p}$ also yields $\mathcal{E}_{\mathbf{M}}(\mathcal{U})$ in the population because $\mathcal{U} = \text{span}\,(\mathbf{u}\mathbf{A}\mathbf{u}^T)$. Similarly, from (A.7) in Corollary A.2, replacing $\mathbf{M}$ with $\mathbf{M}^k$ also yields $\mathcal{E}_{\mathbf{M}}(\mathcal{U})$ in the population. Although the population changes indicated here have no impact on the end result of the algorithm, they will produce different results in application when $\mathbf{u}$ and $\mathbf{M}$ are replaced by $\sqrt{n}$-consistent estimators.

Consider, for example, the problem of estimating the envelope $\mathcal{E}_{\boldsymbol{\Sigma}}(\mathcal{B})$ for response reduction in the multivariate regression of a response vector $\mathbf{Y}$ on a predictor vector $\mathbf{X}$, as discussed in Chapter 1. A seemingly natural choice might be $\hat{\mathbf{u}} = \mathbf{B} = \mathbb{Y}^T\mathbb{X}(\mathbb{X}^T\mathbb{X})^{-1}$ (cf. (1.2)), leading to

$$\hat{\mathbf{U}} = \mathbf{B}\mathbf{B}^T = \mathbb{Y}^T\mathbb{X}(\mathbb{X}^T\mathbb{X})^{-2}\mathbb{X}^T\mathbb{Y}.$$

However, $\hat{\mathbf{U}}$ is not invariant under nonsingular linear transformations $\mathbf{X} \mapsto \mathbf{A}\mathbf{X}$, and consequently the results of the sample algorithm will depend on the scaling of the predictors. Appealing to invariance then leads to a different choice:

$$\hat{\mathbf{U}} = \mathbf{B}(\mathbb{X}^T\mathbb{X})\mathbf{B}^T = \mathbb{Y}^T\mathbb{X}(\mathbb{X}^T\mathbb{X})^{-1}\mathbb{X}^T\mathbb{Y} = \mathbb{Y}^T\mathbf{P}_{\mathbb{X}}\mathbb{Y},$$

which is invariant under transformations $\mathbf{X} \mapsto \mathbf{AX}$. Choosing this form for $\widehat{\mathbf{U}}$ and $\widehat{\mathbf{M}} = \mathbf{S}_{\mathbf{Y}|\mathbf{X}}$ may then lead to a useful envelope algorithm.

A potential advantage of this algorithm is that it does not require that $\mathbf{M}$ be nonsingular, although it still requires $\mathcal{U} \subseteq \text{span}(\mathbf{M})$. This suggests that it may form a starting point for methodology to deal with settings in which the estimator of $\mathbf{M}$ is singular. Nevertheless, experience with simulations and data analyses indicates that on balance the non-Grassmann algorithm described in Section 6.2.2 gives the best results, followed by the 1D algorithm and then the moment-based algorithm.

6.5.2 Krylov Matrices and $\dim(\mathcal{U}) = 1$

The algorithm simplifies considerably when $\dim(\mathcal{U}) = 1$, so $\mathbf{U} = \mathbf{u}\mathbf{u}^T$ with $\mathbf{u} \in \mathbb{R}^r$. In that case, we have that $\mathcal{E}_{\mathbf{M}}(\mathcal{U})$ is the span of a Krylov sequence: $\mathcal{E}_{\mathbf{M}}(\mathcal{U}) = \text{span}(\mathbf{u}, \mathbf{M}^1\mathbf{u}, \dots, \mathbf{M}^{u-1}\mathbf{u})$. We proceed sequentially to see how this conclusion arises. The first direction vector is simply $\mathbf{w}_1 = \ell_1(\mathbf{U}) \propto \mathbf{u}$. The second direction vector is

$$\mathbf{w}_2 = \ell_1(\mathbf{Q}_{\mathcal{E}_1}\mathbf{U}\mathbf{Q}_{\mathcal{E}_1}) \propto \mathbf{Q}_{\mathcal{E}_1}\mathbf{u} = \mathbf{u} - \mathbf{P}_{\mathcal{E}_1}\mathbf{u}.$$

Thus,

$$\begin{aligned}\text{span}(\mathbf{W}_2) &= \text{span}(\mathbf{u}, \mathbf{u} - \mathbf{P}_{\mathcal{E}_1}\mathbf{u}) = \text{span}(\mathbf{u}, \mathbf{P}_{\mathcal{E}_1}\mathbf{u}) \\ &= \text{span}(\mathbf{u}, \mathbf{P}_{\mathbf{Mu}}\mathbf{u}) = \text{span}(\mathbf{u}, \mathbf{M}^1\mathbf{u}).\end{aligned}$$

Proceeding by induction, assume that $\text{span}(\mathbf{W}_k) = \text{span}(\mathbf{u}, \mathbf{M}^1\mathbf{u}, \dots, \mathbf{M}^{k-1}\mathbf{u})$ for $k = 1, \dots, q$. Then next direction vector is given by

$$\mathbf{w}_{q+1} = \ell(\mathbf{Q}_{\mathcal{E}_q}\mathbf{U}\mathbf{Q}_{\mathcal{E}_q}) \propto \mathbf{Q}_{\mathcal{E}_q}\mathbf{u} = \mathbf{u} - \mathbf{P}_{\mathcal{E}_q}\mathbf{u},$$

and therefore

$$\text{span}(\mathbf{W}_{q+1}) = \text{span}(\mathbf{u}, \mathbf{M}^1\mathbf{u}, \dots, \mathbf{M}^{q-1}\mathbf{u}, \mathbf{P}_{\mathcal{E}_q}\mathbf{u}).$$

Since $\mathbf{P}_{\mathcal{E}_q}\mathbf{u} = \mathbf{P}_{\mathbf{MW}_q}\mathbf{u}$, we can write $\mathbf{P}_{\mathcal{E}_q}\mathbf{u}$ as a linear combination of the q spanning vectors $\mathbf{M}^1\mathbf{u}, \mathbf{M}^2\mathbf{u}, \dots, \mathbf{M}^q\mathbf{u}$. It follows that

$$\text{span}(\mathbf{W}_{q+1}) = \text{span}(\mathbf{u}, \mathbf{M}^1\mathbf{u}, \mathbf{M}^2\mathbf{u}, \dots, \mathbf{M}^q\mathbf{u}).$$

6.5.3 Variations on the Basic Algorithm

There are many variations on the choice of $\mathbf{U}$ and $\mathbf{M}$ that lead to the same envelope. As defined in Section A.3.1, let $f : \mathbb{R} \mapsto \mathbb{R}$ with the properties $f(0) = 0$ and $f(x) \neq 0$ whenever $x \neq 0$, and then define $f^* : \mathbb{S}^{r \times r} \mapsto \mathbb{S}^{r \times r}$ as $f^*(\mathbf{M}) = \sum_{i=1}^{q} f(\varphi_i)\mathbf{P}_i$, here $\mathbf{M}$ has spectral decomposition $\sum_{i=1}^{q} \varphi_i \mathbf{P}_i$ with q unique eigenvalues and projections $\mathbf{P}_i$ onto the corresponding eigenspaces.

Then it follows from Corollary A.2 that replacing $\mathbf{U}$ with $f^*(\mathbf{M})\mathbf{U}f^*(\mathbf{M})$ produces the same envelope after u steps, as does replacing $\mathbf{M}$ by $f^*(\mathbf{M})$ for a strictly monotonic f. Such modifications may alter the individual direction vectors $\mathbf{w}_k$, but they will not change $\mathcal{W}_u = \mathcal{E}_{\mathbf{M}}(\mathcal{U})$ in the end.

Variations on the length constraint (6.8) are more involved. Consider replacing the length constraint $\mathbf{w}^T\mathbf{w} = 1$ with $\mathbf{w}^T\mathbf{V}\mathbf{w} = 1$, where $\mathbf{V} \in \mathbb{S}^{r\times r}$ is positive definite, without altering the rest of the algorithm in (6.6) and (6.7). This variation can be studied by changing coordinates to represent it in the form of the original algorithm. Let $\mathbf{h} = \mathbf{V}^{1/2}\mathbf{w}$ and $\mathbf{H}_k = \mathbf{V}^{1/2}\mathbf{W}_k$. Then the algorithm in the $\mathbf{h}$ coordinates is

$$\mathbf{h}_{k+1} = \arg\max_{\mathbf{h}} \mathbf{h}^T\mathbf{V}^{-1/2}\mathbf{U}\mathbf{V}^{-1/2}\mathbf{h}, \text{ subject to} \tag{6.9}$$

$$\mathbf{h}^T\mathbf{V}^{-1/2}\mathbf{M}\mathbf{V}^{-1/2}\mathbf{H}_k = 0, \tag{6.10}$$

$$\mathbf{h}^T\mathbf{h} = 1, \tag{6.11}$$

which produces $\mathcal{E}_{\mathbf{V}^{-1/2}\mathbf{M}\mathbf{V}^{-1/2}}(\mathbf{V}^{-1/2}\mathcal{U})$ after u steps. After transforming back to the $\mathbf{w}$ coordinate, we have the resulting subspace $\mathbf{V}^{-1/2}\mathcal{E}_{\mathbf{V}^{-1/2}\mathbf{M}\mathbf{V}^{-1/2}}(\mathbf{V}^{-1/2}\mathcal{U})$. When is it true that

$$\mathcal{E}_{\mathbf{M}}(\mathcal{U}) = \mathbf{V}^{-1/2}\mathcal{E}_{\mathbf{V}^{-1/2}\mathbf{M}\mathbf{V}^{-1/2}}(\mathbf{V}^{-1/2}\mathcal{U})? \tag{6.12}$$

We know from Proposition A.4 that if $\mathbf{V}$ commutes with $\mathbf{M}$ and $\mathcal{E}_{\mathbf{M}}(\mathcal{U})$ reduces $\mathbf{V}$ then

$$\mathbf{V}^{-1/2}\mathcal{E}_{\mathbf{V}^{-1/2}\mathbf{M}\mathbf{V}^{-1/2}}(\mathbf{V}^{-1/2}\mathcal{U}) = \mathcal{E}_{\mathbf{V}^{-1/2}\mathbf{M}\mathbf{V}^{-1/2}}(\mathcal{U}).$$

From Proposition A.3, we will have $\mathcal{E}_{\mathbf{M}}(\mathcal{U}) = \mathcal{E}_{\mathbf{V}^{-1/2}\mathbf{M}\mathbf{V}^{-1/2}}(\mathcal{U})$ if the eigenspaces of $\mathbf{M}$ that are not orthogonal to $\mathcal{U}$ are the same as the eigenspaces of $\mathcal{E}_{\mathbf{V}^{-1/2}\mathbf{M}\mathbf{V}^{-1/2}}(\mathcal{U})$ that are not orthogonal to $\mathcal{U}$. In reference to Corollary A.2, these sufficient conditions are satisfied when $\mathbf{V} = f^*(\mathbf{M})$, and $f(\cdot)$ is strictly monotonic. In particular, we can use $\mathbf{V} = \mathbf{M}^a$ for $a \in \mathbb{R}^1$ and $a \neq 1$ without altering the end result of the algorithm. The case where $\mathbf{V} = \mathbf{M}$ is nonsingular fails because then $f(\cdot)$ is the identity ($a = 1$), $\mathbf{V}^{-1/2}\mathbf{M}\mathbf{V}^{-1/2} = \mathbf{I}_r$, and consequently its eigenspaces are not the same as those of $\mathbf{M}$.

The envelope algorithm is also invariant under choices of a positive semi-definite $\mathbf{V} \geq 0$ when $\mathbf{V}$ satisfies the previous conditions for the positive definite case: $\mathbf{V}$ commutes with $\mathbf{M}$, $\text{span}(\mathbf{V}) \subseteq \text{span}(\mathbf{M})$, and $\mathcal{E}_{\mathbf{M}}(\mathcal{U})$ reduces $\mathbf{V}$. This can be demonstrated by adapting the justification presented in Section 6.5.1 reasoning as follows. Since $\mathcal{E}_{\mathbf{M}}(\mathcal{U})$ reduces $\mathbf{V}$, we have $\mathbf{V} = \boldsymbol{\Gamma}(\boldsymbol{\Gamma}^T\mathbf{V}\boldsymbol{\Gamma})\boldsymbol{\Gamma}^T + \boldsymbol{\Gamma}_0(\boldsymbol{\Gamma}_0^T\mathbf{V}\boldsymbol{\Gamma}_0)\boldsymbol{\Gamma}_0^T$ and $\boldsymbol{\Gamma}^T\mathbf{V}\boldsymbol{\Gamma} > 0$. Consequently, the rank deficiency in $\mathbf{V}$ is manifested only in $\boldsymbol{\Gamma}_0^T\mathbf{V}\boldsymbol{\Gamma}_0$. The part of $\mathbf{V}$ that is orthogonal to $\mathcal{E}_{\mathbf{M}}(\mathcal{U})$ plays no role in the algorithm, and consequently we can modify $\boldsymbol{\Gamma}_0^T\mathbf{V}\boldsymbol{\Gamma}_0$ to make it nonsingular without affecting the outcome, leading back to the full-rank case.

7

Envelope Extensions

This chapter provides building blocks for developing envelope methodology in general multivariate studies. We describe in Section 7.1 a general paradigm for enveloping a vector-valued parameter (Cook and Zhang 2015a). Matrix-valued parameters are considered in Section 7.2. Section 7.3 contains an introduction to envelopes for regressions with matrix-valued responses. Spatial envelopes (Rekabdarkolaee and Wang 2017) are discussed in Section 7.4, sparse response envelopes (Su et al. 2016) are discussed in Section 7.5, and a cursory introduction to Bayesian response envelopes (Khare et al. 2016) is given in Section 7.6.

7.1 Envelopes for Vector-Valued Parameters

Let $\widetilde{\boldsymbol{\theta}} \in \mathbb{R}^m$ denote an estimator of an unknown parameter vector $\boldsymbol{\theta} \in \mathbb{R}^m$ based on a sample of size n. For example, $\widetilde{\boldsymbol{\theta}}$ might be the maximum likelihood estimator of the coefficient vector in a generalized linear model, a least squares estimator or a robust estimator in linear regression with a univariate response. Assume, as is often the case, that $\sqrt{n}(\widetilde{\boldsymbol{\theta}} - \boldsymbol{\theta})$ converges in distribution to a normal random vector with mean 0 and covariance matrix $\mathbf{V}(\boldsymbol{\theta}) > 0$ as $n \to \infty$. This condition is used to insure that the covariance matrix $\mathbf{V}(\boldsymbol{\theta})$ adequately reflects the asymptotic uncertainty in $\widetilde{\boldsymbol{\theta}}$. A different approach may be necessary when it does not hold. The goal in this context is to use enveloping to reduce estimative variation in $\widetilde{\boldsymbol{\theta}}$.

To accommodate the possible presence of nuisance parameters, we decompose $\boldsymbol{\theta}$ as $\boldsymbol{\theta} = (\boldsymbol{\psi}^T, \boldsymbol{\phi}^T)^T$, where $\boldsymbol{\phi} \in \mathbb{R}^p$ is the parameter vector of interest, and $\boldsymbol{\psi} \in \mathbb{R}^{m-p}$ is the nuisance parameter vector. Let $\mathbf{V}_{\boldsymbol{\phi}}(\boldsymbol{\theta})$ be the $p \times p$ lower right block of $\mathbf{V}(\boldsymbol{\theta})$, which is the asymptotic covariance matrix of $\widetilde{\boldsymbol{\phi}}$. Then the envelope for $\boldsymbol{\phi}$ is defined as follows (Cook and Zhang 2015a).

Definition 7.1 The envelope for the parameter $\boldsymbol{\phi} \in \mathbb{R}^p$ with asymptotically normal estimator $\widetilde{\boldsymbol{\phi}}$, $\sqrt{n}(\widetilde{\boldsymbol{\phi}} - \boldsymbol{\phi}) \to N(0, \mathbf{V}_{\boldsymbol{\phi}}(\boldsymbol{\theta}))$, is $\mathcal{E}_{\mathbf{V}_{\boldsymbol{\phi}}(\boldsymbol{\theta})}(\text{span}(\boldsymbol{\phi})) \subseteq \mathbb{R}^p$.

We think of this as an *asymptotic envelope* since its definition links it to a prespecified method of estimation through the asymptotic covariance matrix $\mathbf{V}_{\boldsymbol{\phi}}(\boldsymbol{\theta})$. The goal of an envelope is to improve that prespecified estimator. The matrix to be reduced – $\mathbf{V}_{\boldsymbol{\phi}}(\boldsymbol{\theta})$ – is dictated by the method of estimation, and it can depend on the parameter being estimated, in addition to perhaps other parameters.

Some insights into the interpretation and potential advantages of envelopes in this context can be gained by considering the estimator $\widetilde{\boldsymbol{\theta}}$ of the whole parameter vector $\boldsymbol{\theta}$ with covariance matrix $\mathbf{V}(\boldsymbol{\theta})$ assuming that the envelope $\mathcal{E}_{\mathbf{V}(\boldsymbol{\theta})}(\boldsymbol{\theta})$ is known. Since $\boldsymbol{\theta} \in \mathcal{E}_{\mathbf{V}(\boldsymbol{\theta})}(\boldsymbol{\theta})$, $\sqrt{n}\mathbf{P}_{\mathcal{E}}(\widetilde{\boldsymbol{\theta}} - \boldsymbol{\theta}) = \sqrt{n}(\widehat{\boldsymbol{\theta}}_{\mathcal{E}} - \boldsymbol{\theta})$, where $\widehat{\boldsymbol{\theta}}_{\mathcal{E}} = \mathbf{P}_{\mathcal{E}}\widetilde{\boldsymbol{\theta}}$ is an envelope estimator of $\boldsymbol{\theta}$ with known envelope. Straightforwardly, $\sqrt{n}(\widehat{\boldsymbol{\theta}}_{\mathcal{E}} - \boldsymbol{\theta})$ converges to a normal random vector with mean $\boldsymbol{\theta}$ and variance $\mathbf{P}_{\mathcal{E}}\mathbf{V}\mathbf{P}_{\mathcal{E}}$, and $\sqrt{n}\mathbf{Q}_{\mathcal{E}}\widetilde{\boldsymbol{\theta}}$ converges to a normal random vector with mean 0 and covariance $\mathbf{Q}_{\mathcal{E}}\mathbf{V}\mathbf{Q}_{\mathcal{E}}$. Since $\mathcal{E}_{\mathbf{V}(\boldsymbol{\theta})}(\boldsymbol{\theta})$ reduces $\mathbf{V}$, we have by Proposition A.1 that $\mathbf{V} = \mathbf{P}_{\mathcal{E}}\mathbf{V}\mathbf{P}_{\mathcal{E}} + \mathbf{Q}_{\mathcal{E}}\mathbf{V}\mathbf{Q}_{\mathcal{E}}$, so that $\widehat{\boldsymbol{\theta}}_{\mathcal{E}}$ and $\mathbf{Q}_{\mathcal{E}}\widetilde{\boldsymbol{\theta}}$ are asymptotically independent. In short, $\widetilde{\boldsymbol{\theta}} = \mathbf{P}_{\mathcal{E}}\widetilde{\boldsymbol{\theta}} + \mathbf{Q}_{\mathcal{E}}\widetilde{\boldsymbol{\theta}} = \widehat{\boldsymbol{\theta}}_{\mathcal{E}} + \mathbf{Q}_{\mathcal{E}}\widetilde{\boldsymbol{\theta}}$. Since $\mathbf{Q}_{\mathcal{E}}\widetilde{\boldsymbol{\theta}}$ is an estimator of 0 and is asymptotically independent of $\widehat{\boldsymbol{\theta}}_{\mathcal{E}}$, it represents immaterial variation in the original estimator $\widetilde{\boldsymbol{\theta}}$ that is eliminated by using the envelope estimator $\widehat{\boldsymbol{\theta}}_{\mathcal{E}}$. Depending on the context, this general notion of immaterial information might be cast in terms of measured variates, as in Chapter 1.

Definition 7.1 indicates how to construct an envelope for a vector-valued parameter, but it does not indicate how that envelope should be estimated. To make use of objective function (6.4) in conjunction with Definition 7.1, define $\mathbf{M} = \mathbf{V}_{\boldsymbol{\phi}}(\boldsymbol{\theta})$ and $\mathbf{U} = \boldsymbol{\phi}\boldsymbol{\phi}^T$ along with corresponding $\sqrt{n}$-consistent estimators such as $\widehat{\mathbf{M}} = \mathbf{V}_{\boldsymbol{\phi}}(\widetilde{\boldsymbol{\theta}})$ and $\widehat{\mathbf{U}} = \widetilde{\boldsymbol{\phi}}\widetilde{\boldsymbol{\phi}}^T$. This setup reproduces many of the estimators that we have already considered and allows us to address new problems, as demonstrated in the illustrations of the next section. The 1D algorithm and the sequential moment-based algorithm discussed in Sections 6.4 and 6.5 can be adapted similarly. Regardless of the algorithm used, we can construct an envelope estimator of $\boldsymbol{\phi}$ by projecting $\widetilde{\boldsymbol{\phi}}$ onto the estimated envelope $\widehat{\mathcal{E}}$ to obtain an envelope estimator $\widehat{\boldsymbol{\phi}} = \mathbf{P}_{\widehat{\mathcal{E}}}\widetilde{\boldsymbol{\phi}}$. This will not necessarily be the same as the estimator obtained when a full likelihood is available, but it should still be a useful $\sqrt{n}$-consistent estimator. Cook and Zhang (2015a) show that if $\widehat{\mathbf{M}}$ and $\widehat{\mathbf{U}}$ are $\sqrt{n}$-consistent estimators of $\mathbf{M}$ and $\mathbf{U}$ and if the 1D algorithm is used to estimate the envelope with a known dimension, then $\widehat{\boldsymbol{\phi}} = \mathbf{P}_{\widehat{\mathcal{E}}}\widetilde{\boldsymbol{\phi}}$ is a $\sqrt{n}$-consistent estimator of $\boldsymbol{\phi}$.

7.1.1 Illustrations

Illustrations 1–4 in this section describe relationships between previously developed envelopes and the envelopes resulting from Definition 7.1. Illustrations 5–7 indicate how Definition 7.1 can be used to address new envelope settings. The goal is to illustrate the use of Definition 7.1 without necessarily relying on a likelihood.

1. *Enveloping a multivariate mean.* As described in Chapter 5, consider the problem of estimating a multivariate mean based on a sample $\mathbf{Y}_1, \dots, \mathbf{Y}_n$ of a random vector $\mathbf{Y}$ with mean $\boldsymbol{\mu}$ and variance $\boldsymbol{\Sigma}$. Set $\boldsymbol{\theta} = (\boldsymbol{\psi}^T, \boldsymbol{\phi}^T)^T = (\text{vech}^T(\boldsymbol{\Sigma}), \boldsymbol{\mu}^T)^T$. If our original estimator of $\boldsymbol{\mu}$ is $\tilde{\boldsymbol{\phi}} = \bar{\mathbf{Y}}$, then $\text{avar}(\sqrt{n}\bar{\mathbf{Y}}) = \mathbf{V} = \boldsymbol{\Sigma}$, and Definition 7.1 leads us to consider $\mathcal{E}_{\mathbf{V}}(\boldsymbol{\mu}) = \mathcal{E}_{\boldsymbol{\Sigma}}(\boldsymbol{\mu})$. Using (6.4) with $\widehat{\mathbf{M}} = \mathbf{S}_{\mathbf{Y}}$ and $\widehat{\mathbf{U}} = \bar{\mathbf{Y}}\bar{\mathbf{Y}}^T$ leads immediately to an equivalent version of objective function (5.4) that we found in Chapter 5.
2. *Univariate linear regression.* Consider estimating the coefficient vector $\boldsymbol{\beta} \in \mathbb{R}^p$ in the univariate linear regression model $Y = \alpha + \boldsymbol{\beta}^T\mathbf{X} + \epsilon$, where ϵ is independent of $\mathbf{X}$, and has mean 0 and variance σ^2. In this case, we have $\boldsymbol{\theta} = (\boldsymbol{\psi}^T, \boldsymbol{\phi}^T)^T = ((\boldsymbol{\alpha}^T, \sigma^2), \boldsymbol{\beta}^T)^T$. The ordinary least squares estimator $\mathbf{b} = \mathbf{S}_{\mathbf{X}}^{-1}\mathbf{S}_{\mathbf{X},Y}$ of $\boldsymbol{\beta}$ has asymptotic variance $\mathbf{V}_{\beta} = \sigma^2\boldsymbol{\Sigma}_{\mathbf{X}}^{-1}$. Direct application of Definition 7.1 then leads to the $\sigma^2\boldsymbol{\Sigma}_{\mathbf{X}}^{-1}$-envelope of span($\boldsymbol{\beta}$). However, it follows from Corollary A.2 that $\mathcal{E}_{\mathbf{V}_\beta}(\boldsymbol{\beta}) = \mathcal{E}_{\boldsymbol{\Sigma}_{\mathbf{X}}}(\boldsymbol{\beta})$, the $\boldsymbol{\Sigma}_{\mathbf{X}}$-envelope of span($\boldsymbol{\beta}$). This is the envelope estimated by the SIMPLS algorithm for partial least squares regression in an effort to improve upon the ordinary least squares estimator for the purpose of prediction, as discussed in Chapter 4.

   To illustrate one possible way to develop an estimator, we use (6.4) with $\mathbf{M} = \mathbf{V}_\beta = \sigma^2\boldsymbol{\Sigma}_{\mathbf{X}}^{-1}, \widehat{\mathbf{M}} = s^2\mathbf{S}_{\mathbf{X}}^{-1}, \mathbf{U} = \boldsymbol{\beta}\boldsymbol{\beta}^T$, and $\widehat{\mathbf{U}} = \mathbf{b}\mathbf{b}^T$, where s^2 is the estimator of σ^2 from the ordinary least squares fit of the model. Then to estimate an envelope of dimension q, we minimize

$$\begin{aligned} L_q(\mathbf{G}) &= \log|s^2\mathbf{G}^T\mathbf{S}_{\mathbf{X}}^{-1}\mathbf{G}| + \log|\mathbf{G}^T(s^2\mathbf{S}_{\mathbf{X}}^{-1} + \mathbf{b}\mathbf{b}^T)^{-1}\mathbf{G}| \\ &= \log|s^2\mathbf{G}^T\mathbf{S}_{\mathbf{X}}^{-1}\mathbf{G}| + \log|\mathbf{G}^T(s^2\mathbf{S}_{\mathbf{X}}^{-1} + \mathbf{S}_{\mathbf{X}}^{-1}\mathbf{S}_{\mathbf{X},Y}\mathbf{S}_{\mathbf{X},Y}^T\mathbf{S}_{\mathbf{X}}^{-1})^{-1}\mathbf{G}| \\ &= \log|\mathbf{G}^T\mathbf{S}_{\mathbf{X}}^{-1}\mathbf{G}| + \log|\mathbf{G}^T(\mathbf{S}_{\mathbf{X}} - \mathbf{S}_{\mathbf{X},Y}\mathbf{S}_{\mathbf{X},Y}^T/s_Y^2)\mathbf{G}|, \end{aligned}$$

   where s_Y^2 is the sample variance of Y and the second equality follows algebraically by using the Woodbury matrix identity to expand $(s^2\mathbf{S}_{\mathbf{X}}^{-1} + \mathbf{S}_{\mathbf{X}}^{-1}\mathbf{S}_{\mathbf{X},Y}\mathbf{S}_{\mathbf{X},Y}^T\mathbf{S}_{\mathbf{X}}^{-1})^{-1}$. This result agrees with the corresponding objective function (4.15) that we found from the normal likelihood in Chapter 4. As indicated previously, although $\mathcal{E}_{\mathbf{V}_\beta}(\boldsymbol{\beta}) = \mathcal{E}_{\boldsymbol{\Sigma}_{\mathbf{X}}}(\boldsymbol{\beta})$, we used $\mathbf{M} = \mathbf{V}_\beta = \sigma^2\boldsymbol{\Sigma}_{\mathbf{X}}^{-1}$ and not $\mathbf{M} = \boldsymbol{\Sigma}_{\mathbf{X}}$ when adapting (6.4).

   Let $\boldsymbol{\Phi}$ be a basis for $\mathcal{E}_{\mathbf{V}_\beta}(\boldsymbol{\beta})$, and let $\widehat{\boldsymbol{\Phi}}$ be a semi-orthogonal matrix that minimizes L_q. In Section 7.1, we indicated that lacking a likelihood we can construct an envelope estimator of $\boldsymbol{\beta}$ as $\widehat{\boldsymbol{\beta}} = \mathbf{P}_{\widehat{\boldsymbol{\Phi}}}\mathbf{b}$. But we saw previously in (4.16) that the normal theory estimator of $\boldsymbol{\beta}$ is $\widehat{\boldsymbol{\beta}} = \mathbf{P}_{\widehat{\boldsymbol{\Phi}}(\mathbf{S}_{\mathbf{X}})}\mathbf{b}$. These two estimators of $\boldsymbol{\beta}$ are different, one using the usual inner product and one using the $\mathbf{S}_{\mathbf{X}}$ inner product. Since $\mathcal{E}_{\mathbf{V}_{\beta}}(\boldsymbol{\beta})$ is a reducing subspace of $\boldsymbol{\Sigma}_{\mathbf{X}}$, we have that $\mathbf{P}_{\boldsymbol{\Phi}} = \mathbf{P}_{\boldsymbol{\Phi}(\boldsymbol{\Sigma}_{\mathbf{X}})}$, and consequently we expect the two estimators of $\boldsymbol{\beta}$ to be close when the sample size n is large relative to p.

3. *Multivariate linear regression with a univariate predictor.* The multivariate linear model $\mathbf{Y} = \boldsymbol{\alpha} + \boldsymbol{\beta}X + \boldsymbol{\varepsilon}$, $\text{var}(\boldsymbol{\varepsilon}) = \boldsymbol{\Sigma}$, with a univariate predictor X provides another illustration on the use of Definition 7.1. In Chapter 1, we based envelope estimation on the $\boldsymbol{\Sigma}$-envelope of $\mathcal{B} = \text{span}(\boldsymbol{\beta})$. To cast this in the context of Definition 7.1, take $\boldsymbol{\psi} = (\boldsymbol{\alpha}^T, \text{vech}^T(\boldsymbol{\Sigma}))^T$, $\boldsymbol{\phi} = \boldsymbol{\beta}$, and $\widetilde{\boldsymbol{\phi}}$ to be the ordinary least squares estimator of $\boldsymbol{\beta}$ (1.2). Let $\sigma_X^2 = \lim_{n\to\infty} s_X^2$. The asymptotic covariance matrix of $\widetilde{\boldsymbol{\phi}}$ is $\sigma_X^{-2}\boldsymbol{\Sigma}$, which does not depend on $\boldsymbol{\alpha}$ or $\boldsymbol{\beta}$. Direct application of Definition 7.1 then leads to the $\sigma_X^{-2}\boldsymbol{\Sigma}$-envelope of span$(\boldsymbol{\beta})$. The equivalence of the envelopes $\mathcal{E}_{\sigma_X^{-2}\boldsymbol{\Sigma}}(\boldsymbol{\beta})$ and $\mathcal{E}_{\boldsymbol{\Sigma}}(\boldsymbol{\beta})$ follows from the construct leading to Corollary A.2. When adapting (6.4), we should again use $\mathbf{M} = \sigma_X^{-2}\boldsymbol{\Sigma}$ and $\mathbf{U} = \boldsymbol{\phi}\boldsymbol{\phi}^T$.
4. *Partial response envelopes.* In Chapter 3, we considered partial response envelopes for $\boldsymbol{\beta}_1 \in \mathbb{R}^r$ in the partitioned multivariate linear regression model

$$\mathbf{Y}_i = \boldsymbol{\mu} + \boldsymbol{\beta}\mathbf{X}_i + \boldsymbol{\varepsilon}_i = \boldsymbol{\mu} + \boldsymbol{\beta}_1 X_{1i} + \boldsymbol{\beta}_{2i}\mathbf{X}_{2i} + \boldsymbol{\varepsilon}_i, \; i = 1, \dots, n, \tag{7.1}$$

where $\mathbf{Y} \in \mathbb{R}^r$, $\boldsymbol{\beta} \in \mathbb{R}^{r\times p}$, the predictor vector $\mathbf{X} \in \mathbb{R}^p$ is nonstochastic and centered in the sample, and the error vectors $\boldsymbol{\varepsilon}_i$ are independent copies of $\boldsymbol{\varepsilon} \sim N(0, \boldsymbol{\Sigma})$. The asymptotic covariance matrix of the least squares estimator of $\boldsymbol{\beta}_1$ is $\mathbf{V}_{\beta_1} = (\boldsymbol{\Sigma}_{\mathbf{X}}^{-1})_{11}\boldsymbol{\Sigma}$, where $(\boldsymbol{\Sigma}_{\mathbf{X}}^{-1})_{11}$ is the $(1,1)$ element of the inverse of $\boldsymbol{\Sigma}_{\mathbf{X}} = \lim_{n\to\infty} n^{-1}\sum_{i=1}^n \mathbf{X}_i\mathbf{X}_i^T$. The $\mathbf{V}_{\beta_1}$-envelope of $\mathcal{B}_1 = \text{span}(\boldsymbol{\beta}_1)$ is then dictated by Definition 7.1, which is the same as the envelope $\mathcal{E}_{\boldsymbol{\Sigma}}(\mathcal{B}_1)$ used in Chapter 3. To construct an envelope estimator, we use (6.4) setting $\mathbf{M} = \mathbf{V}_{\beta_1}$ and $\mathbf{U} = \boldsymbol{\beta}_1\boldsymbol{\beta}_1^T$. Substituting sample versions and letting $\mathbf{b}_1$ denote the first column of the ordinary least squares estimator of $\boldsymbol{\beta}$ gives the following objective function for estimating an envelope of dimension u_1.

$$\begin{aligned} L_{u_1}(\mathbf{G}) &= \log|(\mathbf{S}_{\mathbf{X}}^{-1})_{11}\mathbf{G}^T\mathbf{S}_{\mathbf{Y}|\mathbf{X}}\mathbf{G}| + \log|\mathbf{G}^T\{(\mathbf{S}_{\mathbf{X}}^{-1})_{11}\mathbf{S}_{\mathbf{Y}|\mathbf{X}} + \mathbf{b}_1\mathbf{b}_1^T\}^{-1}\mathbf{G}| \\ &= \log|\mathbf{G}^T\mathbf{S}_{\mathbf{Y}|\mathbf{X}}\mathbf{G}| + \log|\mathbf{G}^T\mathbf{S}_{\mathbf{Y}|\mathbf{X}_2}^{-1}\mathbf{G}|, \end{aligned}$$

which is equivalent to the partially maximized log-likelihood of Chapter 3. Again, although $\mathcal{E}_{\mathbf{V}_{\beta_1}}(\boldsymbol{\beta}_1) = \mathcal{E}_{\boldsymbol{\Sigma}}(\boldsymbol{\beta}_1)$, we used $\mathbf{M} = \mathbf{V}_{\beta_1} = (\boldsymbol{\Sigma}_{\mathbf{X}}^{-1})_{11}\boldsymbol{\Sigma}$ when adapting (6.4) and not $\mathbf{M} = \boldsymbol{\Sigma}$. If $\boldsymbol{\varepsilon}$ is not normal, then the envelope can be viewed as a precursor to improving upon ordinary least squares.
5. *Partial predictor envelopes.* Consider a partitioned form of the univariate linear regression model $Y = \alpha + \boldsymbol{\beta}_1^T\mathbf{X}_1 + \boldsymbol{\beta}_2^T\mathbf{X}_2 + \epsilon$, where $\mathbf{X}_1 \in \mathbb{R}^{p_1}$, $\mathbf{X}_2 \in \mathbb{R}^{p_2}$, and ϵ is independent of $\mathbf{X}$ and has mean 0 and variance σ^2. In

some regressions, we may wish to focus on $\boldsymbol{\beta}_2$, while effectively treating $\boldsymbol{\beta}_1$ as a nuisance parameter. Let $\mathbf{b}_2$ denote the ordinary least squares estimator of $\boldsymbol{\beta}_2$. To construct the envelope for $\boldsymbol{\beta}_2$, we need the asymptotic covariance matrix of $\mathbf{b}_2$: $\mathbf{V}_{\beta_2} = \sigma^2\boldsymbol{\Sigma}^{-1}_{\mathbf{X}_2|\mathbf{X}_1}$, which is the lower $p_2 \times p_2$ diagonal block of $\sigma^2\boldsymbol{\Sigma}^{-1}_{\mathbf{X}}$. The sample version of $\boldsymbol{\Sigma}_{\mathbf{X}_2|\mathbf{X}_1}$ is $\mathbf{S}_{\mathbf{X}_2|\mathbf{X}_1}$, the sample covariance matrix of the residuals from the ordinary least squares regression of $\mathbf{X}_2$ on $\mathbf{X}_1$. The envelope is then $\mathcal{E}_{\sigma^2\boldsymbol{\Sigma}^{-1}_{\mathbf{X}_2|\mathbf{X}_1}}(\mathcal{B}_2) = \mathcal{E}_{\boldsymbol{\Sigma}_{\mathbf{X}_2|\mathbf{X}_1}}(\mathcal{B}_2)$, where $\mathcal{B}_2 = \text{span}(\boldsymbol{\beta}_2)$. To construct an estimator of the envelope of dimension q, we again use (6.4) with $\widehat{\mathbf{M}} = s^2\mathbf{S}^{-1}_{\mathbf{X}_2|\mathbf{X}_1}$ and $\widehat{\mathbf{U}} = \mathbf{b}_2\mathbf{b}_2^T$:

$$\begin{aligned} L_q(\mathbf{G}) &= \log|s^2\mathbf{G}^T\mathbf{S}^{-1}_{\mathbf{X}_2|\mathbf{X}_1}\mathbf{G}| + \log|\mathbf{G}^T(s^2\mathbf{S}^{-1}_{\mathbf{X}_2|\mathbf{X}_1} + \mathbf{b}_2\mathbf{b}_2^T)^{-1}\mathbf{G}| \\ &= \log|\mathbf{G}^T\mathbf{S}^{-1}_{\mathbf{X}_2|\mathbf{X}_1}\mathbf{G}| + \log|\mathbf{G}^T\mathbf{S}_{\mathbf{X}|Y}\mathbf{G}|. \end{aligned} \tag{7.2}$$

The equality (7.2) can be demonstrated by using the orthogonalized model form $Y = \alpha + \boldsymbol{\gamma}^T\mathbf{X}_1 + \boldsymbol{\beta}_2^T\mathbf{r}_{2|1} + \epsilon$, where $\mathbf{r}_{2|1}$ denotes a residual vector from the ordinary least squares fit of $\mathbf{X}_2$ on $\mathbf{X}_1$. The $\boldsymbol{\beta}_2$ coefficient vector is the same as that in the original model form, but $\boldsymbol{\gamma} \neq \boldsymbol{\beta}_1$ unless $\mathbf{X}_1$ and $\mathbf{X}_2$ are uncorrelated in the sample. Since $\mathbf{X}_1$ and $\mathbf{r}_{2|1}$ are uncorrelated in the sample, we can substitute $\mathbf{r}_{2|1}$ for $\mathbf{X}_2$ in $L_q(\mathbf{G})$ and simplify to obtain (7.2). Objective function (7.2) is similar to the normal-theory objective function (4.15) that we found for estimating a predictor envelope for $\mathcal{B}$ in multivariate regression. The main difference is that (7.2) uses $\mathbf{S}_{\mathbf{X}_2|\mathbf{X}_1}$ in place of $\mathbf{S}_{\mathbf{X}}$.

6. *Weighted least squares.* Consider a heteroscedastic linear model with data consisting of n independent copies of $(Y, \mathbf{X}, W)$, where $Y \in \mathbb{R}^1$, $\mathbf{X} \in \mathbb{R}^p$, and $W > 0$ is a weight with $\mathrm{E}(W) = 1$:

$$Y = \mu + \boldsymbol{\beta}^T\mathbf{X} + \varepsilon/\sqrt{W}, \tag{7.3}$$

where $\varepsilon \perp\!\!\!\perp (\mathbf{X}, W)$ and $\mathrm{Var}(\varepsilon) = \sigma^2$. The constraint $\mathrm{E}(W) = 1$ is without loss of generality and serves to normalize the weights and to make subsequent expressions a bit simpler. Fitting under this model is typically done by using weighted least squares:

$$(a, \mathbf{b}) = \arg\min_{\mu,\boldsymbol{\beta}} n^{-1}\sum_{i=1}^{n} W_i(Y_i - \mu - \boldsymbol{\beta}^T\mathbf{X}_i)^2, \tag{7.4}$$

where we normalize the sample weights so that $\bar{W} = 1$. Let

$$\begin{aligned} \boldsymbol{\Sigma}_{\mathbf{X}(W)} &= \mathrm{E}\{W(\mathbf{X} - \mathrm{E}(W\mathbf{X}))(\mathbf{X} - \mathrm{E}(W\mathbf{X}))^T\} \\ \boldsymbol{\Sigma}_{\mathbf{X}Y(W)} &= \mathrm{E}[W\{\mathbf{X} - \mathrm{E}(W\mathbf{X})\}\{Y - \mathrm{E}(WY)\}] \end{aligned}$$

denote the weighted covariance matrix of $\mathbf{X}$ and the weighted covariance between $\mathbf{X}$ and Y, and let $\mathbf{S}_{\mathbf{X}(W)}$ and $\mathbf{S}_{\mathbf{X}Y(W)}$ denote their corresponding sample versions. Then $\boldsymbol{\beta} = \boldsymbol{\Sigma}^{-1}_{\mathbf{X}(W)}\boldsymbol{\Sigma}_{\mathbf{X}Y(W)}$, the prespecified estimator is $\widetilde{\boldsymbol{\beta}} = \mathbf{S}^{-1}_{\mathbf{X}(W)}\mathbf{S}_{\mathbf{X}Y(W)}$, and $\sqrt{n}(\widetilde{\boldsymbol{\beta}} - \boldsymbol{\beta})$ converges to a normal random vector with mean 0 and variance $\mathbf{V}_{\beta} = \sigma^2 \boldsymbol{\Sigma}^{-1}_{\mathbf{X}(W)}$. According to Definition 7.1, we should now strive to estimate the $\sigma^2 \boldsymbol{\Sigma}^{-1}_{\mathbf{X}(W)}$-envelope of span($\boldsymbol{\beta}$).

7. *Nonlinear least squares.* Consider the nonlinear univariate regression model

$$Y = f(\mathbf{X}, \boldsymbol{\beta}) + \epsilon,$$

where f is a known function of the nonstochastic predictor vector $\mathbf{X} \in \mathbb{R}^p$, the parameter vector $\boldsymbol{\beta} \in \mathbb{R}^s$, and the error ϵ is independent of $\mathbf{X}$ and has mean 0 and variance σ^2. Given a sample $(Y_i, \mathbf{X}_i)$, $i = 1, \dots, n$, we take the prespecified estimator to be the usual least squares estimator of $\boldsymbol{\beta}$:

$$\widetilde{\boldsymbol{\beta}} = \arg\min \sum_{i=1}^{n} (Y_i - f(\mathbf{X}_i, \boldsymbol{\beta}))^2.$$

Let $\mathbf{W} \in \mathbb{R}^{n \times s}$ denote the matrix with rows $\mathbf{w}_i^T$ equal to the partial derivative vectors $\mathbf{w}_i^T = \partial f(\mathbf{X}_i, \boldsymbol{\beta})/\partial \boldsymbol{\beta}^T$, $i = 1, \dots, n$, evaluated at the true value of $\boldsymbol{\beta}$ and let $\mathbf{V}_{\beta} = \sigma^2(\lim_{n\to\infty} \mathbf{W}^T\mathbf{W}/n)^{-1}$. Then under the usual regularity conditions, $\sqrt{n}(\widetilde{\boldsymbol{\beta}} - \boldsymbol{\beta})$ is asymptotically normal with mean 0 and variance $\mathbf{V}_{\beta}$, which leads to the $\mathbf{V}_{\beta}$-envelope of span($\boldsymbol{\beta}$). The present case is a bit different from the previous ones since $\mathbf{V}_{\beta}$ can depend on $\boldsymbol{\beta}$, leading to the possibility of iterating over an algorithm of Chapter 6: Evaluate the sample version of $\mathbf{V}_{\beta}$ at $\widetilde{\boldsymbol{\beta}}$, run an algorithm to get the initial envelope estimator $\widehat{\boldsymbol{\beta}}_1$. Then evaluate the sample version of $\mathbf{V}_{\beta}$ at $\widehat{\boldsymbol{\beta}}_1$ and run the algorithm again, continuing until a convergence criterion is met.

7.1.2 Estimation Based on a Complete Likelihood

In this section, we sketch how to use Definition 7.1 when a complete likelihood is available, and we outline the results of a study on the use of envelopes to improve estimates of Darwinian fitness. We expect that the estimators found following the sketch in this section will be superior in large samples to those based on the algorithms of Chapter 6. Additional results are available from Cook and Zhang (2015a) who give details for application in generalized linear models.

7.1.2.1 Likelihood Construction

Let $Y \in \mathbb{R}^1$ and $\mathbf{X} \in \mathbb{R}^p$ have a joint distribution with parameters $\boldsymbol{\theta} := (\boldsymbol{\alpha}^T, \boldsymbol{\beta}^T, \boldsymbol{\psi}^T)^T \in \mathbb{R}^{q+p+s}$, so the joint density can be written as

$$f(Y, \mathbf{X} \mid \boldsymbol{\theta}) = g(Y \mid \boldsymbol{\alpha}, \boldsymbol{\beta}^T\mathbf{X})h(\mathbf{X} \mid \boldsymbol{\psi}).$$

We take $\boldsymbol{\beta}$ to be the parameter vector of interest and, prior to the introduction of envelopes, we restrict the parameters $\boldsymbol{\alpha}$, $\boldsymbol{\beta}$, and $\boldsymbol{\psi}$ to a product space.

The predictors $\mathbf{X}$ are ancillary in most regressions, and thus analysis is often based on the conditional likelihood. Beginning with a random sample $(Y_i, \mathbf{X}_i)$, $i = 1, \ldots, n$, let $L(\boldsymbol{\theta}) = \sum_{i=1}^{n} \log f(Y_i, \mathbf{X}_i \mid \boldsymbol{\theta})$ be the full log-likelihood, let the conditional log-likelihood be represented by $C(\boldsymbol{\alpha}, \boldsymbol{\beta}) = \sum_{i=1}^{n} \log g(Y_i \mid \boldsymbol{\alpha}, \boldsymbol{\beta}^T \mathbf{X}_i)$, and let $M(\boldsymbol{\psi}) = \sum_{i=1}^{n} \log h(\mathbf{X}_i \mid \boldsymbol{\psi})$ be the marginal log-likelihood for $\boldsymbol{\psi}$. Then we can decompose $L(\boldsymbol{\theta}) = C(\boldsymbol{\alpha}, \boldsymbol{\beta}) + M(\boldsymbol{\psi})$. Since our primary interest lies in $\boldsymbol{\beta}$ and $\mathbf{X}$ is ancillary, estimators are often constructed as

$$(\widetilde{\boldsymbol{\alpha}}, \widetilde{\boldsymbol{\beta}}) = \arg \max_{\boldsymbol{\alpha}, \boldsymbol{\beta}} C(\boldsymbol{\alpha}, \boldsymbol{\beta}). \tag{7.5}$$

Our goal here is to improve the prespecified estimator $\widetilde{\boldsymbol{\beta}}$ by introducing the envelope $\mathcal{E}_{\mathbf{V}_\beta}(\boldsymbol{\beta})$, where $\mathbf{V}_\beta = \mathbf{V}_\beta(\boldsymbol{\theta}) = \text{avar}(\sqrt{n}\widetilde{\boldsymbol{\beta}})$, according to Definition 7.1.

Let $(\boldsymbol{\Gamma}, \boldsymbol{\Gamma}_0) \in \mathbb{R}^{p \times p}$ denote an orthogonal matrix where $\boldsymbol{\Gamma} \in \mathbb{R}^{p \times u}$ is a basis for $\mathcal{E}_{\mathbf{V}_\beta}(\boldsymbol{\beta})$. Since $\boldsymbol{\beta} \in \mathcal{E}_{\mathbf{V}_\beta}(\boldsymbol{\beta})$, we can write $\boldsymbol{\beta} = \boldsymbol{\Gamma}\boldsymbol{\eta}$ for some $\boldsymbol{\eta} \in \mathbb{R}^u$. Because $\mathbf{V}_\beta(\boldsymbol{\theta})$ typically depends on the distribution of $\mathbf{X}$ and $\mathcal{E}_{\mathbf{V}_\beta}(\boldsymbol{\beta})$ reduces $\mathbf{V}_\beta(\boldsymbol{\theta})$, the marginal M will depend on $\boldsymbol{\Gamma}$. Then the log-likelihood for fixed u becomes $L(\boldsymbol{\alpha}, \boldsymbol{\eta}, \boldsymbol{\psi}_1, \boldsymbol{\Gamma}) = C(\boldsymbol{\alpha}, \boldsymbol{\eta}, \boldsymbol{\Gamma}) + M(\boldsymbol{\psi}_1, \boldsymbol{\Gamma})$, where $\boldsymbol{\psi}_1$ represents any parameters remaining after incorporating $\boldsymbol{\Gamma}$. Since both C and M depend on $\boldsymbol{\Gamma}$, the predictors are no longer ancillary after incorporating the envelope structure, and estimation must be carried out by maximizing $C(\boldsymbol{\alpha}, \boldsymbol{\eta}, \boldsymbol{\Gamma}) + M(\boldsymbol{\psi}_1, \boldsymbol{\Gamma})$. This is a restatement of the idea expressed in the preamble of Section 4.3.

Writing $L(\boldsymbol{\theta})$ as a function of $\boldsymbol{\Gamma}$ and $\boldsymbol{\eta}$ to construct $L(\boldsymbol{\alpha}, \boldsymbol{\eta}, \boldsymbol{\psi}_1, \boldsymbol{\Gamma})$ could be complicated, depending on the distribution of $\mathbf{X}$. However, the situation simplifies considerably when $\text{E}(\mathbf{X} \mid \boldsymbol{\Gamma}^T\mathbf{X})$ is a linear function of $\boldsymbol{\Gamma}^T\mathbf{X}$ since then $\mathcal{E}_{\mathbf{V}_\beta}(\boldsymbol{\beta}) = \mathcal{E}_{\boldsymbol{\Sigma}_\mathbf{X}}(\boldsymbol{\beta})$ (Cook and Zhang 2015a). The requirement that $\text{E}(\mathbf{X} \mid \boldsymbol{\Gamma}^T\mathbf{X})$ be a linear function of $\boldsymbol{\Gamma}^T\mathbf{X}$ is well known as the linearity condition in the sufficient dimension reduction literature where $\boldsymbol{\Gamma}$ denotes a basis for the central subspace. Background on the linearity condition, which is widely regarded as restrictive but nonetheless rather mild, is available in Section 9.4.

If $\mathcal{E}_{\mathbf{V}_\beta}(\boldsymbol{\beta}) = \mathcal{E}_{\boldsymbol{\Sigma}_\mathbf{X}}(\boldsymbol{\beta})$, then for some positive definite matrices $\boldsymbol{\Omega} \in \mathbb{R}^{u \times u}$ and $\boldsymbol{\Omega}_0 \in \mathbb{R}^{(p-u)\times(p-u)}$, $\boldsymbol{\Sigma}_\mathbf{X} = \boldsymbol{\Gamma}\boldsymbol{\Omega}\boldsymbol{\Gamma}^T + \boldsymbol{\Gamma}_0\boldsymbol{\Omega}_0\boldsymbol{\Gamma}_0^T$, and thus M must depend on $\boldsymbol{\Gamma}$ through the marginal covariance $\boldsymbol{\Sigma}_\mathbf{X}$. Consequently, we can write $M(\boldsymbol{\Sigma}_\mathbf{X}, \boldsymbol{\psi}_2) = M(\boldsymbol{\Gamma}, \boldsymbol{\Omega}, \boldsymbol{\Omega}_0, \boldsymbol{\psi}_2)$, where $\boldsymbol{\psi}_2$ represents any remaining parameters in the marginal function. If $\mathbf{X}$ is normal with mean $\boldsymbol{\mu}_\mathbf{X}$ and variance $\boldsymbol{\Sigma}_\mathbf{X}$, then $\boldsymbol{\psi}_2 = \boldsymbol{\mu}_\mathbf{X}$, and $M(\boldsymbol{\Gamma}, \boldsymbol{\Omega}, \boldsymbol{\Omega}_0, \boldsymbol{\psi}_2) = M(\boldsymbol{\Gamma}, \boldsymbol{\Omega}, \boldsymbol{\Omega}_0, \boldsymbol{\mu}_\mathbf{X})$ is the marginal normal log-likelihood. In this case, it is possible to maximize M over all its parameters, except $\boldsymbol{\Gamma}$:

$$\begin{aligned} M_1(\boldsymbol{\Gamma}) &:= \max_{\boldsymbol{\Omega}, \boldsymbol{\Omega}_0, \boldsymbol{\mu}_\mathbf{X}} M(\boldsymbol{\Gamma}, \boldsymbol{\Omega}, \boldsymbol{\Omega}_0, \boldsymbol{\mu}_\mathbf{X}) \\ &= -\frac{n}{2}\{\log|\boldsymbol{\Gamma}^T\mathbf{S}_\mathbf{X}\boldsymbol{\Gamma}| + \log|\boldsymbol{\Gamma}_0^T\mathbf{S}_\mathbf{X}\boldsymbol{\Gamma}_0|\} \\ &= -\frac{n}{2}\{\log|\boldsymbol{\Gamma}^T\mathbf{S}_\mathbf{X}\boldsymbol{\Gamma}| + \log|\boldsymbol{\Gamma}^T\mathbf{S}_\mathbf{X}^{-1}\boldsymbol{\Gamma}| + \log|\mathbf{S}_\mathbf{X}|\}, \end{aligned}$$

where $(\boldsymbol{\Gamma}, \boldsymbol{\Gamma}_0) \in \mathbb{R}^{p \times p}$ is an orthogonal matrix. This result indicates that if $\mathbf{X}$ is marginally normal, the envelope estimators are

$$(\widehat{\boldsymbol{\alpha}}, \widehat{\boldsymbol{\eta}}, \widehat{\boldsymbol{\Gamma}}) = \arg\max\{C(\boldsymbol{\alpha}, \boldsymbol{\eta}, \boldsymbol{\Gamma}) + M_1(\boldsymbol{\Gamma})\}.$$

In particular, the envelope estimator of $\boldsymbol{\beta}$ is $\widehat{\boldsymbol{\beta}} = \widehat{\boldsymbol{\Gamma}}\widehat{\boldsymbol{\eta}}$, and the envelope estimator of $\boldsymbol{\Sigma}_{\mathbf{X}}$ is $\widehat{\boldsymbol{\Sigma}} = \mathbf{P}_{\widehat{\Gamma}}\mathbf{S}_{\mathbf{X}}\mathbf{P}_{\widehat{\Gamma}} + \mathbf{Q}_{\widehat{\Gamma}}\mathbf{S}_{\mathbf{X}}\mathbf{Q}_{\widehat{\Gamma}}$. The objective function $C(\boldsymbol{\alpha}, \boldsymbol{\eta}, \boldsymbol{\Gamma}) + M_1(\boldsymbol{\Gamma})$ is similar to decompositions of the log-likelihood $L_q(\mathbf{G})$ for predictor reduction that we encountered in Section 4.3.2. They have $M_1(\boldsymbol{\Gamma})$ in common and differ in the contribution of the conditional log-likelihood C. In both cases, M_1 serves to pull the solutions toward the reducing subspaces of $\mathbf{S}_{\mathbf{X}}$ while C tries to insure fidelity to the conditional model.

7.1.2.2 Aster Models

Aster models were developed for use in life history analyses (Geyer et al. 2007; Shaw et al. 2008; Shaw and Geyer 2010) and are suitable for estimating expected Darwinian fitness across covariates and trait values, where Darwinian fitness is the total offspring of a plant or animal over the course of its lifetime. More formally, aster models are directed graphical models that satisfy the following five structural assumptions:

A1. The directed graph is acyclic.
A2. A node has, at most, one predecessor node.
A3. The joint distribution is the product of conditional distributions, one conditional distribution for each arrow in the aster graph.
A4. Predecessor is sample size.
A5. Conditional distributions for arrows are one-parameter exponential families. The exponential families across arrows are not required to be the same.

Figure 7.1 gives the graphical structure for a simulation example to be used for later illustration. The terminal nodes, $C_1, \dots, C_{10}$ in Figure 7.1, correspond to offspring counts, while the preceding nodes, $A_1, \dots, A_{10}$ and $B_1, \dots, B_{10}$, represent important life stages in the plant or animal's life leading up to reproduction. Assumptions A4 and A5 mean for an arrow $y_k \rightarrow y_j$ that y_j is the sum of independent and identically distributed random variables from the exponential

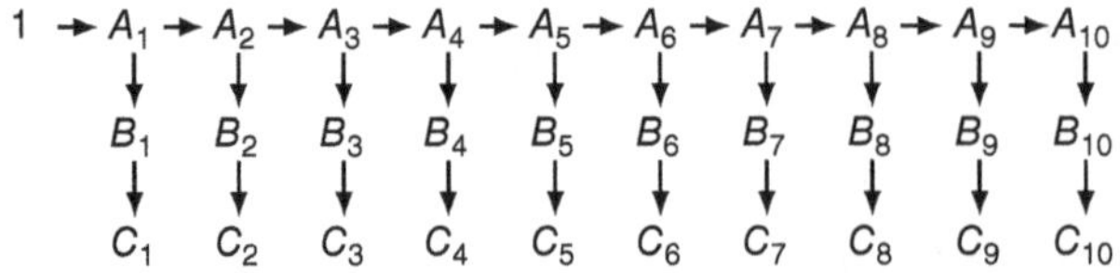

Figure 7.1 Graphical structure of the aster model for simulated data. The top A layer corresponds to survival; these random variables are Bernoulli. The middle B layer corresponds to whether or not an individual reproduced; these random variables are also Bernoulli. The bottom C layer corresponds to offspring count; these random variables are zero-truncated Poisson.

family and there are y_k terms in that sum, where the sum of zero terms is zero (Eck et al. 2017). The structure of an aster model implies that the joint distribution of the node output follows an exponential family (Geyer et al. 2007, Section 2.3). All one-parameter exponential families can be represented in an aster model. However, only the canonical link function can be used or the joint distribution of the aster model will cease to be an exponential family and inferential capabilities of the aster model itself will be lost.

The log-likelihood from the aster model in canonical form is $L(\boldsymbol{\beta}) = \langle \mathbf{M}^T\mathbf{Y}, \boldsymbol{\beta}\rangle - c(\mathbf{a} + \mathbf{M}\boldsymbol{\beta})$, where $\mathbf{Y} \in \mathbb{R}^m$ is the vector of responses consisting of one component for every node in the graph for every individual in the study, $\mathbf{M}$ is the model matrix assumed to have full column rank, $\mathbf{M}^T\mathbf{Y}$ is the canonical statistic, $\mathbf{a}$ is a known offset vector, and $\boldsymbol{\beta}$ is the canonical parameter vector. The vector of mean parameters $\boldsymbol{\theta} = \mathrm{E}(\mathbf{M}^T\mathbf{Y})$ is often of interest in aster modeling and can be estimated using maximum likelihood estimation based on the log-likelihood L.

Eck et al. (2017) studied the advantages of using partial envelopes to estimate subvectors of $\boldsymbol{\theta}$, along the general lines described in Section 7.1. One of their examples involved a simulated dataset formed by generating data for 3000 organisms progressing through the lifecycle is depicted in Figure 7.1. Darwinian fitness for this example is $\sum_{i=1}^{10} C_i$. There are two covariates (z_1, z_2) associated with Darwinian fitness, and the aster model they used supposes that expected Darwinian fitness is a full quadratic model in z_1 and z_2. The aster mean-value parameter vector $\boldsymbol{\theta}$ was partitioned into $(\boldsymbol{\psi}^T, \boldsymbol{\phi}^T)^T$ where $\boldsymbol{\psi} \in \mathbb{R}^4$ are nuisance parameters and $\boldsymbol{\phi} \in \mathbb{R}^5$ are relevant to the estimation of expected Darwinian fitness. The parameter vector of interest $\boldsymbol{\phi}$ was then estimated by maximum likelihood estimation and by envelope estimation using the ideas described earlier in this chapter. To avoid the uncertainty in choice of dimension, Eck et al. (2017) used a version of the weighted envelope estimator (1.49) and assessed variability using the parametric bootstrap procedure suggested by Efron (2014).

The contour plots of the ratios of estimated standard errors for estimated expected Darwinian fitness are displayed in Figure 7.2. The contours show that the envelope estimator of expected Darwinian fitness is less variable than the maximum likelihood estimator for the majority of the observed data. These efficiency gains can be helpful when inferring about trait values that maximize expected Darwinian fitness. Details on this example are available from Eck et al. (2016).

7.2 Envelopes for Matrix-Valued Parameters

Definition 7.1 is sufficiently general to cover a matrix-valued parameter $\boldsymbol{\phi} \in \mathbb{R}^{r\times c}$ by considering the avar$(\sqrt{n}\mathrm{vec}(\widetilde{\boldsymbol{\phi}}))$-envelope of span(vec($\boldsymbol{\phi}$)). However, matrix-valued parameters come with additional structure that is often desirable

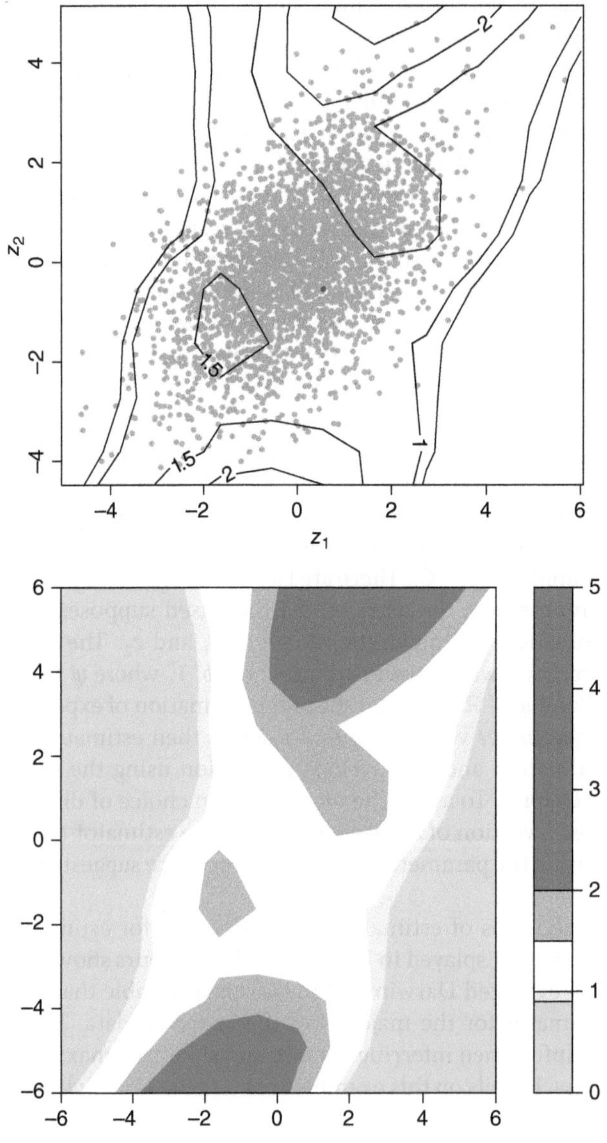

Figure 7.2 Contour plots of the ratios of the bootstrapped standard errors for the maximum likelihood estimator to the bootstrapped standard errors for the corresponding envelope estimator. The point in blue in the top plot corresponds to the highest estimated expected Darwinian fitness value, which is essentially the same using the envelope estimator and the maximum likelihood estimator.

to maintain during estimation. For instance, in the multivariate linear model, our envelope constructions were constrained to reflect separate row and column reduction of the matrix parameter $\boldsymbol{\beta}$, leading to interpretations in terms of material variation in $\mathbf{X}$ and $\mathbf{Y}$. In this section, we describe how Cook and Zhang, (2015a) adapted to a matrix-valued parameter in the spirit of Definition 7.1.

Suppose that $\sqrt{n}(\widetilde{\boldsymbol{\phi}} - \boldsymbol{\phi})$ converges to a matrix normal distribution with mean 0, column variance $\boldsymbol{\Delta}_L \in \mathbb{S}^{r\times r}$, and row variance $\boldsymbol{\Delta}_R \in \mathbb{S}^{c\times c}$. (See Appendix A.8 for background on the matrix normal distribution.) Then

$$\sqrt{n}(\mathrm{vec}(\widetilde{\boldsymbol{\phi}}) - \mathrm{vec}(\boldsymbol{\phi})) \to N(0, \boldsymbol{\Delta}_R \otimes \boldsymbol{\Delta}_L), \tag{7.6}$$

and direct application of Definition 7.1 yields the envelope $\mathcal{E}_{\boldsymbol{\Delta}_R\otimes\boldsymbol{\Delta}_L}(\mathrm{vec}(\boldsymbol{\phi}))$. The least squares estimator of the coefficient matrix $\boldsymbol{\beta}$ in the multivariate linear model satisfies condition (7.6) with $\boldsymbol{\Delta}_R = \boldsymbol{\Sigma}_{\mathbf{X}}^{-1}$ and $\boldsymbol{\Delta}_L = \boldsymbol{\Sigma}$ (Section 1.1). However, the envelope $\mathcal{E}_{\boldsymbol{\Delta}_R\otimes\boldsymbol{\Delta}_L}(\mathrm{vec}(\boldsymbol{\phi}))$ may not preserve the intrinsic row–column structure of $\boldsymbol{\phi}$. In the following definition, we introduce a restricted class of envelopes that maintain the matrix structure of $\boldsymbol{\phi}$. In preparation, recall that the Kronecker product of two subspaces $\mathcal{A}$ and $\mathcal{B}$ is defined as $\mathcal{A} \otimes \mathcal{B} = \{\mathbf{a} \otimes \mathbf{b} \mid \mathbf{a} \in \mathcal{A}, \mathbf{b} \in \mathcal{B}\}$. If $\mathbf{A}$ and $\mathbf{B}$ are basis matrices for $\mathcal{A}$ and $\mathcal{B}$, then $\mathrm{span}(\mathbf{A} \otimes \mathbf{B}) = \mathcal{A} \otimes \mathcal{B}$.

Definition 7.2 Assume that $\widetilde{\boldsymbol{\phi}}$ is asymptotically matrix normal as given in (7.6). Then the tensor envelope for $\boldsymbol{\phi}$, denoted by $\mathcal{K}_{\boldsymbol{\Delta}_R\otimes\boldsymbol{\Delta}_L}(\boldsymbol{\phi})$, is defined as the intersection of all reducing subspaces $\mathcal{E}$ of $\boldsymbol{\Delta}_R \otimes \boldsymbol{\Delta}_L$ that contain $\mathrm{span}(\mathrm{vec}(\boldsymbol{\phi}))$ and can be written as $\mathcal{E} = \mathcal{E}_R \otimes \mathcal{E}_L$ with $\mathcal{E}_R \subseteq \mathbb{R}^c$ and $\mathcal{E}_L \subseteq \mathbb{R}^r$.

We see from this definition that $\mathcal{K}_{\boldsymbol{\Delta}_R\otimes\boldsymbol{\Delta}_L}(\boldsymbol{\phi})$ always exists and is the smallest subspace with the required properties. Let $\boldsymbol{\Gamma}_L \in \mathbb{R}^{r\times d_c}$, $d_c \le r$, and $\boldsymbol{\Gamma}_R \in \mathbb{R}^{c\times d_r}$, $d_r \le c$ be semi-orthogonal matrices such that $\mathcal{K}_{\boldsymbol{\Delta}_R\otimes\boldsymbol{\Delta}_L}(\boldsymbol{\phi}) = \mathrm{span}(\boldsymbol{\Gamma}_R \otimes \boldsymbol{\Gamma}_L)$. By definition, $\mathrm{span}(\boldsymbol{\phi}) \subseteq \mathrm{span}(\boldsymbol{\Gamma}_L)$ and $\mathrm{span}(\boldsymbol{\phi}^T) \subseteq \mathrm{span}(\boldsymbol{\Gamma}_R)$. Hence, we have $\boldsymbol{\phi} = \boldsymbol{\Gamma}_L \boldsymbol{\eta} \boldsymbol{\Gamma}_R^T$ and $\mathrm{vec}(\boldsymbol{\phi}) = (\boldsymbol{\Gamma}_R \otimes \boldsymbol{\Gamma}_L)\mathrm{vec}(\boldsymbol{\eta})$ for some $\eta \in \mathbb{R}^{d_c\times d_r}$.

The next proposition (Cook and Zhang 2015a) shows how to factor $\mathcal{K}_{\boldsymbol{\Delta}_R\otimes\boldsymbol{\Delta}_L}(\boldsymbol{\phi}) \subseteq \mathbb{R}^{rc}$ into the tensor product of envelopes $\mathcal{E}_{\boldsymbol{\Delta}_R}(\boldsymbol{\phi}^T) \subseteq \mathbb{R}^r$ and $\mathcal{E}_{\boldsymbol{\Delta}_L}(\boldsymbol{\phi}) \subseteq \mathbb{R}^c$ for the row and column spaces of $\boldsymbol{\phi}$. These tensor factors are envelopes in smaller spaces that preserve the row and column structure and can facilitate analysis and interpretation.

Proposition 7.1 $\mathcal{K}_{\boldsymbol{\Delta}_R\otimes\boldsymbol{\Delta}_L}(\boldsymbol{\phi}) = \mathcal{E}_{\boldsymbol{\Delta}_R}(\boldsymbol{\phi}^T) \otimes \mathcal{E}_{\boldsymbol{\Delta}_L}(\boldsymbol{\phi})$.

For example, in reference to the multivariate linear model (1.1), the distribution of $\mathbf{B} = \mathbf{S}_{\mathbf{XY}}^T \mathbf{S}_{\mathbf{X}}^{-1}$ satisfies (7.6) with $\boldsymbol{\Delta}_R = \boldsymbol{\Sigma}_{\mathbf{X}}^{-1}$ and $\boldsymbol{\Delta}_L = \boldsymbol{\Sigma}$. The tensor envelope is then

$$\mathcal{K}_{\boldsymbol{\Sigma}_{\mathbf{X}}^{-1}\otimes\boldsymbol{\Sigma}}(\boldsymbol{\beta}) = \mathcal{E}_{\boldsymbol{\Sigma}_{\mathbf{X}}^{-1}}(\boldsymbol{\beta}^T) \otimes \mathcal{E}_{\boldsymbol{\Sigma}}(\boldsymbol{\beta}) = \mathcal{E}_{\boldsymbol{\Sigma}_{\mathbf{X}}}(\boldsymbol{\beta}^T) \otimes \mathcal{E}_{\boldsymbol{\Sigma}}(\boldsymbol{\beta}).$$

If we are interested in reducing only the column space of $\boldsymbol{\beta}$, which corresponds to response reduction, we would use $\mathbb{R}^p \otimes \mathcal{E}_{\boldsymbol{\Sigma}}(\boldsymbol{\beta})$ for constructing an envelope estimator of $\boldsymbol{\beta}$, and then $\boldsymbol{\beta} = \boldsymbol{\Gamma}_L \boldsymbol{\eta} \mathbf{I}_p = \boldsymbol{\Gamma}_L \boldsymbol{\eta}$ where $\boldsymbol{\Gamma}_L$ is a semi-orthogonal basis for $\mathcal{E}_{\boldsymbol{\Sigma}}(\boldsymbol{\beta})$, which reproduces the envelope construction of Chapter 1. Similarly, if we are interested only in predictor reduction, we would take $\mathcal{E}_{\boldsymbol{\Sigma}_{\mathbf{X}}}(\boldsymbol{\beta}^T) \otimes \mathbb{R}^r$, which reproduces the envelope construction in Chapter 4. More generally, the definition of the tensor envelope and Proposition 7.1 connects and combines the envelope models in the predictor space and in the response space and is a basis for the methodology discussed in Section 4.5.

To this point, we have defined envelopes only for improving a $\sqrt{n}$-consistent estimator $\widetilde{\boldsymbol{\phi}}$. Two broad issues now arise: how do we estimate the envelope and how do we use the estimator to improve $\widetilde{\boldsymbol{\phi}}$? When $\widetilde{\boldsymbol{\phi}}$ is a maximum likelihood estimator, we might reparameterize the likelihood in terms of the envelope, following the general logic used for the development of response envelopes in Chapter 1. Lacking a likelihood but having $\sqrt{n}$-consistent estimators of $\boldsymbol{\Delta}_r$ and $\boldsymbol{\Delta}_c$, we can use one of the algorithms described in Chapter 6 to construct estimators $\widehat{\mathcal{E}}_{\boldsymbol{\Delta}_r}(\boldsymbol{\phi}^T)$ and $\widehat{\mathcal{E}}_{\boldsymbol{\Delta}_c}(\boldsymbol{\phi})$, leading to $\widehat{\mathcal{K}}_{\boldsymbol{\Delta}_r \otimes \boldsymbol{\Delta}_c}(\boldsymbol{\phi}) = \widehat{\mathcal{E}}_{\boldsymbol{\Delta}_r}(\boldsymbol{\phi}^T) \otimes \widehat{\mathcal{E}}_{\boldsymbol{\Delta}_c}(\boldsymbol{\phi})$ as an estimator of $\mathcal{K}_{\boldsymbol{\Delta}_r \otimes \boldsymbol{\Delta}_c}(\boldsymbol{\phi})$. We can then project $\text{vec}(\widetilde{\boldsymbol{\phi}})$ onto the estimated envelope, leading to the moment-based envelope estimator

$$\begin{aligned}
\text{vec}(\widehat{\boldsymbol{\phi}}) &= \mathbf{P}_{\widehat{\mathcal{K}}} \text{vec}(\widetilde{\boldsymbol{\phi}}) \\
&= (\mathbf{P}_{\widehat{\mathcal{E}}_r} \otimes \mathbf{P}_{\widehat{\mathcal{E}}_c}) \, \text{vec}(\widetilde{\boldsymbol{\phi}}) \\
&= \text{vec}(\mathbf{P}_{\widehat{\mathcal{E}}_c} \widetilde{\boldsymbol{\phi}} \mathbf{P}_{\widehat{\mathcal{E}}_r}),
\end{aligned} \tag{7.7}$$

where $\mathbf{P}_{\widehat{\mathcal{E}}_c}$ and $\mathbf{P}_{\widehat{\mathcal{E}}_r}$ are the projections onto $\widehat{\mathcal{E}}_{\boldsymbol{\Delta}_c}(\boldsymbol{\phi})$ and $\widehat{\mathcal{E}}_{\boldsymbol{\Delta}_r}(\boldsymbol{\phi}^T)$. This then leads to the envelope estimator $\widehat{\boldsymbol{\phi}} = \mathbf{P}_{\widehat{\mathcal{E}}_c} \widetilde{\boldsymbol{\phi}} \mathbf{P}_{\widehat{\mathcal{E}}_r}$. Adapting this moment-based estimator to response envelopes, we set $\mathcal{E}_{\boldsymbol{\Delta}_r}(\boldsymbol{\phi}^T) = \mathbb{R}^p$ and $\mathcal{E}_{\boldsymbol{\Delta}_c}(\boldsymbol{\phi}) = \mathcal{E}_{\boldsymbol{\Sigma}}(\boldsymbol{\beta})$, which leads back to the maximum likelihood estimator described in Section 1.5.

7.3 Envelopes for Matrix-Valued Responses

We have so far confined attention to regressions in which the response for each experimental unit is a vector. But modern measurement techniques are resulting in an increased prevalence of experiments where the response is naturally a matrix. For instance, a study on the relationship between genetic predisposition and tendency for alcoholism[1] involved collecting electroencephalography data on 121 subjects comprised of 77 alcoholic individuals and 44 controls. Each individual was measured at 256 time points using 64 electrodes placed on the scalp at specific locations, resulting in a 64×256 EEG matrix of voltages.

1 http: //kdd.ics.uci.edu/databases/eeg/eeg.data.html

Similarly, a three-dimensional array of magnetic resonance imaging data[2] was collected for each of 776 subjects during the study of attention-deficit hyperactivity disorder (ADHD). Of the subjects, 285 were classified as having ADHD and 491 were normal controls. In this regression, the response for each subject was an $r_1 \times r_2 \times r_3$ array, and the single predictor was an indicator for ADHT. Ding and Cook (2017) extended envelope methodology to regressions with matrix-valued responses and matrix-valued predictors. Li and Zhang (2017) extended it to regressions in which the response is an $r_1 \times \cdots \times r_m$ array, also called a tensor, and the predictor is a vector. We confine our discussion in this section to matrix-valued responses with vector-valued predictors. This will cover some essential steps in the extension without the need for background on tensor algebra.

7.3.1 Initial Modeling

Consider then a regression with response $\mathbf{Y} \in \mathbb{R}^{r_1 \times r_2}$ and predictor $\mathbf{X} \in \mathbb{R}^p$. To begin, we must decide how to construct a model for the regression of $\mathbf{Y}$ on $\mathbf{X}$. A first possibility is to use the vec operator to convert $\mathbf{Y}$ into a vector, allowing us to form a model covered by our standard construction (1.1):

$$\mathrm{vec}(\mathbf{Y}) = \boldsymbol{\alpha} + \boldsymbol{\nu}\mathbf{X} + \boldsymbol{\epsilon}, \tag{7.8}$$

where $\mathrm{vec}(\mathbf{Y}) \in \mathbb{R}^{r_1 r_2}$, $\boldsymbol{\alpha} \in \mathbb{R}^{r_1 r_2}$, $\boldsymbol{\nu} \in \mathbb{R}^{r_1 r_2 \times p}$, and $\boldsymbol{\epsilon} \sim N_{r_1 r_2}(0, \boldsymbol{\Delta})$ is independent of $\mathbf{X}$. Methods from previous chapters can now be applied without modification. This model, which ignores the matrix structure of $\mathbf{Y}$, allows a different regression for each element of $\mathbf{Y}$. Application of the envelope methodology of Chapter 1 would involve looking for linear combinations of the elements of $\mathbf{Y}$ whose distribution does not depend on the nonstochastic predictor vector. This might be useful as a broad-brush starting point, particularly when $r_1 r_2$ is not too large, but it might also be useful to consider envelope reductions that recognize the matrix structure of $\mathbf{Y}$. One possibility is to allow for linear combinations of the rows or columns of $\mathbf{Y}$ whose distribution does not depend on $\mathbf{X}$. We next consider reductions across the rows of $\mathbf{Y}$; the same reasoning holds for the columns of $\mathbf{Y}$ by reducing the rows of $\mathbf{Y}^T$. There are perhaps many other approaches depending on the application, including the possibility of enveloping rows and columns simultaneously.

Let $(\boldsymbol{\Gamma}, \boldsymbol{\Gamma}_0)$ be an orthogonal matrix with $\boldsymbol{\Gamma} \in \mathbb{R}^{r_1 \times u}$, $u \leq r_1$, so that

$$\boldsymbol{\Gamma}_0^T \mathbf{Y} \mid \mathbf{X} \sim \boldsymbol{\Gamma}_0^T \mathbf{Y} \tag{7.9a}$$

and

$$\boldsymbol{\Gamma}_0^T \mathbf{Y} \perp\!\!\!\perp \boldsymbol{\Gamma}^T \mathbf{Y} \mid \mathbf{X}. \tag{7.9b}$$

2 http://neurobureau.projects.nitrc.org/ADHD200/Data.html

These conditions parallel those used for other envelope constructions and imply that $\mathbf{X}$ affects the distribution of $\mathbf{Y}$ only via $\boldsymbol{\Gamma}^T\mathbf{Y} \in \mathbb{R}^{u\times r_2}$. To see how these conditions impact model (7.8), first partition

$$\boldsymbol{\alpha} = \begin{pmatrix} \boldsymbol{\alpha}_1 \\ \vdots \\ \boldsymbol{\alpha}_{r_2} \end{pmatrix}, \quad \boldsymbol{\nu} = \begin{pmatrix} \boldsymbol{\nu}_1 \\ \vdots \\ \boldsymbol{\nu}_{r_2} \end{pmatrix}, \quad \boldsymbol{\epsilon} = \begin{pmatrix} \boldsymbol{\epsilon}_1 \\ \vdots \\ \boldsymbol{\epsilon}_{r_2} \end{pmatrix},$$

where $\boldsymbol{\alpha}_j \in \mathbb{R}^{r_1}$, $\boldsymbol{\nu}_j \in \mathbb{R}^{r_1\times p}$, and $\boldsymbol{\epsilon}_j \in \mathbb{R}^{r_1}$, $j = 1, \dots, r_2$. The matrices in these block partitions describe the regression of $\mathbf{Y}_j$, the jth column of $\mathbf{Y}$ on $\mathbf{X}$: $\mathbf{Y}_j = \boldsymbol{\alpha}_j + \boldsymbol{\nu}_j\mathbf{X} + \boldsymbol{\epsilon}_j$, $j = 1, \dots, r_2$. In terms of model (7.8), condition (7.9a) implies that the conditional mean $\mathrm{E}(\boldsymbol{\Gamma}_0^T\mathbf{Y} \mid \mathbf{X})$ is a constant function of $\mathbf{X}$. Since

$$\mathrm{E}(\mathrm{vec}(\boldsymbol{\Gamma}_0^T\mathbf{Y}) \mid \mathbf{X}) = (\mathbf{I}_{r_2} \otimes \boldsymbol{\Gamma}_0^T)\mathrm{E}(\mathrm{vec}(\mathbf{Y}) \mid \mathbf{X}) = (\mathbf{I}_{r_2} \otimes \boldsymbol{\Gamma}_0^T)\boldsymbol{\alpha} + (\mathbf{I}_{r_2} \otimes \boldsymbol{\Gamma}_0^T)\boldsymbol{\nu}\mathbf{X},$$

we must have

$$(\mathbf{I}_{r_2} \otimes \boldsymbol{\Gamma}_0^T)\boldsymbol{\nu} = \begin{pmatrix} \boldsymbol{\Gamma}_0^T\boldsymbol{\nu}_1 \\ \vdots \\ \boldsymbol{\Gamma}_0^T\boldsymbol{\nu}_{r_2} \end{pmatrix} = 0,$$

which holds if and only if

$$\sum_{j=1}^{r_2} \mathrm{span}(\boldsymbol{\nu}_j) \subseteq \mathrm{span}(\boldsymbol{\Gamma}). \tag{7.10}$$

To deal with condition (7.9b), block partition $\boldsymbol{\Delta} = (\boldsymbol{\Delta}_{ij})$ to corresponding to the previous partitioning of $\boldsymbol{\epsilon}$, so $\boldsymbol{\Delta}_{ij} = \mathrm{cov}(\mathbf{Y}_i, \mathbf{Y}_j \mid \mathbf{X})$, $i, j = 1, \dots, r_2$. Then condition (7.9b) holds if and only if

$$(\mathbf{I}_{r_2} \otimes \boldsymbol{\Gamma}_0^T)\boldsymbol{\Delta}(\mathbf{I}_{r_2} \otimes \boldsymbol{\Gamma}) = (\boldsymbol{\Gamma}_0^T\boldsymbol{\Delta}_{ij}\boldsymbol{\Gamma}) = 0, \tag{7.11}$$

so $\mathrm{span}(\mathbf{I}_{r_2} \otimes \boldsymbol{\Gamma})$ must reduce $\boldsymbol{\Delta}$ and $\mathrm{span}(\boldsymbol{\Gamma})$ must reduce all elements $\boldsymbol{\Delta}_{ij}$ of the block partition. Consequently,

$$\begin{aligned} \boldsymbol{\Delta} &= (\mathbf{I}_{r_2} \otimes \boldsymbol{\Gamma})\boldsymbol{\Omega}(\mathbf{I}_{r_2} \otimes \boldsymbol{\Gamma}^T) + (\mathbf{I}_{r_2} \otimes \boldsymbol{\Gamma}_0)\boldsymbol{\Omega}_0(\mathbf{I}_{r_2} \otimes \boldsymbol{\Gamma}_0^T), \\ &= (\boldsymbol{\Gamma}\boldsymbol{\Omega}_{ij}\boldsymbol{\Gamma}^T + \boldsymbol{\Gamma}_0\boldsymbol{\Omega}_{0,ij}\boldsymbol{\Gamma}_0^T),\ i, j = 1, \dots, r_2, \end{aligned} \tag{7.12}$$

where $\boldsymbol{\Omega} \in \mathbb{R}^{r_2u\times r_2u}$ and $\boldsymbol{\Omega}_0 \in \mathbb{R}^{(r_1-u)r_2\times(r_1-u)r_2}$ are positive definite with conforming block partitions $\boldsymbol{\Omega}_{ij}$ and $\boldsymbol{\Omega}_{0,ij}$. We can now define the row envelope of $\mathbf{Y}$ as the smallest reducing subspace of $\boldsymbol{\Delta}$ that has the form $\mathrm{span}(\mathbf{I}_{r_2} \otimes \boldsymbol{\Gamma})$ where $\mathrm{span}(\boldsymbol{\Gamma})$ contains $\sum_{j=1}^{r_2} \mathrm{span}(\boldsymbol{\nu}_j)$. This then leads to the row envelope model

$$\mathrm{vec}(\mathbf{Y}) = \boldsymbol{\alpha} + (\mathbf{I}_{r_2} \otimes \boldsymbol{\Gamma})\boldsymbol{\eta}\mathbf{X} + \boldsymbol{\epsilon}, \tag{7.13}$$

where $\mathrm{var}(\boldsymbol{\epsilon})$ has the structure (7.12), and $\boldsymbol{\eta} \in \mathbb{R}^{r_2u\times p}$ is a matrix of coordinates. Derivation of the maximum likelihood estimator can now follow the general steps used for other envelope models.

Of the two conditions (7.10) and (7.12) needed for model (7.13), (7.12) may be the more difficult to imagine. Here are a few special cases to help fix ideas.

- If the columns of $\mathbf{Y}$ are uncorrelated, then condition (7.12) says that span($\boldsymbol{\Gamma}$) must reduce $\boldsymbol{\Delta}_{jj}$, $j = 1, \ldots, r_2$, and we arrive at a setting similar to that described in Definition 5.1. If in addition the columns of $\mathbf{Y}$ have equal variances, then we need only that span($\boldsymbol{\Gamma}$) reduces $\boldsymbol{\Delta}_{11}$ to satisfy condition (7.12).
- Suppose that the elements of $\mathbf{Y}$ have equal variances, say 1, that any two elements within the same column have correlation ρ_w and that any two elements between columns have correlation ρ_b. Then $\boldsymbol{\Delta}_{jj} = (1 - \rho_w + r_2\rho_w)\mathbf{P}_1 + (1 - \rho_w)\mathbf{Q}_1$ and $\boldsymbol{\Delta}_{ij} = r_2\rho_b\mathbf{P}_1$ for $i \neq j$, where $\mathbf{P}_1$ denotes the projection onto span($\mathbf{1}_{r_2}$). In this scenario, we have only two subspaces to consider, span($\mathbf{1}_{r_2}$) and its orthogonal complement.
- If $\boldsymbol{\Delta} = \boldsymbol{\Sigma}_2 \otimes \boldsymbol{\Sigma}_1$ with $\boldsymbol{\Sigma}_2 \in \mathbb{R}^{r_2 \times r_2}$ and $\boldsymbol{\Sigma}_1 \in \mathbb{R}^{r_1 \times r_1}$, then (7.11) reduces to $\boldsymbol{\Gamma}_0^T\boldsymbol{\Sigma}_1\boldsymbol{\Gamma} = 0$, and we need to consider only the reducing subspaces of $\boldsymbol{\Sigma}_1$. This case links to the matrix normal distribution and is considered in more detail in Section 7.3.2.

7.3.2 Models with Kronecker Structure

Recall that the model for the regression of the jth column of $\mathbf{Y}$ on $\mathbf{X}$ is $\mathbf{Y}_j = \boldsymbol{\alpha}_j + \mathbf{v}_j\mathbf{X} + \boldsymbol{\epsilon}_j$, $j = 1, \ldots, r_2$. In many regressions with matrix-valued responses, the columns of $\mathbf{Y}$ represent repeated measures on the same experimental unit, or the same or similar measurements taken at different times or locations. In such cases, it may be reasonable to constrain the coefficient matrix $\mathbf{v}_j$ to change proportionately over the columns of $\mathbf{Y}$: $\mathbf{v}_j = \boldsymbol{\beta}_1\beta_{2j}$, where $\boldsymbol{\beta}_1 \in \mathbb{R}^{r_1 \times p}$, and β_{2j} is a scalar. This leads to the matrix-valued response model

$$\mathbf{Y} = \bar{\boldsymbol{\alpha}} + \boldsymbol{\beta}_1\mathbf{X}\boldsymbol{\beta}_2^T + \bar{\boldsymbol{\epsilon}}, \tag{7.14}$$

where now $\bar{\boldsymbol{\alpha}} = (\boldsymbol{\alpha}_1, \ldots, \boldsymbol{\alpha}_{r_2}) \in \mathbb{R}^{r_1 \times r_2}$, $\boldsymbol{\beta}_2 = (\beta_{21}, \ldots, \beta_{2r_2})^T \in \mathbb{R}^{r_2}$, and the matrix of errors $\bar{\boldsymbol{\epsilon}} = (\boldsymbol{\epsilon}_1, \ldots, \boldsymbol{\epsilon}_{r_2}) \in \mathbb{R}^{r_1 \times r_2}$. Using similar reasoning, we model the covariance between $\boldsymbol{\epsilon}_i$ and $\boldsymbol{\epsilon}_j$ as $\text{cov}(\boldsymbol{\epsilon}_i, \boldsymbol{\epsilon}_j) = (\boldsymbol{\Sigma}_2)_{ij}\boldsymbol{\Sigma}_1$, where $(\boldsymbol{\Sigma}_2)_{ij}$ is the (i,j)th element of the positive definite matrix $\boldsymbol{\Sigma}_2 \in \mathbb{R}^{r_2 \times r_2}$ and similarly $\boldsymbol{\Sigma}_1 \in \mathbb{R}^{r_1 \times r_1}$ is also positive definite. This leads to the overall structure of the error matrix $\text{var}(\boldsymbol{\epsilon}) = \boldsymbol{\Sigma}_2 \otimes \boldsymbol{\Sigma}_1$, which we assume in the remainder of this section. Roy and Khattree (2005) and Lu and Zimmerman (2005) studied testing methods for a Kronecker structure of covariance matrices. Adding normality leads to a matrix normal distribution for $\boldsymbol{\epsilon}$ (see Section A.8). Under (7.14) with matrix normal errors, any linear combination $\mathbf{Yc}$ of the columns of $\mathbf{Y}$ follows the multivariate linear model (1.1) with coefficient matrix $\boldsymbol{\beta} = \boldsymbol{\beta}_1(\boldsymbol{\beta}_2^T\mathbf{c})$ and error covariance matrix $\boldsymbol{\Sigma} = (\mathbf{c}^T\boldsymbol{\Sigma}_1\mathbf{c})\boldsymbol{\Sigma}_2$.

Model (7.14) can be estimated by maximum likelihood assuming non-stochastic predictors, independent observations $\mathbf{Y}_i \mid \mathbf{X}_i$, $i = 1, \ldots, n$, and matrix normal errors. We also center the predictors, so $\bar{\mathbf{X}} = 0$. Then the maximum likelihood estimator of $\bar{\boldsymbol{\alpha}}$ is $\bar{\mathbf{Y}}$. The remaining parameters are not identifiable without further constraints, and so their maximum likelihood estimators are not unique. For instance, multiplying $\boldsymbol{\beta}_1$ by a nonzero constant and dividing $\boldsymbol{\beta}_2$ by the same constant lead to new parameters that also satisfy (7.14). If estimates of $\boldsymbol{\beta}_1$ and $\boldsymbol{\beta}_2$ are required, we could force them to be identifiable by constraining $\boldsymbol{\beta}_2$ to have a positive first element and $\|\boldsymbol{\beta}_2\|_F = 1$, for example. On the other hand, the Kronecker products $\boldsymbol{\beta}_2 \otimes \boldsymbol{\beta}_1$ and $\boldsymbol{\Sigma}_2 \otimes \boldsymbol{\Sigma}_1$ are identifiable without further constraints.

The following alternating algorithm (Ding and Cook 2017) can be used to fit (7.14) by maximizing its likelihood. Let

$$\mathbf{C}_2 = \sum_{i=1}^{n} (\mathbf{Y}_i - \bar{\mathbf{Y}})\boldsymbol{\Sigma}_2^{-1}\boldsymbol{\beta}_2\mathbf{X}_i^T \quad \text{and} \quad \mathbf{M}_2 = \sum_{i=1}^{n} \mathbf{X}_i\boldsymbol{\beta}_2^T\boldsymbol{\Sigma}_2^{-1}\boldsymbol{\beta}_2\mathbf{X}_i^T.$$

Maximizing the log-likelihood function given $\boldsymbol{\beta}_2$ and $\boldsymbol{\Sigma}_2$ fixed, the estimators $\boldsymbol{\beta}_1$ and $\boldsymbol{\Sigma}_1$ are

$$\mathbf{B}_{1|2} = \mathbf{C}_2\mathbf{M}_2^{-1} \tag{7.15}$$

$$\mathbf{S}_{1|2} = \frac{1}{nr_2}\sum_{i=1}^{n} (\mathbf{Y}_i - \bar{\mathbf{Y}} - \mathbf{B}_{1|2}\mathbf{X}_i\boldsymbol{\beta}_2^T)\boldsymbol{\Sigma}_2^{-1}(\mathbf{Y}_i - \bar{\mathbf{Y}} - \mathbf{B}_{1|2}\mathbf{X}_i\boldsymbol{\beta}_2^T)^T. \tag{7.16}$$

Similarly, let

$$\mathbf{C}_1 = \sum_{i=1}^{n} (\mathbf{Y}_i - \bar{\mathbf{Y}})^T\boldsymbol{\Sigma}_1^{-1}\boldsymbol{\beta}_1\mathbf{X}_i \quad \text{and} \quad \mathbf{M}_1 = \sum_{i=1}^{n} \mathbf{X}_i^T\boldsymbol{\beta}_1^T\boldsymbol{\Sigma}_1^{-1}\boldsymbol{\beta}_1\mathbf{X}_i.$$

Given $\boldsymbol{\beta}_1$ and $\boldsymbol{\Sigma}_1$ fixed, the estimators of $\boldsymbol{\beta}_2$ and $\boldsymbol{\Sigma}_2$ are

$$\mathbf{B}_{2|1} = \mathbf{C}_1\mathbf{M}_1^{-1} \tag{7.17}$$

$$\mathbf{S}_{2|1} = \frac{1}{nr_1}\sum_{i=1}^{n} (\mathbf{Y}_i^T - \bar{\mathbf{Y}}^T - \mathbf{B}_{2|1}\mathbf{X}_i^T\boldsymbol{\beta}_1^T)\boldsymbol{\Sigma}_1^{-1}(\mathbf{Y}_i^T - \bar{\mathbf{Y}}^T - \mathbf{B}_{2|1}\mathbf{X}_i^T\boldsymbol{\beta}_1^T)^T. \tag{7.18}$$

With constraints imposed to ensure identifiability, let $\mathbf{B}_1$, $\mathbf{B}_2$, $\mathbf{S}_1$, and $\mathbf{S}_2$ be the maximum likelihood estimators of $\boldsymbol{\beta}_1$, $\boldsymbol{\beta}_2$, $\boldsymbol{\Sigma}_1$, and $\boldsymbol{\Sigma}_2$. Alternating between the conditional estimators until a convergence criterion is met leads to the estimators $\mathbf{B}_1 = \mathbf{B}_{1|2}$, $\mathbf{B}_2 = \mathbf{B}_{2|1}$, $\mathbf{S}_1 = \mathbf{S}_{1|2}$, and $\mathbf{S}_2 = \mathbf{S}_{2|1}$. Additionally, model (7.14) can be tested against model (7.8) by using a likelihood ratio test.

7.3.3 Envelope Models with Kronecker Structure

Although the matrix regression model (7.14) is parsimoniously parameterized relative to the vector model (7.8), there could still be linear combinations of the rows or columns of $\mathbf{Y} \in \mathbb{R}^{r_1 \times r_2}$ whose distribution is invariant to changes in the predictors. The introduction of envelopes into model (7.14) follows the same logic that we have used previously but we consider row and column envelopes simultaneously. Let $\mathcal{B}_1 = \text{span}(\boldsymbol{\beta}_1)$ and $\mathcal{B}_2 = \text{span}(\boldsymbol{\beta}_2)$, let u_j denote the dimensions of the envelope $\mathcal{E}_{\boldsymbol{\Sigma}_j}(\mathcal{B}_j), j = 1, 2$, let $\boldsymbol{\Gamma}_j \in \mathbb{R}^{r_j \times u_j}$ be a semi-orthogonal basis matrix for $\mathcal{E}_{\boldsymbol{\Sigma}_j}(\mathcal{B}_j)$, and let $(\boldsymbol{\Gamma}_j, \boldsymbol{\Gamma}_{j0})$ be an orthogonal matrix, $j = 1, 2$. Reparameterizing model (7.14) in terms of the envelopes, we have

$$\mathbf{Y} = \boldsymbol{\alpha} + \boldsymbol{\Gamma}_1 \boldsymbol{\eta}_1 \mathbf{X} \boldsymbol{\eta}_2^T \boldsymbol{\Gamma}_2^T + \boldsymbol{\varepsilon}, \tag{7.19}$$

with $\text{var}(\text{vec}(\boldsymbol{\varepsilon})) = \boldsymbol{\Sigma}_1 \otimes \boldsymbol{\Sigma}_2$, where $\boldsymbol{\Sigma}_j = \boldsymbol{\Gamma}_j \boldsymbol{\Omega}_j \boldsymbol{\Gamma}_j^T + \boldsymbol{\Gamma}_{j0} \boldsymbol{\Omega}_{j0} \boldsymbol{\Gamma}_{j0}^T$, $\boldsymbol{\Omega}_j \in \mathbb{R}^{u_j \times u_j} > 0$ and $\boldsymbol{\Omega}_{j0} \in \mathbb{R}^{(r_j - u_j) \times (r_j - u_j)}$ are positive definite, and $\boldsymbol{\eta}_1 \in \mathbb{R}^{u_1 \times p}$ and $\boldsymbol{\eta}_2 \in \mathbb{R}^{u_2 \times 1}$ are unconstrained coordinate matrices, $j = 1, 2$.

It can be seen from model (7.19) that

$$\mathbf{Q}_{\boldsymbol{\Gamma}_1} \mathbf{Y} \mid \mathbf{X} = \mathbf{x}_1 \sim \mathbf{Q}_{\boldsymbol{\Gamma}_1} \mathbf{Y} \mid \mathbf{X} = \mathbf{x}_2 \quad \text{and} \quad \mathbf{P}_{\boldsymbol{\Gamma}_1} \mathbf{Y} \perp\!\!\!\perp \mathbf{Q}_{\boldsymbol{\Gamma}_1} \mathbf{Y} \mid \mathbf{X} \tag{7.20}$$

$$\mathbf{Y} \mathbf{Q}_{\boldsymbol{\Gamma}_2} \mid \mathbf{X} = \mathbf{x}_1 \sim \mathbf{Y} \mathbf{Q}_{\boldsymbol{\Gamma}_2} \mid \mathbf{X} = \mathbf{x}_2 \quad \text{and} \quad \mathbf{Y} \mathbf{P}_{\boldsymbol{\Gamma}_2} \perp\!\!\!\perp \mathbf{Y} \mathbf{Q}_{\boldsymbol{\Gamma}_2} \mid \mathbf{X}. \tag{7.21}$$

Conditions (7.20) and (7.21) are the row and column counterparts of conditions (1.18) for envelopes with vector-valued responses. For instance, from (7.21), we see that the distribution of $\mathbf{Y}\mathbf{Q}_{\boldsymbol{\Gamma}_2}$ does not depend on the value of $\mathbf{X}$ and $\mathbf{Y}\mathbf{Q}_{\boldsymbol{\Gamma}_2}$ and $\mathbf{Y}\mathbf{P}_{\boldsymbol{\Gamma}_2}$ are independent given $\mathbf{X}$, with the consequence that, when considering columns, changes in the value of $\mathbf{X}$ can affect only the distribution of $\mathbf{Y}\mathbf{P}_{\boldsymbol{\Gamma}_2}$.

Efficiency gains of envelope model (7.19) over matrix model (7.14) can arise through parameter reduction, particularly when the number of real parameters in (7.14) is large relative to that in (7.19). But the largest efficiency gains are often realized when the immaterial variation in the rows and columns of $\mathbf{Y}$ is large relative to the corresponding material variation $\mathbf{Y}$. Parallel to the discussion in Section 1.4.2, this can happen when $\|\boldsymbol{\Omega}_1\| \ll \|\boldsymbol{\Omega}_{10}\|$ or $\|\boldsymbol{\Omega}_2\| \ll \|\boldsymbol{\Omega}_{20}\|$.

Estimation of the parameters in model (7.19) follows the idea behind the alternating scheme described for model (7.14). Let $\mathbf{S}_{\mathbf{Y}|1} = (nr_1)^{-1} \sum_{i=1}^n (\mathbf{Y}_i - \bar{\mathbf{Y}})^T \boldsymbol{\Sigma}_1^{-1} (\mathbf{Y}_i - \bar{\mathbf{Y}})$ and $\mathbf{S}_{\mathbf{Y}|2} = (nr_2)^{-1} \sum_{i=1}^n (\mathbf{Y}_i - \bar{\mathbf{Y}}) \boldsymbol{\Sigma}_2^{-1} (\mathbf{Y}_i - \bar{\mathbf{Y}})^T$. Given u_1 and values for $\boldsymbol{\beta}_2$ and $\boldsymbol{\Sigma}_2$, the estimator of $\mathcal{E}_{\boldsymbol{\Sigma}_1}(\mathcal{B}_1)$ is

$$\hat{\mathcal{E}}_{\boldsymbol{\Sigma}_1}(\mathcal{B}_1) = \text{span}\{\arg\min_{\mathbf{G}} \log|\mathbf{G}^T \mathbf{S}_{1|2} \mathbf{G}| + \log|\mathbf{G}^T \mathbf{S}_{\mathbf{Y}|2}^{-1} \mathbf{G}|\},$$

where the minimum is over all semi-orthogonal matrices $\mathbf{G} \in \mathbb{R}^{r_1 \times u_1}$ and $\mathbf{S}_{1|2}$ was defined in (7.16). The estimates of $\boldsymbol{\beta}_1$ and $\boldsymbol{\Sigma}_1$ at the current iteration are then

$$\hat{\boldsymbol{\beta}}_1 = \mathbf{P}_{\hat{\mathcal{E}}_1} \mathbf{B}_{1|2} \quad \text{and} \quad \hat{\boldsymbol{\Sigma}}_1 = \mathbf{P}_{\hat{\mathcal{E}}_1} \mathbf{S}_{1|2} \mathbf{P}_{\hat{\mathcal{E}}_1} + \mathbf{Q}_{\hat{\mathcal{E}}_1} \mathbf{S}_{\mathbf{Y}|2} \mathbf{Q}_{\hat{\mathcal{E}}_1},$$

where $\mathbf{B}_{1|2}$ was defined in (7.15). Next, given u_2, $\boldsymbol{\beta}_1 = \widehat{\boldsymbol{\beta}}_1$ and $\boldsymbol{\Sigma}_1 = \widehat{\boldsymbol{\Sigma}}_1$, the log likelihood for $\mathcal{E}_{\boldsymbol{\Sigma}_2}(\mathcal{B}_2)$ is maximized at

$$\widehat{\mathcal{E}}_{\boldsymbol{\Sigma}_2}(\mathcal{B}_2) = \text{span}\{\arg\min_{\mathbf{G}} \log|\mathbf{G}^T\mathbf{S}_{2|1}\mathbf{G}| + \log|\mathbf{G}^T\mathbf{S}_{\mathbf{Y}|1}^{-1}\mathbf{G}|\},$$

where the minimum is over all semi-orthogonal matrices $\mathbf{G} \in \mathbb{R}^{r_2 \times u_2}$ and $\mathbf{S}_{2|1}$ was defined in (7.18). The corresponding estimates of $\boldsymbol{\beta}_2$ and $\boldsymbol{\Sigma}_2$ at the current iteration are then

$$\widehat{\boldsymbol{\beta}}_2 = \mathbf{P}_{\widehat{\mathcal{E}}_2}\mathbf{B}_{2|1} \quad \text{and} \quad \widehat{\boldsymbol{\Sigma}}_2 = \mathbf{P}_{\widehat{\mathcal{E}}_2}\mathbf{S}_{2|1}\mathbf{P}_{\widehat{\mathcal{E}}_2} + \mathbf{Q}_{\widehat{\mathcal{E}}_2}\mathbf{S}_{\mathbf{Y}|1}\mathbf{Q}_{\widehat{\mathcal{E}}_2},$$

where $\mathbf{B}_{2|1}$ was defined in (7.17). These estimates of $\boldsymbol{\beta}_2$ and $\boldsymbol{\Sigma}_2$ can now be used to begin again, with iteration continuing until a convergence criterion is met, at which point the values of the various parameters are taken as maximum likelihood estimators. If desired, constraints can now be introduced to produce unique estimates. The objective functions to be optimized have the same form as (1.25) and many others in this book so the basic optimization is the same, except here iteration is required.

As in previous instances of envelopes in this book, with matrix normal errors, the envelope estimators $\text{vec}(\widehat{\boldsymbol{\beta}}_1)$ and $\text{vec}(\widehat{\boldsymbol{\beta}}_2)$ are asymptotically normal with asymptotic variance no larger than that for the estimators under model (7.14). Lacking matrix normal errors, these estimators are still asymptotically normal under mild technical conditions (Ding and Cook 2017).

7.4 Spatial Envelopes

So far, all envelope methods based on the multivariate linear model (1.1) required that the response vectors be independent, and thus may not be applicable when they are dependent. Emphasizing spatial applications, Rekabdarkolaee and Wang (2017) recently developed an envelope version of the multivariate linear model that allows for the response vectors to be dependent. We sketch highlights of their extension in this section.

Model (1.1) is not adequate to describe settings with dependent responses. Instead, we pass to a variation of it that describes all of the data simultaneously:

$$\mathbb{Y}_0 = \boldsymbol{\alpha}^T \otimes \mathbf{1}_n + \mathbb{X}\boldsymbol{\beta}^T + \mathbb{E}, \tag{7.22}$$

where, in reference to the notation of model (1.1), $\mathbb{Y}_0 \in \mathbb{R}^{n\times r}$ has rows $\mathbf{Y}_i^T$ and $\mathbb{X} \in \mathbb{R}^{n\times p}$ has rows $(\mathbf{X}_i - \bar{\mathbf{X}})^T$, as described in Section 1.1, and $\mathbb{E} \in \mathbb{R}^{n\times r}$ has normally distributed rows $\boldsymbol{\varepsilon}_i^T$. To account for the correlations among the rows of $\mathbb{Y}_0$, Rekabdarkolaee and Wang (2017) adopted a variance of the form $\text{var}\{\text{vec}(\mathbb{Y}_0) \mid \mathbb{X}\} = \text{var}\{\text{vec}(\mathbb{E})\} = \boldsymbol{\Sigma} \otimes \boldsymbol{\rho}(\boldsymbol{\theta})$, where $\boldsymbol{\Sigma} \in \mathbb{R}^{r\times r}$ is positive definite, and $\boldsymbol{\rho}(\boldsymbol{\theta}) \in \mathbb{R}^{n\times n}$ is a parsimoniously parameterized correlation matrix. This covariance structure is similar to that considered in Section 7.3.2. Let $\mathbf{e}_{k,i}$ denote a $k \times 1$ vector with a 1 in position i and zeros elsewhere. Then the variances of the ith and jth response vectors and their covariance can be expressed as

$$\begin{aligned}
\mathrm{var}\{\mathrm{vec}((\mathbf{Y}_i, \mathbf{Y}_j)) \mid \mathbb{X}\} &= \mathrm{var}\{\mathrm{vec}(\mathbb{Y}_0^T(\mathbf{e}_{n,i}, \mathbf{e}_{n,j})) \mid \mathbb{X}\} \\
&= \mathrm{var}\{\mathrm{vec}(\mathbb{E}^T(\mathbf{e}_{n,i}, \mathbf{e}_{n,j}))\} \\
&= ((\mathbf{e}_{n,i}, \mathbf{e}_{n,j})^T \otimes \mathbf{I}_r)\mathrm{var}\{\mathrm{vec}(\mathbb{Y}_0^T) \mid \mathbb{X}\}((\mathbf{e}_{n,i}, \mathbf{e}_{n,j}) \otimes \mathbf{I}_r) \\
&= ((\mathbf{e}_{n,i}, \mathbf{e}_{n,j})^T \otimes \mathbf{I}_r)(\rho(\theta) \otimes \mathbf{\Sigma})((\mathbf{e}_{n,i}, \mathbf{e}_{n,j}) \otimes \mathbf{I}_r) \\
&= \begin{pmatrix} 1 & (\rho)_{ij} \\ (\rho)_{ij} & 1 \end{pmatrix} \otimes \mathbf{\Sigma},
\end{aligned}$$

where the last step follows in part because the diagonals of ρ are all equal to 1. In consequence, we see that the response vectors all have variance $\mathbf{\Sigma}$ and that the covariance between any two different response vectors is proportional to $\mathbf{\Sigma}$.

Rekabdarkolaee and Wang (2017) introduced envelopes onto model (7.22) by using the $\mathbf{\Sigma}$-envelope of $\mathcal{B}$, $\mathcal{E}_{\mathbf{\Sigma}}(\mathcal{B})$. Adopting the notation of Chapter 1, we then have $\mathbf{\Sigma} = \mathbf{\Gamma\Omega\Gamma} + \mathbf{\Gamma}_0\mathbf{\Omega}_0\mathbf{\Gamma}_0$, and thus $\boldsymbol{\beta} = \mathbf{\Gamma}\boldsymbol{\eta}$ and

$$\mathrm{var}\{\mathrm{vec}(\mathbb{Y}_0) \mid \mathbb{X}\} = (\mathbf{\Gamma\Omega\Gamma}) \otimes \rho + (\mathbf{\Gamma}_0\mathbf{\Omega}_0\mathbf{\Gamma}_0) \otimes \rho.$$

Partitioning model (7.22) into the material and X-invariant (immaterial) parts of $\mathbb{Y}_0$,

$$\begin{aligned}
\mathbb{Y}_0\mathbf{\Gamma} &= \boldsymbol{\alpha}^T\mathbf{\Gamma} \otimes \mathbf{1}_n + \mathbb{X}\boldsymbol{\eta}^T + \mathbb{E}\mathbf{\Gamma}, \\
\mathbb{Y}_0\mathbf{\Gamma}_0 &= \boldsymbol{\alpha}^T\mathbf{\Gamma}_0 \otimes \mathbf{1}_n + \mathbb{E}\mathbf{\Gamma}_0.
\end{aligned}$$

Because the error matrices are independent under normality,

$$\mathrm{cov}\{\mathrm{vec}(\mathbb{E}\mathbf{\Gamma}), \mathrm{vec}(\mathbb{E}\mathbf{\Gamma}_0)\} = (\mathbf{\Gamma}^T \otimes \mathbf{I}_n)\mathrm{var}\{\mathrm{vec}(\mathbb{E})\}(\mathbf{\Gamma}_0 \otimes \mathbf{I}_n) = 0,$$

and the marginal distribution of $\mathbb{Y}_0\mathbf{\Gamma}_0$ does not depend on $\mathbb{X}$, we see that the structure is as required to identify the X-invariant part of $\mathbb{Y}_0$. The derivation of the maximum likelihood estimators under this envelope structure can follow the same general logic as used elsewhere in this book.

Rekabdarkolaee and Wang (2017) focused primarily on spatial applications in which they adopted the two-parameter Matérn covariance function as a basis for ρ. That function is commonly used in spatial statistics to define the covariance between measurements as a function of the distance between them. They used their method to predict the air pollution of 270 cites in the Northeast United States. Their response vectors consisted of the concentrations of 10 air pollutants and their predictors were three meteorological variables, wind, temperature, and relative humidity. Using leave-one-out cross-validation, the prediction error for their method was about half of that for the closest competitor, the linear coregionalization model of Zhang (2007), and about a twentieth of that for the other methods they considered. These results seem to give a strong prima facie argument for the use of envelope methods in spatial statistics.

7.5 Sparse Response Envelopes

In Section 1.8.2, we described likelihood-based testing methods for identifying X-variant and X-invariant responses as defined in Section 1.8. In this section, we give an overview of the sparse envelope methodology developed by Su et al. (2016) for the same goal. We divide our discussion into two parts. In the first, we assume that $r \ll n$ so $\mathbf{S}_{\mathbf{Y}}$ and $\mathbf{S}_{\mathbf{Y}|\mathbf{X}}$ are both positive definite. In the second, we allow $r > n$. We assume throughout that $\mathbf{S}_{\mathbf{X}} > 0$.

7.5.1 Sparse Response Envelopes when $r \ll n$

We know from the discussion of Section 1.8 that, in the context of the response envelope model (1.20), a subset of the responses is X-invariant if and only if the corresponding rows of $\boldsymbol{\Gamma}$ are all zero. Consequently, methodology for identifying the X-variant responses can be developed by penalizing the rows of $\boldsymbol{\Gamma}$. However, rather than dealing directly with those rows, Su et al. (2016) proceeded equivalently by penalizing the reparameterized version of the general objective function $L_u(\mathbf{G})$ defined in (6.4). Specifically, they start with the objective function

$$L_u(\mathbf{G}) = \log|\mathbf{G}^T\mathbf{S}_{\mathbf{Y}|\mathbf{X}}\mathbf{G}| + \log|\mathbf{G}^T\mathbf{S}_{\mathbf{Y}}^{-1}\mathbf{G}|$$

and then reparameterize as we did in Section 6.3:

$$\mathbf{G} = \begin{pmatrix} \mathbf{G}_1 \\ \mathbf{G}_2 \end{pmatrix} = \begin{pmatrix} \mathbf{I}_u \\ \mathbf{A} \end{pmatrix} \mathbf{G}_1 = \mathbf{C}_{\mathbf{A}}\mathbf{G}_1,$$

where $\mathbf{G}_1 \in \mathbb{R}^{u\times u}$ is constrained to be nonsingular, $\mathbf{A} = \mathbf{G}_2\mathbf{G}_1^{-1} \in \mathbb{R}^{(r-u)\times u}$ is an unconstrained matrix, and $\mathbf{C}_{\mathbf{A}} = (\mathbf{I}_u, \mathbf{A}^T)^T$. Using these relationships, $L_u(\mathbf{G})$ can be reparameterized as a function of only $\mathbf{A}$:

$$L_u(\mathbf{A}) = -2\log|\mathbf{C}_{\mathbf{A}}^T\mathbf{C}_{\mathbf{A}}| + \log|\mathbf{C}_{\mathbf{A}}^T\mathbf{S}_{\mathbf{Y}|\mathbf{X}}\mathbf{C}_{\mathbf{A}}| + \log|\mathbf{C}_{\mathbf{A}}^T\mathbf{S}_{\mathbf{Y}}^{-1}\mathbf{C}_{\mathbf{A}}|.$$

Since $\mathbf{G}_1$ is nonsingular by construction, any zero rows of $\boldsymbol{\Gamma}$ must correspond to zero rows of $\mathbf{G}_2$ or, equivalently, zero rows of $\mathbf{A}$. In other words, a row of $\mathbf{G}_2$ is zero if and only if the corresponding row of $\mathbf{A}$ is zero. Letting $\mathbf{a}_i^T$ denote the ith row of $\mathbf{A}$, $i = 1, \ldots, r-u$, this connection allowed Su et al. (2016) to penalize via $\mathbf{a}_i$, leading to the penalized objective function

$$L_{\text{pen}}(\mathbf{A}) = -2\log|\mathbf{C}_{\mathbf{A}}^T\mathbf{C}_{\mathbf{A}}| + \log|\mathbf{C}_{\mathbf{A}}^T\mathbf{S}_{\mathbf{Y}|\mathbf{X}}\mathbf{C}_{\mathbf{A}}| + \log|\mathbf{C}_{\mathbf{A}}^T\mathbf{S}_{\mathbf{Y}}^{-1}\mathbf{C}_{\mathbf{A}}| + \lambda\sum_{i=1}^{r-u}\omega_i\|\mathbf{a}_i\|_2, \tag{7.23}$$

where $\|\cdot\|_2$ denotes the L_2-norm, the ω_i's are adaptive weights, and λ is a tuning parameter. Following the determination of $\widehat{\mathbf{A}} = \arg\min_{\mathbf{A}\in\mathbb{R}^{(r-u)\times u}} L_{\text{pen}}(\mathbf{A})$, we have the following sparse envelope estimators:

$$\begin{aligned}
\widehat{\mathbf{\Gamma}} &= \text{any semiorthogonal basis for span}(\mathbf{C}_{\widehat{\mathbf{A}}}) \\
\widehat{\eta} &= \widehat{\mathbf{\Gamma}}^T \mathbf{B} \\
\widehat{\boldsymbol{\beta}} &= \mathbf{P}_{\widehat{\Gamma}} \mathbf{B} \\
\widehat{\boldsymbol{\beta}}_{\emptyset} &= \text{submatrix of } \widehat{\boldsymbol{\beta}} \text{ with nonzero rows} \\
\widehat{\mathbf{\Sigma}} &= \mathbf{P}_{\widehat{\Gamma}} \mathbf{S}_{\mathrm{Y|X}} \mathbf{P}_{\widehat{\Gamma}} + \mathbf{Q}_{\widehat{\Gamma}} \mathbf{S}_{\mathrm{Y}} \mathbf{Q}_{\widehat{\Gamma}}.
\end{aligned}$$

Assuming that the errors $\boldsymbol{\varepsilon}$ on model (1.20) have finite fourth moments and certain mild technical conditions hold, Su et al. (2016) demonstrated that then $\widehat{\boldsymbol{\beta}}$ is a root-n consistent estimator of $\boldsymbol{\beta}$ and that the zero and nonzero rows of $\mathbf{\Gamma}$ are selected with probability approaching 1 as $n \rightarrow \infty$ with r fixed, so the X-variant and X-invariant responses are identified asymptotically in the same way. They further demonstrate that $\sqrt{n}(\mathrm{vec}(\widehat{\boldsymbol{\beta}}_{\emptyset}) - \mathrm{vec}(\boldsymbol{\beta}_{\emptyset}))$ is asymptotically normal with mean zero and variance equal to that of the envelope estimator based on the oracle model where the X-invariant responses are known.

The rather intricate algorithm proposed by Su et al. (2016) involves block-wise coordinate descent, application of the majorization–minimization principle , and a two-stage procedure to determine the adaptive weights ω_i. They proposed a BIC criterion for choosing λ and indicated that in principle any of the methods discussed in Section 1.10 could be used to choose u. Their theoretical and simulation results together indicate that this may be an effective method for fitting a sparse response envelope model when $n \gg r$.

7.5.2 Cattle Weights and Brain Volumes: Sparse Fits

Figure 7.3 shows a plot of the fitted weights from a sparse envelope fit to the cattle data with $u = 1$. The first four elements of $\widehat{\boldsymbol{\beta}}$ from the sparse fit are zero, so the first four fitted weights are identical for the two treatments, with the first nonzero difference occurring at week 10. These results are in qualitative agreement with the initial envelope fit of the cattle data in Section 1.3.4. A sparse envelope fit with $u = 5$ does not lead to any elements of $\widehat{\boldsymbol{\beta}}$ being 0, but using the asymptotic standard errors from Su et al. (2016) resulted in inferences that are consistent with those discussed in Section 1.10.5.

While there is a good agreement between the sparse and nonsparse envelope analyses of the cattle data, such agreement is not always the case. Our study of brain volumes in Section 2.8 leads to the conclusion that 73 regions of the brain regions decreased in volume with age, while all regions either decreased in volume or are correlated with such regions. In contrast, fitting a sparse envelope with $u = 5$, which is the same as the value used previously, led to an estimator of $\boldsymbol{\beta}$ with 21 nonzero and 72 zero elements, suggesting that relatively few regions of the brain change volume over time. This first fit was at the default

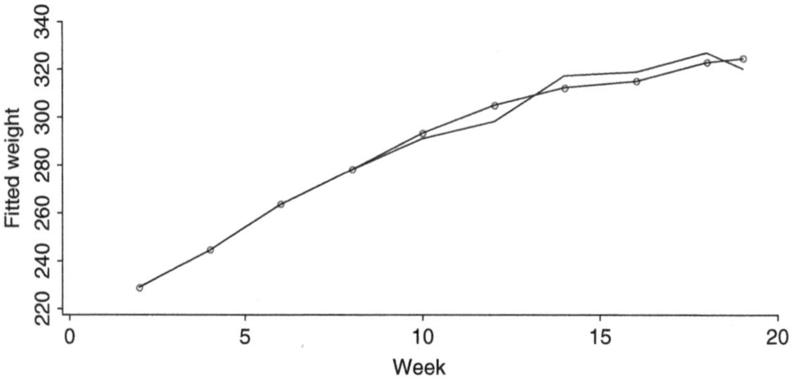

Figure 7.3 Cattle data: Sparse fitted weights with $u = 1$.

values of the tuning parameters (number of iterations, tolerances, etc.) in the code furnished by Su et al. (2016). Increasing the iterations and decreasing tolerances increased the number of coefficients that were estimated to be nonzero from 21 to 34, but that is still substantially less than the 73 found in Section 2.8. The difference could be caused by convergence of the sparse algorithm to a local optimum, or the difference could be a reflection of the nature of sparse methods generally.

7.5.3 Sparse Envelopes when $r > n$

The population structure described in Section 7.5.1 still holds when r is larger than or of the same order as n, but additional problems arise because then $\mathbf{S}_{\mathbf{Y}|\mathbf{X}}$ and $\mathbf{S}_{\mathbf{Y}}$ may be singular or ill conditioned, causing the objective function (7.23) to be unserviceable. Su et al. (2016) got around this issue by replacing $\mathbf{S}_{\mathbf{Y}|\mathbf{X}}$ and $\mathbf{S}_{\mathbf{Y}}$ with sparse permutation invariant covariance estimators (SPICE) $\mathbf{S}_{\mathbf{Y}|\mathbf{X},\mathrm{sp}}$ and $\mathbf{S}_{\mathbf{Y},\mathrm{sp}}$ Rothman et al. (2008), reasoning that SPICE estimators can still perform well when the population quantities $\boldsymbol{\Sigma}$ and $\boldsymbol{\Sigma}_{\mathbf{Y}}$ are not in fact sparse. This reasoning leads to the objective function

$$L_{\mathrm{pen}}^{\mathrm{sp}}(\mathbf{A}) = -2\log|\mathbf{C}_{\mathbf{A}}^T\mathbf{C}_{\mathbf{A}}| + \log|\mathbf{C}_{\mathbf{A}}^T\mathbf{S}_{\mathbf{Y}|\mathbf{X},\mathrm{sp}}\mathbf{C}_{\mathbf{A}}| + \log|\mathbf{C}_{\mathbf{A}}^T\mathbf{S}_{\mathbf{Y},\mathrm{sp}}^{-1}\mathbf{C}_{\mathbf{A}}| + \lambda\sum_{i=1}^{r-u}\omega_i\|\mathbf{a}_i\|_2. \tag{7.24}$$

The algorithm discussed in Section 7.5.1 can also be used to find $\widehat{\mathbf{A}}_{\mathrm{sp}} = \arg\min L_{\mathrm{pen}}^{\mathrm{sp}}(\mathbf{A})$, except Su et al. (2016) express a preference for cross-validation to select λ and u to mitigate issues caused by the relatively small sample size. Following the determination of $\widehat{\mathbf{A}}_{\mathrm{sp}}$, the estimators described in Section 7.5.1

are used here as well; for instance, $\widehat{\boldsymbol{\Gamma}}_{\text{sp}}$ is any semi-orthogonal basis for $\text{span}(\mathbf{C}_{\widehat{\mathbf{A}}_{\text{sp}}})$.

As might be expected, additional theoretical requirements were needed to characterize the asymptotic behavior of the estimators computed based on (7.24):

- To conform to the structure required for the SPICE estimators, they assumed that $\mathbf{X}$ is random with mean $\boldsymbol{\mu}_{\mathbf{X}}$ and variance $\boldsymbol{\Sigma}_{\mathbf{X}} > 0$. They also required that $\mathbf{X} - \boldsymbol{\mu}_{\mathbf{X}}$ follow a sub-Gaussian distribution and that the data $\mathbf{X}_i$, $i = 1, \ldots, n$, are independent copies of $\mathbf{X}$. In contrast, the methodology of Section 7.5.1 treats $\mathbf{X}$ as nonstochastic. Nevertheless, we expect that the estimators based on (7.24) could still be useful when $\mathbf{X}$ is fixed by design.
- The asymptotic results of Section 7.5.1 treated r as fixed while letting $n \to \infty$. For the asymptotics of this section, Su et al. (2016) treated r as a function $r(n)$ of the sample size that is allowed to diverge as $n \to \infty$. This is mainly a technical device to allow r and n to diverge in various alignments.
- Since r is allowed to diverge, it is also necessary to control the asymptotic behavior of $\boldsymbol{\Sigma} \in \mathbb{R}^{r \times r}$. This was accomplished by requiring that the eigenvalues of $\boldsymbol{\Sigma}$ be bound away from 0 and ∞ as $n \to \infty$. This guarantees, for example, that two responses will not be asymptotically collinear.
- Like the predictors, the envelope errors, $\boldsymbol{\varepsilon}_i$, $i = 1, \ldots, n$, were assumed to be independent copies of a sub-Gaussian random vector $\boldsymbol{\varepsilon}$. In contrast, the errors in Section 7.5.1 were required only to have finite fourth moments.

Let s_1 and s_2 denote the number of nonzero elements below the diagonals of $\boldsymbol{\Sigma}$ and $\boldsymbol{\Sigma}_{\mathbf{Y}}$, let $s = \max\{s_1, s_2\}$, and let $\widehat{\boldsymbol{\beta}}_{\text{sp}}$ denote the estimator of $\boldsymbol{\beta}$ stemming from (7.24). Then under the conditions stated above and a couple of additional technical conditions, Su et al. (2016) proved that, as $n \to \infty$, $\|\widehat{\boldsymbol{\beta}}_{\text{sp}} - \boldsymbol{\beta}\|_F = O_p[\{(r+s)\log r/n\}^{1/2}]$ and that the X-invariant responses are identified with probability approaching 1.

7.6 Bayesian Response Envelopes

Khare et al. (2016) developed a Bayesian method for the analysis of data under the response envelope model (1.20). In this section, we mention a few highlights of their approach. Developing a Bayesian approach to an envelope analysis using (1.20) would require in part a prior distribution on the Grassmannian for $\mathcal{E}_{\boldsymbol{\Sigma}}(\mathcal{B})$. To provide a more tractable starting point for their study, they reparameterized model (1.20) in terms of a Stiefel manifold instead of a Grassmannian. The effect of this is to impose a particular coordinate system on $\mathcal{E}_{\boldsymbol{\Sigma}}(\mathcal{B})$ and its orthogonal complement. The reparameterization is accomplished by first constructing the spectral decompositions of $\boldsymbol{\Omega} = \mathbf{U}\mathbf{D}\mathbf{U}^T$ and

$\boldsymbol{\Omega}_0 = \mathbf{U}_0\mathbf{D}_0\mathbf{U}_0^T$, where $\mathbf{D} \in \mathbb{R}^{u\times u}$ and $\mathbf{D}_0 \in \mathbb{R}^{(r-u)\times(r-u)}$ are diagonal matrices of eigenvalues arranged in decreasing order, and $\mathbf{U}$ and $\mathbf{U}_0$ are orthogonal matrices. Define $\boldsymbol{\Gamma}^* = \boldsymbol{\Gamma}\mathbf{U}$, $\boldsymbol{\Gamma}_0^* = \boldsymbol{\Gamma}_0\mathbf{U}_0$, and $\boldsymbol{\eta}^* = \mathbf{U}^T\boldsymbol{\eta}$, subject to the constraint that the maximum entry in each column of $\boldsymbol{\Gamma}^*$ and $\boldsymbol{\Gamma}_0^*$ is positive. These constraints mean that the columns of $\boldsymbol{\Gamma}^*$ and $\boldsymbol{\Gamma}_0^*$ are uniquely defined orthogonal bases for $\mathcal{E}_{\boldsymbol{\Sigma}}(\mathcal{B})$ and $\mathcal{E}_{\boldsymbol{\Sigma}}^{\perp}(\mathcal{B})$. Then the reparameterized model becomes

$$\mathbf{Y} = \boldsymbol{\alpha} + \boldsymbol{\Gamma}^*\boldsymbol{\eta}^*\mathbf{X}, \quad \text{with} \quad \boldsymbol{\Sigma} = \boldsymbol{\Gamma}^*\mathbf{D}\boldsymbol{\Gamma}^{*T} + \boldsymbol{\Gamma}_0^*\mathbf{D}_0\boldsymbol{\Gamma}_0^{*T}.$$

All of the parameters in the model are identifiable by construction. In contrast, recall that $\boldsymbol{\Gamma}$ in model (1.20) is not identifiable. Khare et al. (2016) then proceed to fashion prior distributions for the parameters in their new model, including a matrix Bingham distribution for $(\boldsymbol{\Gamma}^*, \boldsymbol{\Gamma}_0^*)$. The prior for the diagonal elements of $\mathbf{D}$ was taken to be that of the order statistics from a sample of size u from an inverse Gaussian distribution. The prior for the diagonal elements of $\mathbf{D}_0$ was constructed similarly. As one might expect, the posterior density arising out of this construction is intractable, so they developed a Gibbs procedure for sampling from the posterior.

The potential advantages of a Bayesian analysis include, as given by Khare et al. (2016), the ability to address uncertainty by using credible intervals, to incorporate prior information, and to analyze data with $n < r$. For some, the overarching advantage will be simply that the method is Bayesian. In view of the number of ingredients needed for this methodology and the rather intricate nature of its construction, it seems that experience is needed before a clear assessment can be made of its value in practice.

8

Inner and Scaled Envelopes

The envelope constructions considered so far are based on the notion of the *smallest* reducing subspace of a matrix $\mathbf{M} \in \mathbb{S}^{r \times r}$ that *contains* a specified subspace $\mathcal{V} \subseteq \mathbb{R}^r$, which was denoted as $\mathcal{E}_{\mathbf{M}}(\mathcal{V})$ in Chapter 6. The envelopes in this chapter are of a different character since they are intended to provide efficiency gains when the previous methods offer little or no gain, as happens when $\mathcal{E}_{\mathbf{M}}(\mathcal{V}) = \mathbb{R}^r$.

In Section 8.1, we consider models based on the *largest* reducing subspace of $\mathbf{M}$ that is *contained within* a specified subspace $\mathcal{V}$, which is called an inner envelope and indicated by the notation $\mathcal{IE}_{\mathbf{M}}(\mathcal{V})$. Inner envelopes are potentially useful because they may allow for efficiency gains when other methods offer little or no gain. The development in Section 8.1 follows the original work by Su and Cook (2012).

The response and predictor envelopes discussed in Chapters 1 and 4 tend to work best when the responses and predictors are in the same scales since, as mentioned in Section 1.5.4, the methods are not invariant or equivariant under rescaling. In Sections 8.2 and 8.3, we describe how to extend these methods to allow for scales to be estimated along with the corresponding envelope (Cook and Su 2013, 2016). In Section 8.3.3, we discuss also a scaled version of the SIMPLS algorithm.

8.1 Inner Envelopes

Consider a regression in which the response envelope of Chapter 1 is $\mathcal{E}_{\boldsymbol{\Sigma}}(\mathcal{B}) = \mathbb{R}^r$, so the envelope model reduces to the standard linear model (1.1) and offers no advantages. Figure 8.1 illustrates how efficiency might still be gained in such a regression. The axes of the plot represent the eigenvectors $\boldsymbol{\ell}_j$, $j = 1, 2, 3$, of $\boldsymbol{\Sigma}$ with corresponding eigenvalues $\varphi_1 > \varphi_2 > \varphi_3$ for a regression having $r = 3$ responses and $p = 2$ predictors. The two-dimensional coefficient subspace $\mathcal{B} = \text{span}(\boldsymbol{\beta})$ is depicted as a plane in the plot. Since the eigenvalues φ_j are distinct, all three eigenvectors of $\boldsymbol{\Sigma}$ are needed to envelope $\mathcal{B}$ and $\mathcal{E}_{\boldsymbol{\Sigma}}(\mathcal{B}) = \mathbb{R}^3$, so there is

An Introduction to Envelopes: Dimension Reduction for Efficient Estimation in Multivariate Statistics,
First Edition. R. Dennis Cook.

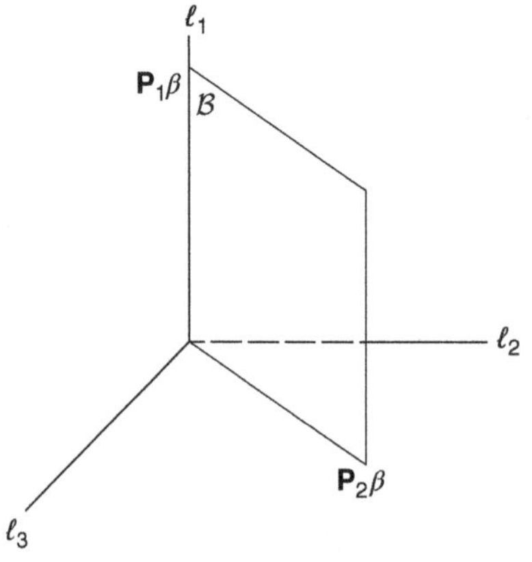

Figure 8.1 Schematic representation of an inner envelope.

no response reduction. However, we see also that span($\boldsymbol{\ell}_1$) $\subset \mathcal{B}$, so $\mathcal{B}$ contains a part of the eigenstructure of $\boldsymbol{\Sigma}$. This illustrates one configuration in which inner response envelopes offer gains where the response envelopes of Chapter 1 do not.

In the context of model (1.1), we assume that $p < r < n$ and rank($\boldsymbol{\beta}$) $= p$ throughout this section.

8.1.1 Definition and Properties of Inner Envelopes

The essential property illustrated in Figure 8.1 is that there is a reducing subspace $\mathcal{S}$ of $\boldsymbol{\Sigma}$ that is contained in $\mathcal{B}$:

$$\text{(i) } \mathcal{S} \subseteq \mathcal{B} \text{ and (ii) } \boldsymbol{\Sigma} = \mathbf{P}\boldsymbol{\Sigma}\mathbf{P} + \mathbf{Q}\boldsymbol{\Sigma}\mathbf{Q}, \tag{8.1}$$

where $\mathbf{P}$ denotes the projection onto $\mathcal{S}$. In Figure 8.1, $\mathcal{S} = \text{span}(\boldsymbol{\ell}_1)$, $\mathbf{P} = \mathbf{P}_1$, $\mathbf{P}_1\boldsymbol{\beta}$ is the projection onto $\mathcal{S}$, and $\mathbf{P}_2$ is the projection onto the part of $\mathcal{B}$ that is orthogonal to $\mathcal{S}$, so $\boldsymbol{\beta} = \mathbf{P}_1\boldsymbol{\beta} + \mathbf{P}_2\boldsymbol{\beta}$. There could be many subspaces $\mathcal{S}$ that satisfy (8.1) and for maximal efficiency gains we require the one with the largest dimension, leading to the following general definition of an inner envelope.

Definition 8.1 Let $\mathbf{M} \in \mathbb{S}^{r\times r}$. The inner $\mathbf{M}$-envelope of the subspace $\mathcal{V} \subseteq \mathbb{R}^r$, denoted by $\mathcal{IE}_{\mathbf{M}}(\mathcal{V})$, is the reducing subspace of $\mathbf{M}$ with maximal dimension that is contained within $\mathcal{V}$.

The existence of inner envelopes is assured because the space with only one element span(0) is a reducing subspace of $\mathbf{M}$ that is contained within $\mathcal{V}$. We use $\mathcal{IE}$ as shorthand in subscripts to denote $\mathcal{IE}_{\mathbf{M}}(\mathcal{V})$. We next state two characterizing propositions.

Proposition 8.1 *Let* $\mathbf{M} \in \mathbb{S}^{r\times r}$*. Then* $\mathcal{IE}_{\mathbf{M}}(\mathcal{V}) = \sum_i \mathcal{V}_i$*, where the sum is over all reducing subspaces* $\mathcal{V}_i$ *of* $\mathbf{M}$ *that are contained in* $\mathcal{V}$*.*

Proof: Because $\mathcal{IE}_{\mathbf{M}}(\mathcal{V})$ itself is a reducing subspace of $\mathbf{M}$ that is contained in $\mathcal{V}$, we have $\mathcal{IE}_{\mathbf{M}}(\mathcal{V}) \subseteq \sum_i \mathcal{V}_i$.

Now we only need to show $\mathcal{IE}_{\mathbf{M}}(\mathcal{V}) \supseteq \sum_i \mathcal{V}_i$. If there is an element $\mathbf{v}$ in $\sum_i \mathcal{V}_i$ but not in $\mathcal{IE}_{\mathbf{M}}(\mathcal{V})$, then there exists a $\mathcal{V}_{i_0}$ so that $\mathbf{v} \in \mathcal{V}_{i_0}$. Let $\mathcal{T} = \mathcal{V}_{i_0} + \mathcal{IE}_{\mathbf{M}}(\mathcal{V})$. Then $\mathcal{T}$ is a reducing subspace in $\mathcal{V}$ that has a bigger dimension than $\mathcal{IE}_{\mathbf{M}}(\mathcal{V})$, which is a contradiction since $\mathcal{IE}_{\mathbf{M}}(\mathcal{V})$ has maximal dimension. □

Proposition 8.1 is a natural consequence of Definition 8.1, which states that the inner envelope contains all the reducing subspaces of $\mathbf{M}$ that are contained in $\mathcal{V}$. The next proposition builds a connection between inner envelopes and envelopes; that is, an inner $\mathbf{M}$-envelope of a subspace is the same as the orthogonal complement of the $\mathbf{M}$-envelope of its orthogonal complement. Propositions 8.1 and 8.2 ensure that the inner envelope is uniquely defined as the largest subspace $\mathcal{S}_1$ that satisfies (8.1).

Proposition 8.2 *Let* $\mathbf{M} \in \mathbb{S}^{r\times r}$*. Then* $\mathcal{IE}_{\mathbf{M}}(\mathcal{V}) = \mathcal{E}_{\mathbf{M}}^{\perp}(\mathcal{V}^{\perp})$*.*

Proof: Let $\mathcal{R} = \mathcal{IE}_{\mathbf{M}}(\mathcal{V}^{\perp})$. For the equality to hold, we need to show that (i) $\mathcal{R}^{\perp}$ is a reducing subspace of $\mathbf{M}$, (ii) $\mathcal{R}^{\perp}$ is contained in $\mathcal{V}$, and (iii) $\mathcal{R}^{\perp}$ is the space with maximum dimension that satisfies (i) and (ii).

For (i), since $\mathcal{R}$ is a reducing subspace of $\mathbf{M}$, we have $\mathbf{M}\mathcal{R} \subseteq \mathcal{R}$ and $\mathbf{M}\mathcal{R}^{\perp} \subseteq \mathcal{R}^{\perp}$, this indicates that $\mathcal{R}^{\perp}$ is also a reducing subspace of $\mathbf{M}$. For (ii), as $\mathcal{R} \supseteq \mathcal{V}^{\perp}$, we have $\mathcal{R}^{\perp} \subseteq \mathcal{V}$. For (iii), if $\mathcal{R}^{\perp}$ does not have maximum dimension, then we can find $\mathcal{R}_0 \supset \mathcal{R}^{\perp}$, and $\mathcal{R}_0$ satisfies (i) and (ii). Then $\mathcal{R}_0^{\perp}$ will be a reducing subspace of $\mathbf{M}$ and $\mathcal{R}_0^{\perp} \supseteq \mathcal{V}^{\perp}$, also $\mathcal{R}_0^{\perp}$ has a smaller dimension than $\mathcal{R}$, which contradicts that $\mathcal{R}$ is the smallest reducing subspace of $\mathbf{M}$ that contains $\mathcal{V}^{\perp}$. So (iii) is also satisfied. □

Consider the implications of Proposition 8.2 for the regression represented in Figure 8.1, where we take the eigenvectors to be those of $\mathbf{M}$ and $\mathcal{B} = \mathcal{V}$. The subspace $\mathcal{B}^{\perp}$ is spanned by a vector (not shown) in the $(\boldsymbol{\ell}_2, \boldsymbol{\ell}_3)$ plane that is orthogonal to $\mathbf{P}_2\boldsymbol{\beta}$, and consequently $\mathcal{E}_{\boldsymbol{\Sigma}}(\mathcal{B}^{\perp}) = \text{span}(\boldsymbol{\ell}_2, \boldsymbol{\ell}_3)$. From this we obtain the inner envelope $\mathcal{IE}_{\boldsymbol{\Sigma}}(\mathcal{B}) = \mathcal{E}_{\boldsymbol{\Sigma}}^{\perp}(\mathcal{B}^{\perp}) = \text{span}(\boldsymbol{\ell}_1)$.

8.1.2 Inner Response Envelopes

Let $d = \dim\{\mathcal{IE}_{\boldsymbol{\Sigma}}(\mathcal{B})\} \leq p$. Like the response envelope model (1.20), the coordinate form of the inner response envelope model is expressed in terms of semi-orthogonal basis matrices $\boldsymbol{\Gamma} \in \mathbb{R}^{r\times d}$ and $\boldsymbol{\Gamma}_0 \in \mathbb{R}^{r\times(r-d)}$ for $\mathcal{IE}_{\boldsymbol{\Sigma}}(\mathcal{B})$

and $\mathcal{IE}_{\Sigma}^{\perp}(\mathcal{B})$. Then we can write $\boldsymbol{\beta} = \mathbf{P}_{\mathcal{IE}}\boldsymbol{\beta} + \mathbf{Q}_{\mathcal{IE}}\boldsymbol{\beta} = \boldsymbol{\Gamma}\boldsymbol{\eta}_1^T + \boldsymbol{\Gamma}_0\mathbf{C}\boldsymbol{\eta}_2^T$, where $\mathbf{C} \in \mathbb{R}^{(r-d)\times(p-d)}$ is a semi-orthogonal matrix so that $\boldsymbol{\Gamma}_0\mathbf{C}$ is a semi-orthogonal basis matrix for $\mathbf{Q}_{\mathcal{IE}}\mathcal{B}$, $\boldsymbol{\eta}_1^T \in \mathbb{R}^{d\times p}$ and $\boldsymbol{\eta}_2^T \in \mathbb{R}^{(p-d)\times p}$, and $(\boldsymbol{\eta}_1, \boldsymbol{\eta}_2) \in \mathbb{R}^{p\times p}$ has full rank. The inner envelope model can now be stated in full as

$$\mathbf{Y} = \boldsymbol{\alpha} + (\boldsymbol{\Gamma}\boldsymbol{\eta}_1^T + \boldsymbol{\Gamma}_0\mathbf{C}\boldsymbol{\eta}_2^T)\mathbf{X} + \boldsymbol{\varepsilon}, \quad \boldsymbol{\Sigma} = \boldsymbol{\Gamma}\boldsymbol{\Omega}\boldsymbol{\Gamma}^T + \boldsymbol{\Gamma}_0\boldsymbol{\Omega}_0\boldsymbol{\Gamma}_0^T, \tag{8.2}$$

where $\boldsymbol{\Omega} \in \mathbb{R}^{d\times d}$ and $\boldsymbol{\Omega}_0 \in \mathbb{R}^{(r-d)\times(r-d)}$ are positive definite matrices. In Figure 8.1, $r = 3$, $p = 2$, $d = 1$, $\boldsymbol{\Gamma}\boldsymbol{\eta}_1^T = \mathbf{P}_1\boldsymbol{\beta}$, $\boldsymbol{\Gamma}_0\mathbf{C}\boldsymbol{\eta}_2^T = \mathbf{P}_2\boldsymbol{\beta}$, and $\text{span}(\boldsymbol{\Gamma}_0) = \text{span}(\boldsymbol{\ell}_2, \boldsymbol{\ell}_3)$. If $d = 0$, then $\mathcal{IE}_{\Sigma}(\mathcal{B}) = \text{span}(0)$, $\boldsymbol{\Gamma}$ and $\boldsymbol{\eta}_1$ do not occur, $\boldsymbol{\Gamma}_0 = \mathbf{I}_r$, $\boldsymbol{\eta}_2 \in \mathbb{R}^{p\times p}$ is full rank, $\boldsymbol{\beta} = \mathbf{C}\boldsymbol{\eta}_2^T$ is full rank and, in consequence, (8.2) reduces to the standard model. On the other extreme, if $d = p$, then $\mathbf{C}$ and $\boldsymbol{\eta}_2$ do not occur, $\boldsymbol{\eta}_1 \in \mathbb{R}^{p\times p}$ is full rank, $\mathcal{IE}_{\Sigma}(\mathcal{B}) = \mathcal{B}$ and, in consequence, (8.2) reduces to an envelope model $\mathcal{IE}_{\Sigma}(\mathcal{B}) = \mathcal{E}_{\Sigma}(\mathcal{B})$ with $\dim(\mathcal{E}_{\Sigma}(\mathcal{B})) = p$.

There are three orthogonal subspaces involved in the statement of the inner envelope model: $\mathcal{IE}_{\Sigma}(\mathcal{B}) = \text{span}(\boldsymbol{\Gamma})$ with dimension d, $\text{span}(\boldsymbol{\Gamma}_0\mathbf{C})$ with dimension $p - d$ and their orthogonal complement $\text{span}^{\perp}(\boldsymbol{\Gamma}, \boldsymbol{\Gamma}_0\mathbf{C})$ with dimension $r - p$. With normal errors, these subspaces are related in the following way.

$$\text{(a) } \mathbf{Q}_{(\Gamma,\Gamma_0 C)}\mathbf{Y} \mid \mathbf{X} \sim \mathbf{Q}_{(\Gamma,\Gamma_0 C)}\mathbf{Y}, \quad \text{(b) } \mathbf{P}_{\Gamma}\mathbf{Y} \perp\!\!\!\perp (\mathbf{P}_{\Gamma_0 C}\mathbf{Y}, \mathbf{Q}_{(\Gamma,\Gamma_0 C)}\mathbf{Y}) \mid \mathbf{X}. \tag{8.3}$$

The distribution of $\mathbf{Q}_{(\Gamma,\Gamma_0 C)}\mathbf{Y} \mid \mathbf{X}$ is independent of $\mathbf{X}$, while the distributions of $\mathbf{P}_{\Gamma}\mathbf{Y} \mid \mathbf{X}$ and $\mathbf{P}_{\Gamma_0 C}\mathbf{Y} \mid \mathbf{X}$ are allowed to depend on $\mathbf{X}$. If we could find such a decomposition with $\mathbf{P}_{\Gamma_0 C}\mathbf{Y} = 0$, then (8.3) would reduce to (8.3) with $\mathbf{P}_{\mathcal{E}} = \mathbf{P}_{\Gamma}$ and $\mathbf{Q}_{\mathcal{E}} = \mathbf{Q}_{\Gamma}$, and we could employ an envelope model (1.20). Otherwise, $\mathbf{P}_{\Gamma_0 C}\mathbf{Y}$ represents a confounder whose distribution depends on $\mathbf{X}$ and is correlated with $\mathbf{Q}_{(\Gamma,\Gamma_0 C)}\mathbf{Y}$; that is, $\text{cov}(\mathbf{P}_{\Gamma_0 C}\mathbf{Y}, \mathbf{Q}_{(\Gamma,\Gamma_0 C)}\mathbf{Y}) \neq 0$.

The potential advantage of an inner envelope analysis arises from the decomposition $\boldsymbol{\beta} = \mathbf{P}_{\mathcal{IE}}\boldsymbol{\beta} + \mathbf{Q}_{\mathcal{IE}}\boldsymbol{\beta}$. Because the first addend $\mathbf{P}_{\mathcal{IE}}\boldsymbol{\beta}$ is connected to the reducing subspaces of $\boldsymbol{\Sigma}$, we may be able to estimate it with greater efficiency than in a standard analysis, while estimating the second addend $\mathbf{Q}_{\mathcal{IE}}\boldsymbol{\beta}$ with about the same precision. Overall then we may be able to estimate $\boldsymbol{\beta}$ with greater precision.

With $\dim\{\mathcal{IE}_{\Sigma}(\mathcal{B})\} = d$, there are $N_d = p^2 + (p - d)(r - p) + r(r + 1)/2$ parameters to be estimated in the inner envelope model (8.2). We need pd parameters for $\boldsymbol{\eta}_1$, $p(p - d)$ for $\boldsymbol{\eta}_2$, $d(d + 1)/2$ and $(r - d)(r - d + 1)/2$ parameters for $\boldsymbol{\Omega}$ and $\boldsymbol{\Omega}_0$ as they are symmetric matrices. Estimation of $\text{span}(\boldsymbol{\Gamma})$ requires $d(r - d)$ parameters. It is the same with estimating $\mathbf{C}$: if we fix an orthogonal basis $(\boldsymbol{\Gamma}, \boldsymbol{\Gamma}_0)$, only $\text{span}(\mathbf{C})$ is estimable, so $\mathbf{C}$ is estimated on a $(r - d) \times (p - d)$ Grassmann manifold and $(p - d)(r - p)$ parameters are needed.

8.1.3 Maximum Likelihood Estimators

As in the analysis using the response envelopes of Chapter 1, we take the predictors to be fixed and assume that, for a sample of size n, the errors $\boldsymbol{\varepsilon}_i$ from (8.2) are independent copies of a $N_r(0, \boldsymbol{\Sigma})$ random variable. As $\mathbf{X}$ is centered, the maximum likelihood estimator of $\boldsymbol{\alpha}$ is $\bar{\mathbf{Y}}$.

Let $(\mathbf{G}, \mathbf{G}_0) \in \mathbb{R}^{r\times r}$ be a partitioned orthogonal matrix with $\mathbf{G} \in \mathbb{R}^{r\times d}$, and let $\mathbf{V}(\mathbf{G}_0)\boldsymbol{\Lambda}(\mathbf{G}_0)\mathbf{V}^T(\mathbf{G}_0)$ denote the spectral decomposition of the $(r-d)\times(r-d)$ matrix

$$\begin{aligned}\widehat{\mathbf{A}}(\mathbf{G}_0) &= \mathbf{S}_{\mathbf{G}_0^T\mathbf{Y}|\mathbf{X}}^{-1/2}\mathbf{S}_{\mathbf{G}_0^T\mathbf{Y}\circ\mathbf{X}}\mathbf{S}_{\mathbf{G}_0^T\mathbf{Y}|\mathbf{X}}^{-1/2}\\ &= (\mathbf{G}_0^T\mathbf{S}_{\mathbf{Y}|\mathbf{X}}\mathbf{G}_0)^{-1/2}(\mathbf{G}_0^T\mathbf{S}_{\mathbf{Y}\circ\mathbf{X}}\mathbf{G}_0)(\mathbf{G}_0^T\mathbf{S}_{\mathbf{Y}|\mathbf{X}}\mathbf{G}_0)^{-1/2},\end{aligned}$$

where $\mathbf{S}_{\mathbf{G}_0^T\mathbf{Y}\circ\mathbf{X}}$ and $\mathbf{S}_{\mathbf{Y}\circ\mathbf{X}}$ are sample covariance matrices of fitted vectors as defined in (1.4). We arrange its eigenvalues $\lambda_i(\mathbf{G}_0)$'s in descending order so

$$\boldsymbol{\Lambda}(\mathbf{G}_0) = \operatorname{diag}\{\lambda_1(\mathbf{G}_0), \dots, \lambda_{r-d}(\mathbf{G}_0)\}.$$

Then a semi-orthogonal basis matrix $\widehat{\boldsymbol{\Gamma}}$ for the maximum likelihood estimator of $\mathcal{IE}_{\boldsymbol{\Sigma}}(\mathcal{B})$ can be obtained by minimizing the following function of $\mathbf{G}$,

$$L_d(\mathbf{G}) := \log|\mathbf{G}^T\mathbf{S}_{\mathbf{Y}|\mathbf{X}}\mathbf{G}| + \log|\mathbf{G}^T\mathbf{S}_{\mathbf{Y}|\mathbf{X}}^{-1}\mathbf{G}| + \sum_{i=p-d+1}^{r-d}\log\{1+\lambda_i(\mathbf{G}_0)\}. \tag{8.4}$$

After obtaining $\widehat{\boldsymbol{\Gamma}}$, $\widehat{\boldsymbol{\Gamma}}_0$ may be constructed as any semi-orthogonal basis matrix for $\operatorname{span}^{\perp}(\widehat{\boldsymbol{\Gamma}})$. Letting

$$\mathbf{K}(\mathbf{G}_0) = \operatorname{diag}\{0, \dots, 0, \lambda_{p-d+1}(\mathbf{G}_0), \dots, \lambda_{r-d}(\mathbf{G}_0)\},$$

the maximum likelihood estimators of the remaining parameters are as given in the following list.

$$\begin{aligned}\widehat{\boldsymbol{\eta}}_1^T &= \widehat{\boldsymbol{\Gamma}}^T\mathbf{B},\\ \widehat{\boldsymbol{\Omega}} &= (\mathbb{Y}\widehat{\boldsymbol{\Gamma}} - \mathbb{X}\widehat{\boldsymbol{\eta}}_1)^T(\mathbb{Y}\widehat{\boldsymbol{\Gamma}} - \mathbb{X}\widehat{\boldsymbol{\eta}}_1)/n,\\ \widehat{\boldsymbol{\Omega}}_0 &= \widehat{\boldsymbol{\Gamma}}_0^T\mathbf{S}_{\mathbf{Y}|\mathbf{X}}\widehat{\boldsymbol{\Gamma}}_0 + (\widehat{\boldsymbol{\Gamma}}_0^T\mathbf{S}_{\mathbf{Y}|\mathbf{X}}\widehat{\boldsymbol{\Gamma}}_0)^{1/2}\mathbf{V}(\widehat{\boldsymbol{\Gamma}}_0)\mathbf{K}(\widehat{\boldsymbol{\Gamma}}_0)\mathbf{V}(\widehat{\boldsymbol{\Gamma}}_0)^T(\widehat{\boldsymbol{\Gamma}}_0^T\mathbf{S}_{\mathbf{Y}|\mathbf{X}}\widehat{\boldsymbol{\Gamma}}_0)^{1/2},\\ \operatorname{span}(\widehat{\mathbf{C}}) &= \widehat{\boldsymbol{\Omega}}_0\mathcal{S}_{p-d}(\widehat{\boldsymbol{\Omega}}_0, \widehat{\boldsymbol{\Gamma}}_0^T\mathbf{S}_{\mathbf{Y}\circ\mathbf{X}}\widehat{\boldsymbol{\Gamma}}_0),\\ \widehat{\boldsymbol{\eta}}_2^T &= (\widehat{\mathbf{C}}^T\widehat{\boldsymbol{\Omega}}_0^{-1}\widehat{\mathbf{C}})^{-1}\widehat{\mathbf{C}}^T\widehat{\boldsymbol{\Omega}}_0^{-1}\widehat{\boldsymbol{\Gamma}}_0^T\mathbf{B},\\ \widehat{\boldsymbol{\beta}} &= \widehat{\boldsymbol{\Gamma}}\widehat{\boldsymbol{\eta}}^T + \widehat{\boldsymbol{\Gamma}}_0\widehat{\mathbf{C}}\widehat{\boldsymbol{\eta}}_2^T = \mathbf{P}_{\widehat{\boldsymbol{\Gamma}}_1}\mathbf{B} + \widehat{\boldsymbol{\Gamma}}_0\mathbf{P}_{\widehat{\mathbf{C}}(\widehat{\boldsymbol{\Omega}}_0^{-1})}\widehat{\boldsymbol{\Gamma}}_0^T\mathbf{B},\\ \widehat{\boldsymbol{\Sigma}} &= \widehat{\boldsymbol{\Gamma}}\widehat{\boldsymbol{\Omega}}\widehat{\boldsymbol{\Gamma}}^T + \widehat{\boldsymbol{\Gamma}}_0\widehat{\boldsymbol{\Omega}}_0\widehat{\boldsymbol{\Gamma}}_0^T,\end{aligned}$$

where $\mathcal{S}_{p-d}(\widehat{\boldsymbol{\Omega}}_0, \widehat{\boldsymbol{\Gamma}}_0^T\mathbf{S}_{\mathbf{Y}\circ\mathbf{X}}\widehat{\boldsymbol{\Gamma}}_0)$ equals $\widehat{\boldsymbol{\Omega}}_0^{-1/2}$ times the span of the first $p-d$ eigenvectors of $\widehat{\boldsymbol{\Omega}}_0^{-1/2}(\widehat{\boldsymbol{\Gamma}}_0^T\mathbf{S}_{\mathbf{Y}\circ\mathbf{X}}\widehat{\boldsymbol{\Gamma}}_0)\widehat{\boldsymbol{\Omega}}_0^{-1/2}$.

The estimation process begins with the minimization of $L_d(\mathbf{G})$. The sum of the first two addends on the right-hand side of (8.4) is minimized when $\mathbf{G} \in \mathbb{R}^{r\times d}$ spans any d-dimensional reducing subspace of $\mathbf{S}_{\mathbf{Y}|\mathbf{X}}$ (see Lemma A.15). The role of these terms is then to pull the solution toward the reducing subspaces of $\mathbf{S}_{\mathbf{Y}|\mathbf{X}}$.

The eigenvalues in the third addend can be interpreted as arising from the standardized version of the response in the regression of $\mathbf{G}_0^T\mathbf{Y}$ on $\mathbf{X}$. It is evidently up to this addend to establish the actual estimator, which will not necessarily correspond to a reducing subspace of $\mathbf{S}_{\mathbf{Y}|\mathbf{X}}$. To gain further intuition, we turn to the population versions $L_d^{\text{pop}}(\mathbf{G})$ and $\mathbf{A}(\mathbf{G}_0)$ of $L_d(\mathbf{G})$ and $\widehat{\mathbf{A}}(\mathbf{G}_0)$. The sample covariance matrices $\mathbf{S}_{\mathbf{Y}}$ and $\mathbf{S}_{\mathbf{Y}|\mathbf{X}}$ converge in probability to their population counterparts as follows:

$$\mathbf{S}_{\mathbf{Y}} \to \boldsymbol{\Sigma}_{\mathbf{Y}} = \boldsymbol{\Gamma}\boldsymbol{\Omega}\boldsymbol{\Gamma}^T + \boldsymbol{\Gamma}_0\boldsymbol{\Omega}_0\boldsymbol{\Gamma}_0^T + (\boldsymbol{\Gamma}\boldsymbol{\eta}_1^T + \boldsymbol{\Gamma}_0\mathbf{C}\boldsymbol{\eta}_2^T)\boldsymbol{\Sigma}_{\mathbf{X}}(\boldsymbol{\Gamma}\boldsymbol{\eta}_1^T + \boldsymbol{\Gamma}_0\mathbf{C}\boldsymbol{\eta}_2^T)^T,$$
$$\mathbf{S}_{\mathbf{Y}|\mathbf{X}} \to \boldsymbol{\Sigma} = \boldsymbol{\Gamma}\boldsymbol{\Omega}\boldsymbol{\Gamma}^T + \boldsymbol{\Gamma}_0\boldsymbol{\Omega}_0\boldsymbol{\Gamma}_0^T,$$

where $\boldsymbol{\Sigma}_{\mathbf{X}} = \lim_{n\to\infty}\mathbf{S}_{\mathbf{X}}$. Since $\text{span}(\boldsymbol{\Gamma}) = \mathcal{IE}_{\boldsymbol{\Sigma}}(\mathcal{B})$ is a reducing subspace of $\boldsymbol{\Sigma}$, it follows that the sum of the first two addends of $L_d^{\text{pop}}(\mathbf{G})$ is minimized at $\mathbf{G} = \boldsymbol{\Gamma}$, although not uniquely. When $\mathbf{G}_0 = \boldsymbol{\Gamma}_0$,

$$\begin{aligned}\mathbf{A}(\boldsymbol{\Gamma}_0) &= (\boldsymbol{\Gamma}_0^T\boldsymbol{\Sigma}\boldsymbol{\Gamma}_0)^{-1/2}\{\boldsymbol{\Gamma}_0^T(\boldsymbol{\Gamma}\boldsymbol{\eta}_1^T + \boldsymbol{\Gamma}_0\mathbf{C}\boldsymbol{\eta}_2^T)\boldsymbol{\Sigma}_{\mathbf{X}}(\boldsymbol{\Gamma}\boldsymbol{\eta}_1^T + \boldsymbol{\Gamma}_0\mathbf{B}\boldsymbol{\eta}_2^T)^T\boldsymbol{\Gamma}_0\}\\ &\quad\times(\boldsymbol{\Gamma}_0^T\boldsymbol{\Sigma}\boldsymbol{\Gamma}_0)^{-1/2}\\ &= \boldsymbol{\Omega}_0^{-1/2}\mathbf{C}\boldsymbol{\eta}_2^T\boldsymbol{\Sigma}_{\mathbf{X}}\boldsymbol{\eta}_2\mathbf{C}^T\boldsymbol{\Omega}_0^{-1/2}.\end{aligned}$$

Because $\mathbf{C}$ is an $(r-d)\times(p-d)$ matrix and $r > p$, the rank of $\mathbf{A}(\boldsymbol{\Gamma}_0)$ is at most $p-d$, so its $(p-d+1)$th to the $(r-d)$th eigenvalues are all 0, and thus the third addend of $L_d^{\text{pop}}(\mathbf{G})$ is minimized at $\mathbf{G}_0 = \boldsymbol{\Gamma}_0$. It follows that $L_d^{\text{pop}}(\mathbf{G})$ is minimized globally at $\mathbf{G} = \boldsymbol{\Gamma}$.

Turning to the estimators listed previously, $\widehat{\boldsymbol{\eta}}_1$, $\widehat{\boldsymbol{\eta}}_2$, $\widehat{\boldsymbol{\Omega}}$, and $\widehat{\boldsymbol{\Omega}}_0$ are all dependent on the selected minimizer of $L_d(\mathbf{G})$, but $\widehat{\boldsymbol{\beta}}$ and $\widehat{\boldsymbol{\Sigma}}$ are coordinate independent. The latter two estimators depend on $\text{span}\{\arg\min L_d(\mathbf{G})\}$ but not on the particular basis selected.

The envelope estimator of $\boldsymbol{\beta}$ has two terms. The first, which is obtained by projecting $\mathbf{B}$ onto the estimated inner envelope $\text{span}(\widehat{\boldsymbol{\Gamma}})$, estimates the part of $\boldsymbol{\beta}$ that is enveloped by a reducing subspace of $\boldsymbol{\Sigma}$. The second term estimates the remaining part of $\boldsymbol{\beta}$: if we multiply both sides of model (8.2) by $\boldsymbol{\Gamma}_0^T$, we have a reduced rank regression with coefficient $\mathbf{C}\boldsymbol{\eta}_2^T$ and covariance matrix $\boldsymbol{\Omega}_0$. The estimator of the coefficients in this reduced rank regression then has the form $\mathbf{P}_{\widehat{\mathbf{C}}(\widehat{\boldsymbol{\Omega}}_0^{-1})}\widehat{\boldsymbol{\Gamma}}_0^T\mathbf{B}$ (Cook and Forzani 2008b), and $\widehat{\boldsymbol{\Gamma}}_0\widehat{\mathbf{C}}\widehat{\boldsymbol{\eta}}_2^T$ becomes $\widehat{\boldsymbol{\Gamma}}_0\mathbf{P}_{\widehat{\mathbf{C}}(\widehat{\boldsymbol{\Omega}}_0^{-1})}\widehat{\boldsymbol{\Gamma}}_0^T\mathbf{B}$. When $d = 0$, we have $\widehat{\boldsymbol{\beta}} = \mathbf{B}$.

In preparation for stating the joint asymptotic distribution of $\text{vec}(\widehat{\boldsymbol{\beta}})$ and $\text{vech}(\widehat{\boldsymbol{\Sigma}})$, recall that the Fisher information matrix $\mathbf{J}$ under the standard model (1.1) was given in (1.14). Let

$$\boldsymbol{\phi} = \{\text{vec}^T(\boldsymbol{\eta}_1), \text{vec}^T(\boldsymbol{\eta}_2), \text{vec}^T(\mathbf{B}), \text{vec}^T(\boldsymbol{\Gamma}), \text{vech}^T(\boldsymbol{\Omega}), \text{vech}^T(\boldsymbol{\Omega}_0)\}^T,$$

and let $\mathbf{H}$ denote the gradient matrix $\partial\{\text{vec}^T(\widehat{\boldsymbol{\beta}}), \text{vech}^T(\widehat{\boldsymbol{\Sigma}})\}^T/\partial\boldsymbol{\phi}^T$,

$$\mathbf{H} = \begin{pmatrix} \mathbf{I}_p\otimes\boldsymbol{\Gamma} & \mathbf{I}_p\otimes\boldsymbol{\Gamma}_0\mathbf{C} & \boldsymbol{\eta}_2\otimes\boldsymbol{\Gamma}_0 & \mathbf{T}_{14} & 0 & 0\\ 0 & 0 & 0 & \mathbf{T}_{24} & \mathbf{C}_r(\boldsymbol{\Gamma}\otimes\boldsymbol{\Gamma})\mathbf{E}_d & \mathbf{C}_r(\boldsymbol{\Gamma}_0\otimes\boldsymbol{\Gamma}_0)\mathbf{E}_{r-d}\end{pmatrix},$$

where $\mathbf{C}$ is as defined in (8.2), $\mathbf{C}_r$ and $\mathbf{E}_d$ denote contraction and expansion matrices,

$$\mathbf{T}_{14} = \boldsymbol{\eta}_1 \otimes \mathbf{I}_r - (\boldsymbol{\eta}_2 \mathbf{C}^T \boldsymbol{\Gamma}_0^T \otimes \boldsymbol{\Gamma})\mathbf{K}_{rd}$$
$$\mathbf{T}_{24} = 2\mathbf{C}_r(\boldsymbol{\Gamma\Omega} \otimes \mathbf{I}_r - \boldsymbol{\Gamma} \otimes \boldsymbol{\Gamma}_0 \boldsymbol{\Omega}_0 \boldsymbol{\Gamma}_0^T).$$

Then under the inner envelope model (8.2) with normal errors,

$$\sqrt{n}\begin{pmatrix} \text{vec}(\widehat{\boldsymbol{\beta}}) - \text{vec}(\boldsymbol{\beta}) \\ \text{vech}(\widehat{\boldsymbol{\Sigma}}) - \text{vech}(\boldsymbol{\Sigma}) \end{pmatrix} \rightarrow N_{rp+r(r+1)/2}(0, \mathbf{V}),$$

in distribution, where $\mathbf{V} = \mathbf{H}(\mathbf{H}^T\mathbf{J}\mathbf{H})^{\dagger}\mathbf{H}^T \leq \mathbf{J}^{-1}$. There is evidently no ponderable closed-form expression for the asymptotic variance of $\text{vec}(\boldsymbol{\beta})$, which is the upper $pr \times pr$ block of $\mathbf{V}$, although given a numerical evaluation of $\mathbf{V}$ it straightforward to obtain.

If the errors for model (8.2) are not normal, but have finite fourth moments, and if the maximum diagonal element of $\mathbf{P}_{\mathbb{X}}$ converges to 0 as $n \rightarrow \infty$, then

$$\sqrt{n}[\{\text{vec}(\widehat{\boldsymbol{\beta}})^T, \text{vech}(\widehat{\boldsymbol{\Sigma}})^T\}^T - \{\text{vec}(\boldsymbol{\beta})^T, \text{vech}(\boldsymbol{\Sigma})^T\}^T]$$

is still asymptotically normally distributed. Although there is no expression for the asymptotic variance in this case, the residual bootstrap still can be used.

8.1.4 Race Times: Inner Envelopes

The data for this example consist of the split times $T_1, \ldots, T_{10}$ on each of ten 10 km segments for the 80 participants who completed a Lincolnshire 100 km race (Jolliffe 2002). Thus, there are 80 observations on 10 variables. Figure 8.2a shows a profile plot of the split times for the 80 participants, the performance of each participant being traced by one line on the plot. The plot indicates, as one might expect, that the average and standard deviations of the split times increase along the race. This can be seen more clearly in Figure 8.2b, which traces the mean split times plus and minus one standard deviation across the segments. We see from the mean profile plot that the mean times decreased in the last two segments, which reflects a runner's tendency to spurt near the end of a race. In this example, we use a multivariate linear regression model to study the dependence of the last seven split times $\mathbf{Y} = (T_4, \ldots, T_{10})^T$ on the first three split times $\mathbf{X} = (T_1, T_2, T_3)^T$. Fitting a response envelope using model (1.20), AIC, BIC, and LRT(.05) indicated dimensions 5, 2, and 4. Instead of trying to sort out the dimension issue, we turn to inner envelopes to see if it might yield a more stable dimension.

As with other envelope methods, an information criterion, likelihood testing, or cross-validation can be used to select the dimension d prior to fitting an inner response envelope model. The maximized log-likelihood function $\hat{L}_d$ under the

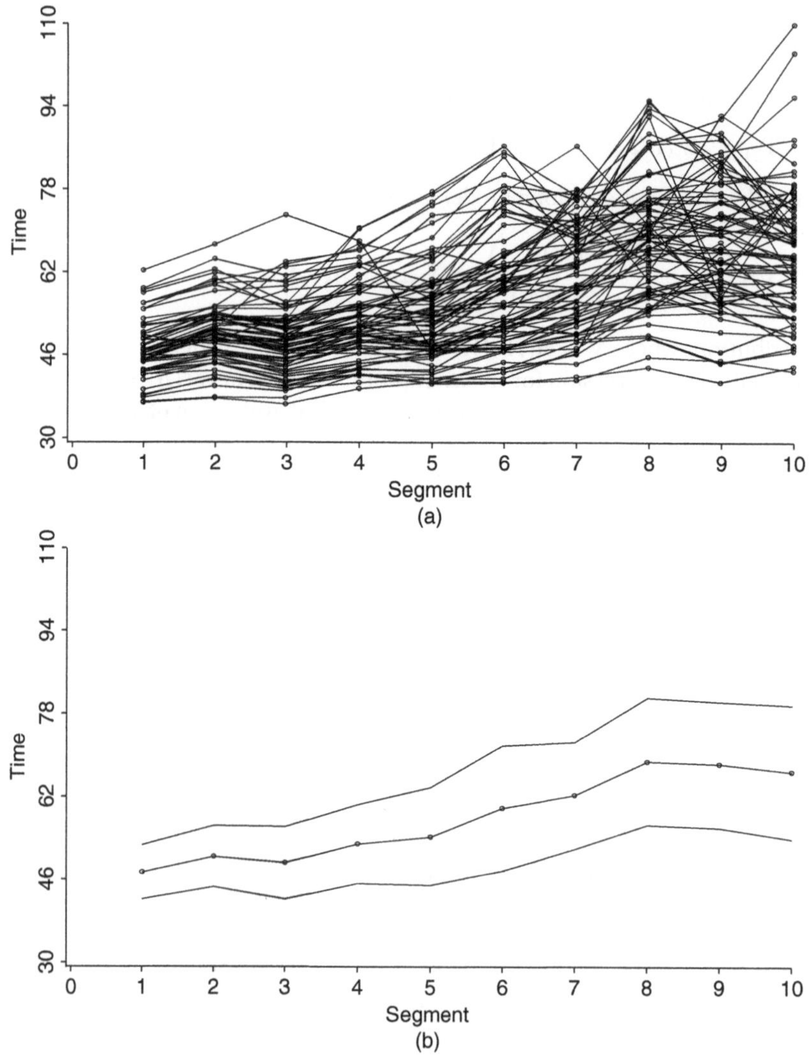

Figure 8.2 Race times: Profile plot of 10 split times for 80 participants in a 100 km race. (a) Profile plots for individual participants, (b) Profile plots of average split times plus and minus one standard deviation.

inner envelope model with dimension d is

$$\hat{L}_d = -(nr/2)\{1 + \log(2\pi)\} - (n/2)\log|\widehat{\mathbf{\Gamma}}^T \mathbf{S}_{\mathbf{Y}|\mathbf{X}} \widehat{\mathbf{\Gamma}}| - (n/2)\log|\widehat{\mathbf{\Gamma}}_0^T \mathbf{S}_{\mathbf{Y}|\mathbf{X}} \widehat{\mathbf{\Gamma}}_0|$$
$$-(n/2) \sum_{i=p-d+1}^{r-d} \log\{1 + \lambda_i(\widehat{\mathbf{\Gamma}}_0)\}.$$

The dimension of the inner envelope can be selected by minimizing over d either Akaike's information criterion $-2\hat{L}_d + 2N_d$ or the Bayes information criterion $-2\hat{L}_d + N_d \log(n)$. It can also be determined by likelihood ratio testing. To test the hypothesis $d = d_0$ $(d_0 \leq p)$, the test statistic $\Lambda(d_0)$ can be constructed as $\Lambda(d_0) = 2\{\hat{L}_0 - \hat{L}_{d_0}\}$. Here $\hat{L}_0$ is the maximized value of the log-likelihood for the standard model (1.1), $\hat{L}_0 = -(nr/2)\{1 + \log(2\pi)\} - (n/2)\log|\mathbf{S}_{\mathrm{Y}|\mathrm{X}}|$. Under the null hypothesis, $\Lambda(d_0)$ is asymptotically distributed as a chi-square random variable with $d_0(r-p)$ degrees of freedom. The testing procedure can be started at $d = p$ with a common significance level and d chosen as the first hypothesized value that is not rejected.

Turning to a fit of the inner envelope model (8.2), AIC, BIC, LRT(.05), and the average prediction error over 10 replications of fivefold cross-validation with random partitions all indicated that $d = 2$. The ratios $\mathrm{se}(\mathbf{B})/\mathrm{se}(\hat{\beta})$ of the asymptotic standard errors ranged between 1.01 and 2.44, indicating improvement over the standard model. The Z-scores from the fit of this model are given in columns 2–4 of Table 8.1. One impression is that the first two split times are not notably related to the response, given the third split time. This is perhaps expected because the third split time is closest to the response times. We might also expect that the absolute Z-scores for the third split time would decrease for the farthest response times. However, it may be surprising that a relationship has evidently been detected between the third split time and the last two response times T_9 and T_{10}, since these are the times that reflect a runner's tendency to spurt near the end of a race. The final column of Table 8.1 is for later use in Section 8.2.3.

Shown in Figure 8.3 is a plot of the fitted times plus and minus one standard deviation of the fitted times. It is shown on approximately the same scale

Table 8.1 Race times: Absolute Z-scores (columns 2–4) and the estimated coefficients $\hat{\beta}_{T_3}$ for T_3 (column 5) from the fit of the inner response envelope model with $d = 2$.

Y\X	T_1	T_2	T_3	$\hat{\beta}_{T_3}$
T_4	0.01	0.17	9.43	0.97
T_5	0.44	0.85	5.53	1.18
T_6	0.97	0.94	4.66	1.32
T_7	0.68	0.84	4.73	0.97
T_8	0.67	0.82	4.84	1.16
T_9	0.79	0.88	4.47	1.09
T_{10}	0.64	0.81	3.90	0.94

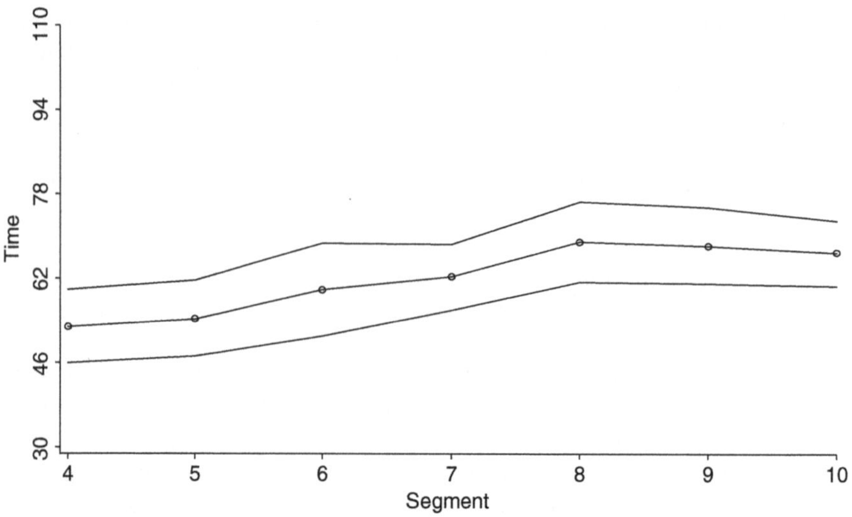

Figure 8.3 Fitted race times.

as Figure 8.2b for ease of comparison. The visual impression is that the fit is reasonable. Similar plots of the residuals could be informative as well.

8.2 Scaled Response Envelopes

As discussed in Section 1.5.4, the estimators based on the response envelope methods introduced in Chapter 1 are not invariant or equivariant under rescaling of the response. Suppose that we rescale $\mathbf{Y}$ by multiplication by a nonsingular diagonal matrix $\mathbf{\Lambda}$. Let $\mathbf{Y}_{\Lambda} = \mathbf{\Lambda Y}$ denote the new response, and let $(\boldsymbol{\beta}_{\Lambda}, \mathbf{\Sigma}_{\Lambda})$ and $(\widehat{\boldsymbol{\beta}}_{\Lambda}, \widehat{\mathbf{\Sigma}}_{\Lambda})$ denote the corresponding parameters and their envelope estimators based on the envelope model for $\mathbf{Y}_{\Lambda}$ on $\mathbf{X}$. Then we do not generally have invariance, $\widehat{\boldsymbol{\beta}}_{\Lambda} = \widehat{\boldsymbol{\beta}}$, $\widehat{\mathbf{\Sigma}}_{\Lambda} = \widehat{\mathbf{\Sigma}}$, or equivariance, $\widehat{\boldsymbol{\beta}}_{\Lambda} = \mathbf{\Lambda}\widehat{\boldsymbol{\beta}}$, $\widehat{\mathbf{\Sigma}}_{\Lambda} = \mathbf{\Lambda}\widehat{\mathbf{\Sigma}}\mathbf{\Lambda}$. In fact, the dimension of the envelope subspace may change because of the transformation. This is illustrated schematically in Figure 8.4 for $r = 2$ responses and a binary predictor X. The left panel, which shows an envelope structure similar to that in Figure 1.3a, has an envelope that aligns with the second eigenvector of $\mathbf{\Sigma}$. Beginning in the left panel, suppose we rescale $\mathbf{Y}$ to $\mathbf{Y}_{\Lambda}$. This may change the relationship between the eigenvectors and $\mathcal{B}$, as illustrated by the distributions shown in Figure 8.4b, where $\mathcal{B}_{\Lambda} = \text{span}(\mathbf{\Lambda}\boldsymbol{\beta})$. Since $\mathcal{B}_{\Lambda}$ aligns with neither eigenvector, the envelope for the regression of $\mathbf{Y}_{\Lambda}$ on X is two dimensional: $\mathcal{E}_{\mathbf{\Sigma}_{\Lambda}}(\mathcal{B}_{\Lambda}) = \mathbb{R}^2$. In this case, all linear combinations of $\mathbf{Y}$ are material to the regression, the envelope model is the same as the standard model and no efficiency gains are achieved. Because envelope methods are not invariant or equivariant under

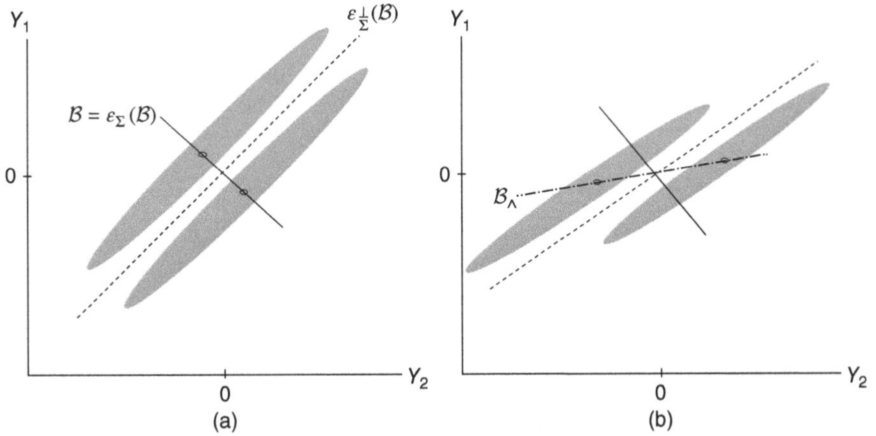

Figure 8.4 Schematic illustration of how rescaling the response can affect an envelope analysis. (a) Original distributions, (b) Rescaled distributions.

rescaling, they tend to work best when the responses are measured in the same or similar units, as in nearly all of the examples in this book.

Suppose now that we begin an analysis with the response $\mathbf{Y}$ distributed as in Figure 8.4b. If we knew the appropriate rescaling, we could transform to Figure 8.4a, perform an envelope analysis, and then transform back to the original scales of Figure 8.4b. In symbols, this means that we first transform $\mathbf{Y}$ to $\mathbf{\Lambda}^{-1}\mathbf{Y}$, then project onto the envelope $\mathbf{P}_{\Gamma}\mathbf{\Lambda}^{-1}\mathbf{Y}$ and then transform back, giving at the end $\mathbf{\Lambda}\mathbf{P}_{\Gamma}\mathbf{\Lambda}^{-1}\mathbf{Y}$. This process results in the material part of $\mathbf{Y}$ being represented as $\mathbf{\Lambda}\mathbf{P}_{\Gamma}\mathbf{\Lambda}^{-1}\mathbf{Y}$, while it is represented as $\mathbf{P}_{\Gamma}\mathbf{Y}$ in an analysis based on standard response envelope. In linear algebra, the transformation matrices $\mathbf{\Lambda}\mathbf{P}_{\Gamma}\mathbf{\Lambda}^{-1}$ and $\mathbf{P}_{\Gamma}$ are said to be similar: A matrix $\mathbf{M} \in \mathbb{R}^{s \times s}$ is similar to $\mathbf{N} \in \mathbb{R}^{s \times s}$ if there exists a nonsingular matrix $\mathbf{T} \in \mathbb{R}^{s \times s}$ such that $\mathbf{N} = \mathbf{T}\mathbf{M}\mathbf{T}^{-1}$ (e.g. Harville 2008). When $\mathbf{M}$ represents a linear transformation from an s-dimensional linear space $\mathcal{S}$ to $\mathcal{S}$, $\mathbf{N}$ is the matrix representation of the same linear transformation but under another basis of $\mathcal{S}$, and $\mathbf{T}^{-1}$ is the matrix representation of the change of basis. Therefore, the process $\mathbf{\Lambda}\mathbf{P}_{\Gamma}\mathbf{\Lambda}^{-1}$ is the same as treating $\mathbf{\Lambda}^{-1}$ as a diagonal similarity transformation to represent $\mathbf{P}_{\Gamma}$ in original coordinate system as $\mathbf{\Lambda}\mathbf{P}_{\Gamma}\mathbf{\Lambda}^{-1}$. This is the essential idea underlying the scaled envelope model proposed by Cook and Su (2013), which we describe in Section 8.2.1. Predictor scaling is discussed in Section 8.3

8.2.1 Scaled Response Model

To represent scaling formally and facilitate estimation, we set the first element of the diagonal scaling matrix $\mathbf{\Lambda}$ to 1, which is necessary for the scale parameters to be identifiable. Accordingly, let $\mathbf{\Lambda} = \text{diag}(1, \lambda_2, \dots, \lambda_r)$, allowing each response to be scaled separately.

We assume that the scaled response vector $\mathbf{\Lambda}^{-1}\mathbf{Y}$ follows the response envelope model (1.20), so the scaled envelope model for the original responses $\mathbf{Y}$ becomes

$$\mathbf{Y} = \boldsymbol{\alpha} + \mathbf{\Lambda}\mathbf{\Gamma}\boldsymbol{\eta}\mathbf{X} + \boldsymbol{\varepsilon}, \text{ with } \mathbf{\Sigma} = \mathbf{\Lambda}\mathbf{\Gamma}\mathbf{\Omega}\mathbf{\Gamma}^T\mathbf{\Lambda} + \mathbf{\Lambda}\mathbf{\Gamma}_0\mathbf{\Omega}_0\mathbf{\Gamma}_0^T\mathbf{\Lambda}. \tag{8.5}$$

The scaled response $\mathbf{\Lambda}^{-1}\mathbf{Y}$ conforms to an envelope model with u-dimensional envelope $\mathcal{E}_{\mathbf{\Lambda}^{-1}\mathbf{\Sigma}\mathbf{\Lambda}^{-1}}(\mathbf{\Lambda}^{-1}\mathcal{B})$ and semi-orthogonal basis matrix $\mathbf{\Gamma}$. The matrices $\mathbf{\Lambda}\mathbf{P}_{\mathbf{\Gamma}}\mathbf{\Lambda}^{-1}$ and $\mathbf{\Lambda}\mathbf{Q}_{\mathbf{\Gamma}}\mathbf{\Lambda}^{-1}$ are orthogonal projection operators,

$$\mathbf{\Lambda}\mathbf{P}_{\mathbf{\Gamma}}\mathbf{\Lambda}^{-1} = \mathbf{P}_{\mathbf{\Lambda}\mathbf{\Gamma}(\mathbf{\Lambda}^{-2})} \text{ and } \mathbf{\Lambda}\mathbf{Q}_{\mathbf{\Gamma}}\mathbf{\Lambda}^{-1} = \mathbf{Q}_{\mathbf{\Lambda}\mathbf{\Gamma}_0(\mathbf{\Lambda}^{-2})},$$

that extract the material and immaterial parts of $\mathbf{Y}$. To facilitate the following discussion, we temporarily use the compact notation $\mathbf{U} = \mathbf{\Lambda}\mathbf{P}_{\mathbf{\Gamma}}\mathbf{\Lambda}^{-1}$ and $\mathbf{V} = \mathbf{\Lambda}\mathbf{Q}_{\mathbf{\Gamma}}\mathbf{\Lambda}^{-1}$. Then the material and immaterial parts of $\mathbf{Y}$ can be expressed as

$$\begin{aligned}\mathbf{U}\mathbf{Y} &= \mathbf{U}\boldsymbol{\alpha} + \mathbf{\Lambda}\mathbf{\Gamma}\boldsymbol{\eta}\mathbf{X} + \mathbf{U}\boldsymbol{\varepsilon}, \quad \text{var}(\mathbf{U}\boldsymbol{\varepsilon}) = \mathbf{\Lambda}\mathbf{\Gamma}\mathbf{\Omega}\mathbf{\Gamma}^T\mathbf{\Lambda} \\ \mathbf{V}\mathbf{Y} &= \mathbf{V}\boldsymbol{\alpha} + \mathbf{V}\boldsymbol{\varepsilon}, \quad \text{var}(\mathbf{V}\boldsymbol{\varepsilon}) = \mathbf{\Lambda}\mathbf{\Gamma}_0\mathbf{\Omega}_0\mathbf{\Gamma}_0^T\mathbf{\Lambda}.\end{aligned}$$

Since the distribution of $\mathbf{V}\mathbf{Y} \mid \mathbf{X}$ does not depend on $\mathbf{X}$ and $\mathbf{U}\boldsymbol{\varepsilon} \perp\!\!\!\perp \mathbf{V}\boldsymbol{\varepsilon}$, we see that

$$\text{(i) } \mathbf{V}\mathbf{Y} \mid (\mathbf{X} = \mathbf{x}_1) \sim \mathbf{V}\mathbf{Y} \mid (\mathbf{X} = \mathbf{x}_2) \text{ and (ii) } \mathbf{U}\mathbf{Y} \perp\!\!\!\perp \mathbf{V}\mathbf{Y} \mid \mathbf{X}, \tag{8.6}$$

so the scaled envelope model (8.5) follows the same independence structure as in (1.18).

Recall from (1.21) that the number of real parameters in the response envelope model (1.20) is $N_u = r + pu + r(r+1)/2$. The incorporation of $\mathbf{\Lambda}$ with $r - 1$ distinct scaling parameters means that the number of parameters in the scaled envelope model is $N_u + r - 1 = 2r - 1 + pu + r(r+1)/2$.

8.2.2 Estimation

Transforming the response to $\mathbf{\Lambda}^{-1}\mathbf{Y}$, it follows from the discussion of Section 1.5.1 that the maximum likelihood estimators $\widehat{\mathbf{\Lambda}}$ and $\widehat{\mathbf{\Gamma}}$ of $\mathbf{\Lambda}$ and $\mathbf{\Gamma}$ are obtained by minimizing the objective function

$$L_u(\mathbf{\Lambda}, \mathbf{\Gamma}) = \log|\mathbf{\Gamma}^T\mathbf{\Lambda}^{-1}\mathbf{S}_{\mathbf{Y}|\mathbf{X}}\mathbf{\Lambda}^{-1}\mathbf{\Gamma}| + \log|\mathbf{\Gamma}^T\mathbf{\Lambda}\mathbf{S}_{\mathbf{Y}}^{-1}\mathbf{\Lambda}\mathbf{\Gamma}|,$$

where the minimization is carried out over the $r - 1$ positive diagonal elements of $\mathbf{\Lambda}$ and the semi-orthogonal matrices $\mathbf{\Gamma} \in \mathbb{R}^{r \times u}$. Following this minimization, the maximum likelihood estimator of $\widehat{\mathbf{\Gamma}}_0$ is any orthogonal basis of the orthogonal complement of span($\widehat{\mathbf{\Gamma}}$). The maximum likelihood estimators of the remaining parameters are as follows:

$$\begin{aligned}\widehat{\boldsymbol{\alpha}} &= \bar{\mathbf{Y}} \\ \widehat{\boldsymbol{\eta}} &= \widehat{\mathbf{\Gamma}}^T\widehat{\mathbf{\Lambda}}^{-1}\mathbf{B}\end{aligned}$$

$$
\begin{aligned}
\widehat{\mathbf{\Omega}} &= \widehat{\mathbf{\Gamma}}^T\widehat{\mathbf{\Lambda}}^{-1}\mathbf{S}_{\mathbf{Y}|\mathbf{X}}\widehat{\mathbf{\Lambda}}^{-1}\widehat{\mathbf{\Gamma}} \\
\widehat{\mathbf{\Omega}}_0 &= \widehat{\mathbf{\Gamma}}_0^T\widehat{\mathbf{\Lambda}}^{-1}\mathbf{S}_{\mathbf{Y}}\widehat{\mathbf{\Lambda}}^{-1}\widehat{\mathbf{\Gamma}}_0 \\
\widehat{\boldsymbol{\beta}} &= \widehat{\mathbf{\Lambda}}\mathbf{P}_{\widehat{\mathbf{\Gamma}}}\widehat{\mathbf{\Lambda}}^{-1}\mathbf{B} \\
\widehat{\mathbf{\Sigma}} &= \widehat{\mathbf{\Lambda}}\mathbf{P}_{\widehat{\mathbf{\Gamma}}}\widehat{\mathbf{\Lambda}}^{-1}\mathbf{S}_{\mathbf{Y}|\mathbf{X}}\widehat{\mathbf{\Lambda}}^{-1}\mathbf{P}_{\widehat{\mathbf{\Gamma}}}\widehat{\mathbf{\Lambda}}^T + \widehat{\mathbf{\Lambda}}\mathbf{P}_{\widehat{\mathbf{\Gamma}}_0}\widehat{\mathbf{\Lambda}}^{-1}\mathbf{S}_{\mathbf{Y}}\widehat{\mathbf{\Lambda}}^{-1}\mathbf{P}_{\widehat{\mathbf{\Gamma}}_0}\widehat{\mathbf{\Lambda}} \\
&= \widehat{\mathbf{\Lambda}}\widehat{\mathbf{\Gamma}}\widehat{\mathbf{\Omega}}\widehat{\mathbf{\Gamma}}^T\widehat{\mathbf{\Lambda}}^T + \widehat{\mathbf{\Lambda}}\widehat{\mathbf{\Gamma}}_0\widehat{\mathbf{\Omega}}_0\widehat{\mathbf{\Gamma}}_0^T\widehat{\mathbf{\Lambda}}.
\end{aligned}
$$

These estimators conform to the projective structure described previously. For example, $\widehat{\boldsymbol{\beta}} = \widehat{\mathbf{\Lambda}}\mathbf{P}_{\widehat{\mathbf{\Gamma}}}\widehat{\mathbf{\Lambda}}^{-1}\mathbf{B}$ is constructed by projecting $\mathbf{B}$ onto span$(\widehat{\mathbf{\Lambda}}\widehat{\mathbf{\Gamma}})$ in the $\widehat{\mathbf{\Lambda}}^{-2}$ inner product. As for other envelopes, the estimators $\widehat{\boldsymbol{\eta}}$, $\widehat{\mathbf{\Omega}}$, and $\widehat{\mathbf{\Omega}}_0$ depend on the selected basis $\widehat{\mathbf{\Gamma}}$ and for that reason are not normally of interest in practice. The estimators of $\boldsymbol{\beta}$ and $\mathbf{\Sigma}$ depend on the estimated envelope but not on the particular basis selected to represent it.

Although the estimators were developed under the assumption of normal errors, Cook and Su (2013, Proposition 3) show that if the errors have finite fourth moments and the diagonal elements of $\mathbf{P}_{\mathbb{X}}$ converge to 0 as $n \to \infty$, then

$$
\sqrt{n}[\{\text{vec}^T(\widehat{\boldsymbol{\beta}}), \text{vech}^T(\widehat{\mathbf{\Sigma}})\}^T - \{\text{vec}^T(\boldsymbol{\beta}), \text{vech}^T(\mathbf{\Sigma})\}^T]
$$

is asymptotically normal. They also give an expression for its asymptotic variance when the errors are normal and show that with normal errors the scaled envelope estimator is at least as efficient as the standard estimator at estimating $\boldsymbol{\beta}$ and $\mathbf{\Sigma}$ when the dimension of $\mathcal{E}_{\mathbf{\Lambda}^{-1}\mathbf{\Sigma}\mathbf{\Lambda}^{-1}}(\mathbf{\Lambda}^{-1}\mathcal{B})$ is known. Whether it is thought that the errors are normal or not, it is typically useful to use the residual bootstrap for standard errors. Additionally, Cook and Su (2013, Corollary 2) show that if $\mathbf{\Sigma} = \sigma^2\mathbf{\Lambda}^2$, then avar$\{\text{vec}(\widehat{\boldsymbol{\beta}})\}$ = avar$\{\text{vec}(\widehat{\boldsymbol{\beta}}_{\Lambda})\}$ = avar$\{\text{vec}(\mathbf{B})\}$, so when the elements of the error vector $\boldsymbol{\varepsilon}$ are independent, there is no asymptotic advantage to the scaled envelope estimator, but there is no asymptotic loss, either. This is the counterpart of Corollary 1.1 for unscaled envelopes. The methods for estimating the dimension of the scaled envelope are direct extensions of those in Chapter 1 for estimating the dimension of the standard response envelope.

8.2.3 Race Times: Scaled Response Envelopes

We use the race time data introduced in Section 8.1.4 to illustrate response scaling. Both AIC and BIC estimated the dimension of $\mathcal{E}_{\mathbf{\Lambda}^{-1}\mathbf{\Sigma}\mathbf{\Lambda}^{-1}}(\mathbf{\Lambda}^{-1}\mathcal{B})$ to be $u = 2$, so we proceed with that dimension. The fit of the scaled response envelope with $u = 2$ gave the scale estimates shown in the second column of Table 8.2. The estimated scale parameters for T_4, T_5, and T_{10} are the most extreme, with the remaining estimates being roughly the same. Evidently, it was necessary to bring the beginning and ending times in line to achieve an envelope structure. Other aspects of the fit of the scaled response envelope were remarkably similar to those for the inner envelope fit of Section 8.1.4, including the values of the maximized log-likelihood. To illustrate the similarity, the third column of

Table 8.2 Race times: Estimated scales $\hat{\lambda}_i$, absolute Z-scores for the coefficients of T_3, and estimated coefficients $\hat{\beta}_{T_3}$ for T_3 from the fit of the scaled response envelope with $u = 2$.

Y	$\hat{\lambda}_i$	Z-score: T_3	$\hat{\beta}_{T_3}$
T_4	1.00	9.44	0.97
T_5	0.05	5.54	1.19
T_6	0.14	5.39	1.46
T_7	0.33	4.69	1.02
T_8	0.31	4.79	1.26
T_9	0.37	4.59	1.17
T_{10}	2.59	4.05	0.96

Table 8.2 gives the Z-scores for the coefficients of T_3, assuming normal errors and constructing standard errors from the plug-in estimate of the asymptotic variances given by Cook and Su (2013). Comparing this with the fourth column of Table 8.1, we see that the scores for the coefficients of T_3 are quite close. The last columns of Tables 8.1 and 8.2 give the estimated coefficients for T_3, and again we see that they are close. The same closeness holds between the Z-scores of other coefficients, between the estimates of $\mathbf{\Sigma}$ and between the fitted vectors. The general conclusion then is that the inner envelope with $d = 2$ and scaled response envelope with $u = 2$ give essentially the same fits.

Simulation results and additional data analyses illustrating the use of scaled envelopes were given by Cook and Su (2013).

8.3 Scaled Predictor Envelopes

Like the response envelopes of Chapter 1, the estimators based on the predictor envelopes of Chapter 4 are not invariant or equivariant under predictor rescaling. This is illustrated schematically in Figure 8.5 for a regression with a univariate response Y and $p = 2$ predictors, X_1 and X_2. According to model (4.5), for envelopes to work effectively in this case, $\boldsymbol{\beta}^T$ must align with an eigenvector $\boldsymbol{\ell}_i$ of $\mathbf{\Sigma}_{\mathbf{X}} \in \mathbb{R}^{2\times 2}$, which is represented in Figure 8.5 by ellipses. The $\boldsymbol{\beta}$ depicted in Figure 8.5a aligns with neither $\boldsymbol{\ell}_1$ or $\boldsymbol{\ell}_2$, and consequently $\mathcal{E}_{\mathbf{\Sigma}_{\mathbf{X}}}(\mathcal{B}') = \mathbb{R}^2$ and the predictor envelope offer no gains. However, in some regressions, progress may be possible by rescaling the predictors $\mathbf{X} \mapsto \mathbf{\Lambda}^{-1}\mathbf{X}$, where $\mathbf{\Lambda} \in \mathbb{R}^{p\times p}$ is a diagonal matrix with positive diagonal elements λ_i, $i = 1, \dots, p$. This possibility is represented in Figure 8.5b. The coefficient vector $\mathbf{\Lambda}\boldsymbol{\beta}^T$ for the regression of

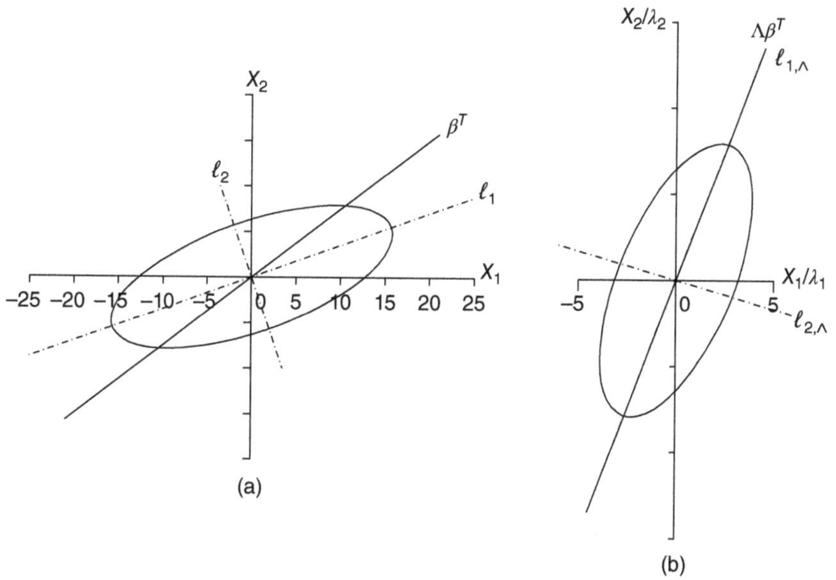

Figure 8.5 Schematic illustration of how rescaling the predictors can affect an envelope analysis (adapted from Cook and Su 2016). (a) Original predictor distribution, (b) Rescaled distribution.

Y on $\mathbf{\Lambda}^{-1}\mathbf{X}$ now aligns with the first eigenvector $\boldsymbol{\ell}_{1,\Lambda}$ of the covariance matrix $\mathbf{\Sigma}_{\Lambda^{-1}\mathbf{X}} = \mathbf{\Lambda}^{-1}\mathbf{\Sigma}_\mathbf{X}\mathbf{\Lambda}^{-1}$ of the scaled predictors. The envelope for the regression of Y on $\mathbf{\Lambda}^{-1}\mathbf{X}$ is $\mathcal{E}_{\Lambda^{-1}\Sigma_X\Lambda^{-1}}(\mathbf{\Lambda}\mathcal{B}')$, which has dimension 1 in Figure 8.5b.

The basic idea of predictor scaling is essentially the same as response scaling: Beginning with Figure 8.5a, rescale and perform an envelope analysis as in Figure 8.5b and then transform back to the original scale in Figure 8.5a. This is the basis for the envelope scaling methods developed by Cook and Su (2016). We describe their proposal in the rest of this section.

8.3.1 Scaled Predictor Model

Predictor scaling can be incorporated into envelope model (4.5) by following the same general steps that we used for response scaling. With response scaling, we allowed each response to have its own scale parameter, but for predictor scaling, we allow for groups of equal scale parameters. Let

$$\mathbf{\Lambda} = \text{diag}(1, \dots, 1, \lambda_2, \dots, \lambda_2, \dots, \lambda_s \dots, \lambda_s) \in \mathbb{R}^{p\times p}$$

be a scaling matrix that has s distinct positive scaling parameters, $(\lambda_1, \lambda_2, \dots, \lambda_s)$, rearranging the predictors as necessary to conform to the group structure of $\mathbf{\Lambda}$. Because $\lambda_1 = 1$, only $s - 1$ of these need to be estimated.

It is assumed that the group structure of $\mathbf{\Lambda}$ is known. Let t_i be the number of repeats of scaling parameter λ_i, so $\sum_{i=1}^{s} t_i = p$. We assume that the regression of $\mathbf{Y}$ on $\mathbf{\Lambda}^{-1}\mathbf{X}$ follows envelope model (4.5) with q-dimensional envelope $\mathcal{E}_{\mathbf{\Lambda}^{-1}\mathbf{\Sigma}_{\mathbf{X}}\mathbf{\Lambda}^{-1}}(\mathbf{\Lambda}\mathcal{B}')$. The resulting extension of envelope model (4.5), which is called the scaled predictor envelope (SPE) model following Cook and Su (2016), is then

$$\begin{aligned} \mathbf{Y} &= \boldsymbol{\alpha} + \boldsymbol{\eta}^T\mathbf{\Phi}^T\mathbf{\Lambda}^{-1}(\mathbf{X} - \boldsymbol{\mu}_{\mathbf{X}}) + \boldsymbol{\varepsilon} \\ \mathbf{\Sigma}_{\mathbf{X}} &= \mathbf{\Lambda}\mathbf{\Phi}\mathbf{\Delta}\mathbf{\Phi}^T\mathbf{\Lambda} + \mathbf{\Lambda}\mathbf{\Phi}_0\mathbf{\Delta}_0\mathbf{\Phi}_0^T\mathbf{\Lambda}, \end{aligned} \tag{8.7}$$

where $\boldsymbol{\beta}^T = \mathbf{\Lambda}^{-1}\mathbf{\Phi}\boldsymbol{\eta}$, $\boldsymbol{\eta} \in \mathbb{R}^{q\times r}$ carries the coordinates of $\mathbf{\Lambda}\boldsymbol{\beta}^T$ with respect to a semi-orthogonal basis matrix $\mathbf{\Phi} \in \mathbb{R}^{p\times q}$ of the envelope of $\mathcal{E}_{\mathbf{\Lambda}^{-1}\mathbf{\Sigma}_{\mathbf{X}}\mathbf{\Lambda}^{-1}}(\mathbf{\Lambda}\mathcal{B}')$, $\mathbf{\Phi}_0 \in \mathbb{R}^{p\times(p-q)}$ is a completion of $\mathbf{\Phi}$ so that $(\mathbf{\Phi}, \mathbf{\Phi}_0)$ is an orthogonal matrix, and $\mathbf{\Omega} \in \mathbb{R}^{q\times q}$ and $\mathbf{\Omega}_0 \in \mathbb{R}^{(p-q)\times(p-q)}$ are positive definite matrices. The number of free parameters in this model is N_q; the number of parameters in model (4.5), plus $s - 1$ scale parameters, giving a total of

$$N_{q,s} = s - 1 + r + p + rq + p(p+1)/2 + r(r+1)/2.$$

The difference between the number of parameters in the full model N_p and $N_{q,s}$ is $N_p - N_{q,s} = r(p - q) - (s - 1)$.

8.3.2 Estimation

We follow the estimation setup in Section 4.3, which is equivalent to assuming that $\mathbf{Y}$ and $\mathbf{X}$ are jointly normal. Normality is not required in the SPE model (8.7), but this assumption produces estimators that perform well when normality does not hold, as discussed in Section 4.3. The maximum likelihood estimators $\widehat{\mathbf{\Lambda}}$ and $\widehat{\mathbf{\Phi}}$ of $\mathbf{\Lambda}$ and $\mathbf{\Phi}$ are obtained by minimizing

$$L_q(\mathbf{\Phi}, \mathbf{\Lambda}) = \log|\mathbf{\Phi}^T\mathbf{\Lambda}^{-1}\mathbf{S}_{\mathbf{X}|\mathbf{Y}}\mathbf{\Lambda}^{-1}\mathbf{\Phi}| + \log|\mathbf{\Phi}^T\mathbf{\Lambda}\mathbf{S}_{\mathbf{X}}^{-1}\mathbf{\Lambda}\mathbf{\Phi}|, \tag{8.8}$$

over the set of $p \times q$ semi-orthogonal matrices for $\mathbf{\Phi}$ and the $s - 1$ positive real numbers for the diagonal elements of $\mathbf{\Lambda}$. Cook and Su (2016) propose minimizing (8.8) by using an alternating algorithm. Given an initial value for $\mathbf{\Lambda}$, perhaps $\mathbf{\Lambda} = \mathbf{I}_p$, minimize L_q over $\mathbf{\Phi}$, and then iterate between $\mathbf{\Lambda}$ and $\mathbf{\Phi}$ until a convergence criterion is met.

Once $\widehat{\mathbf{\Phi}}$ of $\widehat{\mathbf{\Lambda}}$ are obtained, the estimators of the remaining parameters are

$$\begin{aligned} \widehat{\boldsymbol{\mu}}_{\mathbf{X}} &= \bar{\mathbf{X}} \\ \widehat{\boldsymbol{\mu}}_{\mathbf{Y}} &= \bar{\mathbf{Y}} \\ \widehat{\boldsymbol{\eta}} &= (\widehat{\mathbf{\Phi}}^T\widehat{\mathbf{\Lambda}}^{-1}\mathbf{S}_{\mathbf{X}}\widehat{\mathbf{\Lambda}}^{-1}\widehat{\mathbf{\Phi}})^{-1}\widehat{\mathbf{\Phi}}^T\widehat{\mathbf{\Lambda}}^{-1}\mathbf{S}_{\mathbf{XY}} \\ \widehat{\mathbf{\Delta}} &= \widehat{\mathbf{\Phi}}^T\widehat{\mathbf{\Lambda}}^{-1}\mathbf{S}_{\mathbf{X}}\widehat{\mathbf{\Lambda}}^{-1}\widehat{\mathbf{\Phi}} \\ \widehat{\mathbf{\Delta}}_0 &= \widehat{\mathbf{\Phi}}_0^T\widehat{\mathbf{\Lambda}}^{-1}\mathbf{S}_{\mathbf{X}}\widehat{\mathbf{\Lambda}}^{-1}\widehat{\mathbf{\Phi}}_0 \end{aligned}$$

$$\hat{\beta}^T = \hat{\Lambda}^{-1}\hat{\Phi}\hat{\eta}$$
$$\hat{\Sigma}_X = \hat{\Lambda}\hat{\Phi}\hat{\Delta}\hat{\Phi}^T\hat{\Lambda} + \hat{\Lambda}\hat{\Phi}_0\hat{\Delta}_0\hat{\Phi}_0^T\hat{\Lambda}.$$
$$\hat{\Sigma} = (\mathbb{Y} - \mathbb{X}\hat{\beta}^T)^T(\mathbb{Y} - \mathbb{X}\hat{\beta}^T)/n,$$

where $\mathbb{Y}$ and $\mathbb{X}$ were defined in Section 1.1. Let $\mathbf{B}_{\hat{\Lambda}}^T = \mathbf{S}_{\hat{\Lambda}^{-1}\mathbf{X}}^{-1}\mathbf{S}_{\hat{\Lambda}^{-1}\mathbf{X},\mathbf{Y}}$ denote the ordinary least squares estimator of the coefficient matrix based on the scaled predictors. Then we can reexpress $\hat{\beta}^T$ as

$$\hat{\beta}^T = \hat{\Lambda}^{-1}\mathbf{P}_{\hat{\Phi}(\hat{\Lambda}^{-1}\mathbf{S}_\mathbf{X}\hat{\Lambda}^{-1})}\mathbf{B}_{\hat{\Lambda}}^T.$$

Consequently, as described at the outset to this section, the scaled envelope estimator of $\boldsymbol{\beta}^T$ first scales the predictors, then performs envelope estimation, and finally transforms back to the original scale. Cook and Su (2016) showed that if the joint distribution of $\mathbf{Y}$ and $\mathbf{X}$ has finite fourth moments and the SPE model holds with q known, then $\sqrt{n}(\hat{\boldsymbol{\beta}} - \boldsymbol{\beta})$ is asymptotically normally distributed and they give an expression for the asymptotic covariance matrix under normality. They also show that the asymptotic covariance matrix of $\hat{\boldsymbol{\beta}}$ is the same as the ordinary least squares estimator $\mathbf{B}$ when $q \geq p - (s-1)/r$, which places a constraint on the envelope dimensions that can lead to increased efficiency. For instance, if $r = 1$, then the envelope dimension must be $q < p - s + 1$ to avoid reducing to ordinary least squares. If we also consider rescaling all predictors $s = p$, then no asymptotic efficiency gains are possible. In other words, there are no efficiency gains possible from rescaling the predictors in univariate linear regression.

8.3.3 Scaled SIMPLS Algorithm

As indicated in Section 4.3.2, it is customary to standardize the individual predictors marginally $X_j \mapsto X_j/\{\widehat{\text{var}}(X_j)\}^{1/2}, j = 1, \dots, p$, prior to application of SIMPLS, although there is evidently no firm theory to support this operation. In this section, we discuss an alternative method of determining predictor scaling that combines the envelope objective function (8.8) with SIMPLS.

Cook and Su (2016) proposed estimators of $\boldsymbol{\Lambda}$ and $\boldsymbol{\Phi}$ based on a combination of objective function (8.8) and the SIMPLS estimator discussed in Section 4.2.1. The estimators, which can serve effectively as initial values for alternating optimization of (8.8) or as relatively quick standalone estimators, are again based on alternating between $\boldsymbol{\Lambda}$ and $\boldsymbol{\Phi}$.

On the first iteration, set an initial value $\boldsymbol{\Lambda}_0$ for $\boldsymbol{\Lambda}$, perhaps $\boldsymbol{\Lambda}_0 = \mathbf{I}_p$, and then get $\hat{\boldsymbol{\Phi}}_1$ by using SIMPLS to fit the regression of $\mathbf{Y}$ on $\boldsymbol{\Lambda}_0^{-1}\mathbf{X}$. Next, set $\hat{\boldsymbol{\Lambda}}_1 = \arg\min_{\boldsymbol{\Lambda}} L(\hat{\boldsymbol{\Phi}}_1, \boldsymbol{\Lambda})$. This completes the first iteration. To describe subsequent iterations, let $\hat{\boldsymbol{\Lambda}}_i$ and $\hat{\boldsymbol{\Phi}}_i$ denote the values of $\boldsymbol{\Lambda}$ and $\boldsymbol{\Phi}$ after the ith iteration. Then to get $\hat{\boldsymbol{\Lambda}}_{i+1}$ and $\hat{\boldsymbol{\Phi}}_{i+1}$, proceed as follows:

1. Get $\widetilde{\mathbf{\Phi}}$ from the SIMPLS fit of $\mathbf{Y}$ on $\widehat{\mathbf{\Lambda}}_i^{-1}\mathbf{X}$.
2. For $a \in (0,1)$, let $\widehat{\mathbf{\Phi}}_a$ be a semi-orthogonal basis matrix for $\text{span}(a\widehat{\mathbf{\Phi}}_i + (1-a)\widetilde{\mathbf{\Phi}})$. Then find the optimal value a^* for a as $a^* = \arg\min_a L_q(\widehat{\mathbf{\Phi}}_a, \widehat{\mathbf{\Lambda}}_i)$.
3. Set $\widehat{\mathbf{\Phi}}_{i+1}$ to be a semi-orthogonal basis matrix for $\text{span}(a^*\widehat{\mathbf{\Phi}}_i + (1-a^*)\widetilde{\mathbf{\Phi}})$.
4. Set $\widehat{\mathbf{\Lambda}}_{i+1} = \arg\min_{\mathbf{\Lambda}} L(\widehat{\mathbf{\Phi}}_{i+1}, \mathbf{\Lambda})$.

This algorithm can be much faster than the alternating optimization of (8.8) and can produce effective starting values for optimization of (8.8). The estimators may also be useful for assessing the need to pursue estimation via (8.8).

9

Connections and Adaptations

In Chapter 4, we studied the connections between envelopes and partial least squares regression. In this chapter, we discuss relationships between envelopes and other dimension reduction methods, starting with canonical correlations. In each case, we give a brief introduction to the method. Some methods are tied directly to envelopes while for others we indicate how to adapt envelopes to possibly enhance the methods.

9.1 Canonical Correlations

Suppose that we have n independent and identically distributed observations $\mathbf{Z}_i$ on a random vector $\mathbf{Z} \in \mathbb{R}^{p+r}$ partitioned as

$$\mathbf{Z}_i = \begin{pmatrix} \mathbf{X}_i \\ \mathbf{Y}_i \end{pmatrix}, \quad i = 1, \dots, n.$$

Canonical correlations and the corresponding canonical variates are used to characterize the strength of the linear relationships between the two vector-valued random variables $\mathbf{X} \in \mathbb{R}^p$ and $\mathbf{Y} \in \mathbb{R}^r$. Although the XY-notation for these variates is the same as that used for the multivariate linear model (1.1), the context here is not regression, so $\mathbf{Y}$ is not designated as a response vector.

9.1.1 Construction of Canonical Variates and Correlations

Canonical variates are constructed as pairs $(\mathbf{a}_j^T\mathbf{X}, \mathbf{b}_j^T\mathbf{Y})$, where the *canonical vectors* $\mathbf{a}_j \in \mathbb{R}^p$ and $\mathbf{b}_j \in \mathbb{R}^r$, $j = 1, \dots, m$ with $m = \min(p, r)$ are linear transformations of $\mathbf{X}$ and $\mathbf{Y}$ so that $(\mathbf{a}_1^T\mathbf{X}, \mathbf{b}_1^T\mathbf{Y})$ is the most highly correlated pair, $(\mathbf{a}_2^T\mathbf{X}, \mathbf{b}_2^T\mathbf{Y})$ is the second most highly correlated pair, and so on, subject to certain constraints on the construction process. In this way, the pairs of canonical variates give a portrayal of the linear dependencies between $\mathbf{X}$ and $\mathbf{Y}$. We begin by describing how the canonical variates are constructed and then we sketch

An Introduction to Envelopes: Dimension Reduction for Efficient Estimation in Multivariate Statistics, First Edition. R. Dennis Cook.

some of their properties. As in previous chapters, we let $\mathbf{S}_\mathbf{X}$, $\mathbf{S}_\mathbf{Y}$, and $\mathbf{S}_{\mathbf{X},\mathbf{Y}}$ denote the sample versions of $\boldsymbol{\Sigma}_\mathbf{X} = \text{var}(\mathbf{X})$, $\boldsymbol{\Sigma}_\mathbf{Y} = \text{var}(\mathbf{Y})$, and $\boldsymbol{\Sigma}_{\mathbf{X},\mathbf{Y}} = \text{cov}(\mathbf{X}, \mathbf{Y})$.

The correlation structure between $\mathbf{X}$ and $\mathbf{Y}$ is reflected by the matrix of sample cross correlations $\mathbf{C}_{\mathbf{X},\mathbf{Y}} = \mathbf{S}_\mathbf{X}^{-1/2}\mathbf{S}_{\mathbf{X},\mathbf{Y}}\mathbf{S}_\mathbf{Y}^{-1/2}$ between the elements of the standardized vectors $\mathbf{S}_\mathbf{X}^{-1/2}\mathbf{X}$ and $\mathbf{S}_\mathbf{Y}^{-1/2}\mathbf{Y}$. Canonical correlation analysis can be viewed as first constructing the desired canonical vectors in the standardized scale of $\mathbf{C}_{\mathbf{X},\mathbf{Y}}$ and then transforming back to the original scale to obtain the canonical vectors $\{\mathbf{a}_j\}$ and $\{\mathbf{b}_j\}$, followed by forming the canonical variates themselves. Let $\mathbf{C}_{\mathbf{X},\mathbf{Y}} = \mathbf{U}\mathbf{D}\mathbf{V}^T$ denote the compact singular value decomposition of $\mathbf{C}_{\mathbf{X},\mathbf{Y}}$:

$$\mathbf{U} = (\mathbf{u}_1, \dots, \mathbf{u}_m), \quad \mathbf{U}^T\mathbf{U} = \mathbf{I}_m$$

$$\mathbf{V} = (\mathbf{v}_1, \dots, \mathbf{v}_m), \quad \mathbf{V}^T\mathbf{V} = \mathbf{I}_m$$

$$\mathbf{D} = \text{diag}(d_1, \dots, d_m),$$

where we assume that $d_1 > d_2 > \cdots > d_m > 0$. Then the pairs of canonical variates $(\mathbf{a}_j^T\mathbf{X}, \mathbf{b}_j^T\mathbf{Y})$ are constructed from the canonical vectors as

$$\mathbf{a}_j = \mathbf{S}_\mathbf{X}^{-1/2}\mathbf{u}_j, \quad \mathbf{b}_j = \mathbf{S}_\mathbf{Y}^{-1/2}\mathbf{v}_j, \; j = 1, \dots, m. \tag{9.1}$$

The sample variance of the canonical variate $\mathbf{a}_j^T\mathbf{X}$ equals $\mathbf{u}_j^T\mathbf{S}_\mathbf{X}^{-1/2}\mathbf{S}_\mathbf{X}\mathbf{S}_\mathbf{X}^{-1/2}\mathbf{u}_j = 1$, and in the same way, the sample variance of $\mathbf{b}_j^T\mathbf{Y}$ is equal to 1. It can be seen similarly that the pairs of canonical variates $(\mathbf{a}_j^T\mathbf{X}, \mathbf{a}_k^T\mathbf{X})$, $(\mathbf{b}_j^T\mathbf{Y}, \mathbf{b}_k^T\mathbf{Y})$, and $(\mathbf{a}_j^T\mathbf{X}, \mathbf{b}_k^T\mathbf{Y})$ are uncorrelated for $j \neq k$, while the sample correlation between $\mathbf{a}_j^T\mathbf{X}$ and $\mathbf{b}_j^T\mathbf{Y}$ is the jth *sample canonical correlation*:

$$\hat{\rho}(\mathbf{a}_j^T\mathbf{X}, \mathbf{b}_j^T\mathbf{Y}) = \mathbf{u}_j^T\mathbf{S}_\mathbf{X}^{-1/2}\mathbf{S}_{\mathbf{X},\mathbf{Y}}\mathbf{S}_\mathbf{Y}^{-1/2}\mathbf{v}_j = \mathbf{u}_j^T\mathbf{C}_{\mathbf{X},\mathbf{Y}}\mathbf{v}_j = d_j,$$

where $\hat{\rho}(\cdot, \cdot)$ denotes the sample correlation between its arguments. Summarizing these properties,

$$\hat{\rho}(\mathbf{a}_j^T\mathbf{X}, \mathbf{b}_k^T\mathbf{Y}) = \begin{cases} d_j & \text{if } j = k \\ 0 & \text{otherwise} \end{cases}$$

$$\hat{\rho}(\mathbf{a}_j^T\mathbf{X}, \mathbf{a}_k^T\mathbf{X}) = \begin{cases} 1 & \text{if } j = k \\ 0 & \text{otherwise} \end{cases}$$

$$\hat{\rho}(\mathbf{b}_j^T\mathbf{Y}, \mathbf{b}_k^T\mathbf{Y}) = \begin{cases} 1 & \text{if } j = k \\ 0 & \text{otherwise.} \end{cases}$$

Accordingly, the sample canonical correlation for the jth pair of canonical variates is equal to the jth singular value of $\mathbf{C}_{\mathbf{X},\mathbf{Y}}$, the first canonical pair having the largest correlation possible between a linear transformation of $\mathbf{X}$ and a linear transformation of $\mathbf{Y}$. The canonical variates $\{\mathbf{a}_j^T\mathbf{X}\}$ and $\{\mathbf{b}_j^T\mathbf{Y}\}$ are uncorrelated and standardized to have variance 1. In this way, the canonical pairs

$(\mathbf{a}_j^T\mathbf{X}, \mathbf{b}_j^T\mathbf{Y})$ serve to portray the linear relationship between $\mathbf{X}$ and $\mathbf{Y}$ in terms of one-dimensional projections ordered by their correlations.

The population version of a canonical correlation analysis is obtained by replacing the sample matrices $\mathbf{S}_{\mathbf{X}}$, $\mathbf{S}_{\mathbf{Y}}$, and $\mathbf{S}_{\mathbf{X},\mathbf{Y}}$ with their population counterparts $\mathbf{\Sigma}_{\mathbf{X}}$, $\mathbf{\Sigma}_{\mathbf{Y}}$, and $\mathbf{\Sigma}_{\mathbf{X},\mathbf{Y}}$.

9.1.2 Derivation of Canonical Variates

In this section, we sketch how the canonical variates and vectors are derived in the population, starting with the standardized variates $\mathbf{\Sigma}_{\mathbf{X}}^{-1/2}\mathbf{X}$ and $\mathbf{\Sigma}_{\mathbf{Y}}^{-1/2}\mathbf{Y}$. Let

$$\boldsymbol{\rho}_{\mathbf{X},\mathbf{Y}} = \mathbf{\Sigma}_{\mathbf{X}}^{-1/2}\mathbf{\Sigma}_{\mathbf{X},\mathbf{Y}}\mathbf{\Sigma}_{\mathbf{Y}}^{-1/2} = \mathbf{\Upsilon}\mathbf{\Delta}\mathbf{\Psi}^T$$

denote the compact singular value decomposition of the population version $\boldsymbol{\rho}_{\mathbf{X},\mathbf{Y}}$ of the sample cross correlation matrix $\mathbf{C}_{\mathbf{X},\mathbf{Y}}$:

$$\begin{aligned}
&\mathbf{\Upsilon} = (\boldsymbol{\upsilon}_1, \dots, \boldsymbol{\upsilon}_d), \quad \mathbf{\Upsilon}^T\mathbf{\Upsilon} = \mathbf{I}_d \\
&\mathbf{\Psi} = (\boldsymbol{\psi}_1, \dots, \boldsymbol{\psi}_d), \quad \mathbf{\Psi}^T\mathbf{\Psi} = \mathbf{I}_d \\
&\mathbf{\Delta} = \mathrm{diag}(\delta_1, \dots, \delta_d),
\end{aligned}$$

where $d = \mathrm{rank}(\boldsymbol{\rho}_{\mathbf{X},\mathbf{Y}})$, and we assume that the singular values $\delta_1 > \delta_2 > \cdots > \delta_d > 0$ are distinct. Consider finding the first pair of canonical vectors in the standardized scales of $\boldsymbol{\rho}_{\mathbf{X},\mathbf{Y}}$.

For any pair of unit-length vectors $\boldsymbol{\alpha}_1 \in \mathbb{R}^p$ and $\boldsymbol{\beta}_1 \in \mathbb{R}^r$, the correlation between $\boldsymbol{\alpha}_1^T\mathbf{\Sigma}_{\mathbf{X}}^{-1/2}\mathbf{X}$ and $\boldsymbol{\beta}_1^T\mathbf{\Sigma}_{\mathbf{Y}}^{-1/2}\mathbf{Y}$ is

$$\rho(\boldsymbol{\alpha}_1^T\mathbf{\Sigma}_{\mathbf{X}}^{-1/2}\mathbf{X}, \boldsymbol{\beta}_1^T\mathbf{\Sigma}_{\mathbf{Y}}^{-1/2}\mathbf{Y}) = \boldsymbol{\alpha}_1^T\boldsymbol{\rho}_{\mathbf{X},\mathbf{Y}}\boldsymbol{\beta}_1 = \sum_{k=1}^{d}(\boldsymbol{\alpha}_1^T\boldsymbol{\upsilon}_k)(\boldsymbol{\beta}_1^T\boldsymbol{\psi}_k\delta_k). \tag{9.2}$$

The restriction to unit-length vectors $\boldsymbol{\alpha}_1$ and $\boldsymbol{\beta}_1$ serves to standardize the canonical variates:

$$\mathrm{var}(\boldsymbol{\alpha}_1^T\mathbf{\Sigma}_{\mathbf{X}}^{-1/2}\mathbf{X}) = \mathrm{var}(\boldsymbol{\beta}_1^T\mathbf{\Sigma}_{\mathbf{Y}}^{-1/2}\mathbf{Y}) = 1.$$

Using the Cauchy–Schwarz inequality on the right-hand side of (9.2), we have

$$\begin{aligned}
\rho^2(\boldsymbol{\alpha}_1^T\mathbf{\Sigma}_{\mathbf{X}}^{-1/2}\mathbf{X}, \boldsymbol{\beta}_1^T\mathbf{\Sigma}_{\mathbf{Y}}^{-1/2}\mathbf{Y}) &\le \left\{\sum_{k=1}^{d}(\boldsymbol{\alpha}_1^T\boldsymbol{\upsilon}_k)^2\right\}\left\{\sum_{k=1}^{d}(\boldsymbol{\beta}_1^T\boldsymbol{\psi}_k)^2\delta_k^2\right\} \\
&\le \left\{\sum_{k=1}^{d}(\boldsymbol{\beta}_1^T\boldsymbol{\psi}_k)^2\delta_k^2\right\} \\
&\le \delta_1^2\left\{\sum_{k=1}^{d}(\boldsymbol{\beta}_1^T\boldsymbol{\psi}_k)^2\right\} \\
&\le \delta_1^2.
\end{aligned}$$

The second equality follows because

$$\sum_{k=1}^{d} (\boldsymbol{\alpha}_1^T \boldsymbol{v}_k)^2 = \boldsymbol{\alpha}_1^T \boldsymbol{\Upsilon}\boldsymbol{\Upsilon}^T \boldsymbol{\alpha}_1 \leq \boldsymbol{\alpha}_1^T \boldsymbol{\alpha}_1 = 1.$$

The fourth inequality follows similarly. The third inequality follows because $\delta_1^2 > \delta_j^2$, $j = 2, \ldots, d$. The upper bound is attained when we choose $\boldsymbol{\alpha}_1 = \boldsymbol{v}_1$ and $\boldsymbol{\beta}_1 = \boldsymbol{\psi}_1$. Accordingly, the first pair of canonical variates is $(\boldsymbol{v}_1^T \boldsymbol{\Sigma}_{\mathbf{X}}^{-1/2}\mathbf{X}, \boldsymbol{\psi}_1^T \boldsymbol{\Sigma}_{\mathbf{Y}}^{-1/2}\mathbf{Y})$, which implies that the first pair of canonical vectors in the original scale of $\mathbf{X}$ and $\mathbf{Y}$ consists of $\mathbf{a}_1 = \boldsymbol{\Sigma}_{\mathbf{X}}^{-1/2}\boldsymbol{v}_1$ and $\mathbf{b}_1 = \boldsymbol{\Sigma}_{\mathbf{Y}}^{-1/2}\boldsymbol{\psi}_1$, and that δ_1 is the first canonical correlation.

To extend this treatment to all d pairs of population canonical variates, we seek semi-orthogonal matrices $\boldsymbol{\alpha} = (\boldsymbol{\alpha}_1, \ldots, \boldsymbol{\alpha}_d)$ and $\boldsymbol{\beta} = (\boldsymbol{\beta}_1, \ldots, \boldsymbol{\beta}_d)$ so that $\boldsymbol{\alpha}^T \boldsymbol{\rho}_{\mathbf{X},\mathbf{Y}} \boldsymbol{\beta}$ is a diagonal matrix with maximal and ordered diagonal elements. The solution $\boldsymbol{\alpha} = \boldsymbol{\Upsilon}$ and $\boldsymbol{\beta} = \boldsymbol{\Psi}$ follows from the uniqueness of the singular value decomposition of $\boldsymbol{\rho}_{\mathbf{X},\mathbf{Y}}$, recalling that we have assumed the singular values to be distinct. The canonical vectors in the scales of $\mathbf{X}$ and $\mathbf{Y}$ are then the columns of $\boldsymbol{\Sigma}_{\mathbf{X}}^{-1/2}\boldsymbol{\Upsilon}$ and $\boldsymbol{\Sigma}_{\mathbf{Y}}^{-1/2}\boldsymbol{\Psi}$.

9.1.3 Connection to Envelopes

Canonical correlations are related to the response envelope $\mathcal{E}_{\boldsymbol{\Sigma}}(\mathcal{B})$ of Chapter 1, to the predictor envelope $\mathcal{E}_{\boldsymbol{\Sigma}_{\mathbf{X}}}(\mathcal{B}')$ of Chapter 4, and to the simultaneous predictor–response envelope of Chapter 4. To develop this connection, we have from Corollary A.2, Proposition A.5, and the subsequent discussion that, in the context of model (1.1),

$$\begin{aligned}\mathcal{E}_{\boldsymbol{\Sigma}}(\mathcal{B}) = \mathcal{E}_{\boldsymbol{\Sigma}_{\mathbf{Y}}}(\mathcal{B}) &= \mathcal{E}_{\boldsymbol{\Sigma}_{\mathbf{Y}}}(\boldsymbol{\Sigma}_{\mathbf{Y}}^{-1/2}\mathcal{B}\boldsymbol{\Sigma}_{\mathbf{X}}^{1/2}) \\ &= \mathcal{E}_{\boldsymbol{\Sigma}_{\mathbf{Y}}}\{\mathrm{span}(\boldsymbol{\rho}_{\mathbf{YX}})\}.\end{aligned}$$

Similarly, $\mathcal{E}_{\boldsymbol{\Sigma}_{\mathbf{X}}}(\mathcal{B}') = \mathcal{E}_{\boldsymbol{\Sigma}_{\mathbf{X}}}(\mathrm{span}(\boldsymbol{\rho}_{\mathbf{X},\mathbf{Y}}))$. It follows from these relationships that $\mathrm{span}(\boldsymbol{\Psi}) \subseteq \mathcal{E}_{\boldsymbol{\Sigma}}(\mathcal{B})$ and $\mathrm{span}(\boldsymbol{\Upsilon}) \subseteq \mathcal{E}_{\boldsymbol{\Sigma}_{\mathbf{X}}}(\mathcal{B}')$, and consequently

$$\begin{aligned}\boldsymbol{\Sigma}_{\mathbf{Y}}^{-1/2}\mathrm{span}(\boldsymbol{\Psi}) &\subseteq \boldsymbol{\Sigma}_{\mathbf{Y}}^{-1/2}\mathcal{E}_{\boldsymbol{\Sigma}}(\mathcal{B}) = \mathcal{E}_{\boldsymbol{\Sigma}}(\mathcal{B}) \\ \boldsymbol{\Sigma}_{\mathbf{X}}^{-1/2}\mathrm{span}(\boldsymbol{\Upsilon}) &\subseteq \boldsymbol{\Sigma}_{\mathbf{X}}^{-1/2}\mathcal{E}_{\boldsymbol{\Sigma}_{\mathbf{X}}}(\mathcal{B}') = \mathcal{E}_{\boldsymbol{\Sigma}_{\mathbf{X}}}(\mathcal{B}'),\end{aligned}$$

where the final equality in each relation follows from Proposition A.4. While the population canonical vectors $\boldsymbol{\Sigma}_{\mathbf{Y}}^{-1/2}\mathrm{span}(\boldsymbol{\Psi})$ and $\boldsymbol{\Sigma}_{\mathbf{X}}^{-1/2}\mathrm{span}(\boldsymbol{\Upsilon})$ are in $\mathcal{E}_{\boldsymbol{\Sigma}}(\mathcal{B})$ and $\mathcal{E}_{\boldsymbol{\Sigma}_{\mathbf{X}}}(\mathcal{B}')$, respectively, they will not span these envelopes unless $\mathrm{span}(\boldsymbol{\Upsilon})$ and $\mathrm{span}(\boldsymbol{\Psi})$ are themselves reducing subspaces of $\boldsymbol{\Sigma}_{\mathbf{X}}$ and $\boldsymbol{\Sigma}_{\mathbf{Y}}$. Even when $\boldsymbol{\Sigma}_{\mathbf{Y}}^{-1/2}\mathrm{span}(\boldsymbol{\Psi}) = \mathcal{E}_{\boldsymbol{\Sigma}}(\mathcal{B})$ and $\boldsymbol{\Sigma}_{\mathbf{X}}^{-1/2}\mathrm{span}(\boldsymbol{\Upsilon}) = \mathcal{E}_{\boldsymbol{\Sigma}_{\mathbf{X}}}(\mathcal{B}')$, the sample versions of $\boldsymbol{\Sigma}_{\mathbf{Y}}^{-1/2}\mathrm{span}(\boldsymbol{\Psi})$ and $\boldsymbol{\Sigma}_{\mathbf{X}}^{-1/2}\mathrm{span}(\boldsymbol{\Upsilon})$ do not perform as well as the estimators of the corresponding envelopes presented previously (Cook and Zhang 2015b).

9.2 Reduced-Rank Regression

9.2.1 Reduced-Rank Model and Estimation

Reduced-rank regression begins with the multivariate linear model (1.1) and then introduces dimension reduction by constraining $\boldsymbol{\beta}$ to be less than full rank, say $\text{rank}(\boldsymbol{\beta}) = d < m$, where $m = \min(p, r)$ as defined in Section 9.1.1. Consequently, we can write $\boldsymbol{\beta} = \mathbf{H}\mathbf{h}$ as the product of two rank d matrices, $\mathbf{H} \in \mathbb{R}^{r \times d}$ and $\mathbf{h} \in \mathbb{R}^{d \times p}$, leading to the reduced-rank model

$$\mathbf{Y} = \boldsymbol{\alpha} + \mathbf{H}\mathbf{h}\mathbf{X} + \boldsymbol{\varepsilon}, \quad \text{var}(\boldsymbol{\varepsilon}) = \boldsymbol{\Sigma}. \tag{9.3}$$

This is different from an envelope model since there is no required connection between $\boldsymbol{\beta}$ and the error variance $\boldsymbol{\Sigma}$. With $\mathcal{E} = \text{span}(\mathbf{H})$, model (9.3) satisfies condition (1.18(i)) but not necessarily condition (1.18(ii)). This distinction can lead to very different performance in practice, as described later. There are no essential constraints placed on $\mathbf{H}$ and $\mathbf{h}$ as part of the formal model, although constraints might be introduced as part of the estimation process.

Assuming normal errors, maximum likelihood estimators were derived by Anderson (1999), Reinsel and Velu (1998), and Stoica and Viberg (1996) under various constraints on $\mathbf{H}$ and $\mathbf{h}$ for identifiability, such as $\mathbf{h}\mathbf{h}^{\text{T}} = \mathbf{I}_d$ or $\mathbf{H}^T\mathbf{H} = \mathbf{I}_d$. The decomposition $\boldsymbol{\beta} = \mathbf{H}\mathbf{h}$ is not unique even with such constraints: for any orthogonal matrix $\mathbf{O} \in \mathbb{R}^{d \times d}$, $\mathbf{H}_1 = \mathbf{H}\mathbf{O}$ and $\mathbf{h}_1 = \mathbf{O}^T\mathbf{h}$ offer another valid decomposition that satisfies the rank constraint. The parameters of interest, $\boldsymbol{\beta}$ and $\boldsymbol{\Sigma}$, are nevertheless identifiable, as well as $\text{span}(\mathbf{H}) = \text{span}(\boldsymbol{\beta})$ and $\text{span}(\mathbf{h}^T) = \text{span}(\boldsymbol{\beta}^T)$. Cook et al. (2015) gave a unified framework that bypasses the need for constraints $\mathbf{H}$ or $\mathbf{h}$ while providing maximum likelihood estimators of $\boldsymbol{\beta}$ and $\boldsymbol{\Sigma}$.

Recall from Section 9.1.1 that $\mathbf{C}_{\mathbf{X},\mathbf{Y}} = \mathbf{S}_{\mathbf{X}}^{-1/2}\mathbf{S}_{\mathbf{X},\mathbf{Y}}\mathbf{S}_{\mathbf{Y}}^{-1/2}$ is the matrix of sample correlations between the elements of the standardized vectors $\mathbf{S}_{\mathbf{X}}^{-1/2}\mathbf{X}$ and $\mathbf{S}_{\mathbf{Y}}^{-1/2}\mathbf{Y}$, with singular value decomposition $\mathbf{C}_{\mathbf{X},\mathbf{Y}} = \mathbf{U}\mathbf{D}\mathbf{V}^T$. Extending this notation, let $\mathbf{C}_{\mathbf{X},\mathbf{Y}}^{(d)} = \mathbf{U}_d\mathbf{D}_d\mathbf{V}_d^T$, where $\mathbf{U}_d \in \mathbb{R}^{r \times d}$ and $\mathbf{V}_d \in \mathbb{R}^{p \times d}$ consist of the first d columns of $\mathbf{U}$ and $\mathbf{V}$, and $\mathbf{D}_d$ is the diagonal matrix consisting of the first d singular values of $\mathbf{C}_{\mathbf{X},\mathbf{Y}}$. We also use $\mathbf{C}_{\mathbf{Y},\mathbf{X}} = \mathbf{C}_{\mathbf{X},\mathbf{Y}}^T$. Then, assuming normal errors and that d is given, the maximum likelihood estimators of the parameters in the reduced-rank regression model (9.3) are

$$\widehat{\boldsymbol{\alpha}}_{\text{RR}}(\mathbf{Y} \mid \mathbf{X}, d) = \bar{\mathbf{Y}} \tag{9.4}$$

$$\widehat{\boldsymbol{\beta}}_{\text{RR}}(\mathbf{Y} \mid \mathbf{X}, d) = \mathbf{S}_{\mathbf{Y}}^{1/2}\mathbf{C}_{\mathbf{Y},\mathbf{X}}^{(d)}\mathbf{S}_{\mathbf{X}}^{-1/2} \tag{9.5}$$

$$\begin{aligned}\widehat{\boldsymbol{\Sigma}}_{\text{RR}}(\mathbf{Y} \mid \mathbf{X}, d) &= \mathbf{S}_{\mathbf{Y}} - \widehat{\boldsymbol{\beta}}_{\text{RR}}\mathbf{S}_{\mathbf{X},\mathbf{Y}} \\ &= \mathbf{S}_{\mathbf{Y}}^{1/2}(\mathbf{I}_r - \mathbf{C}_{\mathbf{Y},\mathbf{X}}^{(d)}\mathbf{C}_{\mathbf{X},\mathbf{Y}}^{(d)})\mathbf{S}_{\mathbf{Y}}^{1/2}.\end{aligned} \tag{9.6}$$

The nontraditional part of this notation "$(\mathbf{Y} \mid \mathbf{X}, d)$" is meant to indicate that the estimators arise from the rank d linear regression of $\mathbf{Y}$ on $\mathbf{X}$. It will be suppressed until needed in Section 9.2.3 when combining reduced-rank regression and envelopes. The usual estimators are obtained by setting $d = m$, so there is no rank reduction. The asymptotic covariance matrix of the reduced-rank estimator of $\hat{\boldsymbol{\beta}}_{\mathrm{RR}}$ is (Cook et al. 2015, Eq. A17)

$$\begin{aligned}
&\mathrm{avar}\{\sqrt{n}\mathrm{vec}(\hat{\boldsymbol{\beta}}_{\mathrm{RR}})\} \\
&= (\boldsymbol{\Sigma}_{\mathbf{X}}^{-1/2} \otimes \boldsymbol{\Sigma}^{1/2})(\mathbf{I}_{pr} - \mathbf{Q}_{\boldsymbol{\Sigma}_{\mathbf{X}}^{1/2}\mathbf{h}^T} \otimes \mathbf{Q}_{\boldsymbol{\Sigma}^{-1/2}\mathbf{H}})(\boldsymbol{\Sigma}_{\mathbf{X}}^{-1/2} \otimes \boldsymbol{\Sigma}^{1/2}),
\end{aligned}$$

where $\boldsymbol{\Sigma}_{\mathbf{X}}^{-1} \otimes \boldsymbol{\Sigma} = \mathrm{avar}\{\sqrt{n}\mathrm{vec}(\mathbf{B})\}$ is the asymptotic covariance matrix of the maximum likelihood estimator when no rank constraints are imposed (see (1.15)).

Let $\mathbf{R}$ denote the asymptotic variation of the reduced-rank estimator relative to that of the unconstrained maximum likelihood estimator:

$$\begin{aligned}
\mathbf{R} &= \mathrm{avar}^{-1/2}\{\sqrt{n}\mathrm{vec}(\mathbf{B})\}[\mathrm{avar}\{\sqrt{n}\mathrm{vec}(\hat{\boldsymbol{\beta}}_{\mathrm{RR}})\}]\mathrm{avar}^{-1/2}\{\sqrt{n}\mathrm{vec}(\mathbf{B})\} \\
&= \mathbf{I}_{pr} - \mathbf{Q}_{\boldsymbol{\Sigma}_{\mathbf{X}}^{1/2}\mathbf{h}^T} \otimes \mathbf{Q}_{\boldsymbol{\Sigma}^{-1/2}\mathbf{H}}.
\end{aligned}$$

This relationship shows that the gain over the unconstrained maximum likelihood estimator is primarily due to the reduction in the number of parameters being estimated, and not to any internal structure associated with the parameters themselves. For instance, since $\mathbf{R}$ is a projection, its eigenvalues are 0 or 1, and $\mathrm{tr}(\mathbf{R}) = pr - (p-d)(r-d) = d(p+r-d)$, where pr is the number of real parameters needed to determine $\boldsymbol{\beta}$ in multivariate linear model (1.1), and $d(p+r-d)$ is the number of real parameters needed to determine $\boldsymbol{\beta} = \mathbf{Hh}$ in reduced-rank model (9.3). Consequently, the number of parameters needed for $\boldsymbol{\beta}$ in the reduced-rank model is an indicator of its relative gain in efficiency.

The methodology implied by (9.4)–(9.6) requires that $d = \mathrm{rank}(\boldsymbol{\beta})$ be selected first. While there are several ways in which this might be done, including adaptation of information criterions such as AIC and BIC, the relatively straightforward method described in Section 1.10.2 seems to work well in this context.

9.2.2 Contrasts with Envelopes

As mentioned previously, the reduced-rank model (9.3) is distinctly different from the envelope models for the responses (1.20) or predictors (4.5) since there is no connection between $\boldsymbol{\beta}$ and the error covariance matrix $\boldsymbol{\Sigma}$ or the predictor covariance matrix $\boldsymbol{\Sigma}_{\mathbf{X}}$. This distinction has notable consequences. If $r > 1$ and $p = 1$, then $\boldsymbol{\beta} \in \mathbb{R}^r$ and its only possible ranks are 0 and 1. In this case, reduced-rank regression is not useful, while a response envelope can still lead to substantial gain as illustrated by some of the examples in Chapter 2. More generally, reduced-rank regression offers no gain when $\boldsymbol{\beta}$ is full rank, while envelopes can still produce substantial gain. On the other hand, it is also possible to have situations where envelopes offer no gain, while reduced-rank regression provides notable gain. For instance, if $r > 1$, $p > 1$, and $\text{rank}(\boldsymbol{\beta}) = 1$, so reduced-rank regression gives maximal gain, it is still possible that $\mathcal{E}_{\boldsymbol{\Sigma}}(\mathcal{B}) = \mathbb{R}^r$ so that response envelopes produce no gain.

There are also notable differences between the potential gains produced by envelope and reduced-rank regressions. We saw near the end of Section 9.2.1 that the gain from reduced-rank regression is controlled by the reduction in the number of real parameters needed to specify $\boldsymbol{\beta}$. On the other hand, the gain from a response envelope is due to the reduction in parameters and to the structure of $\boldsymbol{\Sigma} = \boldsymbol{\Gamma}\boldsymbol{\Omega}\boldsymbol{\Gamma}^T + \boldsymbol{\Gamma}_0\boldsymbol{\Omega}_0\boldsymbol{\Gamma}_0^T$, with massive gains possible when $\|\boldsymbol{\Omega}\| \ll \|\boldsymbol{\Omega}_0\|$.

These contrasts led to the conclusion that envelopes and reduced-rank regression are distinctly different methods of dimension reduction with different operating characteristics. Cook et al. (2015) combined reduced-rank regression and response envelopes, leading to a new dimension reduction paradigm – *reduced-rank response envelopes* – that can automatically choose the better of the two methods if appropriate and can also give an estimator that does better than both of them.

9.2.3 Reduced-Rank Response Envelopes

The reduced-rank envelope model follows by applying an envelope to the reduced-rank representation (9.3). As in Chapter 1, let $\boldsymbol{\Gamma} \in \mathbb{R}^{r\times u}$ be a semi-orthogonal basis matrix for $\mathcal{E}_{\boldsymbol{\Sigma}}(\mathcal{B}) = \mathcal{E}_{\boldsymbol{\Sigma}}(\mathbf{H})$, where $u = \dim(\mathcal{E}_{\boldsymbol{\Sigma}}(\mathcal{B})) \geq d = \text{rank}(\boldsymbol{\beta})$. Then we can represent $\mathbf{H}$ in terms of $\boldsymbol{\Gamma}$, $\mathbf{H} = \boldsymbol{\Gamma}\boldsymbol{\eta}$, where $\boldsymbol{\eta} \in \mathbb{R}^{u\times d}$, and write the reduced-rank envelope model as

$$\mathbf{Y} = \boldsymbol{\alpha} + \boldsymbol{\Gamma}\boldsymbol{\eta}\mathbf{h}\mathbf{X} + \boldsymbol{\varepsilon}, \quad \boldsymbol{\Sigma} = \boldsymbol{\Gamma}\boldsymbol{\Omega}\boldsymbol{\Gamma}^T + \boldsymbol{\Gamma}_0\boldsymbol{\Omega}_0\boldsymbol{\Gamma}_0^T. \tag{9.7}$$

This model represents $\boldsymbol{\beta}$ as the product of three factors arising from the reduced dimension $d = \text{rank}(\boldsymbol{\beta})$ and the envelope structure. It contains two unknown dimensions d and u. A number of real parameters needed to specify four model variants of (9.7) are

1. Standard model (1.1): $N = r + pr + r(r+1)/2$,
2. Reduced-rank model (9.3): $N_d = r + d(r-d) + pd + r(r+1)/2$,
3. Envelope model (1.20): $N_u = r + pu + r(r+1)/2$,
4. Reduced-rank envelope model (9.7): $N_{d,u} = r + d(u-d) + pd + r(r+1)/2$.

If $u = r$, then $\boldsymbol{\Gamma} = \mathbf{I}_r$, $\mathbf{H} = \boldsymbol{\eta}$, and the model reduces to reduced-rank regression. If $d = u$, then (9.7) reduces to the response envelope model (1.20).

Assuming normal errors in model (9.7) and that $\boldsymbol{\Gamma}$ and d are known, the maximum likelihood estimators of $\boldsymbol{\alpha}$, $\boldsymbol{\beta}$, and $\boldsymbol{\Sigma}$ are as follows:

$$\widehat{\boldsymbol{\alpha}}_{\Gamma} = \bar{\mathbf{Y}}$$

$$\widehat{\boldsymbol{\beta}}_{\Gamma} = \boldsymbol{\Gamma}\widehat{\boldsymbol{\beta}}_{\text{RR}}(\boldsymbol{\Gamma}^T\mathbf{Y} \mid \mathbf{X}, d) \tag{9.8}$$

$$\widehat{\boldsymbol{\Sigma}}_{\Gamma} = \boldsymbol{\Gamma}\{\widehat{\boldsymbol{\Sigma}}_{\text{RR}}(\boldsymbol{\Gamma}^T\mathbf{Y} \mid \mathbf{X}, d)\}\boldsymbol{\Gamma}^T + \mathbf{Q}_{\Gamma}\mathbf{S}_{\mathbf{Y}}\mathbf{Q}_{\Gamma}, \tag{9.9}$$

where we make use of the notation established in (9.5) and (9.6). To construct these estimators, we first reduce to a reduced-rank envelope model (9.7) by transforming both sides by $\boldsymbol{\Gamma}^T$ to obtain the reduced-rank model with coefficient matrix $\boldsymbol{\eta}\mathbf{h}$:

$$\boldsymbol{\Gamma}^T\mathbf{Y} = \boldsymbol{\Gamma}^T\boldsymbol{\alpha} + \boldsymbol{\eta}\mathbf{h}\mathbf{X} + \boldsymbol{\Gamma}^T\boldsymbol{\varepsilon}, \quad \text{var}(\boldsymbol{\Gamma}^T\boldsymbol{\varepsilon}) = \boldsymbol{\Sigma}(\boldsymbol{\Gamma}^T\mathbf{Y} \mid \mathbf{X}, d) = \boldsymbol{\Omega}.$$

This model is then fitted by using the reduced-rank estimators (9.4)–(9.6) and then transformed to the scale of $\mathbf{Y}$. Once the estimator of $\boldsymbol{\Gamma}$ is determined, it can be substituted into (9.8) and (9.9) to obtain the full reduced-rank envelope estimators.

A little additional notation is needed for a compact description of the maximum likelihood estimator of $\boldsymbol{\Gamma}$. Within the class of semi-orthogonal matrices $\mathbf{G} \in \mathbb{R}^{r\times u}$, let $\mathbf{Z}_\mathbf{G} = (\mathbf{G}^T\mathbf{S}_\mathbf{Y}\mathbf{G})^{-1/2}\mathbf{G}^T\mathbf{Y}$ denote the standardized version of $\mathbf{G}^T\mathbf{Y}$ with sample covariance $\mathbf{I}_u$, let $\mathbf{S}^{(d)}_{\mathbf{Z}_\mathbf{G}|\mathbf{X}}$ denote the sample covariance of the residuals from the fit of the reduced-rank model (9.3) with response $\mathbf{Z}_\mathbf{G}$ and predictors $\mathbf{X}$, and let $\hat{\varphi}_j(\mathbf{G})$ denote the jth eigenvalue of $\mathbf{S}^{-1/2}_{\mathbf{G}^T\mathbf{Y}|\mathbf{X}}\mathbf{S}_{\mathbf{G}^T\mathbf{Y}}\mathbf{S}^{-1/2}_{\mathbf{G}^T\mathbf{Y}|\mathbf{X}}$. Define also the objective function

$$L_{d,u}(\mathbf{G}) = \log|\mathbf{G}^T\mathbf{S}_\mathbf{Y}\mathbf{G}| + \log|\mathbf{G}^T\mathbf{S}_\mathbf{Y}^{-1}\mathbf{G}| + \log|\mathbf{S}^{(d)}_{\mathbf{Z}_\mathbf{G}|\mathbf{X}}| \tag{9.10}$$

$$= \log|\mathbf{G}^T\mathbf{S}_{\mathbf{Y}|\mathbf{X}}\mathbf{G}| + \log|\mathbf{G}^T\mathbf{S}_\mathbf{Y}^{-1}\mathbf{G}| + \sum_{j=d+1}^{u} \log(\hat{\varphi}_j(\mathbf{G})), \tag{9.11}$$

which corresponds to the negative of the log likelihood for model (9.7) maximized over all parameters except $\boldsymbol{\Gamma}$. Then a maximum likelihood estimator of $\boldsymbol{\Gamma}$ is obtained as $\hat{\boldsymbol{\Gamma}} = \arg\min_\mathbf{G} L_{d,u}(\mathbf{G})$, where the minimization is over the Grassmannian of dimension u in $\mathbb{R}^r$. These two forms give different insights into the nature of $\hat{\boldsymbol{\Gamma}}$. As in several other objective functions encountered previously, the term $\log|\mathbf{G}^T\mathbf{S}_\mathbf{Y}\mathbf{G}| + \log|\mathbf{G}^T\mathbf{S}_\mathbf{Y}^{-1}\mathbf{G}|$ that occurs in (9.10) is minimized when span($\mathbf{G}$) is any u-dimensional reducing subspace of $\mathbf{S}_\mathbf{Y}$ (Lemma A.15). The remaining term $\log|\mathbf{S}^{(d)}_{\mathbf{Z}_\mathbf{G}|\mathbf{X}}|$ in (9.10) measures the goodness of fit of the reduced-rank regression of the reduced standardized response $\mathbf{Z}_\mathbf{G}$ on $\mathbf{X}$. In this way, the terms in (9.10) balance the requirements of the fit, the first two terms pulling the solution toward the reducing subspaces of $\mathbf{S}_\mathbf{Y}$ and the last term ensuring a reasonable fit of the reduced-rank regression.

Cook et al. (2015) used the second form for $L_{d,u}(\mathbf{G})$ as a basis for computation. They showed that the term $\sum_{j=d+1}^{u}\log(\hat{\varphi}_j(\mathbf{G}))$ converges to 0 uniformly in $\mathbf{G}$, and thus that $L_{d,u}(\mathbf{G})$ converges uniformly to $L_u^{\text{pop}}(\mathbf{G}) := \log|\mathbf{G}^T\boldsymbol{\Sigma}_{\mathbf{Y}|\mathbf{X}}\mathbf{G}| + \log|\mathbf{G}^T\boldsymbol{\Sigma}_\mathbf{Y}^{-1}\mathbf{G}|$, which does not depend on d. The sample version

$$L_u(\mathbf{G}) = \log|\mathbf{G}^T\mathbf{S}_{\mathbf{Y}|\mathbf{X}}\mathbf{G}| + \log|\mathbf{G}^T\mathbf{S}_\mathbf{Y}^{-1}\mathbf{G}|$$

of $L_u^{\text{pop}}(\mathbf{G})$ is the same as objective function (1.25) derived for response envelopes in Chapter 1. The implication of this result is that in large samples the reduced-rank envelope estimator moves toward a two-stage estimator: first estimate the envelope from $L_u(\mathbf{G})$ ignoring the rank d and then obtain the reduced-rank estimator from within the estimated envelope.

The methodology described in this section requires the selection of two dimensions $d = \text{rank}(\boldsymbol{\beta})$ and $u = \dim(\mathcal{E}_{\Sigma}(\mathcal{B}))$ subject to the constraint $0 \leq d \leq u \leq p$. As mentioned at the end of Section 9.2.1, the method outlined in Section 1.10.2 seems to work well for choosing d. Following that, Cook et al. (2015) recommended that u be chosen using BIC over the range $d \leq u \leq p$.

More discussion of this method is available from Cook et al. (2015), who show that the reduced-rank envelope estimators are $\sqrt{n}$-consistent without normality, and give asymptotic comparisons, methods for selecting u and d, simulation results and examples.

9.2.4 Reduced-Rank Predictor Envelopes

The results sketched in the previous section combine reduced-rank regression and response envelopes. In principal, reduced-rank regression can also be combined with the predictor and simultaneous response–predictor envelopes of Chapter 4.

To develop reduced-rank predictor envelopes, we assume as in Chapter 4 that the predictors are random with mean $\boldsymbol{\mu}_{\mathbf{X}}$ and variance $\boldsymbol{\Sigma}_{\mathbf{X}}$. Then we extend the parameterization of the reduced-rank regression model (9.3) to include $\mathcal{E}_{\boldsymbol{\Sigma}_{\mathbf{X}}}(\text{span}(\mathbf{h}^T))$, the $\boldsymbol{\Sigma}_{\mathbf{X}}$-envelope of $\text{span}(\boldsymbol{\beta}^T) = \text{span}(\mathbf{h}^T)$. Let $\boldsymbol{\Phi} \in \mathbb{R}^{p\times q}$ denote a semi-orthogonal basis for this subspace, so that we can write $\mathbf{h}^T = \boldsymbol{\Phi}\boldsymbol{\eta}$, where $\boldsymbol{\eta} \in \mathbb{R}^{q\times d}$. Then, recalling that $\mathbf{H} \in \mathbb{R}^{r\times d}$ and $\mathbf{h} \in \mathbb{R}^{d\times p}$, the model becomes

$$
\begin{aligned}
\mathbf{Y} &= \boldsymbol{\alpha} + \boldsymbol{\beta}(\mathbf{X} - \boldsymbol{\mu}_{\mathbf{X}}) + \boldsymbol{\varepsilon} \\
&= \boldsymbol{\alpha} + \mathbf{H}\mathbf{h}(\mathbf{X} - \boldsymbol{\mu}_{\mathbf{X}}) + \boldsymbol{\varepsilon} \\
&= \boldsymbol{\alpha} + \mathbf{H}\boldsymbol{\eta}^T\boldsymbol{\Phi}^T(\mathbf{X} - \boldsymbol{\mu}_{\mathbf{X}}) + \boldsymbol{\varepsilon} \\
\text{var}(\boldsymbol{\varepsilon}) &= \boldsymbol{\Sigma} \\
\boldsymbol{\Sigma}_{\mathbf{X}} &= \boldsymbol{\Phi}\boldsymbol{\Delta}\boldsymbol{\Phi}^T + \boldsymbol{\Phi}_0\boldsymbol{\Delta}_0\boldsymbol{\Phi}_0^T.
\end{aligned}
$$

This model can be studied following the general steps of previous chapters.

9.3 Supervised Singular Value Decomposition

In this section, we describe how envelopes arose in the development of supervised singular value decompositions by Li et al. (2016). Low-rank approximation of a given data matrix $\mathbb{Y} \in \mathbb{R}^{n\times r}$ is used frequently across the applied sciences, including genomics (Alter et al. 2000), compressed sensing (Candés and Recht 2009), and image analysis (Konstantinides et al. 1997). Let $\widehat{\mathbb{Y}}$ denote a rank $u \leq \min(n, r)$ approximation to $\mathbb{Y}$. It is known that the compact rank u singular value decomposition of $\mathbb{Y}$ minimizes the Frobenius norm $\|\mathbb{Y} - \widehat{\mathbb{Y}}\|_F$ over all matrices $\widehat{\mathbb{Y}} \in \mathbb{R}^{n \times r}$ with rank at most u.

Suppose that, in addition to $\mathbb{Y}$, we observe also an associated *supervision data matrix* $\mathbb{X} \in \mathbb{R}^{n \times p}$, so the total observations are $(\mathbb{Y}, \mathbb{X})$. Li et al. (2016) developed a method that allows using $\mathbb{X}$ to supervise the singular value decomposition of $\mathbb{Y}$. They start by modeling $\mathbb{Y}$ with a rank u structure and isotropic errors:

$$\mathbb{Y} = \mathbf{1}_n \boldsymbol{\alpha}^T + \mathbf{T}\mathbf{V}^T + \mathbf{E}, \tag{9.12}$$

where $\mathbf{1}_n \in \mathbb{R}^n$ is a vector of ones, $\boldsymbol{\alpha} \in \mathbb{R}^r$, $\mathbf{T} \in \mathbb{R}^{n \times u}$, $\mathbf{V} \in \mathbb{R}^{r \times u}$ is a semi-orthogonal matrix , and $\mathbf{E} \in \mathbb{R}^{n \times r}$ is an error matrix with elements that are independent copies of a $N(0, \sigma_e^2)$ variate. This representation effectively adjusts $\mathbb{Y}$ column-wise by the elements of $\boldsymbol{\alpha}$ prior to constructing the primary low-rank approximation $\mathbf{T}\mathbf{V}^T$. The column-wise adjustment can be made more transparent by reparameterizing in terms of $\mathbf{U} = \mathbf{Q}_{1_n}\mathbf{T}$ and substituting $\mathbf{T} = \mathbf{U} + \mathbf{P}_{1_n}\mathbf{T}$, giving

$$\mathbb{Y} = \mathbf{1}_n \boldsymbol{\mu}_{\mathrm{Y}}^T + \mathbf{U}\mathbf{V}^T + \mathbf{E}, \tag{9.13}$$

where $\boldsymbol{\mu}_{\mathrm{Y}} = \boldsymbol{\alpha} + n^{-1}\mathbf{V}\mathbf{T}^T\mathbf{1}_n$. In this representation, the adjustment vector $\boldsymbol{\mu}_{\mathrm{Y}}$ contains the population means of the columns of $\mathbb{Y}$. The goal of Li et al. (2016) was to estimate the terms in model (9.13), particularly the low-rank representation $\mathbf{U}\mathbf{V}^T$ of the column-centered version of $\mathrm{E}(\mathbb{Y})$. Expressing (9.13) in terms of a typical row $\mathbf{Y}_j^T$ of $\mathbb{Y}$, we have

$$\mathbf{Y}_j = \boldsymbol{\mu}_{\mathrm{Y}} + \mathbf{V}\mathbf{u}_j + \mathbf{e}_j, \quad j = 1, \dots, n,$$

where $\mathbf{u}_j^T$ and $\mathbf{e}_j^T$ are the jth rows of $\mathbf{U}$ and $\mathbf{E}$, with $\mathbf{e}_j \sim N(0, \sigma_e^2 \mathbf{I}_r)$. Model (9.13) can now be seen as a multivariate linear regression with isotropic errors, semi-orthogonal coefficient matrix $\mathbf{V}$, and *latent* predictors $\mathbf{u}_j$ with $\sum_{j=1}^n \mathbf{u}_j = 0$. It is essentially the same as the Tipping–Bishop model (Tipping and Bishop 1999) for probabilistic principal components discussed in Section 9.9.2.

Let $\mathbf{t}_j^T$ denote the jth row of $\mathbf{T}$, and let $\mathbf{X}_j^T$ denote the jth row of $\mathbb{X}$, and assume that without the loss of generality $\mathbb{X}$ has been column centered. Li et al. (2016) incorporate the supervisory information by using a multivariate linear regression on the centered *supervision data* $\mathbf{X} \in \mathbb{R}^p$:

$$\mathbf{t}_j = \boldsymbol{\mu}_{\mathrm{t}} + \boldsymbol{\eta}\mathbf{X}_j + \mathbf{f}_j, \quad j = 1, \dots, n,$$

where $\boldsymbol{\eta} \in \mathbb{R}^{u \times p}$ is the unknown coefficient matrix, and the errors $\mathbf{f}_j$ are independent copies of a $N(0, \boldsymbol{\Sigma}_{\mathrm{f}})$ vector. This model can be expressed in matrix form as

$$\mathbf{T} = \mathbf{1}_n \boldsymbol{\mu}_{\mathrm{t}}^T + \mathbb{X}\boldsymbol{\eta}^T + \mathbf{F}, \tag{9.14}$$

where $\mathbf{F} \in \mathbb{R}^{n \times u}$ has rows $\mathbf{f}_j^T$. Substituting this model for $\mathbf{T}$ into (9.12), we get

$$\begin{aligned}\mathbb{Y} &= \mathbf{1}_n \boldsymbol{\alpha}^T + (\mathbf{1}_n \boldsymbol{\mu}_{\mathbf{t}}^T + \mathbb{X}\boldsymbol{\eta}^T + \mathbf{F})\mathbf{V}^T + \mathbf{E} \\ &= \mathbf{1}_n \boldsymbol{\mu}_{\mathbf{Y}}^T + \mathbb{X}\boldsymbol{\eta}^T \mathbf{V}^T + \mathbf{F}\mathbf{V}^T + \mathbf{E}. \end{aligned} \tag{9.15}$$

Let $\boldsymbol{\varepsilon}_j = \mathbf{V}\mathbf{f}_j + \mathbf{e}_j$, $\boldsymbol{\Sigma}_\varepsilon = \text{var}(\boldsymbol{\varepsilon}_j)$, and let $(\mathbf{V}, \mathbf{V}_0)$ be an orthogonal matrix. Then we can write the corresponding vector version as

$$\begin{aligned}\mathbf{Y}_j &= \boldsymbol{\mu}_{\mathbf{Y}} + \mathbf{V}\boldsymbol{\eta}\mathbf{X}_j + \mathbf{V}\mathbf{f}_j + \mathbf{e}_j \\ &= \boldsymbol{\mu}_{\mathbf{Y}} + \mathbf{V}\boldsymbol{\eta}\mathbf{X}_j + \boldsymbol{\varepsilon}_j, \quad j = 1, \ldots, n. \end{aligned} \tag{9.16}$$

Since $\boldsymbol{\Sigma}_\varepsilon = \mathbf{V}(\boldsymbol{\Sigma}_f + \sigma_e^2 \mathbf{I}_u)\mathbf{V}^T + \sigma_e^2 \mathbf{V}_0 \mathbf{V}_0^T$, this is a response envelope model with response $\mathbf{Y}$, predictors $\mathbf{X}$, coefficient matrix $\boldsymbol{\beta} = \mathbf{V}\boldsymbol{\eta}$, $\boldsymbol{\Omega} = \boldsymbol{\Sigma}_f + \sigma_e^2 \mathbf{I}_r$, $\boldsymbol{\Omega}_0 = \sigma_e^2 \mathbf{I}_{r-u}$, and $\mathbf{V}$ denoting a semi-orthogonal basis matrix for $\mathcal{E}_{\boldsymbol{\Sigma}_\varepsilon}(\boldsymbol{\beta})$. Accordingly, $\mathbb{Y}\mathbf{V}$ represents the material variation in $\mathbb{Y}$, and $\mathbb{Y}\mathbf{V}_0$ represents the immaterial variation. The maximum likelihood estimators arising from a fit of envelope model (9.16) are

$$\begin{aligned}\widehat{\mathcal{E}}_{\boldsymbol{\Sigma}_\varepsilon}(\boldsymbol{\beta}) &= \text{span}(\arg\min_{\mathbf{G}}[\log|\mathbf{G}^T \mathbf{S}_{\mathbf{Y}|\mathbf{X}} \mathbf{G}| + (r-u)\log\{\text{tr}(\mathbf{S}_{\mathbf{Y}} \mathbf{Q}_{\mathbf{G}})\}]) \\ \widehat{\boldsymbol{\eta}} &= \widehat{\mathbf{V}}^T \mathbb{Y}^T \mathbb{X}(\mathbb{X}^T \mathbb{X})^{-1} \\ \widehat{\boldsymbol{\beta}} &= \widehat{\mathbf{V}}\widehat{\boldsymbol{\eta}} \\ \widehat{\boldsymbol{\Omega}} &= \widehat{\mathbf{V}}^T \mathbf{S}_{\mathbf{Y}|\mathbf{X}} \widehat{\mathbf{V}} \\ \widehat{\sigma}_e^2 &= (r-u)^{-1}\text{tr}(\mathbf{S}_{\mathbf{Y}} \mathbf{Q}_{\widehat{\mathbf{V}}}), \end{aligned}$$

where $\widehat{\mathbf{V}}$ is any basis for $\widehat{\mathcal{E}}_{\boldsymbol{\Sigma}_\varepsilon}(\boldsymbol{\beta})$.

Although (9.16) is an envelope model, the primary goal is still to estimate the low-rank approximation $\mathbf{U}\mathbf{V}^T$ of the column centered version of $\mathbb{Y}$. Li et al. took the estimator of $\mathbf{V}$ from a fit of (9.16) based on a version of the EM algorithm. The matrix $\mathbf{U}$ is predicted by using

$$\widehat{\text{E}}(\mathbf{U} \mid \mathbb{Y}) = \mathbf{Q}_{\mathbf{1}_n} \mathbb{Y}\widehat{\mathbf{V}} - (\mathbf{Q}_{\mathbf{1}_n} \mathbb{Y}\widehat{\mathbf{V}} - \mathbb{X}\widehat{\boldsymbol{\eta}}^T)\widehat{\sigma}_e^2 \widehat{\boldsymbol{\Omega}}^{-1},$$

which is the estimator of $\text{E}(\mathbf{U} \mid \mathbb{Y})$ constructed by plugging in the estimators of the various unknowns from a fit of (9.16). The term $\mathbf{Q}_{\mathbf{1}_n} \mathbb{Y}\widehat{\mathbf{V}}$ is the estimated centered material part of $\mathbb{Y}$, and $\mathbf{Q}_{\mathbf{1}_n} \mathbb{Y}\widehat{\mathbf{V}} - \mathbb{X}\widehat{\boldsymbol{\eta}}^T$ is the residual matrix from the fit of the estimated material information on $\mathbb{X}$, as in model (9.15). In sum, $\widehat{\mathbf{U}\mathbf{V}^T} = \widehat{\text{E}}(\mathbf{U} \mid \mathbb{Y})\widehat{\mathbf{V}}^T$.

The proposed estimator of $\mathbf{U}\mathbf{V}^T$ depends in part on the assumed error structure in (9.13) and (9.14). For instance, a different envelope model and decomposition results from the assumption that the rows $\mathbf{e}_j^T$ of $\mathbf{E}$ are independent copies of a $N(0, \boldsymbol{\Sigma}_e)$ random vector. This alternative may give an estimated decomposition with a lower dimension since the errors have less structure than those of (9.13).

9.4 Sufficient Dimension Reduction

Sufficient dimension reduction (SDR) began in the mid-1990s with the goal of finding informative low-dimensional graphical representations of data $(Y_i, \mathbf{X}_i), i = 1, \ldots, n$, used to study the regression of a univariate response Y on stochastic predictors $\mathbf{X} \in \mathbb{R}^p$ without requiring a parsimonious parametric model (Cook 1998; Li 2018). It was motivated in part by the understanding that marginal plots of Y versus linear combinations of the predictors are generally imponderable when pursuing information about the full regression of Y on $\mathbf{X}$.

The general goal of SDR is to estimate a low-dimensional subspace $\mathcal{S}$ of the predictor space with the property that Y is independent of $\mathbf{X}$ given the projection $\mathbf{P}_{\mathcal{S}}\mathbf{X}$ of $\mathbf{X}$ onto $\mathcal{S}$ without prespecifying a parametric model. Subspaces with this property are called *dimension reduction subspaces*. If $\mathcal{S}$ is a dimension reduction subspace, then $\mathbf{P}_{\mathcal{S}}\mathbf{X}$ contains all of the information that $\mathbf{X}$ has about Y and, letting $\rho \in \mathbb{R}^{p \times d}$ with $d < p$ denote a basis for $\mathcal{S}$, a plot of Y versus $\rho^T\mathbf{X}$ could be used as a diagnostic guide to the regression. The following definition (Cook 2007) formalizes this idea and gives a basis for many standard linear SDR methods.

Definition 9.1 A projection $\mathbf{P}_{\mathcal{S}} : \mathbb{R}^p \mapsto \mathcal{S} \subseteq \mathbb{R}^p$ onto a q-dimensional subspace $\mathcal{S}$ is a sufficient linear reduction if it satisfies at least one of the following three statements:

(i) inverse reduction, $\mathbf{X} \mid (Y, \mathbf{P}_{\mathcal{S}}\mathbf{X}) \sim \mathbf{X} \mid \mathbf{P}_{\mathcal{S}}\mathbf{X}$,
(ii) forward reduction, $Y \mid \mathbf{X} \sim Y \mid \mathbf{P}_{\mathcal{S}}\mathbf{X}$,
(iii) joint reduction, $Y \perp\!\!\!\perp \mathbf{X} \mid \mathbf{P}_{\mathcal{S}}\mathbf{X}$.

The subspace $\mathcal{S}$ is then called a dimension reduction subspace.

Each of the three conditions in this definition conveys the idea that the reduction $\mathbf{P}_{\mathcal{S}}\mathbf{X}$ carries all the information that $\mathbf{X}$ has about Y, and consequently all the information available to estimate the conditional mean $\mathrm{E}(Y \mid \mathbf{X})$ and variance $\mathrm{var}(Y \mid \mathbf{X})$ functions. They are equivalent when $(Y, \mathbf{X})$ has a joint distribution. We are then free to determine a reduction inversely or jointly and pass it to the forward regression without additional structure; for instance, $\mathrm{E}(Y \mid \mathbf{X}) = \mathrm{E}(Y \mid \mathbf{P}_{\mathcal{S}}\mathbf{X})$. A *sufficient summary plot* of Y versus coordinates of the projection $\mathbf{P}_{\mathcal{S}}\mathbf{X}$ is often an effective tool for guiding the regression.

If $\mathcal{S}$ is a dimension reduction subspace and $\mathcal{S} \subseteq \mathcal{S}_1$, then $\mathcal{S}_1$ is also a dimension reduction subspace. Within the class of linear reductions, we would like to find the smallest dimension reduction subspace (Cook 1994, 1998; Li 2018).

Definition 9.2 The intersection of all dimension reduction subspaces, when it is itself a dimension reduction subspace, is called the central subspace, $\mathcal{S}_{Y|\mathbf{X}}$.

The central subspace does not always exist, but it does so under mild regularity conditions that we assume throughout this section (Cook 1998; Yin et al. 2008). This area is now widely known as *sufficient dimension reduction* because of the similarity between the driving condition $Y \perp\!\!\!\perp \mathbf{X} \mid \mathbf{P}_{S_{Y|\mathbf{X}}}\mathbf{X}$ and Fisher's fundamental notion of sufficiency. The name also serves to distinguish it from other approaches to dimension reduction.

The central subspace turned out to be an effective construct, and over the past 25 years, much work has been devoted to methods for estimating it, the first two methods being sliced inverse regression (Li 1991) and sliced average variance estimation (Cook and Weisberg 1991). These methods, like nearly all of the subsequent methods, require the so called linearity and constant covariance conditions on the marginal distribution of the predictors:

Linearity condition: $\mathrm{E}(\mathbf{X} \mid \boldsymbol{\eta}^T\mathbf{X})$ is a linear function of $\boldsymbol{\eta}^T\mathbf{X}$,
Constant covariance condition: $\mathrm{var}(\mathbf{X} \mid \boldsymbol{\eta}^T\mathbf{X})$ is a nonstochastic matrix,

where $\boldsymbol{\eta} \in \mathbb{R}^{p\times d}$ is a semi-orthogonal basis matrix for $S_{Y|\mathbf{X}}$. These conditions are required only at a basis for $S_{Y|\mathbf{X}}$ and not for all $\boldsymbol{\eta}$. Diaconis and Freedman (1984) and Hall and Li (1993) showed that almost all projections of high-dimensional data are approximately normal, which is often used as partial justification for these conditions. Of the two conditions, the linearity condition seems to be used most often. The following proposition Cook (1998, Proposition 4.2) gives structure that arises from the linearity condition.

Proposition 9.1 *Let* $\mathbf{X} \in \mathbb{R}^p$ *be a random vector with mean* $\mathrm{E}(\mathbf{X}) = 0$ *and covariance matrix* $\boldsymbol{\Sigma}_\mathbf{X} > 0$. *Let* $\boldsymbol{\alpha} \in \mathbb{R}^{p\times q}$, $p \geq q$, *be full rank. Assume that* $\mathrm{E}(\mathbf{X} \mid \boldsymbol{\alpha}^T\mathbf{X} = \mathbf{u}) = \mathbf{M}\mathbf{u}$ *for some nonstochastic matrix* $\mathbf{M} \in \mathbb{R}^{p\times q}$. *Then*

1. $\mathbf{M} = \boldsymbol{\Sigma}_\mathbf{X}\boldsymbol{\alpha}(\boldsymbol{\alpha}^T\boldsymbol{\Sigma}_\mathbf{X}\boldsymbol{\alpha})^{-1}$.
2. $\mathbf{M}^T$ *is a generalized inverse of* $\boldsymbol{\alpha}$.
3. $\boldsymbol{\alpha}\mathbf{M}^T$ *is the orthogonal projection operator for* $\mathrm{span}(\boldsymbol{\alpha})$ *relative to the* $\boldsymbol{\Sigma}_\mathbf{X}$ *inner product.*

Perhaps the most important consequence of this proposition is that if the linearity condition holds for $\mathrm{E}(\mathbf{X} \mid \boldsymbol{\alpha}^T\mathbf{X})$, then we have that

$$\mathrm{E}(\mathbf{X} \mid \boldsymbol{\alpha}^T\mathbf{X}) - \mathrm{E}(\mathbf{X}) = \mathbf{P}^T_{\boldsymbol{\alpha}(\boldsymbol{\Sigma}_\mathbf{X})}(\mathbf{X} - \mathrm{E}(\mathbf{X})). \tag{9.17}$$

See Cook (1998) and Li and Wang (2007) for additional discussion.

Although the linearity and constant covariance conditions are largely seen as mild, they are essentially uncheckable, and thus can be a nag in application. Ma and Zhu (2012) developed a semi-parametric approach that allows modifications of methods so they no longer depend on these conditions. The fundamental restriction to linear reduction $\mathbf{P}_{S_{Y|\mathbf{X}}}\mathbf{X}$ is also a potential limitation. Lee et al. (2013) extended the foundations of sufficient dimension reduction to allow for nonlinear reduction, and Li and Song (2017) developed foundations and methodology for nonlinear sufficient dimension reduction for functional data. These breakthroughs, like that from Ma and Zhu, opened new frontiers in dimension reduction that promise further significant advances. Although SDR methods were originally developed as comprehensive graphical diagnostics, they are now serviceable outside of that context.

The envelope methods described so far in this book require a parametric model or a method to estimate a parametric characteristic of the regression, while most SDR methods are designed specifically to avoid the need for a prespecified model. This distinction has both advantages and disadvantages depending on the context. For instance, if we are dealing with a univariate linear regression, $Y = \alpha + \boldsymbol{\beta}^T\mathbf{X} + \varepsilon$, then $Y \perp\!\!\!\perp \mathbf{X} \mid \boldsymbol{\beta}^T\mathbf{X}$, so SDR leads us back to the estimation of $\boldsymbol{\beta}$, $\mathcal{S}_{Y|\mathbf{X}} = \text{span}(\boldsymbol{\beta})$, while envelopes can offer substantial gain as described in Chapter 4. On the other hand, if we wish to study the regression of Y on $\mathbf{X}$ without specifying a model or estimator of a specific characteristic of the regression, then envelopes at present have little to offer, while SDR methods are applicable. Nevertheless, it is possible to adapt envelopes for use in SDR. The motivating condition underlying SDR is $Y \perp\!\!\!\perp \mathbf{X} \mid \mathbf{P}_{\mathcal{S}}\mathbf{X}$. If we require in addition that $\mathbf{P}_{\mathcal{S}}\mathbf{X} \perp\!\!\!\perp \mathbf{Q}_{\mathcal{S}}\mathbf{X}$, then we arrive at an envelope-type structure that is analogous to that described in Chapter 4 for predictor reduction.

We review sliced inverse regression (SIR; Li 1991) in Section 9.5 and then return to envelopes.

9.5 Sliced Inverse Regression

9.5.1 SIR Methodology

Let $\boldsymbol{\Sigma}_{\mathbf{X}} = \text{var}(\mathbf{X})$, $\boldsymbol{\nu}(y) = \text{E}(\mathbf{X} \mid Y = y) - \text{E}(\mathbf{X})$, and $\boldsymbol{\xi}(y) = \boldsymbol{\Sigma}_{\mathbf{X}}^{-1}\boldsymbol{\nu}(y)$ and recall that $\boldsymbol{\eta} \in \mathbb{R}^{p\times d}$ is a semi-orthogonal basis matrix for $\mathcal{S}_{Y|\mathbf{X}}$. SIR, like many of the early SDR methods, is based on the following result.

Proposition 9.2 *Assume that linearity condition holds. Then for all y in the sample space of Y*

$$\boldsymbol{\nu}(y) = \mathbf{P}^T_{\boldsymbol{\eta}(\boldsymbol{\Sigma}_{\mathbf{X}})}\boldsymbol{\nu}(y).$$

Consequently, $\boldsymbol{\xi}(y) = \mathbf{P}_{\boldsymbol{\eta}(\boldsymbol{\Sigma}_{\mathbf{X}})}\boldsymbol{\xi}(y) \in \mathcal{S}_{Y|\mathbf{X}}$ *for all y.*

Using this result, we should in principle be able to construct an estimator of $\mathcal{S}_{Y|\mathbf{X}}$ if we can find an estimator of $\boldsymbol{\xi}(y)$. If Y is discrete or categorical, a sample version of $\boldsymbol{\xi}(y)$ can be constructed by substituting sample versions of $\boldsymbol{\Sigma}_{\mathbf{X}}$, $\text{E}(\mathbf{X} \mid Y = y)$, and $\text{E}(\mathbf{X})$. When Y is continuous, SIR replaces Y with a discrete version constructed by slicing the range of Y into h slices. Let

$\mathbf{v}_s = \{\text{E}(\mathbf{X} \mid J_s(Y) = 1) - \text{E}(\mathbf{X})\}$, where $J_s(Y)$ is the indicator function for slice s. Then $\boldsymbol{\xi}_s := \boldsymbol{\Sigma}_{\mathbf{X}}^{-1}\mathbf{v}_s \in \mathcal{S}_{Y|\mathbf{X}}, s = 1, \dots, h$, which can again be estimated by substituting sample moments. The fundamental problem is then to estimate $\text{span}(\boldsymbol{\xi}_1, \dots, \boldsymbol{\xi}_h) \subseteq \mathcal{S}_{Y|\mathbf{X}}$, assuming that $h \geq d$, by using the moment estimates of the $\boldsymbol{\xi}_s$'s. For a given dimension $d = \dim(\mathcal{S}_{Y|\mathbf{X}})$ and assuming that Y is categorical possibly induced by slicing, the typical estimation algorithm is as follows.

1. Standardize $\mathbf{X}$ to $\widehat{\mathbf{Z}} = \mathbf{S}_{\mathbf{X}}^{-1/2}(\mathbf{X} - \bar{\mathbf{X}})$.
2. Form $\widehat{\text{var}}\{\widehat{\text{E}}(\widehat{\mathbf{Z}} \mid Y)\}$.
3. Set $\widehat{\mathcal{S}}_{Y|\mathbf{X}} = \mathbf{S}_{\mathbf{X}}^{-1/2} \times$ span of the first d eigenvectors of $\widehat{\text{var}}\{\widehat{\text{E}}(\widehat{\mathbf{Z}} \mid Y)\}$.
4. Select a basis $\widehat{\boldsymbol{\eta}}$ for $\widehat{\mathcal{S}}_{Y|\mathbf{X}}$ and construct a display of Y vs $\widehat{\boldsymbol{\eta}}^T\mathbf{X}$.

The dimension d of $\mathcal{S}_{Y|\mathbf{X}}$ is often selected based on a sequence of hypothesis tests: starting with $d_0 = 0$, test the hypothesis $d = d_0$. If the hypothesis is rejected, increment d_0 by 1 and test again. If not, stop and use d_0 as the value of d. The test statistic for these hypotheses is $\hat{\Lambda}_n = n\sum_{j=d_0+1}^{p} \hat{\varphi}_j$, where the $\hat{\varphi}_j$s are the eigenvalues of $\widehat{\text{var}}\{\widehat{\text{E}}(\widehat{\mathbf{Z}} \mid Y)\}$ ordered from largest to smallest. Under the hypothesis $d = d_0$, $\hat{\Lambda}_n$ converges in distribution to $\sum_{k=1}^{(p-d_0)(h-d_0)} \omega_k C_k$, where the C_ks are independent copies of χ_1^2 and the ω_ks are eigenvalues of a certain covariance matrix (Cook 1998, Proposition 11.4). If the constant covariance condition holds then $\hat{\Lambda}_n$ converges to a chi-square random variable with $(p - d_0)(h - d_0)$ degrees of freedom (Li 1991).

9.5.2 Mussels' Muscles: Sliced Inverse Regression

Generally, for a simple illustration of SIR and SDR methods, we consider the data on horse mussels introduced in Section 4.4.4, but now we take the response to be muscle mass M to allow nonlinearity in the mean function. The predictors $\mathbf{X}$ are still the logarithms of the length, width, height, and mass of the mussel's shell. Fitting with $h = 8$ slices, SIR's dimension tests indicated clearly that $d = 1$, so only a single linear combination of the four predictors is needed to describe the regression of M on $\mathbf{X}$, which is consistent with the finding in Section 4.4.4. Shown in Figure 9.1 is a plot of M versus $\widehat{\boldsymbol{\eta}}^T\mathbf{X}$, the estimated reduction of $\mathbf{X}$ with $\|\widehat{\boldsymbol{\eta}}\| = 1$. The curve on the plot is the estimated mean function from using ordinary least squares to fit a single-index cubic polynomial to the points on the plot; that is, from fitting

$$Y = \mu + \beta_1\widehat{\boldsymbol{\eta}}^T\mathbf{X} + \beta_2(\widehat{\boldsymbol{\eta}}^T\mathbf{X})^2 + \beta_3(\widehat{\boldsymbol{\eta}}^T\mathbf{X})^3 + \epsilon.$$

Standard inferences based on this model will likely be optimistic, since they do not take the variation in $\widehat{\boldsymbol{\eta}}$ into account. To avoid this issue, we can fit the model

$$Y = \mu + \beta_1\boldsymbol{\eta}^T\mathbf{X} + \beta_2(\boldsymbol{\eta}^T\mathbf{X})^2 + \beta_3(\boldsymbol{\eta}^T\mathbf{X})^3 + \epsilon$$

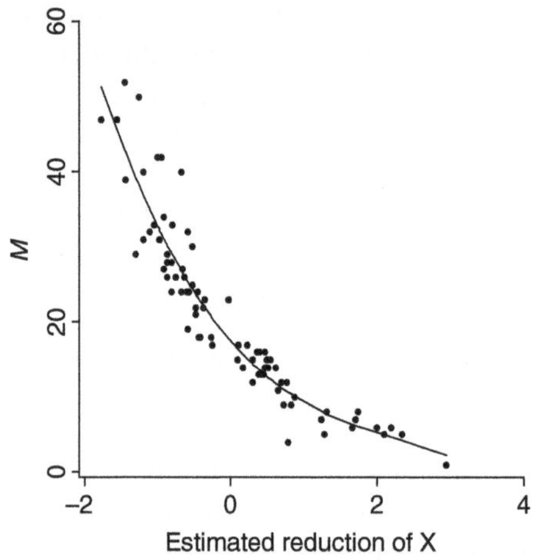

Figure 9.1 Mussels muscles: SIR summary plot of M versus the estimated reduction of $\mathbf{X}$, $\hat{\boldsymbol{\eta}}^T\mathbf{X}$.

with the constraint $\|\boldsymbol{\eta}\| = 1$. This serves to reestimate $\boldsymbol{\eta}$ given the model form inferred from the summary plot. To avoid the constraint $\|\boldsymbol{\eta}\| = 1$, reparameterize the previous model in terms of $\boldsymbol{\alpha} = \boldsymbol{\eta}\beta_1$, $\gamma_2 = \beta_2/\beta_1^2$ and $\gamma_3 = \beta_3/\beta_1^3$, assuming that $\beta_1 \neq 0$:

$$Y = \mu + \boldsymbol{\alpha}^T\mathbf{X} + \gamma_2(\boldsymbol{\alpha}^T\mathbf{X})^2 + \gamma_3(\boldsymbol{\alpha}^T\mathbf{X})^3 + \epsilon.$$

This model can now be fitted and studied using traditional methods for nonlinear models.

Generally, inspection of a low-dimensional summary plot following an SDR reduction can lead to a model specification in the form $Y = f(\boldsymbol{\eta}_1^T\mathbf{X}, \dots, \boldsymbol{\eta}_d^T\mathbf{X}, \boldsymbol{\beta}) + \epsilon$, where f is a function of the SDR variates $\boldsymbol{\eta}_1^T\mathbf{X}, \dots, \boldsymbol{\eta}_d^T\mathbf{X}$ and model parameters $\boldsymbol{\beta}$.

9.5.3 The "Envelope Method"

Recall from our discussion in Section 9.5.1 that $\boldsymbol{\nu}(y) = \mathrm{E}(\mathbf{X} \mid Y = y) - \mathrm{E}(\mathbf{X})$, $\xi(y) = \boldsymbol{\Sigma}_{\mathbf{X}}^{-1}\boldsymbol{\nu}(y)$ and that under the linearity condition $\xi(y) \in \mathcal{S}_{Y|\mathbf{X}}$ for all y in the sample space of Y. This can also be described by saying that the random vector $\xi(Y)$ is in the central subspace almost surely. Guo et al. (2015), following terminology used by Li et al. (2010), referred to the smallest subspace $\mathcal{S} \subseteq \mathbb{R}^p$ to which $\xi(Y)$ belongs almost surely as an "envelope". When Y is discrete or forced to be so by slicing, the envelope according to this usage is simply $\boldsymbol{\Sigma}_{\mathbf{X}}^{-1}\mathrm{span}(\boldsymbol{\nu}_1, \dots, \boldsymbol{\nu}_h)$. To be clear, this use of "envelope" to denote one thing enveloping another differs from the one employed in this book. It is not really needed in the development of SIR, but may be useful in more complicated contexts. For instance, Li et al. (2010) used "Kronecker envelope" in reference to the smallest subspace that preserves a tensorial structure while enclosing an estimator of the central subspace, and Guo et al. (2015) used "direct sum envelope" to denote the smallest subspace that preserves a direct sum structure when performing groupwise dimension reduction. They identified this type of enveloping generically as the "envelope method." Again, this "envelope method" is not the one addressed in this book.

In the next section, we show that envelopes as used in this book have the potential to improve SDR estimation, and that they can be used in conjunction with approaches that use the "envelope method."

9.5.4 Envelopes and SIR

Let $\mathcal{V} = \text{span}(v_1, \dots, v_h)$. The linearity condition guarantees only that $\boldsymbol{\Sigma}_{\mathbf{X}}^{-1}\mathcal{V} \subseteq \mathcal{S}_{Y|\mathbf{X}}$. We assume for simplicity that $\mathcal{S}_{Y|\mathbf{X}} = \boldsymbol{\Sigma}_{\mathbf{X}}^{-1}\mathcal{V}$. SIR's estimator of $\mathcal{S}_{Y|\mathbf{X}}$ might be improved by using $\mathcal{E}_{\boldsymbol{\Sigma}_{\mathbf{X}}}(\mathcal{V})$, the $\boldsymbol{\Sigma}_{\mathbf{X}}$-envelope of $\mathcal{V}$. To see this possibility, let $\boldsymbol{\Gamma} \in \mathbb{R}^{p\times u}$ be a basis for $\mathcal{E}_{\boldsymbol{\Sigma}_{\mathbf{X}}}(\mathcal{V})$, and let $(\boldsymbol{\Gamma}, \boldsymbol{\Gamma}_0) \in \mathbb{R}^{p\times p}$ be an orthogonal matrix. Then, as in other envelope contexts,

$$\begin{aligned}
\boldsymbol{\Sigma}_{\mathbf{X}} &= \mathbf{P}_{\mathcal{E}}\boldsymbol{\Sigma}_{\mathbf{X}}\mathbf{P}_{\mathcal{E}} + \mathbf{Q}_{\mathcal{E}}\boldsymbol{\Sigma}_{\mathbf{X}}\mathbf{Q}_{\mathcal{E}} = \boldsymbol{\Gamma}\boldsymbol{\Omega}\boldsymbol{\Gamma}^T + \boldsymbol{\Gamma}_0\boldsymbol{\Omega}_0\boldsymbol{\Gamma}_0^T \\
\boldsymbol{\Sigma}_{\mathbf{X}}^{-1} &= \boldsymbol{\Gamma}\boldsymbol{\Omega}^{-1}\boldsymbol{\Gamma}^T + \boldsymbol{\Gamma}_0\boldsymbol{\Omega}_0^{-1}\boldsymbol{\Gamma}_0^T \\
\mathcal{S}_{Y|\mathbf{X}} &= \boldsymbol{\Sigma}_{\mathbf{X}}^{-1}\mathcal{V} = \boldsymbol{\Gamma}\boldsymbol{\Omega}^{-1}\boldsymbol{\Gamma}^T\mathcal{V} = \boldsymbol{\Gamma}(\boldsymbol{\Gamma}^T\boldsymbol{\Sigma}_{\mathbf{X}}\boldsymbol{\Gamma})^{-1}\boldsymbol{\Gamma}^T\mathcal{V}.
\end{aligned}$$

Consequently, we may be able to improve the SIR estimator by using the envelope variant $\widehat{\mathcal{S}}_{Y|\mathbf{X}} = \widehat{\boldsymbol{\Gamma}}(\widehat{\boldsymbol{\Gamma}}^T\mathbf{S}_{\mathbf{X}}\widehat{\boldsymbol{\Gamma}})^{-1}\widehat{\boldsymbol{\Gamma}}^T\widehat{\mathcal{V}}$, where $\widehat{\boldsymbol{\Gamma}}$ is an estimator of a semi-orthogonal basis matrix for $\mathcal{E}_{\boldsymbol{\Sigma}_{\mathbf{X}}}(\mathcal{V})$, and $\widehat{\mathcal{V}}$ is an estimator of $\mathcal{V}$. The rationale here is similar to that for predictor envelopes discussed in Chapter 4. The envelope algorithms discussed in Chapter 6 could be used to form $\widehat{\boldsymbol{\Gamma}}$.

Restricting consideration to $d = 1$, Li et al. (2007) estimated $\mathcal{E}_{\boldsymbol{\Sigma}_{\mathbf{X}}}(\mathcal{V})$ by using a Krylov matrix (see Section 6.5.2) because the resulting method does not require inverting $\mathbf{S}_{\mathbf{X}}$ and is thus serviceable when $n < p$. This treatment was extended by Cook et al. (2007) to allow for $d > 1$ and by Yoo (2013) and by Yoo and Im (2014) to allow for multivariate responses. Cook et al. (2013) compared the use of Krylov matrices with other methods of envelope construction.

9.6 Dimension Reduction for the Conditional Mean

Most of the work in sufficient dimension reduction has pursued estimation of the central subspace, which gives a comprehensive characterization of how Y depends on $\mathbf{X}$. But there are also methods for pursuing narrower aspects of a regression. Cook and Li (2002) introduced the central mean subspace (CMS), which is designed for dimension reduction targeted at the conditional mean $\mathrm{E}(Y \mid \mathbf{X})$, again without requiring a prespecified parametric model.

Definition 9.3 A q-dimensional subspace $\mathcal{S} \subseteq \mathbb{R}^p$ is a mean dimension-reduction subspace for the regression of Y on $\mathbf{X}$ if $Y \perp\!\!\!\perp \mathrm{E}(Y \mid \mathbf{X}) \mid \mathbf{P}_{\mathcal{S}}\mathbf{X}$.

The intuition behind this definition is that the projection $\mathbf{P}_{\mathcal{S}}\mathbf{X}$ carries all of the information that $\mathbf{X}$ has about the conditional mean $\mathrm{E}(Y \mid \mathbf{X})$. Let $\boldsymbol{\alpha} \in \mathbb{R}^{p\times q}$ be a basis for a mean dimension reduction subspace $\mathcal{S}$. Then if $\mathcal{S}$ were known, we might expect that $\mathrm{E}(Y \mid \mathbf{X}) = \mathrm{E}(Y \mid \boldsymbol{\alpha}^T\mathbf{X})$, thus reducing the dimension of $\mathbf{X}$ for the purpose of estimating the conditional mean. This expectation is confirmed by the following proposition (Cook and Li 2002).

Proposition 9.3 *The following statements are equivalent.*

(i) $Y \perp\!\!\!\perp \mathrm{E}(Y \mid \mathbf{X}) \mid \boldsymbol{\alpha}^T\mathbf{X}$,
(ii) $\mathrm{cov}\{(Y, \mathrm{E}(Y \mid \mathbf{X})) \mid \boldsymbol{\alpha}^T\mathbf{X}\} = 0$,
(iii) $\mathrm{E}(Y \mid \mathbf{X})$ *is a function of* $\boldsymbol{\alpha}^T\mathbf{X}$.

Statement (i) is equivalent to Definition 9.4, although here it is stated in terms of a basis $\boldsymbol{\alpha}$. Statement (ii) says that the conditional covariance between Y and $\mathrm{E}(Y \mid \mathbf{X})$ is 0, and statement (iii) tells us that $\mathrm{E}(Y \mid \mathbf{X}) = \mathrm{E}(Y \mid \boldsymbol{\alpha}^T\mathbf{X})$. As in our discussion of the central subspace, most of the parsimonious reduction is provided by the smallest mean dimension-reduction subspace:

Definition 9.4 Let $\mathcal{S}_{\mathrm{E}(Y|\mathbf{X})}$ denote the intersection of all mean dimension-reduction subspaces. If $\mathcal{S}_{\mathrm{E}(Y|\mathbf{X})}$ is itself a mean dimension-reduction subspace, then it is called the CMS.

The CMS does not always exist, but it does exist under mild conditions that should not be worrisome in practice (Cook and Li 2002). We assume existence of the CMS throughout this discussion and then $\mathcal{S}_{\mathrm{E}(Y|\mathbf{X})} \subseteq \mathcal{S}_{Y|\mathbf{X}}$.

9.6.1 Estimating One Vector in $\mathcal{S}_{\mathrm{E}(Y|\mathbf{X})}$

There are many methods based on a convex objective function that can be used to estimate the CMS when it is one dimensional. With $\mathbf{b} \in \mathbb{R}^p$, let

$$L(a + \mathbf{b}^T\mathbf{X}, Y) = -Y(a + \mathbf{b}^T\mathbf{X}) + \phi(a + \mathbf{b}^T\mathbf{X})$$

be an objective function based on the natural exponential family for strictly convex ϕ. For instance, to obtain ordinary least squares estimators, we set $\phi(x) = x^2/2$. Also, let

$$(\boldsymbol{\alpha}, \boldsymbol{\beta}) = \arg\min_{a,\mathbf{b}} \mathrm{E}\{L(a + \mathbf{b}^T\mathbf{X}, Y)\},$$

where the expectation is with respect to both Y and $\mathbf{X}$. This use of an objective function is not meant to imply that any associated model is true or even provides an adequate fit of the data. Nevertheless, there is still a close connection between $\boldsymbol{\beta}$ and $\mathcal{S}_{\mathrm{E}(Y|\mathbf{X})}$ as described in the following proposition whose proof may provide some intuition (Cook and Li 2002).

Proposition 9.4 *Let $\boldsymbol{\gamma} \in \mathbb{R}^{p\times q}$ be a basis matrix for $\mathcal{S}_{\mathrm{E}(Y|\mathbf{X})}$, and assume that* $\mathrm{E}(\mathbf{X} \mid \boldsymbol{\gamma}^T\mathbf{X})$ *satisfies the linearity condition. Then $\boldsymbol{\beta} \in \mathcal{S}_{\mathrm{E}(Y|\mathbf{X})}$.*

Proof: Let $R(a, \mathbf{b}) = \mathrm{E}\{L(a + \mathbf{b}^T\mathbf{X}, Y)\}$. Then

$$
\begin{aligned}
R(a,\mathbf{b}) &= \mathrm{E}\{-Y(a+\mathbf{b}^T\mathbf{X}) + \phi(a+\mathbf{b}^T\mathbf{X})\}\\
&= \mathrm{E}\{-\mathrm{E}(Y \mid \boldsymbol{\gamma}^T\mathbf{X})(a+\mathbf{b}^T\mathbf{X}) + \phi(a+\mathbf{b}^T\mathbf{X})\}\\
&= \mathrm{E}_{\boldsymbol{\gamma}^T\mathbf{X}}\mathrm{E}_{\mathbf{X}|\boldsymbol{\gamma}^T\mathbf{X}}\{-\mathrm{E}(Y \mid \boldsymbol{\gamma}^T\mathbf{X})(a+\mathbf{b}^T\mathbf{X}) + \phi(a+\mathbf{b}^T\mathbf{X})\}\\
&> \mathrm{E}_{\boldsymbol{\gamma}^T\mathbf{X}}\{-\mathrm{E}(Y \mid \boldsymbol{\gamma}^T\mathbf{X})(a+\mathbf{b}^T\mathrm{E}(\mathbf{X} \mid \boldsymbol{\gamma}^T\mathbf{X})) + \phi(a+\mathbf{b}^T\mathrm{E}(\mathbf{X} \mid \boldsymbol{\gamma}^T\mathbf{X}))\}\\
&= \mathrm{E}\{-Y(a+\mathbf{b}^T\mathrm{E}(\mathbf{X} \mid \boldsymbol{\gamma}^T\mathbf{X})) + \phi(a+\mathbf{b}^T\mathrm{E}(\mathbf{X} \mid \boldsymbol{\gamma}^T\mathbf{X}))\}\\
&= \mathrm{E}\{-Y(a^*+\mathbf{b}^T\mathbf{P}^T_{\boldsymbol{\gamma}(\boldsymbol{\Sigma}_\mathbf{X})}\mathbf{X}) + \phi(a^*+\mathbf{b}^T\mathbf{P}^T_{\boldsymbol{\gamma}(\boldsymbol{\Sigma}_\mathbf{X})}\mathbf{X})\},
\end{aligned}
$$

where $a^* = a + \mathbf{b}^T\mathbf{P}^T_{\boldsymbol{\gamma}(\boldsymbol{\Sigma}_\mathbf{X})}\mathrm{E}(\mathbf{X})$. The second equality follows because $\boldsymbol{\gamma}$ is a basis for $\mathcal{S}_{\mathrm{E}(Y|\mathbf{X})}$. The third equality is just for clarity of the next step that follows because ϕ is strictly convex. The final equality follows from the linearity condition and Proposition 9.1. As a consequence, we see that if $\mathbf{b}$ is not in $\mathcal{S}_{\mathrm{E}(Y|\mathbf{X})}$, then $\min_{a,\mathbf{b}} R(a, \mathbf{b}) > \min_{a,\mathbf{b}} R(a, \mathbf{P}_{\boldsymbol{\gamma}(\boldsymbol{\Sigma}_\mathbf{X})}\mathbf{b})$ and thus $\boldsymbol{\beta} \in \mathcal{S}_{\mathrm{E}(Y|\mathbf{X})}$. □

If the regression satisfies the linearity condition, then $\boldsymbol{\beta} \in \mathcal{S}_{\mathrm{E}(Y|\mathbf{X})}$ even if $q > 1$. This says in effect that we can estimate a vector in the CMS by minimizing the sample version $n^{-1}\sum_{i=1}^n L(a + \mathbf{b}^T\mathbf{X}_i, Y_i)$ of $\mathrm{E}\{L(a + \mathbf{b}^T\mathbf{X}, Y)\}$. Envelopes can come into play by combining this result with the envelope extensions discussed in Section 7.1 to achieve more efficient estimation of $\boldsymbol{\beta}$. A more immediate connection with envelopes is described in Section 9.6.2.

9.6.2 Estimating $\mathcal{S}_{\mathrm{E}(Y|\mathbf{X})}$

A little buildup is needed before describing the method of estimating $\mathcal{S}_{\mathrm{E}(Y|\mathbf{X})}$ proposed by Cook and Li (2002) and its connection with envelopes. Let $q = \dim(\mathcal{S}_{\mathrm{E}(Y|\mathbf{X})})$ and define the standardized predictor $\mathbf{Z} = \boldsymbol{\Sigma}_\mathbf{X}^{-1/2}(\mathbf{X} - \mathrm{E}(\mathbf{X}))$. The CMS is equivariant in the sense that $\mathcal{S}_{\mathrm{E}(Y|\mathbf{X})} = \boldsymbol{\Sigma}_\mathbf{X}^{-1/2}\mathcal{S}_{\mathrm{E}(Y|\mathbf{Z})}$, so there is no loss of generality when working with the standardized predictor. Also define the standardized regression coefficients $\boldsymbol{\beta}_{Y\mathbf{Z}} = \mathrm{E}(Y\mathbf{Z})$ and the third moment matrix $\boldsymbol{\Sigma}_{Y\mathbf{Z}\mathbf{Z}} = \mathrm{E}\{(Y - \mathrm{E}(Y))\mathbf{Z}\mathbf{Z}^T\}$.

The following proposition shows one way to get CMS vectors in the population. We again present its proof to aid understanding.

Proposition 9.5 *Let $\boldsymbol{\gamma}$ be a basis matrix for $\mathcal{S}_{\mathrm{E}(Y|\mathbf{Z})}$. If the linearity condition holds for* $\mathrm{E}(\mathbf{Z} \mid \boldsymbol{\gamma}^T\mathbf{Z})$ *and if* $\mathrm{var}(\mathbf{Z} \mid \boldsymbol{\gamma}^T\mathbf{Z})$ *is uncorrelated with Y, then* $\mathrm{span}(\boldsymbol{\beta}_{Y\mathbf{Z}}, \boldsymbol{\Sigma}_{Y\mathbf{Z}\mathbf{Z}}) \subseteq \mathcal{S}_{\mathrm{E}(Y|\mathbf{Z})}$.

Proof: The proof follows by direct calculation.

$$\begin{aligned}
\boldsymbol{\Sigma}_{Y\mathbf{ZZ}} &= \mathrm{E}_{Y,\boldsymbol{\gamma}^T\mathbf{Z}}\mathrm{E}_{\mathbf{Z}|(Y,\boldsymbol{\gamma}^T\mathbf{Z})}\{(Y-\mathrm{E}(Y))\mathbf{ZZ}^T\} \\
&= \mathrm{E}\{(Y-\mathrm{E}(Y))\mathrm{E}(\mathbf{ZZ}^T \mid \boldsymbol{\gamma}^T\mathbf{Z})\} \\
&= \mathrm{E}\{(Y-\mathrm{E}(Y))(\mathrm{E}(\mathbf{Z} \mid \boldsymbol{\gamma}^T\mathbf{Z})\mathrm{E}(\mathbf{Z}^T \mid \boldsymbol{\gamma}^T\mathbf{Z}) + \mathrm{var}(\mathbf{Z} \mid \boldsymbol{\gamma}^T\mathbf{Z})\} \\
&= \mathrm{E}\{(Y-\mathrm{E}(Y))(\mathbf{P}_{\mathcal{S}_{\mathrm{E}(Y|\mathbf{Z})}}\mathbf{ZZ}^T\mathbf{P}_{\mathcal{S}_{\mathrm{E}(Y|\mathbf{Z})}} + \mathrm{var}(\mathbf{Z} \mid \boldsymbol{\gamma}^T\mathbf{Z})\} \\
&= \mathrm{E}\{(Y-\mathrm{E}(Y))\mathbf{P}_{\mathcal{S}_{\mathrm{E}(Y|\mathbf{Z})}}\mathbf{ZZ}^T\mathbf{P}_{\mathcal{S}_{\mathrm{E}(Y|\mathbf{Z})}}\} \\
&= \mathrm{E}\{\mathbf{P}_{\mathcal{S}_{\mathrm{E}(Y|\mathbf{Z})}}\boldsymbol{\Sigma}_{Y\mathbf{ZZ}}\mathbf{P}_{\mathcal{S}_{\mathrm{E}(Y|\mathbf{Z})}}\},
\end{aligned}$$

where the third equality follows from the linearity condition and the fourth from the requirement that Y and $\mathrm{var}(\mathbf{Z} \mid \boldsymbol{\gamma}^T\mathbf{Z})$ are uncorrelated. The result here shows that $\mathrm{span}(\boldsymbol{\Sigma}_{Y\mathbf{ZZ}}) \subseteq \mathcal{S}_{\mathrm{E}(Y|\mathbf{Z})}$. The conclusion follows because we know from Proposition 9.4 that $\boldsymbol{\beta}_{Y\mathbf{Z}} \in \mathcal{S}_{\mathrm{E}(Y|\mathbf{Z})}$. □

Building methodology based on Proposition 9.5 would require the stated conditions on the conditional mean and the conditional variance. However, Cook and Li (2002) show also that, *under the linearity condition alone,*

$$\mathrm{span}(\boldsymbol{\Sigma}^j_{Y\mathbf{ZZ}}\boldsymbol{\beta}_{Y\mathbf{Z}}, j = 0, 1, \dots, p-1) \subseteq \mathcal{S}_{\mathrm{E}(Y|\mathbf{Z})}. \tag{9.18}$$

They then used this and a Krylov construction as a basis for methodology – called iterative Hessian transformations – by inferring about the rank of $\widehat{\mathbf{K}}\widehat{\mathbf{K}}^T$, where

$$\widehat{\mathbf{K}} = (\widehat{\boldsymbol{\beta}}_{Y\mathbf{Z}}, \widehat{\boldsymbol{\Sigma}}_{Y\mathbf{ZZ}}\widehat{\boldsymbol{\beta}}_{Y\mathbf{Z}}, \dots, \widehat{\boldsymbol{\Sigma}}^{p-1}_{Y\mathbf{ZZ}}\widehat{\boldsymbol{\beta}}_{Y\mathbf{Z}})$$

is a sample Krylov matrix. The span of the eigenvectors of $\widehat{\mathbf{K}}\widehat{\mathbf{K}}^T$ corresponding to its inferred nonzero eigenvalues is an estimator of $\mathcal{S}_{\mathrm{E}(Y|\mathbf{Z})}$. Premultiplying these eigenvectors by $\widehat{\boldsymbol{\Sigma}}_{\mathbf{X}}^{-1/2}$ then provides an estimator of $\mathcal{S}_{\mathrm{E}(Y|\mathbf{X})}$.

We know from the discussion of Section 6.5.2 that the subspace on the left-hand side of (9.18) equals the $\boldsymbol{\Sigma}_{Y\mathbf{ZZ}}$-envelope of $\mathrm{span}(\boldsymbol{\beta}_{Y\mathbf{Z}})$. Consequently,

Proposition 9.6 *Under the linearity condition for* $\mathrm{E}(\mathbf{Z} \mid \boldsymbol{\gamma}^T\mathbf{Z})$, *we have*

$$\mathcal{E}_{\boldsymbol{\Sigma}_{Y\mathbf{ZZ}}}(\mathrm{span}(\boldsymbol{\beta}_{Y\mathbf{Z}})) \subseteq \mathcal{S}_{\mathrm{E}(Y|\mathbf{Z})}.$$

The envelope algorithms discussed in Chapter 6 are all available for estimation of the subspace $\mathcal{E}_{\boldsymbol{\Sigma}_{Y\mathbf{ZZ}}}(\mathrm{span}(\boldsymbol{\beta}_{Y\mathbf{Z}}))$ of $\mathcal{S}_{\mathrm{E}(Y|\mathbf{Z})}$. In view of the Krylov construction associated with SIMPLS (Section 4.2.1) and the general superiority of envelope methods over SIMPLS (Section 4.3.2), we would expect envelope methodology to generally dominate the Krylov-based methodology originally proposed by Cook and Li (2002).

Several extensions and applications of the CMS have been developed since 2002, including the CMS for time series (Park et al. 2009), Fourier methods for estimation of the CMS (Zhu and Zeng 2006), extensions to multivariate regressions (Zhu and Wei 2015), locally efficient estimators (Ma and Zhu 2014), in addition to many others. Many of these methods could likely be improved by use of envelopes.

9.7 Functional Envelopes for SDR

We sketch in this section the main ideas behind the functional envelopes for SDR developed by Wang (2017) and Zhang et al. (2017).

9.7.1 Functional SDR

Consider the regression of a univariate response $Y \in \mathbb{R}$ on a random function $X(t)$, where t lies in a compact interval T and X is restricted to a real separable Hilbert space $\mathcal{H}$ with inner product $\langle f, g\rangle = \int_T f(t)g(t)\mathrm{d}t$ and norm $\|f\|_{\mathcal{H}} = \langle f, f\rangle^{1/2}$. Assume that X is centered, $\mathrm{E}\{X(t)\} = 0$ for all $t \in T$ and has finite fourth moment, $\int_T \mathrm{E}\{X^4(t)\}\mathrm{d}t < \infty$. Let $\Sigma_X(s,t) = \mathrm{E}\{X(s)X(t)\}$ denote the covariance operator, and let $\mathcal{B}(\mathcal{H})$ denote the set of all bounded linear operators from $\mathcal{H}$ to $\mathcal{H}$.

In the finite dimensional case discussed in Section 9.4, the central subspace $\mathcal{S}_{Y|\mathbf{X}}$, with semi-orthogonal basis matrix $\boldsymbol{\eta} \in \mathbb{R}^{p\times d}$, is the intersection of all subspaces $\mathcal{S}$ of $\mathbb{R}^p$ having the property that $Y \perp\!\!\!\perp \mathbf{X}|\mathbf{P}_{\mathcal{S}}\mathbf{X}$. Letting $\boldsymbol{\eta}_j$ denote the jth column of $\boldsymbol{\eta}$, this can be stated in terms of $\mathcal{S}_{Y|\mathbf{X}}$ as $Y \perp\!\!\!\perp \mathbf{X} \mid (\boldsymbol{\eta}_1^T\mathbf{X}, \dots, \boldsymbol{\eta}_d^T\mathbf{X})$. The corresponding goal in terms of the random function X is to find the smallest collection of linearly independent functions $\eta_j(t)$, $j = 1, \dots, d$, so that

$$Y \perp\!\!\!\perp X \mid (\langle \eta_1, X\rangle, \dots, \langle \eta_d, X\rangle),$$

with corresponding central subspace $\mathcal{S}_{Y|X} = \mathrm{span}(\eta_1, \dots, \eta_d)$. Although directed at reducing a functional predictor, this is still regarded as a linear reduction of X. This general setup has been developed for longitudinal data by Jiang et al. (2014) and Yao et al. (2015) (see also Hsing and Ren 2009). A complication that arises in these and other extensions of SDR to functional regressions is that the inverse covariance operator Σ_X^{-1} will not normally exist except under rather stringent conditions (Cook et al. 2010). In consequence, pursuing methodology based on straightforward extensions of the logic in Section 9.5.1 may not be reasonable because of the demanding conditions necessary to ensure functional analogs of finite-dimensional relationships like $\xi(y) = \boldsymbol{\Sigma}_{\mathbf{X}}^{-1}\nu(y)$.

9.7.2 Functional Predictor Envelopes

As in the finite-dimensional case, the definition of a reducing subspace for the regression of Y on the functional predictor X requires first a "functional reducing subspace." Reducing subspaces of bounded linear operators can be defined in a manner similar to Definition 1.1 for the finite-dimensional case (Wang 2017; Zhang et al. 2017):

Definition 9.5 Let $\mathcal{R} \subseteq \mathcal{H}$ be a subspace of $\mathcal{H}$ and let $M \in \mathcal{B}(\mathcal{H})$. Then $\mathcal{R}$ is a reducing subspace of M if it decomposes M as

$$M = P_{\mathcal{R}} M P_{\mathcal{R}} + Q_{\mathcal{R}} M Q_{\mathcal{R}},$$

where $P_{\mathcal{R}} = P_{\mathcal{R}}(t,s) \in \mathcal{B}(\mathcal{H})$ and $Q_{\mathcal{R}} = Q_{\mathcal{R}}(t,s) \in \mathcal{B}(\mathcal{H})$ are projections onto $\mathcal{R}$ and $\mathcal{R}^{\perp}$.

With this Zhang et al. (2017) formulated a definition of envelopes like Definition 1.2 for the finite-dimensional case:

Definition 9.6 Let $\mathcal{R} \subseteq \mathcal{H}$ be a subspace of $\mathcal{H}$ and let $M \in \mathcal{B}(\mathcal{H})$. Then the M-envelope of $\mathcal{R}$, denoted $\mathcal{E}_M(\mathcal{R})$, is the intersection of all reducing subspaces of M that contain $\mathcal{R}$.

Zhang et al. (2017) pursued SDR via $\mathcal{E}_{\Sigma_X}(\mathcal{S}_{Y|X})$, the smallest reducing subspace of Σ_X that contains the functional central subspace $\mathcal{S}_{Y|X}$ as represented in Section 9.7.1. Since $\mathcal{S}_{Y|X} \subseteq \mathcal{E}_{\Sigma_X}(\mathcal{S}_{Y|X})$, pursuing this functional envelope does not lose any information on the central subspace. Additionally, $\mathcal{E}_{\Sigma}(\mathcal{S}_{Y|X})$ has the following useful properties. Let $\gamma_1(t), \ldots, \gamma_u(t)$ denote linearly independent functions that form a basis for $\mathcal{E}_{\Sigma_X}(\mathcal{S}_{Y|X})$. Then $Y \perp\!\!\!\perp X \mid (\langle \gamma_1, X \rangle, \ldots, \langle \gamma_u, X \rangle)$ and $\langle \alpha, X \rangle$ and $\langle \alpha_0, X \rangle$ are uncorrelated for all $\alpha \in \mathcal{E}_{\Sigma_X}(\mathcal{S}_{Y|X})$ and $\alpha_0 \in \mathcal{E}^{\perp}_{\Sigma_X}(\mathcal{S}_{Y|X})$.

Let

$$\Lambda(s,t) = \mathrm{E}[\mathrm{E}\{X(s) \mid Y\}\mathrm{E}\{X(t) \mid Y\}]$$

denote the functional counterpart of $\mathrm{var}\{\mathrm{E}(\mathbf{X} \mid Y)\}$ in the finite-dimensional case and define the functional Krylov spaces

$$\mathcal{S}_k = \mathrm{span}(\Lambda, \Sigma_X \Lambda, \ldots, \Sigma_X^{k-1}\Lambda), \; k = 1, 2, \ldots.$$

Then Wang (2017) showed that there exists an integer K so that

$$\mathcal{S}_1 \subset \mathcal{S}_2 \subset \cdots \subset \mathcal{S}_K = \mathcal{E}_{\Sigma_X}(\mathcal{S}_{Y|X}) = \mathcal{S}_{K+1} = \ldots,$$

and used this result to develop a Krylov-based method for estimating $\mathcal{E}_{\Sigma_X}(\mathcal{S}_{Y|X})$, and subsequently an envelope-based method for estimating $\mathcal{S}_{Y|X}$.

In the next section, we return to the finite-dimensional setting.

9.8 Comparing Covariance Matrices

Most of the advances in SDR have been on estimation of the central subspace or the CMS, but there is SDR methodology for other multivariate problems as well. In this section, we consider the problem of comparing positive definite covariance matrices $\boldsymbol{\Sigma}_g = \mathrm{cov}(\mathbf{X} \mid g)$, $g = 1, \ldots, h$, of a random vector $\mathbf{X} \in \mathbb{R}^p$ observed in each of h populations. We assume that the data $\mathbf{X}_{gj}$ consist of n_g independent observations from population $g = 1, \ldots, h$, $j = 1, \ldots, n_g$. In practice, sampling may or may not be stratified by population, but in either case, we condition on the observed sample sizes, n_g. A regression with response Y and predictor $\mathbf{X}$ can be cast into this mold by slicing the response into h slices and then g becomes the slice counter. Let $n = \sum_{g=1}^{h} n_g$, let $\mathbf{S}_g$ denote the sample version of $\boldsymbol{\Sigma}_g$, and let $\mathbf{T}_g = n_g \mathbf{S}_g$ denote the sum of squares matrix for population $g = 1, \ldots, h$.

There are standard methods that can aid in comparing the $\boldsymbol{\Sigma}_g$'s, including tests for equality and proportionality (Muirhead 2005; Flury and Riedwyl 1988; Flury 1988a; Jensen and Madsen 2004). But more intricate methods are needed when such simple characterizations are insufficient. Perhaps the most common methods for comparing covariance matrices stem from spectral models for common principal components. Developed primarily by Flury (1988b, 1987), Flury and Riedwyl (1988), Schott (1991, 1999, 2003), and Boik (2002), many of these approaches are based on fitting versions of the general spectral structure

$$\boldsymbol{\Sigma}_g = \mathbf{V}_{1,g} \boldsymbol{\Phi}_{1,g} \mathbf{V}_{1,g}^T + \mathbf{V}_{2,g} \boldsymbol{\Phi}_{2,g} \mathbf{V}_{2,g}^T, \quad g = 1, \ldots, h, \tag{9.19}$$

where $\boldsymbol{\Phi}_{1,g} > 0$ and $\boldsymbol{\Phi}_{2,g} > 0$ are diagonal matrices of eigenvalues, and $(\mathbf{V}_{1,g}, \mathbf{V}_{2,g})$ is an orthogonal matrix of eigenvectors of $\boldsymbol{\Sigma}_g$ with $\mathbf{V}_{1,g} \in \mathbb{R}^{p \times q}$, $q \leq p - 1$, so the dimensions of $\mathbf{V}_{1,g}$ are constant across populations. If we further require $\mathbf{V}_{1,g}$ itself to be constant across groups, so $\mathbf{V}_{1,g} = \mathbf{V}_1$, $g = 1, \ldots, h$, then we get a model of partial common principal components (Flury 1987), the linear combinations $\mathbf{V}_1^T \mathbf{X}$ giving the principal components that are common to all h populations. If $q = p - 1$, then (9.19) reduces to the model of common principal components proposed by Flury (1988b). Requiring only that $\mathrm{span}(\mathbf{V}_{1,g}) = \mathrm{span}(\mathbf{V}_1)$, we arrive at a model of common subspaces (Flury 1987). Boik (2002) proposed perhaps the most comprehensive methodology by allowing the $\boldsymbol{\Sigma}_g$'s to share multiple eigenspaces without sharing eigenvectors and permitting sets of homogeneous eigenvalues. While these approaches can be useful for dimension reduction, their motivation seems to rest, not primarily with statistical reasoning, but with the convenience of spectral algebra.

We next turn to a rather different approach proposed by Cook and Forzani (2008a) for comparing covariance matrices. Their starting point is based on SDR, which differs from (9.19), and the consequent methodologies can produce quite different conclusions. However, we will see later in this section that envelopes provide a link between the approach proposed by Cook and Forzani (2008a) and (9.19).

9.8.1 SDR for Covariance Matrices

Let $(\boldsymbol{\alpha}, \boldsymbol{\alpha}_0)$ be an orthogonal matrix with $\boldsymbol{\alpha} \in \mathbb{R}^{p \times d}$. Our goal is now to find the fewest linear combinations of the predictors $\boldsymbol{\alpha}^T \mathbf{X}$ so that $\boldsymbol{\alpha}^T(\mathbf{X} - \mathrm{E}(\mathbf{X} \mid g))$

accounts for all differences in $\boldsymbol{\Sigma}_g$ across populations. More specifically, we require that for any two populations j and k,

$$\mathbf{T}_j \mid (\boldsymbol{\alpha}^T\mathbf{T}_j\boldsymbol{\alpha}, n_j = m) \sim \mathbf{T}_k \mid (\boldsymbol{\alpha}^T\mathbf{T}_k\boldsymbol{\alpha}, n_k = m). \tag{9.20}$$

This condition implies that, apart from differences in sample size, the quadratic reduction $\boldsymbol{\alpha}^T\mathbf{T}_j\boldsymbol{\alpha}$ is sufficient to account for all differences in the covariance matrices. As in our discussion of the central subspace $\mathcal{S}_{Y|\mathbf{X}}$ for regression, if (9.20) holds, then it also holds for any full rank transformation $\boldsymbol{\alpha}\mathbf{A}$ of the columns $\boldsymbol{\alpha}$. In consequence, this condition for covariance matrices is again a requirement on the subspace span($\boldsymbol{\alpha}$).

Cook and Forzani (2008a) developed corresponding methodology for covariance reduction (CORE) by assuming that the sample sum of squares matrices $\mathbf{T}_j$ follow independent Wishart distributions, in which case span($\boldsymbol{\alpha}$) satisfies (9.20) if and only if, in addition to n_g,

$$\mathbf{P}_{\alpha(\boldsymbol{\Sigma}_g)} \tag{9.21a}$$

and

$$\boldsymbol{\Sigma}_g\{\mathbf{I}_p - \mathbf{P}_{\alpha(\boldsymbol{\Sigma}_g)}\} \tag{9.21b}$$

are constant in g. When $\mathbf{X}$ is normal, $\text{var}(\mathbf{X} \mid \boldsymbol{\alpha}^T\mathbf{X}, g) = \boldsymbol{\Sigma}_g\{\mathbf{I}_p - \mathbf{P}_{\alpha(\boldsymbol{\Sigma}_g)}\}$, so requirement (b) means that the conditional variances are constant in g; that is,

$$\text{var}(\mathbf{X} \mid \boldsymbol{\alpha}^T\mathbf{X}, g = i) = \text{var}(\mathbf{X} \mid \boldsymbol{\alpha}^T\mathbf{X}, g = j), \quad i,j = 1, \dots, h.$$

The central subspace C also exists with Wishart matrices, since then the intersection of all subspaces span($\boldsymbol{\alpha}$) that satisfy (9.20) also satisfies (9.20), so we can now take C = span($\boldsymbol{\alpha}$). When the Wishart assumption fails, their methodology estimates a CMS for covariance matrices.

Let $\boldsymbol{\Sigma}_* = \sum_{h=1}^{g}(n_g/n)\boldsymbol{\Sigma}_g$ denote the pooled population covariance matrix. Conditions (9.21) give the essential population implications of (9.20). Without requiring Wishart matrices, these conditions are equivalent to the following four conditions for $g = 1, \dots, h$,

I. $\boldsymbol{\Sigma}_g^{-1}\boldsymbol{\alpha}_0 = \boldsymbol{\Sigma}_*^{-1}\boldsymbol{\alpha}_0$,
II. $\mathbf{P}_{\alpha(\boldsymbol{\Sigma}_g)} = \mathbf{P}_{\alpha(\boldsymbol{\Sigma}_*)}$ and $\boldsymbol{\Sigma}_g\{\mathbf{I}_p - \mathbf{P}_{\alpha(\boldsymbol{\Sigma}_g)}\} = \boldsymbol{\Sigma}_*\{\mathbf{I}_p - \mathbf{P}_{\alpha(\boldsymbol{\Sigma}_*)}\}$,
III. $\boldsymbol{\Sigma}_g = \boldsymbol{\Sigma}_* + \mathbf{P}^T_{\alpha(\boldsymbol{\Sigma}_*)}(\boldsymbol{\Sigma}_g - \boldsymbol{\Sigma}_*)\mathbf{P}_{\alpha(\boldsymbol{\Sigma}_*)}$,
IV. $\boldsymbol{\Sigma}_g^{-1} = \boldsymbol{\Sigma}_*^{-1} + \boldsymbol{\alpha}\{(\boldsymbol{\alpha}^T\boldsymbol{\Sigma}_g\boldsymbol{\alpha})^{-1} - (\boldsymbol{\alpha}^T\boldsymbol{\Sigma}_*\boldsymbol{\alpha})^{-1}\}\boldsymbol{\alpha}^T$.

The equivalent condition I says that $\boldsymbol{\Sigma}_*^{-1/2}C^\perp$ is an eigenspace of $\boldsymbol{\Sigma}_*^{-1/2}\boldsymbol{\Sigma}_g\boldsymbol{\Sigma}_*^{-1/2}$ with eigenvalue 1, $g = 1, \dots, h$. Condition II gives the constant values for (a) and (b) in (9.21). Condition III tells us that $\text{span}(\boldsymbol{\Sigma}_g - \boldsymbol{\Sigma}_*) \subseteq \text{span}(\boldsymbol{\Sigma}_*\boldsymbol{\alpha})$, and condition V shows the relationship in the inverse scale.

Let $\mathbf{S}_* = \sum_{g=1}^{h}(n_g/n)\mathbf{S}_g$ and

$$L_d(\mathcal{S}) = n\log|\mathbf{P}_\mathcal{S}\mathbf{S}_*\mathbf{P}_\mathcal{S}|_0 - \sum_{g=1}^{h} n_g|\mathbf{P}_\mathcal{S}\mathbf{S}_g\mathbf{P}_\mathcal{S}|_0.$$

Then Cook and Forzani (2008a) show that, under the assumption of Wishart matrices, the maximum likelihood estimator of C with fixed dimension d is $\widehat{C} = \arg\max_S L_d(S)$, where the maximum is over all d-dimensional subspaces in $\mathbb{R}^p$. The addends in L_d are similar to those in (1.28), and similar steps can be taken to reexpress it in terms of a basis matrix for C. The derivation of L_d is similar to the derivation of objective function (5.11) for enveloping multivariate means with heteroscedastic errors. The maximum likelihood estimator of $\mathbf{\Sigma}_g$ is then

$$\widehat{\mathbf{\Sigma}}_g = \mathbf{S}_* + \mathbf{P}^T_{\widehat{C}(\mathbf{S}_*)}(\mathbf{S}_g - \mathbf{S}_*)\mathbf{P}_{\widehat{C}(\mathbf{S}_*)}.$$

Cook and Forzani (2008a) also described methodology for selecting d based on likelihood ratio testing, information criteria, and cross-validation and for testing hypotheses of the form $\mathbf{P}_{\mathcal{H}}C = 0_p$, where $\mathcal{H}$ is a user-specified subspace.

9.8.2 Connections with Envelopes

A connection between SDR for covariance matrices and envelopes arises by considering $\mathcal{E}_{\mathbf{\Sigma}_*}(C)$, the smallest reducing subspace of $\mathbf{\Sigma}_*$ that contains the central subspace C. Let $u = \dim(\mathcal{E}_{\mathbf{\Sigma}_*}(C))$ and let $(\boldsymbol{\gamma}, \boldsymbol{\gamma}_0) \in \mathbb{R}^{p\times p}$ be an orthogonal matrix so that $\boldsymbol{\gamma} \in \mathbb{R}^{p\times u}$ is a basis matrix for $\mathcal{E}_{\mathbf{\Sigma}_*}(C)$, leading to the representation $\mathbf{\Sigma}_* = \boldsymbol{\gamma}\mathbf{V}\boldsymbol{\gamma}^T + \boldsymbol{\gamma}_0\mathbf{V}_0\boldsymbol{\gamma}_0^T$, where $\mathbf{V} \in \mathbb{S}^{u\times u}$ and $\mathbf{V}_0 \in \mathbb{S}^{(p-u)\times(p-u)}$ are positive definite. Since $C \subseteq \mathcal{E}_{\mathbf{\Sigma}_*}(C)$, the envelope provides an upper bound on the central subspace, $u \geq d$, and we can write $\boldsymbol{\alpha} = \boldsymbol{\gamma}\boldsymbol{\eta}$ for some semi-orthogonal matrix $\boldsymbol{\eta} \in \mathbb{R}^{u\times d}$. Substituting this structure into the right side of condition III leads to the representation

$$\mathbf{\Sigma}_g = \boldsymbol{\gamma}\mathbf{M}_g\boldsymbol{\gamma}^T + \boldsymbol{\gamma}_0\mathbf{M}_0\boldsymbol{\gamma}_0^T,$$

where $\mathbf{M}_g \in \mathbb{S}^{u\times u}$ and $\mathbf{M}_0 \in \mathbb{S}^{(p-u)\times(p-u)}$. This relationship, which is an instance of the common space model of Flury (1987), shows that $\mathcal{E}_{\mathbf{\Sigma}_*}(C)$ and $\mathcal{E}^{\perp}_{\mathbf{\Sigma}_*}(C)$ reduce all $\mathbf{\Sigma}_g$ and that the eigenvectors of $\mathbf{\Sigma}_g$ can be constructed to fall in either $\mathcal{E}_{\mathbf{\Sigma}_*}(C)$ or $\mathcal{E}^{\perp}_{\mathbf{\Sigma}_*}(C)$. The eigenvectors that fall in $\mathcal{E}^{\perp}_{\mathbf{\Sigma}_*}(C)$ can be constructed to be common to all $\mathbf{\Sigma}_g$ with the same eigenvalues.

Another characteristic of the spectral structure arises when we rewrite condition III in terms of the standardized variable $\mathbf{Z} = \mathbf{\Sigma}_*^{-1/2}\mathbf{X}$ that has central subspace $C_Z = \mathbf{\Sigma}_*^{1/2}C$ with semi-orthogonal basis matrix $\mathbf{\Gamma}$. Then $\mathbf{P}_{\alpha(\mathbf{\Sigma}_*)} = \mathbf{\Sigma}_*^{-1/2}\mathbf{P}_{\mathbf{\Gamma}}\mathbf{\Sigma}_*^{1/2}$ and condition III can be written as

$$\mathbf{\Sigma}_*^{-1/2}\mathbf{\Sigma}_g\mathbf{\Sigma}_*^{-1/2} = \mathbf{Q}_{\mathbf{\Gamma}} + \mathbf{P}_{\mathbf{\Gamma}}\mathbf{\Sigma}_*^{-1/2}\mathbf{\Sigma}_g\mathbf{\Sigma}_*^{-1/2}\mathbf{P}_{\mathbf{\Gamma}}. \tag{9.22}$$

Consequently, we see that the group covariance matrices in the $\mathbf{Z}$ scale, $\mathbf{\Sigma}_*^{-1/2}\mathbf{\Sigma}_g\mathbf{\Sigma}_*^{-1/2}$, share an eigenspace $\text{span}^{\perp}(\mathbf{\Gamma})$ with eigenvalue 1 and also share the complementary space $\text{span}(\mathbf{\Gamma})$.

9.8.3 Illustrations

Cook and Forzani (2008a) gave an example in which they compare the covariance matrices for six traits of female garter snakes at coastal and inland populations in northern California. Their illustration includes dimension selection, hypothesis tests of the form $\mathbf{P}_{\mathcal{H}}C = 0$, and a comparison with a spectral analysis. Here we give two brief examples to emphasize the potential advantages of graphical comparisons. Let $\widehat{\boldsymbol{\alpha}} = (\widehat{\boldsymbol{\alpha}}_1, \dots, \widehat{\boldsymbol{\alpha}}_d)$ denote the estimated basis matrix for C with individual basis vectors $\widehat{\boldsymbol{\alpha}}_j$. Then a d-dimensional plot of $(\widehat{\boldsymbol{\alpha}}_1^T\mathbf{X}, \dots, \widehat{\boldsymbol{\alpha}}_d^T\mathbf{X})$ marked by populations may give useful information on the nature of the differences between the $\boldsymbol{\Sigma}_g$'s.

Using CORE to compare the covariance matrices from the data on genuine and counterfeit banknotes introduced in Section 2.3, we first used the likelihood ratio procedure to select d, concluding that $d = 2$, and then determined $\widehat{\boldsymbol{\alpha}} = (\widehat{\boldsymbol{\alpha}}_1, \widehat{\boldsymbol{\alpha}}_2)$ by maximizing $L_2(S)$. A plot of $\widehat{\boldsymbol{\alpha}}_1^T\mathbf{X}$ versus $\widehat{\boldsymbol{\alpha}}_2^T\mathbf{X}$ with points marked by note type is shown in Figure 9.2a. Although the plot shows location separation between the counterfeit and genuine banknotes, location separation per se is not a goal of this methodology.

Our second illustration is from a pilot study to assess the possibility of distinguishing birds, planes, and cars by the sounds they make. A two-hour recording was made in Ermont, France, and then five seconds snippets of sounds were selected, resulting in 58 recordings identified as birds, 43 as cars, and 64 as planes. Each recording was processed and ultimately represented by 13 scale dependent mel-frequency cepstrum coefficients, which we represent

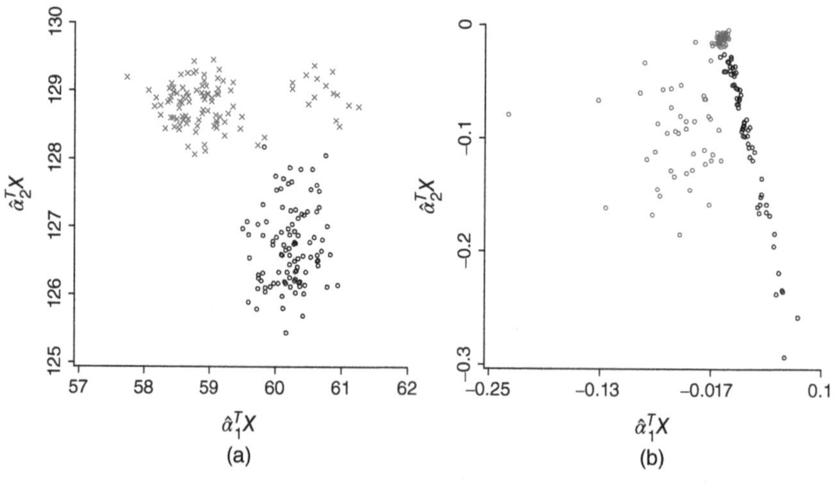

Figure 9.2 Plots of the first two CORE predictors $\hat{\alpha}_1^T\mathbf{X}$ and $\hat{\alpha}_2^T\mathbf{X}$. (a) Banknote data: ∘, genuine; ×, counterfeit. (b) Birds–planes–cars data: (b) Blue, birds; black, planes; red, cars.

as **X**. In terms of our basic structure then we have observations on $\mathbf{X} \in \mathbb{R}^{13}$ from three populations, birds, planes, and cars. Further background on these data is available from Cook and Forzani (2009) who used them to illustrate simultaneous dimension reduction for group means and covariance matrices. Here we use CORE to compare the covariance matrices for birds, planes, and cars. Likelihood ratio testing, BIC and AIC all indicated that d is large, perhaps 11 or 12. Nevertheless, a 2D or 3D marked plot from fitting with $d = 2$ or 3 can still give good clues about properties of the covariance matrices. Fitting with $d = 2$, a plot of $\hat{\boldsymbol{\alpha}}_1^T\mathbf{X}$ versus $\hat{\boldsymbol{\alpha}}_2^T\mathbf{X}$ with points marked by population is given in Figure 9.2b. The plot shows quite different variance structures for birds, planes, and cars.

9.8.4 SDR for Means and Covariance Matrices

Cook and Forzani (2009) extended CORE methodology to allow simultaneous dimension reduction for group means and covariance matrices. Let $\boldsymbol{\mu}_g = \mathrm{E}(\mathbf{X} \mid g)$, and let $\boldsymbol{\mu}$ denote the grand mean. Write the deviations of the group means from the grand mean in the **Z** scale as $\boldsymbol{\delta}_g = \boldsymbol{\Sigma}_*^{-1/2}(\boldsymbol{\mu}_g - \boldsymbol{\mu})$, and let $\mathcal{M}_{\mathbf{Z}} = \mathrm{span}\{\boldsymbol{\delta}_1, \dots, \boldsymbol{\delta}_h\}$. Then adding dimension reduction for the conditional means to CORE is equivalent in the **Z** scale to requiring the CORE condition (9.22) and the additional condition that $\mathcal{M}_{\mathbf{Z}} \subseteq \mathrm{span}(\boldsymbol{\Gamma})$. In effect, $\mathrm{span}(\boldsymbol{\Gamma})$ is the smallest subspace that contains $\mathcal{M}_{\mathbf{Z}}$ and that reduces each of the group covariance matrices in the **Z** scale.

9.9 Principal Components

9.9.1 Introduction

Principal component analysis (PCA) has a long tradition as a data-analytic method of unsupervised reduction of multivariate observations, including predictors in regression. Its first uses have been traced back to Adcock (1878), Pearson (1901), and Hotelling (1933), and it may now be the most widely used dimension reduction method in the applied sciences. PCA has been justified on various grounds (Jolliffe 2002), but mainly users sought a few uncorrelated linear combinations of the original variables that capture maximal variation, with the hope that those linear combinations would, in some sense, preserve relevant information.

Suppose that we have n observations on the random vector $\mathbf{X} \in \mathbb{R}^p$ with mean $\boldsymbol{\mu}$ and variance $\boldsymbol{\Sigma}_{\mathbf{X}} > 0$. Let $q = \mathrm{rank}(\mathbf{S}_{\mathbf{X}}) \leq p$. Additionally, let $\hat{\varphi}_1 > \hat{\varphi}_2 > \cdots > \hat{\varphi}_q > 0$ and $\hat{\boldsymbol{\ell}}_1, \hat{\boldsymbol{\ell}}_2, \dots, \hat{\boldsymbol{\ell}}_q$ be the nonzero eigenvalues and corresponding eigenvectors of $\mathbf{S}_{\mathbf{X}}$. The linear combinations $\hat{\boldsymbol{\ell}}_j^T\mathbf{X}$ are the *sample principal components of* **X**; we refer to the eigenvectors $\hat{\boldsymbol{\ell}}_j$, $j = 1, \dots, q$, as *principal component directions*. The population principal components are computed in the same way with $\mathbf{S}_{\mathbf{X}}$ replaced by $\mathbf{\Sigma}_{\mathbf{X}}$. The word "principal" in principal components indicates a natural ordering on importance; $\hat{\ell}_1^T\mathbf{X}$ being the most important component, $\hat{\ell}_2^T\mathbf{X}$ being the second most important component, and so on.

The sample variance of the jth principal component $\hat{\ell}_j^T\mathbf{X}$ is $\hat{\varphi}_j$. If the fraction of explained variation $f_d = \sum_{j=1}^{d}\hat{\varphi}_j / \sum_{j=1}^{q}\hat{\varphi}_j$ is sufficiently close to 1, then the first $d \leq q$ principal components $\hat{\ell}_j^T\mathbf{X}, j = 1, \dots, d$, are uncorrelated linear combinations of the original variables that capture most of the variation in the data. There are many methods for selecting the number of principal components d, depending on application specific requirements. Jolliffe (2002) gave a comprehensive account of principal components, including methods for selecting d. Despite its popularity, it seems that PCA was not based on a probability model until Tipping and Bishop (1999) introduced a context in which the principal component directions arise through maximum likelihood estimation.

The developments in this section are based on the latent variable model

$$\mathbf{X}_i = \boldsymbol{\mu} + \boldsymbol{\theta}\boldsymbol{v}_i + \boldsymbol{\varepsilon}_i, \quad i = 1, \dots, n, \tag{9.23}$$

where $\boldsymbol{\mu} \in \mathbb{R}^p$, $\boldsymbol{\theta} \in \mathbb{R}^{p\times d}$ with $d \leq p$, and the errors $\boldsymbol{\varepsilon}_i$ are independent copies of $\boldsymbol{\varepsilon} \sim N_p(0, \boldsymbol{\Delta})$. The $\boldsymbol{v}_i \in \mathbb{R}^d$ are latent random vectors that are assumed to be independent copies of a random vector $\boldsymbol{v} \sim N_d(0, \mathbf{I}_d)$ that is independent of $\boldsymbol{\varepsilon}$, but we will also consider in Section 9.9.3 the case where the $\boldsymbol{v}_i$'s are nonstochastic. The identity covariance matrix in the former assumption can always be achieved by normalization: If $\text{var}(\boldsymbol{v}) = \mathbf{A}$, then write $\boldsymbol{\theta}\mathbf{A}^{1/2}\mathbf{A}^{-1/2}\boldsymbol{v}$ and redefine $\boldsymbol{\theta}$ and $\boldsymbol{v}$ accordingly. The parameter $\boldsymbol{\theta}$ is not identified since $\boldsymbol{\theta}\boldsymbol{v} = (\boldsymbol{\theta}\mathbf{O})(\mathbf{O}^T\boldsymbol{v})$ for any orthogonal matrix $\mathbf{O} \in \mathbb{R}^{d\times d}$, resulting in an equivalent model. Model (9.23) is like the multivariate linear model (1.1), except for the crucial difference that the "predictors" $\boldsymbol{v}$ are unobserved random vectors.

We think of $\boldsymbol{v}$ as representing variation that is caused by latent extrinsic factors, while $\boldsymbol{\varepsilon}$ represents intrinsic variation that would be present if all extrinsic factors were held fixed. The goal of an analysis is to extract the part of $\mathbf{X}$ that is affected by the extrinsic factors. More specifically, under model (9.23), it can be shown that

$$\mathbf{X} \perp\!\!\!\perp \boldsymbol{v} \mid \boldsymbol{\theta}^T\boldsymbol{\Delta}^{-1}\mathbf{X}, \tag{9.24}$$

and thus, given $\boldsymbol{\theta}^T\boldsymbol{\Delta}^{-1}\mathbf{X}$, $\mathbf{X}$ is unaffected by the extrinsic vector $\boldsymbol{v}$. Consequently, $\mathbf{R}(\mathbf{X}) := \boldsymbol{\theta}^T\boldsymbol{\Delta}^{-1}\mathbf{X}$ is the part of $\mathbf{X}$ affected by $\boldsymbol{v}$ and is the minimal reduction we would like to estimate (Cook and Forzani 2008b, Thm 2.1), reasoning that the intrinsic variation is generally of little relevance to the dimension reduction goals. Full rank linear transformations of $\mathbf{R}$ do not matter in this context: for any full rank matrix $\mathbf{A} \in \mathbb{R}^{d\times d}$, $\mathbf{X} \perp\!\!\!\perp \boldsymbol{v} \mid \boldsymbol{\theta}^T\boldsymbol{\Delta}^{-1}\mathbf{X}$ if and only if $\mathbf{X} \perp\!\!\!\perp \boldsymbol{v} \mid \mathbf{A}^T\boldsymbol{\theta}^T\boldsymbol{\Delta}^{-1}\mathbf{X}$ and so the fundamental goal is to estimate $\mathcal{T} := \text{span}(\boldsymbol{\Delta}^{-1}\boldsymbol{\theta})$. Because $\boldsymbol{v}$ is not observable, only the marginal distribution of $\mathbf{X}$ is available to estimate $\mathcal{T}$. Under model (9.23), $\mathbf{X}$ is normal with mean $\boldsymbol{\mu}$ and variance $\boldsymbol{\Sigma}_{\mathbf{X}} = \boldsymbol{\Delta} + \boldsymbol{\theta}\boldsymbol{\theta}^T$. The maximum likelihood estimator of $\boldsymbol{\mu}$ is simply the sample mean of $\mathbf{X}$. However, $\boldsymbol{\Delta}$ and $\boldsymbol{\theta}$ are confounded, and thus $\mathcal{T}$ cannot be estimated without additional structure.

An additional assumption of *isotropic variation* $\boldsymbol{\Delta} = \sigma^2\mathbf{I}_p$ might be reasonable in some applications. Principal components have been used in face recognition to construct eigenface representations of individual images. Beginning with a set of n $r \times c$ grayscale images normalized to have eyes and mouths aligned, the images are then vectorized so each face is represented as a vector $\mathbf{X}$ of length $p = rc$. The mean $\boldsymbol{\mu}$ in model (9.23) represents the average face and $\boldsymbol{\theta}\mathbf{v}_i$ models the deviations of individual faces from the average. Isotropic variation might be reasonable in this case since the error $\boldsymbol{\varepsilon}$ reflects the intrinsic variation in grayscale measurements across images of the same face. It could be reasonable also in applications involving measurement error, microarray data, and calibration. Cavalli-Sforza et al. (1994) used PCA to produce acclaimed continental maps summarizing human genetic variation. When used in the analysis of microarray data, principal components have been called "eigengenes" (Alter et al. 2000).

There is a long history of uneven attempts to use PCA as a method for reducing the predictors in regression (Cook 2007). It is now generally recognized that PCA is not a reliable method for this because it ignores the response, but see Artemiou and Li (2009) for a different perspective. In Section 9.10, we describe how principal components arise naturally in predictor reduction when the response is taken into account.

9.9.2 Random Latent Variables

Tipping and Bishop (1999) studied the version of model (9.23) with *isotropic errors*, $\boldsymbol{\Delta} = \sigma^2\mathbf{I}_p$. In this case, $\mathcal{T} = \text{span}(\boldsymbol{\theta})$ is identified and Tipping and Bishop showed that its maximum likelihood estimator is the span of the first d sample principal component directions of $\mathbf{X}$, leading to their notion of *probabilistic principal components.*

The maximum likelihood estimator of $\mathcal{T}$ is perhaps easiest to find by first reformulating the isotropic version of model (9.23) as

$$\mathbf{X} = \boldsymbol{\mu} + \boldsymbol{\Theta}\boldsymbol{\delta}\boldsymbol{v} + \boldsymbol{\varepsilon}, \tag{9.25}$$

where $\boldsymbol{\theta} = \boldsymbol{\Theta}\boldsymbol{\delta}$, $\boldsymbol{\Theta} \in \mathbb{R}^{p\times d}$ is a semi-orthogonal basis matrix for $\mathcal{T}$, $\boldsymbol{\delta} \in \mathbb{R}^{d\times d}$ is a full rank coordinate matrix, $\boldsymbol{v} \sim N_d(0, \mathbf{I}_d)$, and $\boldsymbol{\varepsilon} \sim N_p(0, \sigma^2\mathbf{I}_p)$. The variance $\boldsymbol{\Delta} = \sigma^2\mathbf{I}_p$ says that, conditional on the extrinsic variables $\boldsymbol{v}$, the measurements are independent and have the same variance. The covariance matrix of $\mathbf{X}$ can be expressed under (9.25) as

$$\boldsymbol{\Sigma}_{\mathbf{X}} = \boldsymbol{\Theta}\boldsymbol{\delta}\boldsymbol{\delta}^T\boldsymbol{\Theta}^T + \sigma^2\mathbf{I}_p = \boldsymbol{\Theta}\mathbf{V}\boldsymbol{\Theta}^T + \sigma^2\boldsymbol{\Theta}_0\boldsymbol{\Theta}_0^T, \tag{9.26}$$

where $\mathbf{V} = \boldsymbol{\delta}\boldsymbol{\delta}^T + \sigma^2 \mathbf{I}_d$, and $(\boldsymbol{\Theta}, \boldsymbol{\Theta}_0)$ is an orthogonal matrix. From this we see that (9.25) is a version of an envelope model since $\mathcal{T} = \mathcal{E}_{\boldsymbol{\Delta}}(\mathcal{T})$. Since $\boldsymbol{\delta}$ is full rank, we parameterize the model in terms of $\mathbf{V}$, $\mathcal{T}$, and σ^2.

The maximum likelihood estimator of $\boldsymbol{\mu}$ is $\hat{\boldsymbol{\mu}} = \bar{\mathbf{X}}$. With d known, the log-likelihood $L_d(\mathcal{T}, \mathbf{V}, \sigma^2)$ maximized over $\boldsymbol{\mu}$ is then

$$\begin{aligned}(2/n)L_d(\mathcal{T}, \mathbf{V}, \sigma^2) &= c - \log|\mathbf{V}| - \mathrm{tr}(\boldsymbol{\Theta}^T \mathbf{S}_{\mathbf{X}} \boldsymbol{\Theta} \mathbf{V}^{-1}) \\ &\quad - (p-d)\log(\sigma^2) - \sigma^{-2}\mathrm{tr}(\boldsymbol{\Theta}_0^T \mathbf{S}_{\mathbf{X}} \boldsymbol{\Theta}_0),\end{aligned}$$

where $c = -p\log(2\pi)$. This log-likelihood, like others encountered in this book, depends only on $\mathcal{T}$ and not on a particular basis $\boldsymbol{\Theta}$. Maximizing over $\mathbf{V}$ and σ^2 separately leads to the same partially maximized likelihood function as found by Tipping and Bishop (1999). However, the parameters $\mathbf{V}$ and σ^2 are not in a proper product space because the eigenvalues of $\mathbf{V}$ are bounded below by σ^2. Thus, it seems inappropriate to maximize over $\mathbf{V}$ and σ^2 separately and instead we reexpress the log likelihood as

$$\begin{aligned}(2/n)L_d(\mathcal{T}, \mathbf{V}, \sigma^2) &= c - \log|\mathbf{V}| - (p-d)\log(\sigma^2) - \sigma^{-2}\mathrm{tr}(\mathbf{S}_{\mathbf{X}}) \\ &\quad + \mathrm{tr}\{\boldsymbol{\Theta}^T \mathbf{S}_{\mathbf{X}} \boldsymbol{\Theta}(\sigma^{-2}\mathbf{I}_p - \mathbf{V}^{-1})\},\end{aligned}$$

where $(\sigma^{-2}\mathbf{I}_p - \mathbf{V}^{-1}) > 0$. It now follows immediately from Lemma A.19 that this likelihood is maximized over $\mathcal{T}$ by the span of the first d eigenvectors of $\mathbf{S}_{\mathbf{X}}$ regardless of the particular value of $(\sigma^{-2}\mathbf{I}_p - \mathbf{V}^{-1}) > 0$. We then have

Proposition 9.7 *The maximum likelihood estimator of $\mathcal{T}$ under model (9.25) is the subspace spanned by the first d principal component directions of $\mathbf{S}_{\mathbf{X}}$.*

The result of Proposition 9.7 is the same as that given by Tipping and Bishop (1999), but the justification here is different. The proposition gives no notion of the relative importance of the principal component directions, but they can still be ordered using traditional reasoning (Jolliffe 2002).

Figure 9.3 is a schematic representation of how principal components work in the isotropic model (9.25) based on identity (9.26). $\mathcal{T}$ is a reducing subspace of $\boldsymbol{\Sigma}_{\mathbf{X}}$, and consequently the eigenvectors of $\boldsymbol{\Sigma}_{\mathbf{X}}$ are in either $\mathcal{T}$ or $\mathcal{T}^{\perp}$. And the eigenvalues of $\boldsymbol{\Sigma}_{\mathbf{X}}$ corresponding to $\mathcal{T}$ are uniformly larger than the remaining eigenvalues. As represented in Figure 9.3a, we begin with the circular contours of $\mathrm{var}(\mathbf{X}|\boldsymbol{\nu}) = \boldsymbol{\Delta} = \sigma^2\mathbf{I}_p$. Adding the signal $\boldsymbol{\Theta}\boldsymbol{\delta}\boldsymbol{\delta}^T\boldsymbol{\Theta}^T$ then distorts the circular contours into ellipses, as seen in Figure 9.3b, with the span of its first axis being equal to $\mathcal{T}$.

We next use envelopes to expand the probabilistic principal component model (9.25).

9.9.2.1 Envelopes

Instead of assuming an isotropic error structure, consider model (9.23) where $\boldsymbol{\Delta} > 0$. We still think of the latent variable $\boldsymbol{\nu}$ as representing extrinsic variation in $\mathbf{X}$ and the error $\boldsymbol{\varepsilon}$ as representing intrinsic variation, but now the intrinsic

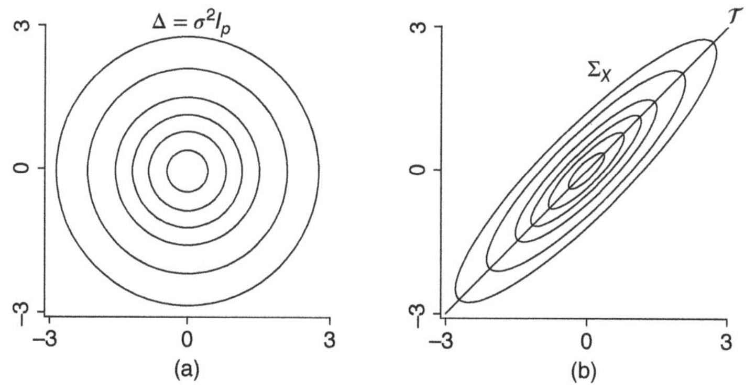

Figure 9.3 Schematic representation showing why principal components work in isotropic models. (a) Circular contour of $\boldsymbol{\Delta}$. (b) Elliptical contours of $\boldsymbol{\Sigma}_{\mathbf{X}}$.

variation is anisotropic and might have correlated elements. For instance, are nearby measurements on an $r \times c$ grayscale image correlated across replicate images of the same face? This leads us back to the general confounding between $\boldsymbol{\Delta}$ and $\boldsymbol{\theta}$ mentioned previously. Nevertheless, useful results might still be possible if we can estimate an upper bound on $\mathcal{T}$. By doing so, we would not lose any information on its extrinsic variation, but we might retain some of its intrinsic variation. Following Chen (2010), we construct an upper bound on $\mathcal{T}$ by using an envelope.

To apply these ideas in the context of model (9.23), consider the $\boldsymbol{\Delta}$-envelope of $\mathcal{T}$, $\mathcal{E}_{\boldsymbol{\Delta}}(\mathcal{T})$. It follows from Proposition A.5 that $\mathcal{E}_{\boldsymbol{\Delta}}(\mathcal{T}) = \mathcal{E}_{\boldsymbol{\Delta}}(\boldsymbol{\theta})$, so we can equivalently consider the $\boldsymbol{\Delta}$-envelope of span($\boldsymbol{\theta}$). Let the columns of the semi-orthogonal matrix $\boldsymbol{\Gamma} \in \mathbb{R}^{p \times u}$ be a basis for $\mathcal{E}_{\boldsymbol{\Delta}}(\boldsymbol{\theta})$ and let $(\boldsymbol{\Gamma}, \boldsymbol{\Gamma}_0)$ be an orthogonal matrix, where $u = \dim(\mathcal{E}_{\boldsymbol{\Delta}}(\boldsymbol{\theta})) \geq d$. We have $\text{span}(\boldsymbol{\theta}) \subseteq \mathcal{E}_{\boldsymbol{\Delta}}(\boldsymbol{\theta})$. Consequently, there is a coordinate matrix $\boldsymbol{\eta} \in \mathbb{R}^{u \times d}$ with rank d so that $\boldsymbol{\theta} = \boldsymbol{\Gamma}\boldsymbol{\eta}$. From this, we obtain an *envelope component model*

$$\begin{aligned} \mathbf{X} &= \boldsymbol{\mu} + \boldsymbol{\Gamma}\boldsymbol{\eta}\boldsymbol{\nu} + \boldsymbol{\varepsilon}, \\ \boldsymbol{\Delta} &= \boldsymbol{\Gamma}\boldsymbol{\Omega}\boldsymbol{\Gamma}^T + \boldsymbol{\Gamma}_0\boldsymbol{\Omega}_0\boldsymbol{\Gamma}_0^T. \end{aligned} \tag{9.27}$$

The interpretation of this model follows from the interpretation of (1.20): $\boldsymbol{\Gamma}^T\mathbf{X} \perp\!\!\!\perp \boldsymbol{\Gamma}_0^T\mathbf{X}|\boldsymbol{\nu}$ and $\boldsymbol{\Gamma}_0^T\mathbf{X} \perp\!\!\!\perp \boldsymbol{\nu}$, and thus $\boldsymbol{\Gamma}_0^T\mathbf{X} \perp\!\!\!\perp (\boldsymbol{\Gamma}^T\mathbf{X}, \boldsymbol{\nu})$ and $\mathbf{X} \perp\!\!\!\perp \boldsymbol{\nu}|\boldsymbol{\Gamma}^T\mathbf{X}$. Consequently, we think of $\boldsymbol{\Gamma}_0^T\mathbf{X}$ with $\text{var}(\boldsymbol{\Gamma}_0^T\mathbf{X}|\boldsymbol{\nu}) = \text{var}(\boldsymbol{\Gamma}_0^T\mathbf{X}) = \boldsymbol{\Omega}_0$ as capturing just intrinsic variation, while $\boldsymbol{\Gamma}^T\mathbf{X}$ with $\text{var}(\boldsymbol{\Gamma}^T\mathbf{X}|\boldsymbol{\nu}) = \boldsymbol{\Omega}$ and $\boldsymbol{\Psi} := \text{var}(\boldsymbol{\Gamma}^T\mathbf{X}) = \boldsymbol{\Omega} + \boldsymbol{\eta}\boldsymbol{\eta}^T$ holds all of the extrinsic variation. The intrinsic and extrinsic variations are both represented in the marginal covariance matrix of $\mathbf{X}$: $\boldsymbol{\Sigma}_{\mathbf{X}} = \boldsymbol{\Gamma}\boldsymbol{\Psi}\boldsymbol{\Gamma}^T + \boldsymbol{\Gamma}_0\boldsymbol{\Omega}_0\boldsymbol{\Gamma}_0^T$. Under model (9.27), $\boldsymbol{\Delta}^{-1}\boldsymbol{\theta} = \boldsymbol{\Gamma}\boldsymbol{\Omega}^{-1}\boldsymbol{\eta}$, and so the desired reduction is $\mathbf{R} = \boldsymbol{\eta}^T\boldsymbol{\Omega}^{-1}\boldsymbol{\Gamma}^T\mathbf{X} \in \mathbb{R}^d$. However, $\boldsymbol{\eta}$ is not estimable from the marginal of $\mathbf{X}$ as it is confounded with $\boldsymbol{\Omega}$, and thus we pursue the envelope reduction $\mathbf{R} = \mathbf{\Gamma}^T\mathbf{X} \in \mathbb{R}^u$. If $u = d$, then $\mathcal{T}$ and the envelope coincide, $\mathcal{T} = \mathcal{E}_{\mathbf{\Delta}}(\mathcal{T})$.

An estimate of the envelope $\mathcal{E}_{\mathbf{\Delta}}(\mathcal{T}) = \text{span}(\mathbf{\Gamma})$ then provides an estimated upper bound on $\mathcal{T}$, which is the target of our inquiry. In short, we need to estimate $\mathcal{E}_{\mathbf{\Delta}}(\mathcal{T})$ based on a sample $\mathbf{X}_1, \dots, \mathbf{X}_n$ from a $N_p(\boldsymbol{\mu}, \mathbf{\Sigma}_{\mathbf{X}})$ distribution. The estimated reduction is then $\widehat{\mathbf{R}} = \widehat{\mathbf{\Gamma}}^T\mathbf{X}$, where the columns of $\widehat{\mathbf{\Gamma}}$ form a basis for the estimated envelope. We turn next to maximum likelihood estimation.

After some algebra, the partially maximized log-likelihood $L_u^{(1)}$ with $\hat{\boldsymbol{\mu}} = \bar{\mathbf{X}}$ can be represented in the form

$$\begin{aligned}(2/n)L_u^{(1)}(\mathcal{E}, \mathbf{\Psi}, \mathbf{\Omega}_0) = &- \log|\mathbf{\Psi}| - \text{tr}(\mathbf{\Gamma}^T\mathbf{S}_{\mathbf{X}}\mathbf{\Gamma}\mathbf{\Psi}^{-1}) \\ &- \log|\mathbf{\Omega}_0| - \text{tr}(\mathbf{\Gamma}_0^T\mathbf{S}_{\mathbf{X}}\mathbf{\Gamma}_0\mathbf{\Omega}_0^{-1}),\end{aligned}$$

where $\mathcal{E}_{\mathbf{\Delta}}(\mathcal{T})$ has been shortened to $\mathcal{E}$ for use as an argument. Maximizing over $\mathbf{\Psi}$ and $\mathbf{\Omega}_0$, which are defined on a product space, we have the partially maximized log-likelihood function

$$\begin{aligned}(2/n)L_u^{(11)}(\mathcal{E}) &= -\log|\mathbf{\Gamma}^T\mathbf{S}_{\mathbf{X}}\mathbf{\Gamma}| - \log|\mathbf{\Gamma}_0^T\mathbf{S}_{\mathbf{X}}\mathbf{\Gamma}_0| - p \\ &= -\log|\mathbf{\Gamma}^T\mathbf{S}_{\mathbf{X}}\mathbf{\Gamma}| - \log|\mathbf{\Gamma}^T\mathbf{S}_{\mathbf{X}}^{-1}\mathbf{\Gamma}| - p - \log|\mathbf{S}_{\mathbf{X}}|,\end{aligned}$$

which requires $n > p$ as $\mathbf{S}_{\mathbf{X}}$ must not be singular. Also, as expected, it is invariant under right orthogonal transformations of $\mathbf{\Gamma}$, and thus it determines only a subspace and not a particular coordinate system. It follows immediately from Proposition A.15 that the span of any u sample principal component directions is a maximum likelihood estimator of $\mathcal{E}_{\mathbf{\Delta}}(\mathcal{T})$. Model (9.27) represents a demarcation point for the effectiveness of PCA. Because any subset of u principal component directions is equally supported by the likelihood, there seems little hope of obtaining useful estimates of $\mathcal{T}$ or its envelope $\mathcal{E}_{\mathbf{\Delta}}(\mathcal{T})$ with less structure. However, useful results might be obtained when additional structure is imposed. For instance, suppose that $u = d$, $\text{var}(\mathbf{\Gamma}^T\mathbf{X}|\nu) = \mathbf{\Omega} = \sigma^2\mathbf{I}_d$, and $\text{var}(\mathbf{\Gamma}_0^T\mathbf{X}) = \mathbf{\Omega}_0 = \sigma^2\mathbf{I}_{p-d}$, so that the conditional extrinsic and intrinsic variations have the same structure. Then $\mathbf{\Delta} = \sigma^2\mathbf{I}_p$, and model (9.27) reduces to probabilistic principal component model (9.25). In effect, models (9.25) and (9.27) represent extremes of a modeling environment for multivariate reduction. Perhaps there is sufficient flexibility between these extremes for application-specific requirements. Two possibilities are described in the next two sections.

9.9.2.2 Envelopes with Isotropic Intrinsic and Extrinsic Variation

We begin with a relatively small generalization of model (9.23) that nevertheless directs us away from the leading principal components as the desired reduction of $\mathbf{X}$. Assume that the marginal variation $\mathbf{\Sigma}_{\mathbf{X}} = \sigma^2\mathbf{\Gamma}\mathbf{\Gamma}^T + \sigma_0^2\mathbf{\Gamma}_0\mathbf{\Gamma}_0^T$, where the intrinsic variation $\mathbf{\Omega}_0 = \sigma_0^2\mathbf{I}_{p-u}$. Then the log-likelihood $L_u^{(2)}$ can be written as

$$\begin{aligned}(2/n)L_u^{(2)}(\mathcal{E}, \sigma^2, \sigma_0^2) = &-u\log(\sigma^2) - \sigma^{-2}\text{tr}(\mathbf{\Gamma}^T\mathbf{S}_{\mathbf{X}}\mathbf{\Gamma}) \\ &- (p-u)\log(\sigma_0^2) - \sigma_0^{-2}\text{tr}(\mathbf{\Gamma}_0^T\mathbf{S}_{\mathbf{X}}\mathbf{\Gamma}_0).\end{aligned}$$

Maximizing over σ^2 and σ_0^2, we have the partially maximized log-likelihood function

$$(2/n)L_u^{(21)}(\mathcal{E}) = -u\log\{\mathrm{tr}(\boldsymbol{\Gamma}^T\mathbf{S}_{\mathbf{X}}\boldsymbol{\Gamma})\} - (p-u)\log\{\mathrm{tr}(\mathbf{S}_{\mathbf{X}}) - \mathrm{tr}(\boldsymbol{\Gamma}^T\mathbf{S}_{\mathbf{X}}\boldsymbol{\Gamma})\}$$
$$-p + u\log(u) + (p-u)\log(p-u),$$

which requires $n > p - u + 1$ to ensure that $\mathrm{tr}(\boldsymbol{\Gamma}^T\mathbf{S}_{\mathbf{X}}\boldsymbol{\Gamma}) > 0$ for all $\boldsymbol{\Gamma}$.

Proposition 9.8 *Assume envelope model (9.27) with $\boldsymbol{\Sigma}_{\mathbf{X}} = \sigma^2\boldsymbol{\Gamma}\boldsymbol{\Gamma}^T + \sigma_0^2\boldsymbol{\Gamma}_0\boldsymbol{\Gamma}_0^T$, where $\boldsymbol{\Omega}_0 = \sigma_0^2\mathbf{I}_{p-u}$. Then $L_u^{(21)}(\mathcal{E})$ is maximized by the span of either the first u principal component directions or the last u principal component directions.*

Proof: Maximizing $L_u^{(21)}(\mathcal{E})$ is equivalent to minimizing

$$\log\{\mathrm{tr}(\boldsymbol{\Gamma}^T\mathbf{S}_{\mathbf{X}}\boldsymbol{\Gamma})\} + \frac{p-u}{u}\log\{\mathrm{tr}(\mathbf{S}_{\mathbf{X}}) - \mathrm{tr}(\boldsymbol{\Gamma}^T\mathbf{S}_{\mathbf{X}}\boldsymbol{\Gamma})\}.$$

We next apply Lemma A.20 with $x = \mathrm{tr}(\boldsymbol{\Gamma}^T\mathbf{S}_{\mathbf{X}}\boldsymbol{\Gamma})$, $C = (p-u)/u$, and $K = \mathrm{tr}(\mathbf{S}_{\mathbf{X}})$. Then, with probability one, we have

$$\min\{\mathrm{tr}(\boldsymbol{\Gamma}^T\mathbf{S}_{\mathbf{X}}\boldsymbol{\Gamma})\} < \frac{K}{1+C} = \frac{u}{p}\mathrm{tr}(\mathbf{S}_{\mathbf{X}}) < \max\{\mathrm{tr}(\boldsymbol{\Gamma}^T\mathbf{S}_{\mathbf{X}}\boldsymbol{\Gamma})\}$$

subject to $\boldsymbol{\Gamma}^T\boldsymbol{\Gamma} = \mathbf{I}_u$. Following Lemma A.20, we know that the maximum value of $L_u^{(21)}(\mathcal{E})$ is reached at either $\max\{\mathrm{tr}(\boldsymbol{\Gamma}^T\mathbf{S}_{\mathbf{X}}\boldsymbol{\Gamma})\}$ or $\min\{\mathrm{tr}(\boldsymbol{\Gamma}^T\mathbf{S}_{\mathbf{X}}\boldsymbol{\Gamma})\}$. That is, the maximum likelihood estimator of $\mathcal{E}_{\boldsymbol{\Delta}}(\mathcal{T})$ is the span of either the first u principal component directions or the last u principal component directions. □

Using this proposition in combination with the likelihood function $L_u^{(21)}$, it can be seen that the maximum likelihood estimator is the span of the first u principal component directions if (a) $f_u > 1/2$ and $r - 2u > 0$ or if (b) $f_u < 1/2$ and $r - 2u < 0$, where f_u is the fraction of explained variation as defined in Section 9.9.1. Otherwise, the maximum likelihood estimator is the span of the last u principal component directions.

9.9.2.3 Envelopes with Isotropic Intrinsic Variation

Additional flexibility can be incorporated by requiring only that $\boldsymbol{\Omega}_0 = \sigma_0^2\mathbf{I}_{p-u}$, leaving the extrinsic component $\boldsymbol{\Psi} > 0$ arbitrary. In this case, model (9.27) reduces to

$$\mathbf{X} = \boldsymbol{\mu} + \boldsymbol{\Gamma}\boldsymbol{\eta}\boldsymbol{\nu} + \boldsymbol{\varepsilon} \qquad (9.28)$$
$$\boldsymbol{\Delta} = \boldsymbol{\Gamma}\boldsymbol{\Omega}\boldsymbol{\Gamma}^T + \sigma_0^2\boldsymbol{\Gamma}_0\boldsymbol{\Gamma}_0^T.$$

The log-likelihood function $L_u^{(3)}$ for model (9.28) can be written as

$$(2/n)L_u^{(3)}(\mathcal{E}, \boldsymbol{\Delta}, \sigma_0^2) = -\log|\boldsymbol{\Psi}| - \mathrm{tr}(\boldsymbol{\Gamma}^T\mathbf{S}_{\mathbf{X}}\boldsymbol{\Gamma}\boldsymbol{\Psi}^{-1})$$
$$-(p-u)\log(\sigma_0^2) - \sigma_0^{-2}\mathrm{tr}(\boldsymbol{\Gamma}_0^T\mathbf{S}_{\mathbf{X}}\boldsymbol{\Gamma}_0).$$

Maximizing over $\boldsymbol{\Psi}$ and σ_0^2, we have the partially maximized log-likelihood function

$$(2/n)L_u^{(31)}(\mathcal{E}) = -\log|\boldsymbol{\Gamma}^T\mathbf{S}_{\mathbf{X}}\boldsymbol{\Gamma}| - (p-u)\log(\mathrm{tr}(\mathbf{S}_{\mathbf{X}}) - \mathrm{tr}(\boldsymbol{\Gamma}^T\mathbf{S}_{\mathbf{X}}\boldsymbol{\Gamma})) \\ -p + (p-u)\log(p-u).$$

The next proposition (Chen 2010) describes how to maximize $L_u^{(31)}$.

Proposition 9.9 *Under model (9.28), the maximum likelihood estimator of $\mathcal{E}_{\Delta}(\mathcal{T})$ is the span of the first k and last $u-k$ principal component directions for some $k \in \{0, 1, \dots, u\}$.*

Proof: Maximizing $L_u^{(31)}(\mathcal{E})$ is equivalent to minimizing

$$L_u^{(32)}(\mathcal{E}) = \log|\boldsymbol{\Gamma}^T\mathbf{S}_{\mathbf{X}}\boldsymbol{\Gamma}| + (p-u)\log\{\mathrm{tr}(\mathbf{S}_{\mathbf{X}}) - \mathrm{tr}(\boldsymbol{\Gamma}^T\mathbf{S}_{\mathbf{X}}\boldsymbol{\Gamma})\} \tag{9.29}$$

subject to $\boldsymbol{\Gamma}^T\boldsymbol{\Gamma} = \mathbf{I}_u$. Using Lagrange multipliers, a solution can be obtained by finding the stationary points of the unconstrained function

$$\log|\boldsymbol{\Gamma}^T\mathbf{S}_{\mathbf{X}}\boldsymbol{\Gamma}| + (p-u)\log\{\mathrm{tr}(\mathbf{S}_{\mathbf{X}}) - \mathrm{tr}(\boldsymbol{\Gamma}^T\mathbf{S}_{\mathbf{X}}\boldsymbol{\Gamma})\} + \mathrm{tr}\{\mathbf{U}(\boldsymbol{\Gamma}^T\boldsymbol{\Gamma} - \mathbf{I}_u)\},$$

where $\mathbf{U}$ is a $u \times u$ matrix of Lagrange multipliers. Let

$$\mathbf{A}_{\boldsymbol{\Gamma}} = (\boldsymbol{\Gamma}^T\mathbf{S}_{\mathbf{X}}\boldsymbol{\Gamma})^{-1} - \frac{(p-u)}{\mathrm{tr}(\mathbf{S}_{\mathbf{X}}) - \mathrm{tr}(\boldsymbol{\Gamma}^T\mathbf{S}_{\mathbf{X}}\boldsymbol{\Gamma})}\mathbf{I}_u.$$

Then the Lagrange analysis leads to the conclusion that the minimizers of (9.29) must satisfy $\mathbf{Q}_{\boldsymbol{\Gamma}}\mathbf{S}_{\mathbf{X}}\boldsymbol{\Gamma}\mathbf{A}_{\boldsymbol{\Gamma}} = 0$. Chen (2010) showed that $\mathbf{A}_{\boldsymbol{\Gamma}}$ must be of full rank, which implies that $\mathrm{span}(\mathbf{S}_{\mathbf{X}}\boldsymbol{\Gamma}) = \mathrm{span}(\boldsymbol{\Gamma})$, and thus that $\boldsymbol{\Gamma}$ spans a reducing subspace of $\mathbf{S}_{\mathbf{X}}$. Consequently, we can take the columns of $\boldsymbol{\Gamma}$ to be a subset of eigenvectors of $\mathbf{S}_{\mathbf{X}}$ with eigenvalues $\hat{\varphi}_1, \dots, \hat{\varphi}_u$, which now are still not necessarily ordered. Let $\hat{\varphi}_{(u+1)}, \dots, \hat{\varphi}_{(p)}$ denote the complement of $\hat{\varphi}_1, \dots, \hat{\varphi}_u$. Then objective function (9.29) becomes

$$L_{32}(\mathcal{E}) = \sum_{i=1}^{u}\log(\hat{\varphi}_i) + (p-u)\log\{\hat{\varphi}_{(u+1)} + \cdots + \hat{\varphi}_{(p)}\}. \tag{9.30}$$

Suppose there exists a configuration so that $\hat{\varphi}_{(i)} < \hat{\varphi}_l < \hat{\varphi}_{(j)}$. It follows from Lemma A.20 that (9.30) can be reduced by replacing $\hat{\varphi}_l$ with either $\hat{\varphi}_{(i)}$ or $\hat{\varphi}_{(j)}$. This tells us that the complementary set of eigenvalues $\hat{\varphi}_{(u+1)}, \dots, \hat{\varphi}_{(p)}$ must form a contiguous block. In other words, there exists an integer $1 \le k \le p$ so that the maximum likelihood estimator of $\mathcal{E}$ is the span of the first k and last $u-k$ principal component directions. □

Let σ_i, $i = 1, 2, \dots, u$ be the eigenvalues of $\boldsymbol{\Psi}$, $\sigma_1 \ge \sigma_2 \ge \cdots \ge \sigma_u$. The setting of model (9.28) basically says that the extrinsic signals can have different scales and be correlated, both conditionally $\mathrm{var}(\boldsymbol{\Gamma}^T\mathbf{X}|\nu) = \boldsymbol{\Omega}$ and unconditionally $\mathrm{var}(\boldsymbol{\Gamma}^T\mathbf{X}) = \boldsymbol{\Psi}$, but the intrinsic noise has an isotropic normal distribution

$\text{var}(\boldsymbol{\Gamma}_0) = \sigma_0^2\mathbf{I}$. If $\sigma_i > \sigma_0$ for all $i = 1, 2, \dots, u$, then in the population we have the same solution as the usual PCA. In effect, if the extrinsic signal is strong enough, the usual PCA is doing a sensible thing. If $\sigma_0 > \sigma_i$ for all $i = 1, 2, \dots, u$, then we have the last u principal component directions as the solution. If σ_0 lies among σ_i for $i = 1, 2, \dots, u$, then the solution is the span of the first k principal component directions and $u - k$ last principal component directions for some k, where k ranges from 0 to u. This provides a fast algorithm to search for the maximizer of $L_u^{(31)}$. If $u = d$, so that $\mathcal{T}$ is itself a reducing subspace of $\boldsymbol{\Delta}$ and $\mathcal{T} = \mathcal{E}_{\boldsymbol{\Delta}}(\mathcal{T})$, then this reduces to the *extreme components* studied by Welling et al. (2004).

9.9.2.4 Selection of the Dimension u

Sequential likelihood ratio testing can be used to help determine the dimensionality u of the envelope. For example, consider model (9.28). The hypothesis $u = u_0$ can be tested by using the likelihood ratio statistic $\Lambda(u_0) = 2(\hat{L}_p - \hat{L}_{u_0})$, where $\hat{L}_p = -(np)/2 - (n/2)\log|\mathbf{S_X}|$ denotes the maximum value of the log likelihood for the full model ($u = p$), and $\hat{L}_{u_0}$ denotes the maximum value of the log likelihood when $u = u_0$. The total number of parameters needed to estimate model (9.28) is $p + u(u+1)/2 + u(p-u) + 1$. The first term corresponds to the estimation of the grand mean $\boldsymbol{\mu}$. The second term corresponds to the estimation of the unconstrained symmetric matrix $\boldsymbol{\Psi}$. The third term corresponds to the number of parameters needed to describe the subspace $\mathcal{E}_{\boldsymbol{\Delta}}(\mathcal{T})$. The last term corresponds to σ_0^2. Following standard likelihood theory, under the null hypothesis, $\Lambda(u_0)$ is distributed asymptotically as a chi-squared random variable with $(p - u_0 + 2)(p - u_0 - 1)/2$ degrees of freedom. To choose u, start with $u_0 = 0$, test the hypothesis $u = u_0$ against the alternative $u > u_0$. If the test is rejected, increment u_0 by 1 and test again, stopping at the first hypothesis that is not rejected.

9.9.3 Fixed Latent Variables and Isotropic Errors

Model (9.25) is based on the assumption that $\mathbf{v} \sim N_d(0, \mathbf{I}_d)$. However, we obtain the same reduction if we condition on the realized (but unobserved) values $\mathbf{v}_1, \dots, \mathbf{v}_n$ of $\mathbf{v}$ and treat them as fixed centered vectors so that $\sum_{i=1}^n \mathbf{v}_i = 0$. Since the $\mathbf{v}_i$'s are now fixed, we absorb $\boldsymbol{\delta}$ into them and write the isotropic model as

$$\mathbf{X}_i = \boldsymbol{\mu} + \boldsymbol{\Theta}\mathbf{v}_i + \boldsymbol{\varepsilon}_i, \quad i = 1, \dots, n. \tag{9.31}$$

The log likelihood for this setting is

$$L_d(\boldsymbol{\mu}, \mathcal{T}, \sigma^2, \mathbf{v}_1, \dots, \mathbf{v}_n) = -(np/2)\log(\sigma^2) - (1/2\sigma^2)\sum_{i=1}^n \|\mathbf{X}_i - \boldsymbol{\mu} - \boldsymbol{\Theta}\mathbf{v}_i\|^2.$$

Because $\sum_{i=1}^n \mathbf{v}_i = 0$, it follows that $\hat{\boldsymbol{\mu}} = \bar{\mathbf{X}}$. To estimate $\mathbf{v}_i$ with the remaining parameters held fixed, we need to minimize $\|\mathbf{X}_i - \bar{\mathbf{X}} - \boldsymbol{\Theta}\mathbf{v}_i\|^2$. This is just ordinary least squares and consequently we minimize by setting $\mathbf{v}_i = \boldsymbol{\Theta}^T(\mathbf{X}_i - \bar{\mathbf{X}})$.

These automatically satisfy the constraint $\sum_{i=1}^{n} \boldsymbol{v}_i = 0$. Substituting back, we get the partially maximized log likelihood

$$L_d^{(1)}(\mathcal{T}, \sigma^2) = -(np/2)\log(\sigma^2) - (n/2\sigma^2)\mathrm{tr}(\mathbf{S}_{\mathbf{X}}) + (n/2\sigma^2)\mathrm{tr}(\mathbf{P}_{\mathcal{T}}\mathbf{S}_{\mathbf{X}}).$$

It follows from Lemma A.17 that $\hat{\mathcal{T}} = \mathrm{span}(\hat{\ell}_1, \dots, \hat{\ell}_d)$. Again there is no notion of relative importance of vectors in $\hat{\mathcal{T}}$. Substituting this estimator into $L_u^{(1)}$, we get

$$L_d^{(2)}(\sigma^2) = -(np/2)\log(\sigma^2) - (n/2\sigma^2)\sum_{i=d+1}^{p} \hat{\varphi}_i,$$

which is maximized at $\hat{\sigma}^2 = \sum_{i=d+1}^{p} \hat{\varphi}_i/p$.

It follows from the above results that the estimated reduction evaluated at the data points $\widehat{\mathbf{R}}(\mathbf{X}_i) = (\hat{\ell}_1, \hat{\ell}_2, \dots, \hat{\ell}_d)^T\mathbf{X}_i$ consists of the first d sample principal components of $\mathbf{X}$. Consequently, we now have two related models that produce principal components as maximum likelihood estimators.

9.9.4 Numerical Illustrations

The following simulation example may help to further fix ideas. Observations on $\mathbf{X}$ were generated from model (9.23) as $\mathbf{X}_i = \boldsymbol{\theta}\boldsymbol{v}_i + \boldsymbol{\varepsilon}_i$, where $\boldsymbol{v}_i$ was sampled uniformly from the boundary of the square $[-1, 1]^2$, and the elements of the $p \times 2$ matrix $\boldsymbol{\theta}$ were sampled independently from a standard normal distribution. Then $\boldsymbol{\theta}$ was normalized to give $\boldsymbol{\Theta} = \boldsymbol{\theta}(\boldsymbol{\theta}^T\boldsymbol{\theta})^{-1/2}$, leading to the representation

$$\mathbf{X}_i = \boldsymbol{\Theta}(\boldsymbol{\theta}^T\boldsymbol{\theta})^{1/2}\boldsymbol{v}_i + \boldsymbol{\varepsilon}_i = \boldsymbol{\Theta}\boldsymbol{\delta}\boldsymbol{v}_i + \boldsymbol{\varepsilon}_i,$$

where $\boldsymbol{\delta} = (\boldsymbol{\theta}^T\boldsymbol{\theta})^{1/2}$, and the error vector $\boldsymbol{\varepsilon}$ was sampled from a normal distribution with mean 0 and variance matrix $\mathbf{I}_p$. The sampling used to construct $\boldsymbol{\Theta}$ is for convenience only; the model is still conditional on $\boldsymbol{\Theta}$ regardless of how it was obtained. This sampling process was repeated $n = 80$ times for various values of p. If PCA works as predicted by the theory, then a plot of the first two sample principal components should recover the square that holds the $\boldsymbol{v}$'s.

Figure 9.4 shows plots of the first two principal components for four values of p. We see that for small p, the square is not recognizable, but for larger values of p, the square is quite clear. In Figure 9.4d, $p = 500$, while the number of observations is still $n = 80$. The sides of the estimated square in Figure 9.4d do not align with the coordinate axes because the method is designed to estimate only the subspace $\mathcal{T}$ with isotropic errors.

Intuition about why principal components are apparently working in this example can be found in the work of Johnstone and Lu (2009) who studied the asymptotic behavior of the first principal component direction under model (9.23) with $d = 1$ and $\boldsymbol{v} \sim N(0, 1)$. In this case, $\boldsymbol{\delta}$ is simply the length

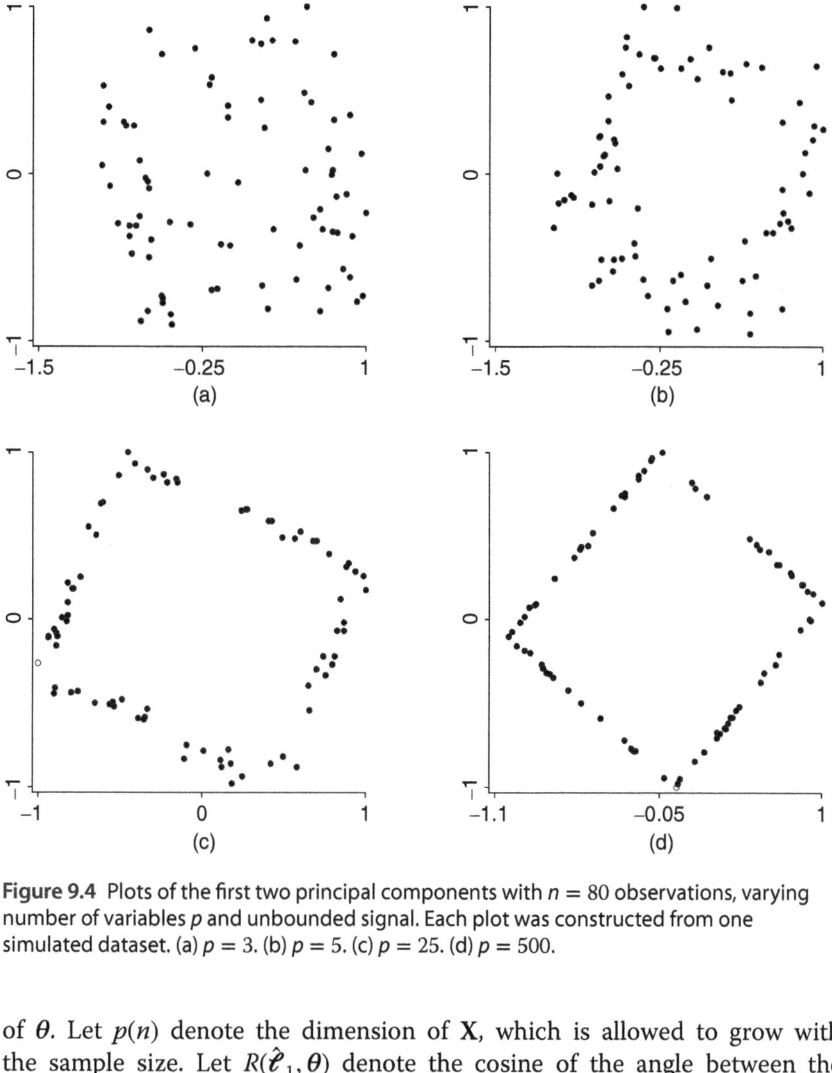

Figure 9.4 Plots of the first two principal components with $n = 80$ observations, varying number of variables p and unbounded signal. Each plot was constructed from one simulated dataset. (a) $p = 3$. (b) $p = 5$. (c) $p = 25$. (d) $p = 500$.

of $\boldsymbol{\theta}$. Let $p(n)$ denote the dimension of $\mathbf{X}$, which is allowed to grow with the sample size. Let $R(\hat{\boldsymbol{\ell}}_1, \boldsymbol{\theta})$ denote the cosine of the angle between the first principal component direction $\hat{\boldsymbol{\ell}}_1$ and $\boldsymbol{\theta}$, let $\omega = \lim_{n\to\infty} \delta^2/\sigma^2$, and let $c = \lim_{n\to\infty} p(n)/n$. Then (Johnstone and Lu 2009, Theorem 1)

$$\lim_{n\to\infty} R^2(\hat{\boldsymbol{\ell}}_1, \boldsymbol{\theta}) = R^2_\infty(\omega, c) = \frac{(\omega^2 - c)_+}{\omega^2 + c\omega}.$$

This implies that if ω is finite, then $R^2_\infty < 1$ if and only if $c > 0$, and consequently $\hat{\boldsymbol{\ell}}_1$ is a consistent estimator of span($\boldsymbol{\theta}$) if and only if $p(n)/n \to 0$. These results can be taken to imply that the sample size should be large relative to the number of variables for principal components to work well in practice when ω is finite.

However, if ω is large and $\omega \gg c$, then $R^2 \approx 1$, and principal components can be expected to work well. Suppose for example that $p(n)/n$ is bounded above and that the elements of $\boldsymbol{\theta}$ are sampled from a distribution with mean 0 and finite fourth moment. Then ω^2 will diverge, and principal components should work well. This is effectively what seems to have happened in the illustration of Figure 9.4, although $d = 2$ in that illustration. Because the elements of $\boldsymbol{\theta} \in \mathbb{R}^{p\times 2}$ were sampled from a standard normal distribution, $\boldsymbol{\delta}$ will diverge, leading to an ever-increasing signal. This suggests that principal components should work well if information accumulates as p increases.

The previous results also suggest that principal components will not work as well when only finitely many variables are relevant; that is, $\boldsymbol{\theta}$ has only finitely many nonzero rows or the signal grows very slowly, unless $n \gg p$. For example, Figure 9.5 shows the results of a simulation conducted like that for Figure 9.4,

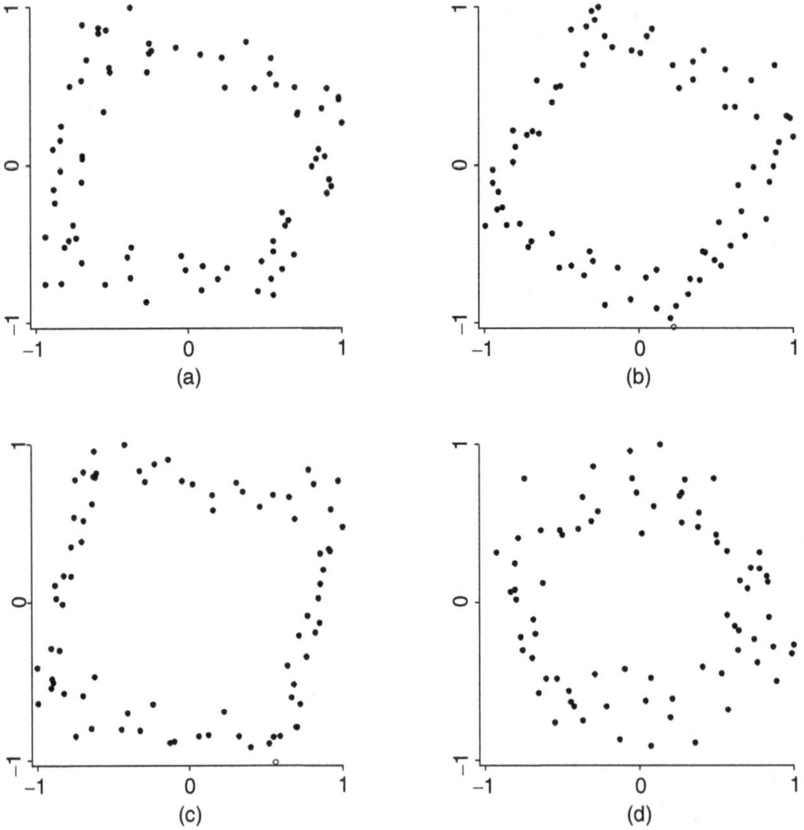

Figure 9.5 Plots of the first two principal components with $n = 80$ observations, varying number of predictors p and bounded signal. Each plot was constructed from one simulated dataset. (a) $p = 5$. (b) $p = 25$. (c) $p = 250$. (d) $p = 1500$.

except the ith row of θ was sampled from a normal distribution with mean 0 and variance $(10/i)\mathbf{I}_2$. This suggests that $\hat{\boldsymbol{\ell}}_1$ will be an inconsistent estimator of span(θ), unless $c = 0$. The patterns shown in Figure 9.5 seem to support this conclusion. Although p has been increased from 5 to 1500, the signal does not seem to be increasing. Indeed, the best of the results shown are for $p = 250$, suggesting that there could be value in screening to remove the less informative variables. A screening method for $d = 1$ was developed by Johnstone and Lu (2009).

9.10 Principal Fitted Components

Principal fitted components (PFC) can be seen as representing a transition between PCA and sufficient dimension reduction for the regression of response $Y \in \mathbb{R}^1$ on predictors $\mathbf{X} \in \mathbb{R}^p$. In model (9.23), $\boldsymbol{\nu}$ is a random vector that is intended to capture extrinsic latent structure that contributes to the variation in $\mathbf{X}$. The extrinsic structure of interest in regression stems from the response, so we are specifically interested in extracting the part of $\boldsymbol{\nu}$ that is associated with Y. Since the observed responses are known, we can adopt a regression formulation starting with the conditional mean of (9.23): $\mathrm{E}(\mathbf{X} \mid Y) = \boldsymbol{\mu} + \boldsymbol{\Theta}\boldsymbol{\delta}\mathrm{E}(\boldsymbol{\nu} \mid Y)$, where, in reference to (9.23), $\theta = \boldsymbol{\Theta}\boldsymbol{\delta}$ as defined for (9.25). We next model $\boldsymbol{\delta}\mathrm{E}(\boldsymbol{\nu} \mid Y) = \mathbf{hf}(Y)$, where $\mathbf{f} \in \mathbb{R}^r$ is a known user-specified vector-valued function of the response and $\mathbf{h} \in \mathbb{R}^{d \times r}$ with $r \geq d$, leading to the inverse regression model

$$\mathbf{X}_i = \boldsymbol{\mu} + \boldsymbol{\Theta}\mathbf{hf}(Y_i) + \boldsymbol{\epsilon}_i, \quad i = 1, \ldots, n, \tag{9.32}$$

where the $\boldsymbol{\epsilon}_i$s are independent copies of a $N_p(0, \boldsymbol{\Sigma}_{\mathbf{X}|Y})$ random variable. If $\mathrm{E}(\boldsymbol{\nu} \mid Y)$ spans $\mathbb{R}^d$ as the value of Y varies in its marginal sample space, then $\mathrm{rank}(\mathbf{h}) = d$ and we can work with model (9.32). However, if $\mathrm{E}(\boldsymbol{\nu} \mid Y)$ spans a lower-dimensional space, then the rank of $\mathbf{h}$ is less than d and model (9.32) is overparameterized. We continue the discussion assuming that $\mathrm{rank}(\mathbf{h}) = d$, understanding that (9.32) can be reparameterized to remove the overparameterization when $\mathrm{rank}(\mathbf{h}) < d$. Model (9.32) can be seen as a targeted form of (9.23) that is conditioned on the responses, still assuming that $\boldsymbol{\epsilon} \sim N(0, \boldsymbol{\Sigma}_{\mathbf{X}|Y})$, although here $\boldsymbol{\Sigma}_{\mathbf{X}|Y}$ need not be the same as the $\boldsymbol{\Delta}$ in (9.23). Methods for selecting $\mathbf{f}$ were discussed by Cook (2007), Cook and Forzani (2008b), and Adragni (2009). Some ways of selecting $\mathbf{f}$ are discussed in Section 9.10.3.

Under (9.32), $\mathcal{S}_{Y|\mathbf{X}} = \mathrm{span}(\boldsymbol{\Sigma}_{\mathbf{X}|Y}^{-1}\boldsymbol{\Theta})$; that is,

$$\mathbf{X} \mid (Y, \boldsymbol{\Theta}^T\boldsymbol{\Sigma}_{\mathbf{X}|Y}^{-1}\mathbf{X}) \sim \mathbf{X} \mid \boldsymbol{\Theta}^T\boldsymbol{\Sigma}_{\mathbf{X}|Y}^{-1}\mathbf{X}, \tag{9.33}$$

which agrees with the conclusion of Proposition 9.2 that the centered conditional means are contained in $\boldsymbol{\Sigma}_{\mathbf{X}}\mathcal{S}_{Y|\mathbf{X}}$.

Let $\boldsymbol{\beta} = \boldsymbol{\Theta}\mathbf{h}$ denote the coefficient matrix for the regression of $\mathbf{X}$ on $\mathbf{f}$. Then

$$\text{span}(\boldsymbol{\Sigma}_{\mathbf{X}|Y}^{-1}\boldsymbol{\Theta}) = \text{span}(\boldsymbol{\Sigma}_{\mathbf{X}|Y}^{-1}\boldsymbol{\Theta}\mathbf{h}) = \text{span}(\boldsymbol{\Sigma}_{\mathbf{X}|Y}^{-1}\boldsymbol{\beta}) = \mathcal{S}_{Y|\mathbf{X}}. \tag{9.34}$$

Accordingly, once d is available, we can estimate the central subspace from a fit of model (9.32) by substituting the estimates of $\boldsymbol{\Sigma}_{\mathbf{X}|Y}$, $\boldsymbol{\Theta}$ and $\mathbf{h}$ into the left-hand side of (9.34). We next consider estimation of $\mathcal{S}_{Y|\mathbf{X}}$ under two versions of model (9.32).

9.10.1 Isotropic Errors, $\boldsymbol{\Sigma}_{\mathbf{X}|Y} = \sigma^2\mathbf{I}_p$

With isotropic errors, model (9.32) is like model (9.25), but with the latent effects modeled as $\boldsymbol{\nu}_i = \mathbf{hf}(Y_i)$. From (9.34), the central subspace is simply $\mathcal{S}_{Y|\mathbf{X}} = \text{span}(\boldsymbol{\Theta})$, so we need to estimate only $\text{span}(\boldsymbol{\Theta})$. Assuming normal errors and known dimension d, the maximum likelihood estimator of $\mathcal{S}_{Y|\mathbf{X}}$ is the span of the first d eigenvectors of the sample covariance matrix $\mathbf{S}_{\mathbf{X}\circ\mathbf{f}}$ of the fitted vectors from the linear regression of $\mathbf{X}$ on $\mathbf{f}$ (Cook 2007; Cook and Forzani 2008b). More specifically, for a sample $(\mathbf{X}_1, Y_1) \dots (\mathbf{X}_n, Y_n)$, let $\mathbb{X}$ denote the $n \times p$ matrix with rows $(\mathbf{X}_i - \bar{\mathbf{X}})^T$, and let $\mathbb{F}$ denote the $n \times r$ matrix with rows $(\mathbf{f}(y_i) - \bar{\mathbf{f}})^T$. Then the maximum likelihood estimator of $\mathcal{S}_{Y|\mathbf{X}}$ is the subspace spanned by the first d eigenvectors $\boldsymbol{\ell}_{i,\text{fit}}$ of $\mathbf{S}_{\mathbf{X}\circ\mathbf{f}} = \mathbb{X}^T\mathbf{P}_{\mathbb{F}}\mathbb{X}/n$. The corresponding estimator of σ^2 is

$$\hat{\sigma}^2 = \left(\sum_{j=1}^{p} \hat{\varphi}_i - \sum_{j=1}^{d} \hat{\varphi}_i^{\text{fit}}\right) \Big/ p,$$

where the $\hat{\varphi}_j^{\text{fit}}$'s are the eigenvalues of $\mathbf{S}_{\mathbf{X}\circ\mathbf{f}}$, and the $\hat{\varphi}_j$'s are the eigenvalues of $\mathbf{S}_{\mathbf{X}}$.

The estimated reductions from evaluating at the data points $\boldsymbol{\ell}_{i,\text{fit}}^T\mathbf{X}$ are called *principal fitted components* since they consist of the first d sample principal components of the fitted vectors $\widehat{\mathbf{X}}_i$. The rank of the sample covariance matrix $\mathbf{S}_{\mathbf{X}\circ\mathbf{f}}$ will typically equal r and so it will have only r positive eigenvalues. This means that the choice of $\mathbf{f}(y)$ automatically bounds the dimension of $\mathcal{S}_{Y|\mathbf{X}}$, $d \leq r$. Theoretical properties of principal fitted components were studied by Johnson (2008) who argued that they should outperform principal components.

We can arrive at an envelope model by using a slightly different line of reasoning. Write $\boldsymbol{\delta\nu} = \boldsymbol{\delta}\text{E}(\boldsymbol{\nu} \mid Y) + \boldsymbol{\epsilon}$, where we assume that $\boldsymbol{\epsilon} \perp\!\!\!\perp (\boldsymbol{\varepsilon}, Y)$. Substituting this into model (9.25), we get

$$\begin{aligned}\mathbf{X} &= \boldsymbol{\mu} + \boldsymbol{\Theta\delta\nu} + \boldsymbol{\varepsilon} \\ &= \boldsymbol{\mu} + \boldsymbol{\Theta\delta}\text{E}(\boldsymbol{\nu} \mid Y) + \boldsymbol{\Theta\epsilon} + \boldsymbol{\varepsilon} \\ &= \boldsymbol{\mu} + \boldsymbol{\Theta}\mathbf{hf}(Y) + \boldsymbol{\Theta\epsilon} + \boldsymbol{\varepsilon},\end{aligned}$$

where in the last step we modeled $\delta \mathrm{E}(\boldsymbol{\nu} \mid Y = y) = \mathbf{hf}(y)$ as before. Since $\mathrm{span}(\boldsymbol{\Theta})$ is a reducing subspace of $\mathrm{var}(\boldsymbol{\Theta}\boldsymbol{\epsilon} + \boldsymbol{\varepsilon}) = \boldsymbol{\Theta}(\boldsymbol{\Sigma}_{\epsilon} + \sigma^2 \mathbf{I})\boldsymbol{\Theta} + \sigma^2 \boldsymbol{\Theta}_0 \boldsymbol{\Theta}_0^T$, we have an envelope model. Estimation under this model can be developed following the general steps outlined elsewhere in this book.

9.10.2 Anisotropic Errors, $\boldsymbol{\Sigma}_{\mathbf{X}|Y} > 0$

With general errors $\boldsymbol{\Sigma}_{\mathbf{X}|Y} > 0$, model (9.32) becomes an instance of the reduced-rank regression (9.3), with response $\mathbf{X}$ and predictor $\mathbf{f}$. Adapting the estimators of $\boldsymbol{\beta}$ and $\boldsymbol{\Sigma}$ given at (9.5) and (9.6), we have the maximum likelihood estimators for model (9.32),

$$\widehat{\boldsymbol{\beta}} = \widehat{\mathbf{H}}\widehat{\mathbf{h}} = \mathbf{S}_{\mathbf{X}}^{1/2} \mathbf{C}_{\mathbf{X},\mathbf{f}}^{(d)} \mathbf{S}_{\mathbf{f}}^{-1/2} \tag{9.35}$$

$$\widehat{\boldsymbol{\Sigma}}_{\mathbf{X}|Y} = \mathbf{S}_{\mathbf{X}}^{1/2}(\mathbf{I}_p - \mathbf{C}_{\mathbf{X},\mathbf{f}}^{(d)} \mathbf{C}_{\mathbf{f},\mathbf{X}}^{(d)}) \mathbf{S}_{\mathbf{X}}^{1/2}. \tag{9.36}$$

The estimator of the central subspace with known dimension d is then

$$\widehat{\mathcal{S}}_{Y|\mathbf{X}} = \mathrm{span}(\widehat{\boldsymbol{\Sigma}}_{\mathbf{X}|Y}^{-1} \widehat{\boldsymbol{\beta}}) = \mathrm{span}(\mathbf{S}_{\mathbf{X}}^{-1/2} \mathbf{C}_{\mathbf{X},\mathbf{f}}^{(d)}), \tag{9.37}$$

where the second equality follows from reexpressing $(\mathbf{I}_p - \mathbf{C}_{\mathbf{X},\mathbf{f}}^{(d)} \mathbf{C}_{\mathbf{f},\mathbf{X}}^{(d)})^{-1}$ by using the Woodbury matrix identity.

As in Section 9.2.3, the envelope version of model (9.32) is obtained by parameterizing it in terms of the $\boldsymbol{\Sigma}_{\mathbf{X}|Y}$ envelope of $\mathrm{span}(\boldsymbol{\Theta})$, leading to the adaptation of model (9.7),

$$\mathbf{X} = \boldsymbol{\alpha} + \boldsymbol{\Gamma}\boldsymbol{\eta}\mathbf{hf}(y) + \boldsymbol{\varepsilon}, \quad \boldsymbol{\Sigma}_{\mathbf{X}|Y} = \boldsymbol{\Gamma}\boldsymbol{\Omega}\boldsymbol{\Gamma}^T + \boldsymbol{\Gamma}_0 \boldsymbol{\Omega}_0 \boldsymbol{\Gamma}_0^T. \tag{9.38}$$

Accordingly, the central subspace is given by

$$\mathcal{S}_{Y|\mathbf{X}} = \mathrm{span}(\boldsymbol{\Sigma}_{\mathbf{X}|Y}^{-1} \boldsymbol{\beta}) = \mathrm{span}(\boldsymbol{\Gamma}\boldsymbol{\Omega}^{-1}\boldsymbol{\eta}\mathbf{h}) = \mathcal{E}_{\boldsymbol{\Sigma}_{\mathbf{X}|Y}}(\boldsymbol{\beta}).$$

The central subspace can then be estimated by using the following steps:

1. Select d and u using one of the methods referenced at the end of Section 9.2.3.
2. Estimate $\boldsymbol{\Gamma}$ by finding an argument $\widehat{\boldsymbol{\Gamma}}$ that minimizes objective function (9.10) with $\mathbf{Y}$ replaced by $\mathbf{X}$ and $\mathbf{X}$ replaced by $\mathbf{f}$.
3. Regress the reduced predictor vector $\widehat{\boldsymbol{\Gamma}}^T\mathbf{X}$ on $\mathbf{f}$, and then use (9.35) and (9.36) to estimate $\boldsymbol{\Omega}$ and $\boldsymbol{\eta}\mathbf{h}$:
$$\widehat{\boldsymbol{\eta}}\widehat{\mathbf{h}} = \mathbf{S}_{\widehat{\boldsymbol{\Gamma}}^T\mathbf{X}}^{1/2} \mathbf{C}_{\widehat{\boldsymbol{\Gamma}}^T\mathbf{X},\mathbf{f}}^{(d)} \mathbf{S}_{\mathbf{f}}^{-1/2}$$
$$\widehat{\boldsymbol{\Omega}} = \mathbf{S}_{\widehat{\boldsymbol{\Gamma}}^T\mathbf{X}}^{1/2}(\mathbf{I}_u - \mathbf{C}_{\widehat{\boldsymbol{\Gamma}}^T\mathbf{X},\mathbf{f}}^{(d)} \mathbf{C}_{\mathbf{f},\widehat{\boldsymbol{\Gamma}}^T\mathbf{X}}^{(d)}) \mathbf{S}_{\widehat{\boldsymbol{\Gamma}}^T\mathbf{X}}^{1/2}.$$
4. The central subspace is then estimated as
$$\widehat{\mathcal{S}}_{Y|\mathbf{X}} = \mathrm{span}(\widehat{\boldsymbol{\Gamma}}\widehat{\boldsymbol{\Omega}}^{-1}\widehat{\boldsymbol{\eta}}) = \mathrm{span}(\widehat{\boldsymbol{\Gamma}}\mathbf{S}_{\widehat{\boldsymbol{\Gamma}}^T\mathbf{X}}^{-1/2} \mathbf{C}_{\widehat{\boldsymbol{\Gamma}}^T\mathbf{X},\mathbf{f}}^{(d)}).$$

9.10.3 Nonnormal Errors and the Choice of f

The estimators in the preceding sections were developed under the assumptions that $\mathbf{f}(Y)$ correctly models $\mathrm{E}(\mathbf{X} \mid Y)$ and that the errors in model (9.32) are normal. Neither of these assumptions is crucial for the success of those methods.

We relax normality by assuming that the errors $\boldsymbol{\varepsilon}$ for model (9.32) have finite fourth moments and are independent of Y. We allow for misspecification of $\mathbf{f}$ by assuming that $\mathrm{E}(\mathbf{X} \mid Y) = \boldsymbol{\alpha} + \boldsymbol{\Theta}\boldsymbol{\phi}(Y)$ for $\boldsymbol{\phi} \in \mathbb{R}^d$. Let $\boldsymbol{\rho}_{\phi,\mathbf{f}} \in \mathbb{R}^{d\times r}$ denote the matrix of cross correlations between the elements of $\boldsymbol{\phi}$ and $\mathbf{f}$. Then the estimators in Section 9.10.2 are $\sqrt{n}$-consistent if and only if $\boldsymbol{\rho}_{\phi,\mathbf{f}}$ has rank d (Cook and Forzani 2008b, Theorem 3.5). As a consequence, there is considerable flexibility in the choice of $\mathbf{f}$, and some misspecification is not necessarily worrisome. As long as $\mathbf{f}$ is sufficiently correlated with the true function $\boldsymbol{\phi}$, we may expect useful results. We next describe some possibilities for the choice of $\mathbf{f}$.

9.10.3.1 Graphical Choices

Under model (9.32), the individual elements X_j of $\mathbf{X}$ each follow a univariate linear model with predictor $\mathbf{f}$:

$$X_j = \alpha_j + \boldsymbol{\beta}_j^T \mathbf{f} + \varepsilon_j.$$

Consequently, we are able to use inverse response plots (Cook 1998, Chapter 10) of X_j versus Y, $j = 1, \ldots, p$, to gain graphical information about suitable choices for $\mathbf{f}$, which is an ability that is not generally available in the forward regression of Y on $\mathbf{X}$.

9.10.3.2 Basis Functions

There are several generic possibilities for the choice of $\mathbf{f}$, perhaps guided by graphics. Polynomials deriving from a Taylor approximation,

$$\mathbf{f}(y) = \{y, y^2, y^3, \ldots, y^r\}^T,$$

are one possibility. Periodic behavior could be modeled using a Fourier series form

$$\mathbf{f}(y) = \{\cos(2\pi y), \sin(2\pi y), \ldots, \cos(2\pi k y), \sin(2\pi k y)\}^T$$

as perhaps in signal processing applications. Here, k is a user-selected integer and $r = 2k$. Splines and other types of nonparametric constructions could also be used to form a suitable $\mathbf{f}$. A variety of basis functions are available from Adragni (2009).

9.10.3.3 Categorical Response

In some regressions, there may be a natural choice for $\mathbf{f}$. Suppose for instance that Y is categorical, taking values in one of h categories C_k, $k = 1, \ldots, c$. We can then set $r = c - 1$ and specify the kth element of $\mathbf{f}$ to be $J(y \in C_k)$, where J is the indicator function. Here there is no approximation in $\mathbf{f}$.

It may help fix ideas to see how we arrive at an inverse model with a categorical response starting from the beginning. Let $\mathrm{E}(\mathbf{X}|Y \in C_k) = \boldsymbol{\mu}_k = \boldsymbol{\mu}\mathbf{J}(y)$, where the columns of the $p \times h$ matrix $\boldsymbol{\mu}$ are $\boldsymbol{\mu}_k$, $k = 1, \ldots, c$, and $\mathbf{J}(y)$ is the $c \times 1$ indicator vector with elements $J(y \in C_k)$, $k = 1, \ldots, c$. Centering the columns of $\boldsymbol{\mu}$ we have

$$\begin{aligned}\mathrm{E}(\mathbf{X}|Y \in C_k) &= (\boldsymbol{\mu} - \bar{\boldsymbol{\mu}}\mathbf{1}_c^T + \bar{\boldsymbol{\mu}}\mathbf{1}_c^T)\mathbf{J}(Y)\\ &= \bar{\boldsymbol{\mu}}\mathbf{1}_c^T\mathbf{J}(Y) + (\boldsymbol{\mu} - \bar{\boldsymbol{\mu}}\mathbf{1}_c^T)\mathbf{J}(Y)\\ &= \bar{\boldsymbol{\mu}} + (\boldsymbol{\mu} - \bar{\boldsymbol{\mu}}\mathbf{1}_c^T)\mathbf{J}(Y)\\ &= \bar{\boldsymbol{\mu}} + \boldsymbol{\Theta}\mathbf{h}\mathbf{J}(Y),\end{aligned}$$

where $\mathbf{1}_c^T\mathbf{J}(Y) = 1$ and $\boldsymbol{\Theta}$ is a semi-orthogonal basis matrix for $\mathrm{span}(\boldsymbol{\mu} - \bar{\boldsymbol{\mu}}\mathbf{1}_c^T)$. This model is not estimable because the elements of $\mathbf{J}$ always sum to 1. To remove this linear dependency, we can just remove the last element of $\mathbf{J}$ and define

$$\mathbf{f}(Y) = \{J(Y \in C_1), \ldots, J(Y \in C_{c-1})\}^T. \tag{9.39}$$

9.10.3.4 Sliced Inverse Regression

Another option with continuous responses consists of "slicing" the observed values of Y into h bins (categories) C_k, $k = 1, \ldots, c$, and then specifying the kth coordinate of $\mathbf{f}$ as for the case of a categorical Y. This has the effect of approximating each conditional mean $\mathrm{E}(X_j \mid Y)$ as a step function of Y with c steps,

$$\mathrm{E}(X_j \mid Y) \approx \mu_j + \sum_{k=1}^{c-1} \boldsymbol{\Theta}_j^T\mathbf{h}_k J(Y \in C_k),$$

where $\boldsymbol{\Theta}_j^T$ is the jth row of $\boldsymbol{\Theta}$ and $\mathbf{h}_k$ is the kth column of $\mathbf{h}$.

The estimator of the central subspace (9.37) is the same as that given by sliced inverse regression (Section 9.5) when the response is categorical or slicing is used to categorize a continuous response. Additionally, if the errors are normal and the response is categorical, then sliced inverse regression gives the maximum likelihood estimator of the central subspace (Cook and Forzani 2008b).

9.10.4 High-Dimensional PFC

Our discussion of PFCs with general $\boldsymbol{\Sigma}_{\mathbf{X}|Y}$ has so far required $\mathbf{S}_{Y|\mathbf{X}} > 0$, and thus we need $n > p$. Versions of PFCs for high-dimensional regressions in which we may have $n < p$ were studied by Cook et al. (2012). Their treatment allows for a variety of different estimators, depending on the assumed form of $\boldsymbol{\Sigma}_{\mathbf{X}|Y}$. They obtained particularly promising results using the sparse permutation invariant covariance estimator of Rothman et al. (2008) to estimate $\boldsymbol{\Sigma}_{\mathbf{X}|Y}$.

Appendix A

Envelope Algebra

We consider the linear algebra and statistical foundations of envelopes in this appendix. Section A.1 contains a discussion of invariant and reducing subspaces, which are precursors to envelopes introduced in Section A.2. Relationships between envelopes are discussed in Section A.3. Kronecker products, commutation and related operators are discussed in Sections A.4 and A.5. Derivatives related to envelope derivations are summarized in Section A.6. Miscellaneous results and the matrix normal distribution are discussed in Sections A.7 and A.8. Notes on the literature related to this appendix are given in Section A.9.

A.1 Invariant and Reducing Subspaces

The following definition of invariant and reducing subspaces is a common construction in linear algebra (Conway, 1990).

Definition A.1 A subspace $\mathcal{R}$ of $\mathbb{R}^r$ is an *invariant subspace* of $\mathbf{M} \in \mathbb{R}^{r \times r}$ if $\mathbf{M}\mathcal{R} \subseteq \mathcal{R}$; so $\mathbf{M}$ maps $\mathcal{R}$ to a subset of itself. $\mathcal{R}$ is a *reducing subspace* of $\mathbf{M}$ if $\mathcal{R}$ and $\mathcal{R}^{\perp}$ are both invariant subspaces of $\mathbf{M}$. If $\mathcal{R}$ is a reducing subspace of $\mathbf{M}$, we say that $\mathcal{R}$ reduces $\mathbf{M}$.

The notion of reduction conveyed by this definition is incompatible with how reduction is usually understood in statistics. Nevertheless, it is a recognized term in linear algebra and functional analysis, and it forms a convenient starting point for envelopes. The following results provide characterizations that may facilitate intuition and will be helpful later.

The next lemma describes a matrix equation that characterizes invariant subspaces.

Lemma A.1 *Let $\mathcal{R}$ be a u dimensional subspace of $\mathbb{R}^r$ and let $\mathbf{M} \in \mathbb{R}^{r \times r}$. Then $\mathcal{R}$ is an invariant subspace of $\mathbf{M}$ if and only if, for any $\mathbf{A} \in \mathbb{R}^{r \times s}$ with* $\text{span}(\mathbf{A}) = \mathcal{R}$, *there exists a $\mathbf{B} \in \mathbb{R}^{s \times s}$ such that $\mathbf{MA} = \mathbf{AB}$.*

An Introduction to Envelopes: Dimension Reduction for Efficient Estimation in Multivariate Statistics, First Edition. R. Dennis Cook.

Proof: Suppose there is a $\mathbf{B}$ that satisfies $\mathbf{MA} = \mathbf{AB}$. For every $\mathbf{v} \in \mathcal{R}$, there is a $\mathbf{t} \in \mathbb{R}^s$ so that $\mathbf{v} = \mathbf{At}$. Consequently, $\mathbf{Mv} = \mathbf{MAt} = \mathbf{ABt} \in \mathcal{R}$, which implies that $\mathcal{R}$ is an invariant subspace of $\mathbf{M}$.

Suppose that $\mathcal{R}$ is an invariant subspace of $\mathbf{M}$, and let $\mathbf{a}_j$, $j = 1, \ldots, s$ denote the columns of $\mathbf{A}$. Then $\mathbf{Ma}_j \in \mathcal{R}$, $j = 1, \ldots, s$. Consequently, $\text{span}(\mathbf{MA}) \subseteq \mathcal{R}$, which implies there is a $\mathbf{B} \in \mathbb{R}^{s\times s}$ such that $\mathbf{MA} = \mathbf{AB}$. □

The next result gives a connection between reducing subspaces and the eigenvectors of a symmetric $\mathbf{M}$. It does not require that the eigenvalues of $\mathbf{M}$ be unique.

Lemma A.2 *Suppose that $\mathcal{R}$ reduces $\mathbf{M} \in \mathbb{S}^{r\times r}$. Then $\mathbf{M}$ has a spectral decomposition with eigenvectors in $\mathcal{R}$ or in $\mathcal{R}^\perp$.*

Proof: Let $\mathbf{A}_0 \in \mathbb{R}^{r\times u}$ be a semi-orthogonal matrix whose columns span $\mathcal{R}$ and let $\mathbf{A}_1$ be its completion, such that $(\mathbf{A}_0, \mathbf{A}_1) \equiv \mathbf{A}$ is an orthogonal matrix. Because $\mathbf{M}\mathcal{R} \subseteq \mathcal{R}$ and $\mathbf{M}\mathcal{R}^\perp \subseteq \mathcal{R}^\perp$, it follows from Lemma A.1 there exist matrices $\mathbf{B}_0 \in \mathbb{R}^{u\times u}$ and $\mathbf{B}_1 \in \mathbb{R}^{(r-u)\times(r-u)}$ such that $\mathbf{MA}_0 = \mathbf{A}_0\mathbf{B}_0$ and $\mathbf{MA}_1 = \mathbf{A}_1\mathbf{B}_1$. Hence

$$\mathbf{M}\left(\mathbf{A}_0\ \mathbf{A}_1\right) = \left(\mathbf{A}_0\ \mathbf{A}_1\right)\begin{pmatrix}\mathbf{B}_0 & 0\\ 0 & \mathbf{B}_1\end{pmatrix} \Leftrightarrow \mathbf{M} = \mathbf{A}\begin{pmatrix}\mathbf{B}_0 & 0\\ 0 & \mathbf{B}_1\end{pmatrix}\mathbf{A}^T.$$

Because $\mathbf{M}$ is symmetric, $\mathbf{B}_0$ and $\mathbf{B}_1$ are also symmetric. Hence, $\mathbf{B}_0$ and $\mathbf{B}_1$ have spectral decompositions $\mathbf{C}_0\boldsymbol{\Lambda}_0\mathbf{C}_0^T$ and $\mathbf{C}_1\boldsymbol{\Lambda}_1\mathbf{C}_1^T$ for some diagonal matrices $\boldsymbol{\Lambda}_0$ and $\boldsymbol{\Lambda}_1$ and orthogonal matrices $\mathbf{C}_0$ and $\mathbf{C}_1$. Let $\mathbf{C} = \text{bdiag}(\mathbf{C}_0, \mathbf{C}_1)$ and $\boldsymbol{\Lambda} = \text{bdiag}(\boldsymbol{\Lambda}_0, \boldsymbol{\Lambda}_1)$. Then,

$$\mathbf{M} = \mathbf{AC}\boldsymbol{\Lambda}\mathbf{C}^T\mathbf{A}^T \equiv \mathbf{D}\boldsymbol{\Lambda}\mathbf{D}^T, \tag{A.1}$$

where $\mathbf{D} = \mathbf{AC}$. The first u columns of $\mathbf{D}$, which form the matrix $\mathbf{A}_0\mathbf{C}_0$, span $\mathcal{R}$. Moreover, $\mathbf{D}$ is an orthogonal matrix, and thus (A.1) is a spectral decomposition of $\mathbf{M}$ with eigenvectors in $\mathcal{R}$ or $\mathcal{R}^\perp$. □

The next proposition gives a characterization of $\mathbf{M}$ in terms of its projections onto its reducing subspaces. It shows in effect that $\mathbf{M}$ can be decomposed into a sum of orthogonal matrices, and it was used in forming the parameters in the envelope models of Section 1.4. Representation (A.2) was used previously in Definition 1.1 to define a reducing subspace, since it is equivalent to the definition of a reducing subspace given in Definition A.1. Definition 1.1 is related directly to the required envelope properties stated in (1.18), while Definition A.1 is often easier to employ in proofs.

Proposition A.1 *$\mathcal{R}$ reduces $\mathbf{M} \in \mathbb{R}^{r\times r}$ if and only if $\mathbf{M}$ can be written in the form*

$$\mathbf{M} = \mathbf{P}_\mathcal{R}\mathbf{M}\mathbf{P}_\mathcal{R} + \mathbf{Q}_\mathcal{R}\mathbf{M}\mathbf{Q}_\mathcal{R}. \tag{A.2}$$

Proof: Assume that $\mathbf{M}$ can be written as in (A.2). Then for any $\mathbf{v} \in \mathcal{R}$, $\mathbf{Mv} \in \mathcal{R}$, and for $\mathbf{v} \in \mathcal{R}^{\perp}$, $\mathbf{Mv} \in \mathcal{R}^{\perp}$. Consequently, $\mathcal{R}$ reduces $\mathbf{M}$.

Next, assume that $\mathcal{R}$ reduces $\mathbf{M}$. We must show that $\mathbf{M}$ satisfies (A.2). Let $u = \dim(\mathcal{R})$. It follows from Lemma A.1 that there is a $\mathbf{B} \in \mathbb{R}^{u\times u}$ that satisfies $\mathbf{MA} = \mathbf{AB}$, where $\mathbf{A} \in \mathbb{R}^{r\times u}$ and $\text{span}(\mathbf{A}) = \mathcal{R}$. This implies $\mathbf{Q}_{\mathcal{R}}\mathbf{MA} = 0$, which is equivalent to $\mathbf{Q}_{\mathcal{R}}\mathbf{MP}_{\mathcal{R}} = 0$. By the same logic applied to $\mathcal{R}^{\perp}$, $\mathbf{P}_{\mathcal{R}}\mathbf{MQ}_{\mathcal{R}} = 0$. Consequently,

$$\mathbf{M} = (\mathbf{P}_{\mathcal{R}} + \mathbf{Q}_{\mathcal{R}})\mathbf{M}(\mathbf{P}_{\mathcal{R}} + \mathbf{Q}_{\mathcal{R}}) = \mathbf{P}_{\mathcal{R}}\mathbf{MP}_{\mathcal{R}} + \mathbf{Q}_{\mathcal{R}}\mathbf{MQ}_{\mathcal{R}}.$$

□

An eigenspace of $\mathbf{M} \in \mathbb{S}^{r\times r}$ is a subspace of $\mathbb{R}^r$ spanned by the eigenvectors of $\mathbf{M}$ with the same eigenvalue and the zero vector. The dimension of an eigenspace is then equal to the multiplicity of the corresponding eigenvalue. An eigenspace of $\mathbf{M}$ reduces $\mathbf{M}$. However, a reducing subspace of $\mathbf{M}$ need not be an eigenspace of $\mathbf{M}$. For example, if $\mathbf{M}$ has an eigenvalue λ_1 with multiplicity 1 and an eigenvalue λ_2 with multiplicity 2, then $\mathbf{M}$ is reduced by the span of the eigenvector associated λ_1 and any vector in the eigenspace of λ_2, but this reducing subspace is not an eigenspace. The span of any vector in the eigenspace of λ_2 also reduces $\mathbf{M}$ but is not an eigenspace of $\mathbf{M}$.

Corollary A.1 describes consequences of Proposition A.1, including a relationship between reducing subspaces of $\mathbf{M}$ and $\mathbf{M}^{-1}$ when $\mathbf{M}$ is nonsingular. Results 3 and 4 in this corollary were used during the analysis of the log likelihood L in Section 1.5. Connecting the notation of that section with the corollary, $\mathbf{\Omega} = \mathbf{A}^T\mathbf{MA}$ and $\mathbf{\Omega}_0 = \mathbf{A}_0\mathbf{MA}_0$.

Corollary A.1 *Let $\mathcal{R}$ reduce $\mathbf{M} \in \mathbb{R}^{r\times r}$, let $\mathbf{A} \in \mathbb{R}^{r\times u}$ be a semi-orthogonal basis matrix for $\mathcal{R}$, and let $\mathbf{A}_0$ be a semi-orthogonal basis matrix for $\mathcal{R}^{\perp}$. Then*

1. *$\mathbf{M}$ and $\mathbf{P}_{\mathcal{R}}$, and $\mathbf{M}$ and $\mathbf{Q}_{\mathcal{R}}$ commute.*
2. *$\mathcal{R} \subseteq \text{span}(\mathbf{M})$ if and only if $\mathbf{A}^T\mathbf{MA}$ is full rank.*
3. *$|\mathbf{M}| = |\mathbf{A}^T\mathbf{MA}| \times |\mathbf{A}_0^T\mathbf{MA}_0|$.*
4. *If $\mathbf{M}$ is full rank, then*

$$\begin{aligned}\mathbf{M}^{-1} &= \mathbf{A}(\mathbf{A}^T\mathbf{MA})^{-1}\mathbf{A}^T + \mathbf{A}_0(\mathbf{A}_0^T\mathbf{MA}_0)^{-1}\mathbf{A}_0^T \\ &= \mathbf{P}_{\mathcal{R}}\mathbf{M}^{-1}\mathbf{P}_{\mathcal{R}} + \mathbf{Q}_{\mathcal{R}}\mathbf{M}^{-1}\mathbf{Q}_{\mathcal{R}}.\end{aligned} \tag{A.3}$$

5. *If $\mathcal{R} \subseteq \text{span}(\mathbf{M})$, then*

$$\mathbf{M}^{\dagger} = \mathbf{A}(\mathbf{A}^T\mathbf{MA})^{-1}\mathbf{A}^T + \mathbf{A}_0(\mathbf{A}_0^T\mathbf{MA}_0)^{\dagger}\mathbf{A}_0^T.$$

Proof: The first conclusion follows immediately from Proposition A.1.

To show the second conclusion, first assume that $\mathbf{A}^T\mathbf{MA}$ is full rank. Then, from Lemma A.1, $\mathbf{B}$ must be full rank in the representation $\mathbf{MA} = \mathbf{AB}$. Consequently, any vector in $\mathcal{R}$ can be written as a linear combination of the columns of $\mathbf{M}$ and thus $\mathcal{R} \subseteq \text{span}(\mathbf{M})$. Next, assume that $\mathcal{R} \subseteq \text{span}(\mathbf{M})$. Then there is a full rank matrix $\mathbf{V} \in \mathbb{R}^{r \times u}$ such that $\mathbf{MV} = \mathbf{A}$ and thus that $\mathbf{A}^T\mathbf{MV} = \mathbf{I}_u$. Substituting $\mathbf{M}$ from Proposition A.1, we have $(\mathbf{A}^T\mathbf{MA})(\mathbf{A}^T\mathbf{V}) = \mathbf{I}_u$. It follows that $\mathbf{A}^T\mathbf{MA}$ is of full rank.

For the fourth conclusion, since $\mathbf{M}$ is full rank $\mathcal{R} \subseteq \text{span}(\mathbf{M})$ and $\mathcal{R}^\perp \subseteq \text{span}(\mathbf{M})$. Consequently, both $\mathbf{A}^T\mathbf{MA}$ and $\mathbf{A}_0^T\mathbf{MA}_0$ are full rank. Thus, the right-hand side of (A.3) is defined. Meanwhile, note that $\mathbf{P}_\mathcal{R} = \mathbf{AA}^T$ and $\mathbf{Q}_\mathcal{R} = \mathbf{A}_0\mathbf{A}_0^T$. Hence, by (A.2), $\mathbf{M} = \mathbf{AA}^T\mathbf{MAA}^T + \mathbf{A}_0\mathbf{A}_0^T\mathbf{MA}_0\mathbf{A}_0^T$. Multiply this and the right-hand side of (A.3) to complete the proof. The final conclusion follows similarly: Since $\mathcal{R} \subseteq \text{span}(\mathbf{M})$, $\mathbf{A}^T\mathbf{MA}$ is full rank. The conclusion follows by checking the conditions for the Moore-Penrose inverse. □

Lemma A.3 *Let $\mathcal{R}$ reduce $\mathbf{M} \in \mathbb{R}^{r \times r}$. Then $\mathbf{M}\mathcal{R} = \mathcal{R}$ if and only if $\mathcal{R} \subseteq$ span($\mathbf{M}$).*

Proof: Assume that $\mathbf{M}\mathcal{R} = \mathcal{R}$. Then, with $\mathbf{A}$ as defined in Corollary A.1, $\mathbf{MA} = \mathbf{AB}$ for some full rank matrix $\mathbf{B} \in \mathbb{R}^{u \times u}$. Consequently, $\mathbf{A}^T\mathbf{MA}$ is full rank. It follows from Corollary A.1 that $\mathcal{R} \subseteq \text{span}(\mathbf{M})$.

Assume that $\mathcal{R} \subseteq \text{span}(\mathbf{M})$. Then it follows from Corollary A.1 that $\mathbf{A}^T\mathbf{MA}$ is of full rank. Thus, $\mathbf{B}$ must have full rank in the representation $\mathbf{MA} = \mathbf{AB}$, which implies $\mathbf{M}\mathcal{R} = \mathcal{R}$. □

Lemma A.3 gives necessary and sufficient conditions for $\mathbf{M}$ to map $\mathcal{R}$ to itself, rather than to a proper subset of $\mathcal{R}$. We will nearly always use reducing subspaces in contexts where $\mathbf{M}$ is a full rank covariance matrix. Lemma A.3 holds trivially in such cases since then it is always true that $\mathcal{R} \subseteq \text{span}(\mathbf{M}) = \mathbb{R}^r$.

As mentioned in connection with Lemma A.2, there is a relationship between the eigenstructure of a symmetric matrix $\mathbf{M}$ and its reducing subspaces. By definition, any invariant subspace of $\mathbf{M} \in \mathbb{S}^{r \times r}$ is also a reducing subspace of $\mathbf{M}$. In particular, it follows from Proposition A.1 that the subspace spanned by any set of eigenvectors of $\mathbf{M}$ is a reducing subspace of $\mathbf{M}$. This connection is formalized as follows.

Proposition A.2 *Let $\mathcal{R}$ be a subspace of $\mathbb{R}^r$ and let $\mathbf{M} \in \mathbb{S}^{r \times r}$. Assume that $\mathbf{M}$ has $q \leq r$ distinct eigenvalues, and let $\mathbf{P}_i$, $i = 1, \dots, q$, indicate the projections on the corresponding eigenspaces. Then the following statements are equivalent:*

1. *$\mathcal{R}$ reduces $\mathbf{M}$,*
2. *$\mathcal{R} = \sum_{i=1}^q \mathbf{P}_i\mathcal{R}$,*
3. *$\mathbf{P}_\mathcal{R} = \sum_{i=1}^q \mathbf{P}_i\mathbf{P}_\mathcal{R}\mathbf{P}_i$,*
4. *$\mathbf{M}$ and $\mathbf{P}_\mathcal{R}$ commute.*

Proof: **Equivalence of 1 and 4:** If $\mathcal{R}$ reduces $\mathbf{M}$, then it follows immediately from Corollary A.1 that $\mathbf{M}$ and $\mathbf{P}_\mathcal{R}$ commute. If $\mathbf{M}$ and $\mathbf{P}_\mathcal{R}$ commute, then $\mathbf{P}_\mathcal{R}\mathbf{M}\mathbf{Q}_\mathcal{R} = \mathbf{M}\mathbf{P}_\mathcal{R}\mathbf{Q}_\mathcal{R} = 0$, and thus it follows from Proposition A.1 that $\mathcal{R}$ reduces $\mathbf{M}$.

1 implies 2: If $\mathbf{v} \in \mathcal{R}$, then

$$\mathbf{v} = \mathbf{I}_p\mathbf{v} = \left(\sum_{i=1}^{q} \mathbf{P}_i\right)\mathbf{v} = \sum_{i=1}^{q} \mathbf{P}_i\mathbf{v} \in \sum_{i=1}^{q} \mathbf{P}_i\mathcal{R}.$$

Hence $\mathcal{R} \subseteq \sum_{i=1}^{q} \mathbf{P}_i\mathcal{R}$. Conversely, if $\mathbf{v} \in \sum_{i=1}^{q} \mathbf{P}_i\mathcal{R}$, then $\mathbf{v}$ can be written as a linear combination of $\mathbf{P}_1\mathbf{v}_1, \dots, \mathbf{P}_q\mathbf{v}_q$ where $\mathbf{v}_1, \dots, \mathbf{v}_q$ belong to $\mathcal{R}$. By Lemma A.2, $\mathbf{P}_i\mathbf{w} \in \mathcal{R}$ for any $\mathbf{w} \in \mathcal{R}$. Hence, any linear combination of $\mathbf{P}_1\mathbf{v}_1, \dots, \mathbf{P}_q\mathbf{v}_q$, with $\mathbf{v}_1, \dots, \mathbf{v}_q$ belonging to $\mathcal{R}$, belongs to $\mathcal{R}$. That is, $\sum_{i=1}^{q} \mathbf{P}_i\mathcal{R} \subseteq \mathcal{R}$.

2 implies 3: If $\mathbf{v} \in \mathcal{R}$, then, from the previous step, $\mathbf{P}_i\mathbf{v} \in \mathcal{R}$, $i = 1, \dots, q$. Hence,

$$\left(\sum_{i=1}^{q} \mathbf{P}_i\mathbf{P}_{\mathcal{R}}\mathbf{P}_i\right)\mathbf{v} = \sum_{i=1}^{q} \mathbf{P}_i\mathbf{v} = \mathbf{v} = \mathbf{P}_{\mathcal{R}}\mathbf{v}.$$

Now let $\mathbf{v} \in \mathcal{R}^{\perp}$. Then, $\mathbf{v} \perp \mathbf{P}_i\mathcal{R}$ for each i. Because $\mathbf{P}_i$ is self-adjoint, we see that $\mathbf{P}_i\mathbf{v} \perp \mathcal{R}$ for each i. Consequently,

$$\left(\sum_{i=1}^{q} \mathbf{P}_i\mathbf{P}_{\mathcal{R}}\mathbf{P}_i\right)\mathbf{v} = 0 = \mathbf{P}_{\mathcal{R}}\mathbf{v}.$$

It follows that $(\sum \mathbf{P}_i\mathbf{P}_{\mathcal{R}}\mathbf{P}_i)\mathbf{v} = \mathbf{P}_{\mathcal{R}}\mathbf{v}$ for all $\mathbf{v} \in \mathbb{R}^r$. Hence, the two matrices are the same.

3 implies 1: Again, if $\mathbf{v} \in \mathcal{R}$, then $\mathbf{P}_i\mathbf{v} \in \mathcal{R}$, $i = 1, \dots, q$. Hence, indicating with m_i, $i = 1, \dots, q$ the distinct eigenvalues of M we have

$$\mathbf{P}_{\mathcal{R}}\mathbf{M}\mathbf{v} = \sum_{i=1}^{q} m_i\mathbf{P}_i\mathbf{P}_{\mathcal{R}}\mathbf{P}_i\mathbf{P}_{\mathcal{R}}\mathbf{P}_i\mathbf{v} = \sum_{i=1}^{q} m_i\mathbf{P}_i\mathbf{P}_{\mathcal{R}}\mathbf{P}_i\mathbf{v} = \mathbf{M}\mathbf{v}.$$

It follows that $\mathbf{M}\mathcal{R} \subseteq \mathcal{R}$. □

Lemma A.4 *The intersection of any two reducing subspaces of* $\mathbf{M} \in \mathbb{R}^{r\times r}$ *is also a reducing subspace of* $\mathbf{M}$.

Proof: Let $\mathcal{R}_1$ and $\mathcal{R}_2$ be reducing subspaces of $\mathbf{M}$. Then by definition $\mathbf{M}\mathcal{R}_1 \subseteq \mathcal{R}_1$ and $\mathbf{M}\mathcal{R}_2 \subseteq \mathcal{R}_2$. Clearly, if $\mathbf{v} \in \mathcal{R}_1 \cap \mathcal{R}_2$, then $\mathbf{M}\mathbf{v} \in \mathcal{R}_1 \cap \mathcal{R}_2$, and it follows that the intersection is an invariant subspace of $\mathbf{M}$. The same argument shows that if $\mathbf{v} \in (\mathcal{R}_1 \cap \mathcal{R}_2)^{\perp} = \mathcal{R}_1^{\perp} + \mathcal{R}_2^{\perp}$ then $\mathbf{M}\mathbf{v} \in \mathcal{R}_1^{\perp} + \mathcal{R}_2^{\perp}$: If $\mathbf{v} \in \mathcal{R}_1^{\perp} + \mathcal{R}_2^{\perp}$, then it can be written as $\mathbf{v} = \mathbf{v}_1 + \mathbf{v}_2$, where $\mathbf{v} \in \mathcal{R}_1^{\perp}$ and $\mathbf{v} \in \mathcal{R}_2^{\perp}$. Then $\mathbf{M}\mathbf{v} = \mathbf{M}\mathbf{v}_1 + \mathbf{M}\mathbf{v}_2 \in \mathcal{R}_1^{\perp} + \mathcal{R}_2^{\perp}$. □

Lemma A.5 *Let* $\mathbf{M}_j \in \mathbb{R}^{r_j\times r_j}$, *and let* $\mathcal{R} \subseteq \mathbb{R}^{r_j}$, $j = 1, 2$. *Then* $\mathcal{R} = \mathcal{R}_1 \otimes \mathcal{R}_2 := \{\mathbf{v}_1 \otimes \mathbf{v}_2 | \mathbf{v}_1 \in \mathcal{R}_1, \mathbf{v}_2 \in \mathcal{R}_2\}$ *reduces* $\mathbf{M} = \mathbf{M}_1 \otimes \mathbf{M}_2$ *if and only if* $\mathcal{R}_j$ *reduce* $\mathbf{M}_j$, $j = 1, 2$.

Proof: If $\mathcal{R}_j$ reduce $\mathbf{M}_j, j = 1, 2$. Then

$$\begin{aligned}
\mathbf{M} &= (\mathbf{P}_{\mathcal{R}_1}\mathbf{M}_1\mathbf{P}_{\mathcal{R}_1} + \mathbf{Q}_{\mathcal{R}_1}\mathbf{M}_1\mathbf{Q}_{\mathcal{R}_1}) \otimes (\mathbf{P}_{\mathcal{R}_2}\mathbf{M}_2\mathbf{P}_{\mathcal{R}_2} + \mathbf{Q}_{\mathcal{R}_2}\mathbf{M}_1\mathbf{Q}_{\mathcal{R}_2}) \\
&= \mathbf{P}_{\mathcal{R}_1}\mathbf{M}_1\mathbf{P}_{\mathcal{R}_1} \otimes \mathbf{P}_{\mathcal{R}_2}\mathbf{M}_2\mathbf{P}_{\mathcal{R}_2} + \mathbf{P}_{\mathcal{R}_1}\mathbf{M}_1\mathbf{P}_{\mathcal{R}_1} \otimes \mathbf{Q}_{\mathcal{R}_2}\mathbf{M}_2\mathbf{Q}_{\mathcal{R}_2} \\
&\quad + \mathbf{Q}_{\mathcal{R}_1}\mathbf{M}_1\mathbf{Q}_{\mathcal{R}_1} \otimes \mathbf{P}_{\mathcal{R}_2}\mathbf{M}_2\mathbf{P}_{\mathcal{R}_2} + \mathbf{Q}_{\mathcal{R}_1}\mathbf{M}_1\mathbf{Q}_{\mathcal{R}_1} \otimes \mathbf{Q}_{\mathcal{R}_2}\mathbf{M}_2\mathbf{Q}_{\mathcal{R}_2} \\
&= (\mathbf{P}_{\mathcal{R}_1} \otimes \mathbf{P}_{\mathcal{R}_2})\mathbf{M}(\mathbf{P}_{\mathcal{R}_1} \otimes \mathbf{P}_{\mathcal{R}_2}) + (\mathbf{P}_{\mathcal{R}_1} \otimes \mathbf{Q}_{\mathcal{R}_2})\mathbf{M}(\mathbf{P}_{\mathcal{R}_1} \otimes \mathbf{Q}_{\mathcal{R}_2}) \\
&\quad + (\mathbf{Q}_{\mathcal{R}_1} \otimes \mathbf{P}_{\mathcal{R}_2})\mathbf{M}(\mathbf{Q}_{\mathcal{R}_1} \otimes \mathbf{P}_{\mathcal{R}_2}) + (\mathbf{Q}_{\mathcal{R}_1} \otimes \mathbf{Q}_{\mathcal{R}_2})\mathbf{M}(\mathbf{Q}_{\mathcal{R}_1} \otimes \mathbf{Q}_{\mathcal{R}_2}) \\
&= (\mathbf{P}_{\mathcal{R}_1} \otimes \mathbf{P}_{\mathcal{R}_2})\mathbf{M}(\mathbf{P}_{\mathcal{R}_1} \otimes \mathbf{P}_{\mathcal{R}_2}) \\
&\quad + (\mathbf{I}_{r_1 r_2} - \mathbf{P}_{\mathcal{R}_1} \otimes \mathbf{P}_{\mathcal{R}_2})\mathbf{M}(\mathbf{I}_{r_1 r_2} - \mathbf{P}_{\mathcal{R}_1} \otimes \mathbf{P}_{\mathcal{R}_2}),
\end{aligned}$$

where the last step follows because $\mathcal{R}_j$ reduces $\mathbf{M}_j, j = 1, 2$.

If $\mathcal{R}$ reduces $\mathbf{M}$, then by Proposition A.1, we have $\mathbf{P}_{\mathcal{R}}\mathbf{M}\mathbf{Q}_{\mathcal{R}} = 0$. Since $\mathbf{P}_{\mathcal{R}} = \mathbf{P}_{\mathcal{R}_1} \otimes \mathbf{P}_{\mathcal{R}_2}$,

$$\begin{aligned}
\mathbf{P}_{\mathcal{R}}\mathbf{M}\mathbf{Q}_{\mathcal{R}} &= (\mathbf{P}_{\mathcal{R}_1} \otimes \mathbf{P}_{\mathcal{R}_2})(\mathbf{M}_1 \otimes \mathbf{M}_2)(\mathbf{I}_{r_1 r_2} - \mathbf{P}_{\mathcal{R}_1} \otimes \mathbf{P}_{\mathcal{R}_2}) \\
&= \mathbf{P}_{\mathcal{R}_1}\mathbf{M}_1 \otimes \mathbf{P}_{\mathcal{R}_2}\mathbf{M}_2 - \mathbf{P}_{\mathcal{R}_1}\mathbf{M}_1\mathbf{P}_{\mathcal{R}_1} \otimes \mathbf{P}_{\mathcal{R}_2}\mathbf{M}_2\mathbf{P}_{\mathcal{R}_2}.
\end{aligned}$$

Now, $\mathbf{P}_{\mathcal{R}}\mathbf{M}\mathbf{Q}_{\mathcal{R}} = 0$ if and only if $\mathbf{P}_{\mathcal{R}_j}\mathbf{M}_j = \mathbf{P}_{\mathcal{R}_j}\mathbf{M}_j\mathbf{P}_{\mathcal{R}_j}$, $j = 1, 2$. Equivalently, we must have $\mathbf{P}_{\mathcal{R}_j}\mathbf{M}_j\mathbf{Q}_{\mathcal{R}_j} = 0$, $j = 1, 2$. The conclusion follows from Proposition A.1. □

A.2 M-Envelopes

Lemma A.4, which states that the intersection of two reducing subspaces of $\mathbf{M} \in \mathbb{S}^{r \times r}$ is itself a reducing subspace of $\mathbf{M}$, is important for envelope construction since it allows us to define the smallest reducing subspace of $\mathbf{M}$ that contains a specified subspace $\mathcal{S}$. This leads then to the definition of an envelope given in Definition 1.2. That definition requires that $\mathcal{S} \subseteq \text{span}(\mathbf{M})$. This containment guarantees existence of the $\mathbf{M}$-envelope and it holds trivially if $\mathbf{M}$ is full rank; that is, if $\text{span}(\mathbf{M}) = \mathbb{R}^r$. Moreover, closure under intersection guarantees that the $\mathbf{M}$-envelope is in fact a reducing subspace of $\mathbf{M}$. Thus, the $\mathbf{M}$-envelope of $\mathcal{S}$ represents a well-defined parameter in some statistical problems.

Consider the case where the eigenvalues of $\mathbf{M}$ are distinct, and let $\mathcal{S}$ be a d-dimensional subspace of $\mathbb{R}^r$. It will take at least d eigenvectors of $\mathbf{M}$ to construct a subspace that contains $\mathcal{S}$, and in some cases, it may take all r eigenvectors of $\mathbf{M}$. The envelope by definition is spanned by the smallest number u, $d \le u \le r$, of eigenvectors needed to construct a subspace that contains $\mathcal{S}$. The situation can become more complicated if $\mathbf{M}$ has less than r distinct eigenvalues. However, reducing subspaces get around this potential difficulty since they do not require distinct eigenvalues, which is why they were used in Definition 1.2.

Consider next the application of this definition in terms of Figure 1.3. We set $\mathbf{M} = \boldsymbol{\Sigma}$ and $\mathcal{S} = \mathcal{B} = \text{span}(\boldsymbol{\beta})$. Then the envelope $\mathcal{E}_{\boldsymbol{\Sigma}}(\mathcal{B})$ shown in the figure is the smallest reducing subspace of $\boldsymbol{\Sigma}$ that contains $\mathcal{B}$. We know from Proposition A.2 that the reducing subspaces of $\boldsymbol{\Sigma}$ correspond to eigenspaces of $\boldsymbol{\Sigma}$. Thus, in the example, we find the smallest eigenspace of $\boldsymbol{\Sigma}$ that contains $\mathcal{B}$. In the figure, $\mathcal{B}$ is equal to the eigenspace corresponding to the smaller eigenvalue of $\boldsymbol{\Sigma}$.

The following proposition, derived from Proposition A.2 and Definition 1.2, gives a constructive characterization of $\mathbf{M}$-envelopes and provided a connection between $\mathcal{E}_{\mathbf{M}}(\mathcal{S})$ and the eigenspaces of $\mathbf{M}$.

Proposition A.3 *Let* $\mathbf{M} \in \mathbb{S}^{r\times r}$, *let* $\mathbf{P}_i$, $i = 1, \dots, q \leq r$, *be the projections onto the eigenspaces of* $\mathbf{M}$, *and let* $\mathcal{S}$ *be a subspace of* $\text{span}(\mathbf{M})$. *Then* $\mathcal{E}_{\mathbf{M}}(\mathcal{S}) = \sum_{i=1}^{q} \mathbf{P}_i\mathcal{S}$ *and, as a special case,* $\mathcal{E}_{\mathbf{I}_p}(\mathcal{S}) = \mathcal{S}$.

Proof: To prove that $\sum_{i=1}^{q} \mathbf{P}_i\mathcal{S}$ is the smallest reducing subspace of $\mathbf{M}$ that contains $\mathcal{S}$, it suffices to prove the following statements:

1. $\sum_{i=1}^{q} \mathbf{P}_i\mathcal{S}$ reduces $\mathbf{M}$.
2. $\mathcal{S} \subseteq \sum_{i=1}^{q} \mathbf{P}_i\mathcal{S}$.
3. If $\mathcal{T}$ reduces $\mathbf{M}$ and $\mathcal{S} \subseteq \mathcal{T}$, then $\sum_{i=1}^{q} \mathbf{P}_i\mathcal{S} \subseteq \mathcal{T}$.

Statement 1 follows from Proposition A.2, as applied to $\mathcal{R} \equiv \sum_{i=1}^{q} \mathbf{P}_i\mathcal{S}$. Statement 2 holds because $\mathcal{S} = \{\mathbf{P}_1\mathbf{v} + \cdots + \mathbf{P}_q\mathbf{v} : \mathbf{v} \in \mathcal{S}\} \subseteq \sum_{i=1}^{q} \mathbf{P}_i\mathcal{S}$. Turning to statement 3, if $\mathcal{T}$ reduces $\mathbf{M}$, it can be written as $\mathcal{T} = \sum_{i=1}^{q} \mathbf{P}_i\mathcal{T}$ by Proposition A.2. If, in addition, $\mathcal{S} \subseteq \mathcal{T}$, then we have $\mathbf{P}_i\mathcal{S} \subseteq \mathbf{P}_i\mathcal{T}$ for $i = 1, \dots, q$. Statement 3 follows since $\sum_{i=1}^{q} \mathbf{P}_i\mathcal{S} \subseteq \sum_{i=1}^{q} \mathbf{P}_i\mathcal{T} = \mathcal{T}$. □

A.3 Relationships Between Envelopes

In this section, we consider relationships between various types of envelopes. We first investigate how the $\mathbf{M}$-envelope is modified by linear transformations of $\mathcal{S}$, keeping $\mathbf{M}$ fixed.

A.3.1 Invariance and Equivariance

An envelope does not transform equivariantly for all linear transformations, but it does so for symmetric linear transformations that commute with $\mathbf{M}$:

Proposition A.4 *Let* $\mathbf{K} \in \mathbb{S}^{r\times r}$ *commute with* $\mathbf{M} \in \mathbb{S}^{r\times r}$, *and let* $\mathcal{S}$ *be a subspace of* $\text{span}(\mathbf{M})$. *Then* $\mathbf{K}\mathcal{S} \subseteq \text{span}(\mathbf{M})$ *and*

$$\mathcal{E}_{\mathbf{M}}(\mathbf{K}\mathcal{S}) = \mathbf{K}\mathcal{E}_{\mathbf{M}}(\mathcal{S}). \tag{A.4}$$

If, in addition, $\mathcal{S} \subseteq \text{span}(\mathbf{K})$ *and* $\mathcal{E}_{\mathbf{M}}(\mathcal{S})$ *reduces* $\mathbf{K}$*, then*

$$\mathcal{E}_{\mathbf{M}}(\mathbf{K}\mathcal{S}) = \mathcal{E}_{\mathbf{M}}(\mathcal{S}). \tag{A.5}$$

Proof: Since $\mathcal{S} \subseteq \text{span}(\mathbf{M})$ and $\mathbf{M}$ and $\mathbf{K}$ commute, we have

$$\mathbf{K}\mathcal{S} \subseteq \mathbf{K}\text{span}(\mathbf{M}) = \text{span}(\mathbf{KM}) = \text{span}(\mathbf{MK}) \subseteq \text{span}(\mathbf{M}).$$

Turning to (A.4), because $\mathbf{K}$ and $\mathbf{M}$ commute, they can be diagonalized simultaneously by an orthogonal matrix, say $\mathbf{U}$. Recall that $\mathbf{P}_i$ is the projection on the ith eigenspace of $\mathbf{M}$, and let $d_i = \text{rank}(\mathbf{P}_i)$. Partition $\mathbf{U} = (\mathbf{U}_1, \dots, \mathbf{U}_q)$, where $\mathbf{U}_i$ contains d_i columns, $i = 1, \dots, q$. Without loss of generality, we can assume that $\mathbf{U}_i\mathbf{U}_i^T = \mathbf{P}_i$ for $i = 1, \dots, q$. Then $\mathbf{K}$ can be written as $\mathbf{U}_1\boldsymbol{\Lambda}_1\mathbf{U}_1^T + \cdots \mathbf{U}_q\boldsymbol{\Lambda}_q\mathbf{U}_q^T$, where the $\boldsymbol{\Lambda}_i$'s are diagonal matrices of dimension $d_i \times d_i$. It follows that

$$\begin{aligned}\mathbf{K}\mathbf{P}_i &= (\mathbf{U}_1\boldsymbol{\Lambda}_1\mathbf{U}_1^T + \cdots + \mathbf{U}_q\boldsymbol{\Lambda}_q\mathbf{U}_q^T)\mathbf{U}_i\mathbf{U}_i^T \\ &= \mathbf{U}_i\boldsymbol{\Lambda}_i\mathbf{U}_i^T = \mathbf{U}_i\mathbf{U}_i^T(\mathbf{U}_1\boldsymbol{\Lambda}_1\mathbf{U}_1^T + \cdots \mathbf{U}_q\boldsymbol{\Lambda}_q\mathbf{U}_q^T) = \mathbf{P}_i\mathbf{K}.\end{aligned}$$

That is, $\mathbf{K}$ and $\mathbf{P}_i$ commute. Now, by Proposition A.3, $\mathcal{E}_{\mathbf{M}}(\mathcal{S}) = \sum_i^q \mathbf{P}_i\mathcal{S}$. Hence,

$$\begin{aligned}\mathbf{K}\mathcal{E}_{\mathbf{M}}(\mathcal{S}) &= \{\mathbf{K}\mathbf{P}_1\mathbf{h}_1 + \cdots + \mathbf{K}\mathbf{P}_q\mathbf{h}_q : \mathbf{h}_1, \dots, \mathbf{h}_q \in \mathcal{S}\} \\ &= \{\mathbf{P}_1\mathbf{K}\mathbf{h}_1 + \cdots + \mathbf{P}_q\mathbf{K}\mathbf{h}_q : \mathbf{h}_1, \dots, \mathbf{h}_q \in \mathcal{S}\} = \sum_{i=1}^{q} \mathbf{P}_i\mathbf{K}\mathcal{S}.\end{aligned}$$

By Proposition A.3 again, the right-hand side is $\mathcal{E}_{\mathbf{M}}(\mathbf{K}\mathcal{S})$.

Now suppose, in addition, that $\mathcal{S} \subseteq \text{span}(\mathbf{K})$ and $\mathcal{E}_{\mathbf{M}}(\mathcal{S})$ reduces $\mathbf{K}$. Since $\mathbf{K}$ commutes with $\mathbf{M}$, span($\mathbf{K}$) is an invariant subspace of $\mathbf{M}$: for all $\mathbf{h} \in \mathbb{R}^r$, $\mathbf{MKh} = \mathbf{KMh} \subseteq \text{span}(\mathbf{K})$. Since $\mathbf{M}$ is symmetric, we further have that span($\mathbf{K}$) reduces $\mathbf{M}$. In sum then, span($\mathbf{K}$) is a reducing subspace of $\mathbf{M}$ that contains $\mathcal{S}$, and thus it must contain the smallest reducing substance of $\mathbf{M}$ that contains $\mathcal{S}$; that is, $\mathcal{E}_{\mathbf{M}}(\mathcal{S}) \subseteq \text{span}(\mathbf{K})$. By Lemma A.3, then, $\mathbf{K}\mathcal{E}_{\mathbf{M}}(\mathcal{S}) = \mathcal{E}_{\mathbf{M}}(\mathcal{S})$, which, in conjunction with (A.4), implies (A.5). □

To explore a useful consequence of Proposition A.4, start with a function $f : \mathbb{R} \to \mathbb{R}$ with the properties $f(0) = 0$ and $f(x) \neq 0$ whenever $x \neq 0$. Let m_i and $\mathbf{P}_i$, $i = 1, \dots, q$ indicate the distinct eigenvalues and the projections onto the corresponding eigenspaces for a matrix $\mathbf{M} \in \mathbb{S}^{r \times r}$, and let $f^*(\mathbf{M}) = \sum_{i=1}^q f(m_i)\mathbf{P}_i$. Then it is straightforward to verify that (i) $f^*(\mathbf{M})$ commutes with $\mathbf{M}$, (ii) any subspace $\mathcal{S} \subseteq \text{span}(\mathbf{M})$ satisfies $\mathcal{S} \subseteq \{f^*(\mathbf{M})\}$, and (iii) $\mathcal{E}_{\mathbf{M}}(\mathcal{S})$ reduces $f^*(\mathbf{M})$. Hence, by Proposition A.4 we have $\mathcal{E}_{\mathbf{M}}(f^*(\mathbf{M})\mathcal{S}) = \mathcal{E}_{\mathbf{M}}(\mathcal{S})$. Additionally, if $f(\cdot)$ is a strictly monotonic function, then $\mathbf{M}$ and $f^*(\mathbf{M})$ have the same eigenspaces. It follows from Proposition A.3 that $\mathcal{E}_{f^*(\mathbf{M})}(\mathcal{S}) = \mathcal{E}_{\mathbf{M}}(\mathcal{S})$. We summarize these results in the following corollary.

Corollary A.2 *With f and f^* as previously defined, $\mathcal{E}_{\mathbf{M}}(f^*(\mathbf{M})\mathcal{S}) = \mathcal{E}_{\mathbf{M}}(\mathcal{S})$. If f is strictly monotonic, then $\mathcal{E}_{f^*(\mathbf{M})}(\mathcal{S}) = \mathcal{E}_{\mathbf{M}}(\mathcal{S})$. In particular,*

$$\mathcal{E}_{\mathbf{M}}(\mathbf{M}^k\mathcal{S}) = \mathcal{E}_{\mathbf{M}}(\mathcal{S}) \text{ for all } k \in \mathbb{R} \tag{A.6}$$

$$\mathcal{E}_{\mathbf{M}^k}(\mathcal{S}) = \mathcal{E}_{\mathbf{M}}(\mathcal{S}) \text{ for all } k \in \mathbb{R} \text{ with } k \neq 0. \tag{A.7}$$

We next consider settings in which envelopes are invariant under certain simultaneous changes in $\mathbf{M}$ and $\mathcal{S}$. Let $\boldsymbol{\Delta} \in \mathbb{S}^{r\times r}$ be a positive definite matrix and let $\mathcal{S}$ be a u-dimensional subspace of $\mathbb{R}^r$. Let $\mathbf{G} \in \mathbb{R}^{r\times u}$ be a semi-orthogonal basis matrix for $\mathcal{S}$ and let $\mathbf{V} \in \mathbb{S}^{u\times u}$ be positive definite. Define $\boldsymbol{\Psi} = \boldsymbol{\Delta} + \mathbf{G}\mathbf{V}\mathbf{G}^T$. Then

Proposition A.5 $\boldsymbol{\Delta}^{-1}\mathcal{S} = \boldsymbol{\Psi}^{-1}\mathcal{S}$ *and*

$$\mathcal{E}_{\boldsymbol{\Delta}}(\mathcal{S}) = \mathcal{E}_{\boldsymbol{\Psi}}(\mathcal{S}) = \mathcal{E}_{\boldsymbol{\Delta}}(\boldsymbol{\Delta}^{-1}\mathcal{S}) = \mathcal{E}_{\boldsymbol{\Psi}}(\boldsymbol{\Psi}^{-1}\mathcal{S}) = \mathcal{E}_{\boldsymbol{\Psi}}(\boldsymbol{\Delta}^{-1}\mathcal{S}) = \mathcal{E}_{\boldsymbol{\Delta}}(\boldsymbol{\Psi}^{-1}\mathcal{S}).$$

Proof: Using a variant of the Woodbury identity for matrix inverses, we have

$$\boldsymbol{\Psi}^{-1} = \boldsymbol{\Delta}^{-1} - \boldsymbol{\Delta}^{-1}\mathbf{G}(\mathbf{V}^{-1} + \mathbf{G}^T\boldsymbol{\Delta}^{-1}\mathbf{G})^{-1}\mathbf{G}^T\boldsymbol{\Delta}^{-1},$$
$$\boldsymbol{\Delta}^{-1} = \boldsymbol{\Psi}^{-1} - \boldsymbol{\Psi}^{-1}\mathbf{G}(-\mathbf{V}^{-1} + \mathbf{G}^T\boldsymbol{\Psi}^{-1}\mathbf{G})^{-1}\mathbf{G}^T\boldsymbol{\Psi}^{-1}.$$

Multiplying both equations on the right by $\mathbf{G}$, The first implies $\text{span}(\boldsymbol{\Psi}^{-1}\mathbf{G}) \subseteq \text{span}(\boldsymbol{\Delta}^{-1}\mathbf{G})$; the second implies $\text{span}(\boldsymbol{\Delta}^{-1}\mathbf{G}) \subseteq \text{span}(\boldsymbol{\Psi}^{-1}\mathbf{G})$. Hence, $\boldsymbol{\Psi}^{-1}\mathcal{S} = \boldsymbol{\Delta}^{-1}\mathcal{S}$. From this we have also that $\mathcal{E}_{\boldsymbol{\Psi}}(\boldsymbol{\Psi}^{-1}\mathcal{S}) = \mathcal{E}_{\boldsymbol{\Psi}}(\boldsymbol{\Delta}^{-1}\mathcal{S})$ and $\mathcal{E}_{\boldsymbol{\Delta}}(\boldsymbol{\Psi}^{-1}\mathcal{S}) = \mathcal{E}_{\boldsymbol{\Delta}}(\boldsymbol{\Delta}^{-1}\mathcal{S})$.

We next show that $\mathcal{E}_{\boldsymbol{\Delta}}(\mathcal{S}) = \mathcal{E}_{\boldsymbol{\Psi}}(\mathcal{S})$ by demonstrating that $\mathcal{R} \subseteq \mathbb{R}^r$ is a reducing subspace of $\boldsymbol{\Delta}$ that contains $\mathcal{S}$ if and only if it is a reducing subspace of $\boldsymbol{\Psi}$ that contains $\mathcal{S}$. Suppose $\mathcal{R}$ is a reducing subspace of $\boldsymbol{\Delta}$ that contains $\mathcal{S}$. Let $\boldsymbol{\alpha} \in \mathcal{R}$. Then $\boldsymbol{\Psi}\boldsymbol{\alpha} = \boldsymbol{\Delta}\boldsymbol{\alpha} + \mathbf{G}\mathbf{V}\mathbf{G}^T\boldsymbol{\alpha}$. $\boldsymbol{\Delta}\boldsymbol{\alpha} \in \mathcal{R}$ because $\mathcal{R}$ reduces $\boldsymbol{\Delta}$; the second term on the right is a vector in $\mathcal{R}$ because $\mathcal{S} \subseteq \mathcal{R}$. Thus, $\mathcal{R}$ is a reducing subspace of $\boldsymbol{\Psi}$ and by construction it contains $\mathcal{S}$. Next, suppose $\mathcal{R}$ is a reducing subspace of $\boldsymbol{\Psi}$ that contains $\mathcal{S}$. The reverse implication follows similarly by reasoning in terms of $\boldsymbol{\Delta}\boldsymbol{\alpha} = \boldsymbol{\Psi}\boldsymbol{\alpha} - \mathbf{G}\mathbf{V}\mathbf{G}^T\boldsymbol{\alpha}$. We have $\boldsymbol{\Psi}\boldsymbol{\alpha} \in \mathcal{R}$ because $\mathcal{R}$ reduces $\boldsymbol{\Psi}$; the second term on the right is a vector in $\mathcal{R}$ because $\mathcal{S} \subseteq \mathcal{R}$. The remaining equalities follow immediately from (A.6). □

For a first application of Proposition A.5, consider the multivariate linear regression model of Chapter 1 but now assume that the predictors $\mathbf{X}$ are random with covariance matrix $\boldsymbol{\Sigma}_{\mathbf{X}}$, so that $\mathbf{Y}$ and $\mathbf{X}$ have a joint distribution. The envelope that we wish to estimate is still $\mathcal{E}_{\boldsymbol{\Sigma}}(\mathcal{B})$ since the analysis is conditional on the observed predictors, but because the predictors are random, we can express the marginal variance of $\mathbf{Y}$ as

$$\boldsymbol{\Sigma}_{\mathbf{Y}} = \boldsymbol{\Sigma} + \boldsymbol{\beta}\boldsymbol{\Sigma}_{\mathbf{X}}\boldsymbol{\beta}^T = \boldsymbol{\Sigma} + \mathbf{G}\mathbf{V}\mathbf{G}^T, \tag{A.8}$$

where $\mathbf{G}$ is a semi-orthogonal basis matrix for $\mathcal{B} = \text{span}(\boldsymbol{\beta})$, $\boldsymbol{\beta} = \mathbf{GA}$ and $\mathbf{V} = \mathbf{A\Sigma A}^T$ is positive definite since $\mathbf{A}$ must have full row rank. The second form for $\boldsymbol{\Sigma}$ matches the decomposition required for Proposition A.5 with $\boldsymbol{\Psi} = \boldsymbol{\Sigma}_{\mathbf{Y}}$ and $\boldsymbol{\Delta} = \boldsymbol{\Sigma}$. Consequently, we have

$$\mathcal{E}_{\boldsymbol{\Sigma}}(\mathcal{B}) = \mathcal{E}_{\boldsymbol{\Sigma}_{\mathbf{Y}}}(\mathcal{B}) = \mathcal{E}_{\boldsymbol{\Sigma}}(\boldsymbol{\Sigma}^{-1}\mathcal{B}) = \mathcal{E}_{\boldsymbol{\Sigma}_{\mathbf{Y}}}(\boldsymbol{\Sigma}_{\mathbf{Y}}^{-1}\mathcal{B}) = \mathcal{E}_{\boldsymbol{\Sigma}_{\mathbf{Y}}}(\boldsymbol{\Sigma}^{-1}\mathcal{B}) = \mathcal{E}_{\boldsymbol{\Sigma}}(\boldsymbol{\Sigma}_{\mathbf{Y}}^{-1}\mathcal{B}).$$

These relationships were used to connect envelopes with canonical correlations in Section 9.1.1 and elsewhere.

A.3.2 Direct Sums of Envelopes

The direct sum of $\mathbf{A} \in \mathbb{R}^{m\times n}$ and $\mathbf{B} \in \mathbb{R}^{p\times q}$ is defined as the block diagonal matrix $\mathbf{A} \oplus \mathbf{B} = \text{bdiag}(\mathbf{A}, \mathbf{B}) \in \mathbb{R}^{(m+p)\times(n+q)}$. The direct sum of two subspaces $\mathcal{S} \subseteq \mathbb{R}^p$ and $\mathcal{R} \subseteq \mathbb{R}^q$ is $\mathcal{S} \oplus \mathcal{R} = \text{span}(\mathbf{S} \oplus \mathbf{R})$, where $\mathbf{S}$ and $\mathbf{R}$ are basis matrices for $\mathcal{S}$ and $\mathcal{R}$.

The next lemma concerns the direct sum of envelopes. It was used during the discussion simultaneous reduction of responses and predictors in Section 4.5. Let $\mathbf{M}_j \in \mathbb{S}^{p_j\times p_j}$ and let $\mathcal{S}_j \subseteq \text{span}(\mathbf{M}_j)$, $j = 1, 2$. Then

Lemma A.6 $\mathcal{E}_{\mathbf{M}_1}(\mathcal{S}_1) \oplus \mathcal{E}_{\mathbf{M}_2}(\mathcal{S}_2) = \mathcal{E}_{\mathbf{M}_1\oplus\mathbf{M}_2}(\mathcal{S}_1 \oplus \mathcal{S}_2)$.

Proof: Let $\mathbf{P}_{ij}$ denote the projection onto the jth eigenspace of $\mathbf{M}_i$, $j = 1, \dots, q_i$, $i = 1, 2$. It then follows from Proposition A.3 that $\mathcal{E}_{\mathbf{M}_i}(\mathcal{S}_i) = \sum_{j=1}^{q_i} \mathbf{P}_{ij}\mathcal{S}_i$, $i = 1, 2$. The eigen-projections of $\mathbf{M}_1 \oplus \mathbf{M}_2$ are $\mathbf{P}_{1j} \oplus \mathbf{O}_{p_2\times p_2}$, $j = 1, \dots, q_1$, and $\mathbf{O}_{p_1\times p_1} \oplus \mathbf{P}_{2k}$, $k = 1, \dots, q_2$. Applying Proposition A.3 again we have

$$\begin{aligned}\mathcal{E}_{\mathbf{M}_1\oplus\mathbf{M}_2}(\mathcal{S}_1 \oplus \mathcal{S}_2) &= \sum_{j=1}^{q_1}\{(\mathbf{P}_{1j} \oplus \mathbf{O})(\mathcal{S}_1 \oplus \mathcal{S}_2)\} + \sum_{k=1}^{q_2}\{(\mathbf{O} \oplus \mathbf{P}_{2k})(\mathcal{S}_1 \oplus \mathcal{S}_2)\} \\ &= \{\mathcal{S}_{\mathbf{M}_1}(\mathcal{S}_1 \oplus \mathbf{O}\} + \{\mathbf{O} \oplus \mathcal{S}_{\mathbf{M}_2}(\mathcal{S}_2)\} \\ &= \mathcal{E}_{\mathbf{M}_1}(\mathcal{S}_1) \oplus \mathcal{E}_{\mathbf{M}_2}(\mathcal{S}_2).\end{aligned}$$

□

Direct sums of envelopes were used in Section 4.5 when developing predictor–response envelopes.

A.3.3 Coordinate Reduction

The next proposition shows a way to express an envelope in terms of reduced coordinates.

Proposition A.6 *Let* $\mathbf{B}$ *be a semi-orthogonal basis matrix for a reducing subspace of* $\mathbf{M} \in \mathbb{S}^{r\times r}$ *that contains* $\mathcal{S} \subseteq \mathbb{R}^r$. *Then* $\mathcal{E}_{\mathbf{M}}(\mathcal{S}) = \mathbf{B}\mathcal{E}_{\mathbf{B}^T\mathbf{MB}}(\mathbf{B}^T\mathcal{S})$.

Proof: For the conclusion to be reasonable, we need to have $\mathbf{B}^T\mathcal{S} \subseteq \text{span}(\mathbf{B}^T\mathbf{MB})$ (cf. Definition 1.2). To see that this requirement holds, let

$\mathbf{A}$ be a basis matrix for $\mathcal{S}$. Then since $\mathcal{S} \subseteq \text{span}(\mathbf{M})$ there is a matrix $\mathbf{C}$ so that

$$\mathbf{A} = \mathbf{MC} = \mathbf{P_B M P_B C} + \mathbf{Q_B M Q_B C},$$

where the second equality follows from Proposition A.1 since span($\mathbf{B}$) reduces $\mathbf{M}$. This implies that $\mathbf{B}^T\mathbf{A} = \mathbf{B}^T\mathbf{MB}(\mathbf{B}^T\mathbf{C})$ from which the requirement follows.

Let $\mathcal{D} = \mathcal{E}_{\mathbf{B}^T\mathbf{MB}}(\mathbf{B}^T\mathcal{S})$ for notational convenience. The conclusion can be deduced from the following quantities:

$$\begin{aligned}
\mathbf{M} &= \mathbf{P_B M P_B} + \mathbf{Q_B M Q_B} \\
&= \mathbf{B}(\mathbf{B}^T\mathbf{MB})\mathbf{B}^T + \mathbf{Q_B M Q_B} \\
&= \mathbf{B}\{\mathbf{P}_\mathcal{D}(\mathbf{B}^T\mathbf{MB})\mathbf{P}_\mathcal{D} + \mathbf{Q}_\mathcal{D}(\mathbf{B}^T\mathbf{MB})\mathbf{Q}_\mathcal{D}\}\mathbf{B}^T + \mathbf{Q_B M Q_B} \\
&= (\mathbf{BP}_\mathcal{D}\mathbf{B}^T)\mathbf{M}(\mathbf{BP}_\mathcal{D}\mathbf{B}^T) + (\mathbf{BQ}_\mathcal{D}\mathbf{B}^T + \mathbf{Q_B})\mathbf{M}(\mathbf{BQ}_\mathcal{D}\mathbf{B}^T + \mathbf{Q_B}),
\end{aligned}$$

where the third equality follows from Proposition A.1 because $\mathcal{D}$ reduces $\mathbf{B}^T\mathbf{MB}$. The final equation holds because $\mathbf{BQ}_\mathcal{D}\mathbf{B}^T\mathbf{MQ_B} = 0$ since span($\mathbf{B}$) reduces $\mathbf{M}$, and therefore $\mathbf{Q_B}$ and $\mathbf{M}$ commute. Since $\mathbf{BP}_\mathcal{D}\mathbf{B}^T$ and $\mathbf{BQ}_\mathcal{D}\mathbf{B}^T + \mathbf{Q_B}$ are orthogonal projections that sum to $\mathbf{I}_r$, it follows from Proposition A.1 that $\mathbf{BP}_\mathcal{D}\mathbf{B}^T$ reduces $\mathbf{M}$. Additionally, since $\mathbf{B}^T\mathcal{S} \subseteq \mathcal{D}$, we have $\mathcal{S} = \mathbf{BB}^T\mathcal{S} \subseteq \mathbf{B}\mathcal{D}$. Consequently,

$$\text{span}(\mathbf{BP}_\mathcal{D}\mathbf{B}^T) = \text{span}(\mathbf{BP}_\mathcal{D}) = \mathbf{B}\mathcal{D}$$

is a reducing subspace of $\mathbf{M}$ that contains $\mathcal{S}$. The equality

$$\mathcal{E}_\mathbf{M}(\mathcal{S}) = \mathbf{B}\mathcal{D} = \mathbf{B}\mathcal{E}_{\mathbf{B}^T\mathbf{MB}}(\mathbf{B}^T\mathcal{S})$$

follows from the minimality of $\mathcal{D}$. □

Proposition A.6 says that we never need to deal with a singular $\mathbf{M} \in \mathbb{S}^{r\times r}$ because we can always choose $\mathbf{B}$ to be a semi-orthogonal basis matrix for span($\mathbf{M}$), construct the envelope $\mathcal{E}_{\mathbf{B}^T\mathbf{MB}}(\mathbf{B}^T\mathcal{S})$ in reduced coordinates and then transform back to the full coordinates. The nonsingularity of $\mathbf{B}^T\mathbf{MB}$ follows from conclusion 2 of Corollary A.1 because $\mathcal{S} \subseteq \text{span}(\mathbf{M})$ by construction.

Proposition A.6 posits information about a superset of $\mathcal{E}_\mathbf{M}(\mathcal{S})$. In contrast, the next proposition starts with a subset of $\mathcal{E}_\mathbf{M}(\mathcal{S})$. It was used in Chapter 6 as a basis of sequential algorithms for determining $\mathcal{E}_\mathbf{M}(\mathcal{S})$.

Proposition A.7 *Let* $\mathbf{B}$ *be a semi-orthogonal basis matrix for a subspace* $\mathcal{B}$ *of* $\mathcal{E}_\mathbf{M}(\mathcal{S})$, *where* $\mathbf{M} \in \mathbb{S}^{r\times r}$ *and* $\mathcal{S} \subseteq \text{span}(\mathbf{M})$. *Let* $(\mathbf{B}, \mathbf{B}_0)$ *be an orthogonal matrix. Then*

$$\mathbf{B}_0\mathcal{E}_{\mathbf{B}_0^T\mathbf{MB}_0}(\mathbf{B}_0^T\mathcal{S}) \subseteq \mathcal{E}_\mathbf{M}(\mathcal{S}).$$

Proof: Since $\mathcal{S} \subseteq \text{span}(\mathbf{M})$, we have $\mathbf{B}_0^T\mathcal{S} \subseteq \text{span}(\mathbf{B}_0^T\mathbf{MB}_0)$, so the envelope $\mathcal{E}_{\mathbf{B}_0^T\mathbf{MB}_0}(\mathbf{B}_0^T\mathcal{S})$ is well defined. The next step is to use Proposition A.1 to show that $\mathbf{B}_0^T\mathcal{E}_\mathbf{M}(\mathcal{S})$ is a reducing subspace of $\mathbf{B}_0^T\mathbf{MB}_0$ that contains $\mathbf{B}_0^T\mathcal{S}$.

Since $\mathcal{E}_{\mathbf{M}}(\mathcal{S})$ reduces $\mathbf{M}$, $\mathbf{M} = \mathbf{P}_{\mathcal{E}}\mathbf{M}\mathbf{P}_{\mathcal{E}} + \mathbf{Q}_{\mathcal{E}}\mathbf{M}\mathbf{Q}_{\mathcal{E}}$, and thus

$$\mathbf{B}_0^T\mathbf{M}\mathbf{B}_0 = \mathbf{B}_0^T\mathbf{P}_{\mathcal{E}}\mathbf{M}\mathbf{P}_{\mathcal{E}}\mathbf{B}_0 + \mathbf{B}_0^T\mathbf{Q}_{\mathcal{E}}\mathbf{M}\mathbf{Q}_{\mathcal{E}}\mathbf{B}_0.$$

Since $\mathcal{B} \subseteq \mathcal{E}_{\mathbf{M}}(\mathcal{S})$, we have the decomposition $\mathbf{P}_{\mathcal{E}} = \mathbf{P}_{\mathcal{B}} + \mathbf{P}_{\mathbf{Q}_{\mathcal{B}}\mathcal{E}}$, and thus

$$\mathbf{B}_0^T\mathbf{P}_{\mathcal{E}} = \mathbf{B}_0^T\mathbf{P}_{\mathbf{Q}_{\mathcal{B}}\mathcal{E}} = \mathbf{B}_0^T\mathbf{P}_{\mathbf{B}_0\mathbf{B}_0^T\mathcal{E}} = \mathbf{B}_0^T\mathbf{B}_0\mathbf{P}_{\mathbf{B}_0^T\mathcal{E}}\mathbf{B}_0^T = \mathbf{P}_{\mathbf{B}_0^T\mathcal{E}}\mathbf{B}_0^T.$$

Also, since $\mathcal{E}_{\mathbf{M}}^{\perp}(\mathcal{S}) \subseteq \mathcal{B}^{\perp}$, $\mathbf{Q}_{\mathcal{E}}\mathbf{Q}_{\mathcal{B}} = \mathbf{Q}_{\mathcal{E}}$. Substituting these relationships and $\mathbf{Q}_{\mathcal{B}} = \mathbf{B}_0\mathbf{B}_0^T$, we have

$$\mathbf{B}_0^T\mathbf{M}\mathbf{B}_0 = \mathbf{P}_{\mathbf{B}_0^T\mathcal{E}}(\mathbf{B}_0^T\mathbf{M}\mathbf{B}_0)\mathbf{P}_{\mathbf{B}_0^T\mathcal{E}} + \mathbf{B}_0^T\mathbf{Q}_{\mathcal{E}}\mathbf{B}_0(\mathbf{B}_0^T\mathbf{M}\mathbf{B}_0)\mathbf{B}_0^T\mathbf{Q}_{\mathcal{E}}\mathbf{B}_0,$$

where $\mathbf{P}_{\mathbf{B}_0^T\mathcal{E}}$ and $\mathbf{B}_0^T\mathbf{Q}_{\mathcal{E}}\mathbf{B}_0$ are orthogonal projections, as can be seen by direct calculation:

$$\mathbf{B}_0^T\mathbf{Q}_{\mathcal{E}}\mathbf{B}_0\mathbf{B}_0^T\mathbf{Q}_{\mathcal{E}}\mathbf{B}_0 = \mathbf{B}_0^T\mathbf{Q}_{\mathcal{E}}\mathbf{B}_0 \quad \text{and}$$
$$\mathbf{B}_0^T\mathbf{Q}_{\mathcal{E}}\mathbf{B}_0\mathbf{B}_0^T\mathcal{E}_{\mathbf{M}}(\mathcal{S}) = \mathbf{B}_0^T\mathbf{Q}_{\mathcal{E}}\mathcal{E}_{\mathbf{M}}(\mathcal{S}) = 0.$$

Since $\text{span}(\mathbf{B}_0^T\mathbf{P}_{\mathcal{E}})$ and $\text{span}(\mathbf{B}_0^T\mathbf{Q}_{\mathcal{E}})$ are orthogonal subspaces, we can find their joint span as $\text{span}(\mathbf{B}_0^T\mathbf{P}_{\mathcal{E}}) + \text{span}(\mathbf{B}_0^T\mathbf{Q}_{\mathcal{E}}) = \text{span}(\mathbf{B}_0^T)$. It follows that $\text{span}(\mathbf{B}_0^T\mathbf{P}_{\mathcal{E}}) = \mathbf{B}_0^T\mathcal{E}_{\mathbf{M}}(\mathcal{S})$ is a reducing subspace of $\mathbf{B}_0^T\mathbf{M}\mathbf{B}_0$ that contains $\mathbf{B}_0^T\mathcal{S}$, since $\dim(\text{span}(\mathbf{B}_0^T))$ is the same as the dimension of the matrix $\mathbf{B}_0^T\mathbf{M}\mathbf{B}_0$.

Since $\mathcal{E}_{\mathbf{B}_0^T\mathbf{M}\mathbf{B}_0}(\mathbf{B}_0^T\mathcal{S})$ is the smallest reducing subspace of $\mathbf{B}_0^T\mathbf{M}\mathbf{B}_0$ that contains $\mathbf{B}_0^T\mathcal{S}$, we have

$$\mathcal{E}_{\mathbf{B}_0^T\mathbf{M}\mathbf{B}_0}(\mathbf{B}_0^T\mathcal{S}) \subseteq \mathbf{B}_0^T\mathcal{E}_{\mathbf{M}}(\mathcal{S}). \tag{A.9}$$

Consequently, if $\mathbf{v} \in \mathcal{E}_{\mathbf{B}_0^T\mathbf{M}\mathbf{B}_0}(\mathbf{B}_0^T\mathcal{S})$, then $\mathbf{B}_0\mathbf{v} \in \mathbf{Q}_{\mathcal{B}}\mathcal{E}_{\mathbf{M}}(\mathcal{S}) \subseteq \mathcal{E}_{\mathbf{M}}(\mathcal{S})$, since $\mathcal{B} \subseteq \mathcal{E}_{\mathbf{M}}(\mathcal{S})$: letting $\boldsymbol{\Gamma}$ be a semi-orthogonal basis matrix for $\mathcal{E}_{\mathbf{M}}(\mathcal{S})$ and writing $\mathbf{B} = \boldsymbol{\Gamma}\mathbf{A}$, where $\mathbf{A}$ is a semi-orthogonal matrix, we have

$$\mathbf{Q}_{\mathcal{B}}\boldsymbol{\Gamma} = \boldsymbol{\Gamma} - \mathbf{P}_{\mathcal{B}}\boldsymbol{\Gamma} = \boldsymbol{\Gamma} - \boldsymbol{\Gamma}\mathbf{P}_{\mathbf{A}} = \boldsymbol{\Gamma}\mathbf{Q}_{\mathbf{A}}.$$

□

Proposition A.6 was used in Chapters 1, 4, and 6, and Proposition A.7 was used in Chapters 6 and 7.

A.4 Kronecker Products, vec and vech

The Kronecker product $\otimes$ between two matrices $\mathbf{A} \in \mathbb{R}^{r\times s}$ and $\mathbf{B} \in \mathbb{R}^{t\times u}$ is the $rt \times su$ matrix defined in blocks as

$$\mathbf{A} \otimes \mathbf{B} = (\mathbf{A})_{ij}\mathbf{B}, \quad i = 1, \dots, r,\ j = 1, \dots s.$$

Kronecker products are not in general commutative, $\mathbf{A} \otimes \mathbf{B} \neq \mathbf{B} \otimes \mathbf{A}$. Lemma A.7 reviews some basic properties of Kronecker products.

Lemma A.7 *Let* $\mathbf{A}$, $\mathbf{B}$, $\mathbf{C}$, *and* $\mathbf{D}$ *be matrices of arbitrary but finite dimension. Then*

1. $\mathbf{A} \otimes (\mathbf{B} \otimes \mathbf{C}) = (\mathbf{A} \otimes \mathbf{B}) \otimes \mathbf{C}$
2. $(\mathbf{A} \otimes \mathbf{B})^T = \mathbf{A}^T \otimes \mathbf{B}^T$
3. $(\mathbf{A} \otimes \mathbf{B})(\mathbf{C} \otimes \mathbf{D}) = (\mathbf{AC} \otimes \mathbf{BD})$
4. *If* $\mathbf{A}$ *and* $\mathbf{B}$ *are nonsingular, then* $(\mathbf{A} \otimes \mathbf{B})^{-1} = (\mathbf{A}^{-1} \otimes \mathbf{B}^{-1})$
5. $\text{rank}(\mathbf{A} \otimes \mathbf{B}) = \text{rank}(\mathbf{A})\text{rank}(\mathbf{B})$
6. $\text{tr}(\mathbf{A} \otimes \mathbf{B}) = \text{tr}(\mathbf{A})\text{tr}(\mathbf{B})$
7. *If* $\mathbf{A} \in \mathbb{R}^{r\times r}$ *and* $\mathbf{B} \in \mathbb{R}^{s\times s}$, *then* $|\mathbf{A} \otimes \mathbf{B}| = |\mathbf{A}|^s|\mathbf{B}|^r$
8. $\mathbf{P}_{\mathbf{A}\otimes\mathbf{B}} = \mathbf{P}_{\mathbf{A}} \otimes \mathbf{P}_{\mathbf{B}}$.

The vec operator transforms a matrix $\mathbf{A} \in \mathbb{R}^{r\times u}$ to a vector $\text{vec}(\mathbf{A}) \in \mathbb{R}^{ru}$ by stacking its columns. If $\mathbf{A} = (\mathbf{a}_1, \dots, \mathbf{a}_u)$, then

$$\text{vec}(\mathbf{A}) = \begin{pmatrix} \mathbf{a}_1 \\ \mathbf{a}_2 \\ \vdots \\ \mathbf{a}_u \end{pmatrix}.$$

The vech operator transforms a symmetric matrix $\mathbf{A} \in \mathbb{S}^{r\times r}$ to a vector $\text{vech}(\mathbf{A}) \in \mathbb{R}^{r(r+1)/2}$ by stacking its unique elements on and below the diagonal. If

$$\mathbf{A} = \begin{pmatrix} \mathbf{a}_{11} & \mathbf{a}_{12} & \mathbf{a}_{13} \\ \mathbf{a}_{12} & \mathbf{a}_{22} & \mathbf{a}_{23} \\ \mathbf{a}_{13} & \mathbf{a}_{23} & \mathbf{a}_{33} \end{pmatrix}$$

then

$$\text{vech}(\mathbf{A}) = \left(\mathbf{a}_{11}\ \mathbf{a}_{12}\ \mathbf{a}_{13}\ \mathbf{a}_{22}\ \mathbf{a}_{23}\ \mathbf{a}_{33}\right)^T.$$

The following lemma summarizes some useful properties of these operators.

Lemma A.8 *Let* $\mathbf{A}$, $\mathbf{B}$, *and* $\mathbf{C}$ *be matrices whose dimensions allow the required multiplication or addition. Then*

1. $\text{vec}(\mathbf{A} + \mathbf{B}) = \text{vec}(\mathbf{A}) + \text{vec}(\mathbf{B})$
2. $\text{vec}(\mathbf{ABC}) = (\mathbf{C}^T \otimes \mathbf{A})\text{vec}(\mathbf{B})$
3. $\text{tr}(\mathbf{A}^T\mathbf{B}) = \text{vec}^T(\mathbf{A})\text{vec}(\mathbf{B})$
4. *For nonsingular* $\mathbf{A}$, $\text{vec}(\mathbf{A}^{-1}) = (\mathbf{A}^{-T} \otimes \mathbf{A}^{-1})\text{vec}(\mathbf{A})$.

A.5 Commutation, Expansion, and Contraction Matrices

The commutation matrix $\mathbf{K}_{pm} \in \mathbb{R}^{pm \times pm}$ is the unique matrix that transforms the vec of a matrix into the vec of its transpose: For $\mathbf{A} \in \mathbb{R}^{p \times m}$, $\text{vec}(\mathbf{A}^T) = \mathbf{K}_{pm}\text{vec}(\mathbf{A})$. The expansion $\mathbf{E}_r \in \mathbb{R}^{r^2 \times r(r+1)/2}$ and contraction matrices $\mathbf{C}_r \in \mathbb{R}^{r(r+1)/2 \times r^2}$ are unique matrices that connect the vec and vech operators: If $\mathbf{A} \in \mathbb{S}^{r \times r}$, then $\text{vech}(\mathbf{A}) = \mathbf{C}_r\text{vec}(\mathbf{A})$ and $\text{vec}(\mathbf{A}) = \mathbf{E}_r\text{vech}(\mathbf{A})$. We summarize properties of these operators in this section, focusing on those used in various derivations related to envelope models.

The next lemma gives properties of commutation matrices.

Lemma A.9 *The following properties hold:*

1. $\mathbf{K}_{pm}^T = \mathbf{K}_{mp}$
2. $\mathbf{K}_{pm}^T\mathbf{K}_{pm} = \mathbf{K}_{pm}\mathbf{K}_{pm}^T = \mathbf{I}_{pm}$
3. $\mathbf{K}_{p1} = \mathbf{K}_{1p} = \mathbf{I}_p$
4. *Let* $\mathbf{A} \in \mathbb{R}^{r_1 \times r_2}$ *and* $\mathbf{B} \in \mathbb{R}^{r_3 \times r_4}$. *Then* $\mathbf{K}_{r_3 r_1}(\mathbf{A} \otimes \mathbf{B})\mathbf{K}_{r_2 r_4} = \mathbf{B} \otimes \mathbf{A}$ *and* $\mathbf{K}_{r_3 r_1}(\mathbf{A} \otimes \mathbf{B}) = (\mathbf{B} \otimes \mathbf{A})\mathbf{K}_{r_4 r_2}$
5. *Let* $\mathbf{a} \in \mathbb{R}^r$ *and* $\mathbf{b} \in \mathbb{R}^u$. *Then* $\mathbf{a} \otimes \mathbf{b} = \mathbf{K}_{ru}(\mathbf{b} \otimes \mathbf{a})$.

The next lemma gives properties of expansion and contraction matrices. Let $\mathbf{P}_{\mathbf{E}_r}$ be the projection onto span($\mathbf{E}_r$).

Lemma A.10 *Let* $\mathbf{A} \in \mathbb{R}^{r \times u}$. *Then*

1. $\mathbf{C}_r\mathbf{E}_r = \mathbf{I}_{r(r+1)/2}$
2. $\mathbf{E}_r\mathbf{C}_r(\mathbf{A} \otimes \mathbf{A})\mathbf{E}_u = (\mathbf{A} \otimes \mathbf{A})\mathbf{E}_u$
3. $\mathbf{E}_r\mathbf{C}_r(\mathbf{A} \otimes \mathbf{A})\mathbf{C}_u^T = (\mathbf{A} \otimes \mathbf{A})\mathbf{C}_u^T$
4. $\mathbf{P}_{\mathbf{E}_r}(\mathbf{A} \otimes \mathbf{A})\mathbf{P}_{\mathbf{E}_u} = \mathbf{P}_{\mathbf{E}_r}(\mathbf{A} \otimes \mathbf{A}) = (\mathbf{A} \otimes \mathbf{A})\mathbf{P}_{\mathbf{E}_u}$.

The next corollary gives relationships that involve commutation, expansions, and/or contraction matrices.

Lemma A.11 *The following properties hold:*

1. $\mathbf{C}_r\mathbf{K}_{rr} = \mathbf{C}_r$
2. $\mathbf{E}_r\mathbf{C}_r = (\mathbf{E}_r\mathbf{C}_r)^T = \frac{1}{2}(\mathbf{I}_{r^2} + \mathbf{K}_{rr}) = \mathbf{P}_{\mathbf{E}_r}$
3. *Let* $\mathbf{C} \in \mathbb{R}^{s \times r}$, $\mathbf{D} \in \mathbb{R}^{t \times r}$, $\mathbf{A} \in \mathbb{R}^{r \times u}$ *and* $\mathbf{B} \in \mathbb{R}^{r \times v}$. *Then*

$$2(\mathbf{C} \otimes \mathbf{D})\mathbf{P}_{\mathbf{E}_r}(\mathbf{A} \otimes \mathbf{B}) = (\mathbf{CA} \otimes \mathbf{DB}) + (\mathbf{CB} \otimes \mathbf{DA})\mathbf{K}_{vu}.$$

If either $\mathbf{CB} = \mathbf{0}$ *or* $\mathbf{DA} = \mathbf{0}$, *then*

$$2(\mathbf{C} \otimes \mathbf{D})\mathbf{P}_{\mathbf{E}_r}(\mathbf{A} \otimes \mathbf{B}) = (\mathbf{CA} \otimes \mathbf{DB}).$$

A.6 Derivatives

The derivatives given in Sections A.6.1 and A.6.2 are needed in deriving the asymptotic variances of $\text{vec}(\widehat{\boldsymbol{\beta}})$ and $\text{vech}(\widehat{\boldsymbol{\Sigma}})$ for various envelope models. The notation used here conforms to that used in Section 1.6 for envelope model (1.20). We enforce the condition that $\boldsymbol{\Gamma}^T\boldsymbol{\Gamma}_0 = 0$, but we do not enforce the condition that $\boldsymbol{\Gamma}$ and $\boldsymbol{\Gamma}_0$ are semi-orthogonal, since the overparameterization is taken onto account in the asymptotic calculations. All derivatives are to be evaluated at the true values of the parameters.

In Section A.6.3, we turn to Grassmann derivatives, which may be needed in optimization algorithms. In this section, we do enforce the condition that $\boldsymbol{\Gamma}$ is semi-orthogonal.

A.6.1 Derivatives for $\boldsymbol{\eta}$, $\boldsymbol{\Omega}$, and $\boldsymbol{\Omega}_0$

Lemma A.12 gives a general tool that is used frequently in derivatives associated with asymptotic variances.

Lemma A.12 *Let* $\mathbf{X} \in \mathbb{R}^{a\times b}$, *and let* $\mathbf{F}(\mathbf{X}) \in \mathbb{R}^{m\times p}$ *and* $\mathbf{G}(\mathbf{X}) \in \mathbb{R}^{p\times q}$ *be matrix-valued differentiable function of* $\mathbf{X}$. *Then*

$$\frac{\partial \text{vec}[\mathbf{F}(\mathbf{X})\mathbf{G}(\mathbf{X})]}{\partial \text{vec}^T(\mathbf{X})} = (\mathbf{G}^T \otimes \mathbf{I}_m)\frac{\partial \text{vec}[\mathbf{F}(\mathbf{X})]}{\partial \text{vec}^T(\mathbf{X})} + (\mathbf{I}_q \otimes \mathbf{F})\frac{\partial \text{vec}[\mathbf{G}(\mathbf{X})]}{\partial \text{vec}^T(\mathbf{X})}.$$

In particular,

$$\frac{\partial \text{vec}(\mathbf{X})}{\partial \text{vec}^T(\mathbf{X})} = \mathbf{I}_{ab}.$$

It follows immediately from Lemma A.12 that

$$\frac{\partial \text{vec}(\boldsymbol{\Gamma\eta})}{\partial \text{vec}^T(\boldsymbol{\eta})} = \frac{\partial[(\mathbf{I}_p \otimes \boldsymbol{\Gamma})\text{vec}(\boldsymbol{\eta})]}{\partial \text{vec}^T(\boldsymbol{\eta})} = \mathbf{I}_p \otimes \boldsymbol{\Gamma} \in \mathbb{R}^{pr\times pu}$$

and that

$$\begin{aligned}\frac{\partial \text{vech}(\boldsymbol{\Gamma\Omega\Gamma}^T)}{\partial \text{vech}^T(\boldsymbol{\Omega})} &= \mathbf{C}_r\frac{\partial \text{vec}(\boldsymbol{\Gamma\Omega\Gamma}^T)}{\partial \text{vech}^T(\boldsymbol{\Omega})}\\ &= \mathbf{C}_r(\boldsymbol{\Gamma}\otimes\boldsymbol{\Gamma})\mathbf{E}_u\frac{\partial \text{vech}(\boldsymbol{\Omega})}{\partial \text{vech}^T(\boldsymbol{\Omega})}\\ &= \mathbf{C}_r(\boldsymbol{\Gamma}\otimes\boldsymbol{\Gamma})\mathbf{E}_u,\end{aligned}$$

where $\mathbf{C}_r$ and $\mathbf{E}_u$ are contraction and expansion matrices described in Section A.5. Similarly,

$$\frac{\partial \text{vech}(\boldsymbol{\Gamma\Omega}_0\boldsymbol{\Gamma}^T)}{\partial \text{vech}^T(\boldsymbol{\Omega}_0)} = \mathbf{C}_r(\boldsymbol{\Gamma}_0\otimes\boldsymbol{\Gamma}_0)\mathbf{E}_{(r-u)}.$$

A.6.2 Derivatives with Respect to $\boldsymbol{\Gamma}$

From Lemma A.12,

$$\frac{\partial \text{vec}(\boldsymbol{\Gamma}\boldsymbol{\eta})}{\partial \text{vec}^T(\boldsymbol{\Gamma})} = \boldsymbol{\eta}^T \otimes \mathbf{I}_r \in \mathbb{R}^{pr \times ur}.$$

Again applying Lemma A.12,

$$\begin{aligned}
\frac{\partial \text{vec}\{(\boldsymbol{\Gamma}\boldsymbol{\Omega})\boldsymbol{\Gamma}^T\}}{\partial \text{vec}^T(\boldsymbol{\Gamma})} &= (\boldsymbol{\Gamma} \otimes \mathbf{I}_r)\frac{\partial \text{vec}(\boldsymbol{\Gamma}\boldsymbol{\Omega})}{\partial \text{vec}^T(\boldsymbol{\Gamma})} + (\mathbf{I}_r \otimes \boldsymbol{\Gamma}\boldsymbol{\Omega})\frac{\partial \text{vec}(\boldsymbol{\Gamma}^T)}{\partial \text{vec}^T(\boldsymbol{\Gamma})} \\
&= (\boldsymbol{\Gamma} \otimes \mathbf{I}_r)(\boldsymbol{\Omega} \otimes \mathbf{I}_r)\frac{\partial \text{vec}(\boldsymbol{\Gamma})}{\partial \text{vec}^T(\boldsymbol{\Gamma})} + (\mathbf{I}_r \otimes \boldsymbol{\Gamma}\boldsymbol{\Omega})\mathbf{K}_{ru}\frac{\partial \text{vec}(\boldsymbol{\Gamma})}{\partial \text{vec}^T(\boldsymbol{\Gamma})} \\
&= \{(\boldsymbol{\Gamma}\boldsymbol{\Omega} \otimes \mathbf{I}_r) + (\mathbf{I}_r \otimes \boldsymbol{\Gamma}\boldsymbol{\Omega})\mathbf{K}_{ru}\}\frac{\partial \text{vec}(\boldsymbol{\Gamma})}{\partial \text{vec}^T(\boldsymbol{\Gamma})} \qquad \text{(A.10)} \\
&= (\boldsymbol{\Gamma}\boldsymbol{\Omega} \otimes \mathbf{I}_r) + (\mathbf{I}_r \otimes \boldsymbol{\Gamma}\boldsymbol{\Omega})\mathbf{K}_{ru}. \qquad \text{(A.11)}
\end{aligned}$$

In the second equality, the commutation matrix $\mathbf{K}_{ru}$ is used to convert $\text{vec}(\boldsymbol{\Gamma}^T)$ to $\text{vec}(\boldsymbol{\Gamma})$. Next, multiplying each side of (A.11) by the contraction matrix $\mathbf{C}_r$, we have

$$\begin{aligned}
\frac{\partial \text{vech}\{(\boldsymbol{\Gamma}\boldsymbol{\Omega})\boldsymbol{\Gamma}^T\}}{\partial \text{vec}^T(\boldsymbol{\Gamma})} &= \mathbf{C}_r\frac{\partial \text{vec}\{(\boldsymbol{\Gamma}\boldsymbol{\Omega})\boldsymbol{\Gamma}^T\}}{\partial \text{vec}^T(\boldsymbol{\Gamma})} \\
&= \mathbf{C}_r(\boldsymbol{\Gamma}\boldsymbol{\Omega} \otimes \mathbf{I}_r) + \mathbf{C}_r(\mathbf{I}_r \otimes \boldsymbol{\Gamma}\boldsymbol{\Omega})\mathbf{K}_{ru} \\
&= \mathbf{C}_r(\boldsymbol{\Gamma}\boldsymbol{\Omega} \otimes \mathbf{I}_r) + \mathbf{C}_r\mathbf{K}_{rr}(\mathbf{I}_r \otimes \boldsymbol{\Gamma}\boldsymbol{\Omega})\mathbf{K}_{ru} \\
&= 2\mathbf{C}_r(\boldsymbol{\Gamma}\boldsymbol{\Omega} \otimes \mathbf{I}_r), \qquad \text{(A.12)}
\end{aligned}$$

where the last equality follows from conclusion 4 of Lemma A.9.

To find $\partial \text{vech}\{(\boldsymbol{\Gamma}_0\boldsymbol{\Omega}_0)\boldsymbol{\Gamma}_0^T\}/\partial \text{vec}^T(\boldsymbol{\Gamma})$, we follow the steps leading to (A.12). Beginning with (A.10),

$$\frac{\partial \text{vec}\{(\boldsymbol{\Gamma}_0\boldsymbol{\Omega}_0)\boldsymbol{\Gamma}_0^T\}}{\partial \text{vec}^T(\boldsymbol{\Gamma})} = \{(\boldsymbol{\Gamma}_0\boldsymbol{\Omega}_0 \otimes \mathbf{I}_r) + (\mathbf{I}_r \otimes \boldsymbol{\Gamma}_0\boldsymbol{\Omega}_0)\mathbf{K}_{r(r-u)}\}\frac{\partial \text{vec}(\boldsymbol{\Gamma}_0)}{\partial \text{vec}^T(\boldsymbol{\Gamma})}. \qquad \text{(A.13)}$$

Multiplying both sides of (A.13) by $\mathbf{C}_r$ and proceed as before, we have

$$\frac{\partial \text{vech}(\boldsymbol{\Gamma}_0\boldsymbol{\Omega}_0\boldsymbol{\Gamma}_0^T)}{\partial \text{vec}^T(\boldsymbol{\Gamma})} = 2\mathbf{C}_r(\boldsymbol{\Gamma}_0\boldsymbol{\Omega}_0 \otimes \mathbf{I}_r)\frac{\partial \text{vec}(\boldsymbol{\Gamma}_0)}{\partial \text{vec}^T(\boldsymbol{\Gamma})}. \qquad \text{(A.14)}$$

To complete the calculation, we need to evaluate $\partial \text{vec}(\boldsymbol{\Gamma}_0)/\partial \text{vec}^T(\boldsymbol{\Gamma})$, which requires that we define $\boldsymbol{\Gamma}_0$ so that it is uniquely associated with $\boldsymbol{\Gamma}$. Since all derivatives are to be evaluated at true bases for the envelope and its orthogonal complement, we need to define a unique association only in a neighborhood of the true basis for the envelope. Accordingly, temporarily let $\boldsymbol{\gamma}$ and $\boldsymbol{\gamma}_0$ denote true semi-orthogonal basis matrices for $\mathcal{E}_{\boldsymbol{\Sigma}}(\mathcal{B})$ and its orthogonal complement. Then in a neighborhood of $\boldsymbol{\gamma}$, $\mathbf{Q}_{\boldsymbol{\Gamma}}\boldsymbol{\gamma}_0$ and $\mathbf{Q}_{\boldsymbol{\Gamma}}$ have the same column space, and we can take $\boldsymbol{\Gamma}_0 = \mathbf{Q}_{\boldsymbol{\Gamma}}\boldsymbol{\gamma}_0$.

Then the required derivative becomes

$$\begin{aligned}\frac{\partial \mathrm{vec}(\boldsymbol{\Gamma}_0)}{\partial \mathrm{vec}^T(\boldsymbol{\Gamma})} &= \frac{\partial \mathrm{vec}(\mathbf{Q}_{\boldsymbol{\Gamma}}\boldsymbol{\gamma}_0)}{\partial \mathrm{vec}^T(\boldsymbol{\Gamma})} = -\frac{\partial \mathrm{vec}(\mathbf{P}_{\boldsymbol{\Gamma}}\boldsymbol{\gamma}_0)}{\partial \mathrm{vec}^T(\boldsymbol{\Gamma})} = -(\boldsymbol{\gamma}_0^T \otimes \mathbf{I}_r)\frac{\partial \mathrm{vec}(\mathbf{P}_{\boldsymbol{\Gamma}})}{\partial \mathrm{vec}^T(\boldsymbol{\Gamma})}\\
&= -(\boldsymbol{\gamma}_0^T \otimes \mathbf{I}_r)\frac{\partial \mathrm{vec}(\boldsymbol{\Gamma}(\boldsymbol{\Gamma}^T\boldsymbol{\Gamma})^{-1}\boldsymbol{\Gamma}^T)}{\partial \mathrm{vec}^T(\boldsymbol{\Gamma})}\\
&= -(\boldsymbol{\gamma}_0^T \otimes \mathbf{I}_r)(\boldsymbol{\Gamma}(\boldsymbol{\Gamma}^T\boldsymbol{\Gamma})^{-1} \otimes \mathbf{I}_r)\frac{\partial \mathrm{vec}(\boldsymbol{\Gamma})}{\partial \mathrm{vec}^T(\boldsymbol{\Gamma})}\\
&\quad - (\boldsymbol{\gamma}_0^T \otimes \mathbf{I}_r)(\mathbf{I}_r \otimes \boldsymbol{\Gamma})\frac{\partial \mathrm{vec}((\boldsymbol{\Gamma}^T\boldsymbol{\Gamma})^{-1}\boldsymbol{\Gamma}^T)}{\partial \mathrm{vec}^T(\boldsymbol{\Gamma})}\\
&= -(\boldsymbol{\gamma}_0^T \otimes \boldsymbol{\Gamma})\frac{\partial \mathrm{vec}((\boldsymbol{\Gamma}^T\boldsymbol{\Gamma})^{-1}\boldsymbol{\Gamma}^T)}{\partial \mathrm{vec}^T(\boldsymbol{\Gamma})},\end{aligned}$$

where the last step follows because $\boldsymbol{\gamma}_0^T\boldsymbol{\Gamma} = 0$ when evaluated at $\boldsymbol{\Gamma} = \boldsymbol{\gamma}$.

To complete the calculation, we need to evaluate

$$\begin{aligned}\frac{\partial \mathrm{vec}((\boldsymbol{\Gamma}^T\boldsymbol{\Gamma})^{-1}\boldsymbol{\Gamma}^T)}{\partial \mathrm{vec}^T(\boldsymbol{\Gamma})} &= (\boldsymbol{\Gamma} \otimes \mathbf{I}_u)\frac{\partial \mathrm{vec}((\boldsymbol{\Gamma}^T\boldsymbol{\Gamma})^{-1})}{\partial \mathrm{vec}^T(\boldsymbol{\Gamma})} + (\mathbf{I}_r \otimes (\boldsymbol{\Gamma}^T\boldsymbol{\Gamma})^{-1})\frac{\partial \mathrm{vec}(\boldsymbol{\Gamma}^T)}{\partial \mathrm{vec}^T(\boldsymbol{\Gamma})}\\
&= (\boldsymbol{\Gamma} \otimes \mathbf{I}_u)\frac{\partial \mathrm{vec}((\boldsymbol{\Gamma}^T\boldsymbol{\Gamma})^{-1})}{\partial \mathrm{vec}^T(\boldsymbol{\Gamma})} + (\mathbf{I}_r \otimes (\boldsymbol{\Gamma}^T\boldsymbol{\Gamma})^{-1})\mathbf{K}_{ru}.\end{aligned}$$

The first term on the right-hand side is again 0 when multiplied by $(\boldsymbol{\gamma}_0^T \otimes \mathbf{I}_r)$, and therefore

$$\frac{\partial \mathrm{vec}(\boldsymbol{\Gamma}_0)}{\partial \mathrm{vec}^T(\boldsymbol{\Gamma})} = -(\boldsymbol{\gamma}_0^T \otimes \boldsymbol{\Gamma}(\boldsymbol{\Gamma}^T\boldsymbol{\Gamma})^{-1})\mathbf{K}_{ru} = -(\boldsymbol{\gamma}_0^T \otimes \boldsymbol{\gamma})\mathbf{K}_{ru},$$

where the final term is explicitly evaluated at the true values, recalling that $\boldsymbol{\gamma}^T\boldsymbol{\gamma} = \mathbf{I}_u$.

Substituting this result into (A.14) and reverting to the original notation, we obtain

$$\begin{aligned}\frac{\partial \mathrm{vech}(\boldsymbol{\Gamma}_0\boldsymbol{\Omega}_0\boldsymbol{\Gamma}_0^T)}{\partial \mathrm{vec}^T(\boldsymbol{\Gamma})} &= -2\mathbf{C}_r(\boldsymbol{\Gamma}_0\boldsymbol{\Omega}_0 \otimes \mathbf{I}_r)(\boldsymbol{\Gamma}_0^T \otimes \boldsymbol{\Gamma})\mathbf{K}_{ru}\\
&= -2\mathbf{C}_r(\boldsymbol{\Gamma}_0\boldsymbol{\Omega}_0\boldsymbol{\Gamma}_0^T \otimes \boldsymbol{\Gamma})\mathbf{K}_{ru}.\end{aligned}$$

A.6.3 Derivatives of Grassmann Objective Functions

Let $\mathbf{G} \in \mathbb{R}^{r\times u}$, $r > u$, and suppose that we wish to minimize a generic real objective function $f(\mathbf{G})$ subject to the constraint $\mathbf{G}^T\mathbf{G} = \mathbf{I}_u$. We require further that $f(\mathbf{G})$ depend only on span($\mathbf{G}$); that is, $f(\mathbf{GO}) = \mathbf{f}(\mathbf{G})$ for all orthogonal matrices $\mathbf{O} \in \mathbb{R}^{u\times u}$. Since f depends only on the subspace spanned by the columns of $\mathbf{G}$, the minimization is effectively over the Grassmannian $\mathcal{G}(u, r)$. The likelihood objective function (1.25) is an instance of this more general setup.

Algorithms for the minimization of $f(\mathbf{G})$ over $\mathcal{G}(u,r)$ require the gradient matrix $\nabla f(\mathbf{G}) \in \mathbb{R}^{r\times u}$ of $f(\mathbf{G})$ along a Grassmann geodesic. Let $\partial f(\mathbf{G})/\partial \mathbf{G} \in \mathbb{R}^{r\times u}$ denote the ordinary derivative of f with respect to the elements of $\mathbf{G}$. Then the Grassmann derivative of $\mathbf{f}(\mathbf{G})$ can be expressed as (Edelman et al., 1998, Section 2.5.3).

$$\nabla f(\mathbf{G}) = \mathbf{Q}_{\mathbf{G}} \frac{\partial f(\mathbf{G})}{\partial \mathbf{G}} \in \mathbb{R}^{r\times u}. \tag{A.15}$$

Optimization of $f(\mathbf{G})$ is guided by the condition $\nabla \mathbf{f}(\mathbf{G}) = 0$.

The form of the gradient (A.15) is the same as that obtained when using Lagrange multipliers to induce the constraint $\mathbf{G}^T\mathbf{G} = \mathbf{I}_u$: Let $\mathbf{U} \in \mathbb{R}^{u\times u}$ be a matrix of Lagrange multipliers and differentiate $f(\mathbf{G}) + \text{tr}(\mathbf{U}^T(\mathbf{G}^T\mathbf{G} - \mathbf{I}_u))$ to get

$$\frac{\partial f(\mathbf{G})}{\partial \mathbf{G}} + \mathbf{G}(\mathbf{U} + \mathbf{U}^T) = 0.$$

Solving for $\mathbf{U} + \mathbf{U}^T$ by using the constraint $\mathbf{G}^T\mathbf{G} = \mathbf{I}_u$ yields $\mathbf{U} + \mathbf{U}^T = -\mathbf{G}^T(\partial f(\mathbf{G})/\partial \mathbf{G})$, which reduces the derivative to

$$\frac{\partial f(\mathbf{G})}{\partial \mathbf{G}} - \mathbf{P}_{\mathbf{G}} \frac{\partial f(\mathbf{G})}{\partial \mathbf{G}} = \mathbf{Q}_{\mathbf{G}} \frac{\partial f(\mathbf{G})}{\partial \mathbf{G}} = 0.$$

Edelman et al. (1998) gave a nice account of optimization over Grassmann and Stiefel manifolds.

Applying these results to the likelihood objective function (1.25)

$$L(\mathbf{G}) = \log|\mathbf{G}^T\mathbf{S}_{\mathbf{Y}|\mathbf{X}}\mathbf{G}| + \log|\mathbf{G}^T\mathbf{S}_{\mathbf{Y}}^{-1}\mathbf{G}|$$

we get

$$\nabla L(\mathbf{G}) = 2\mathbf{Q}_{\mathbf{G}}\{\mathbf{S}_{\mathbf{Y}|\mathbf{X}}\mathbf{G}(\mathbf{G}^T\mathbf{S}_{\mathbf{Y}|\mathbf{X}}\mathbf{G})^{-1} + \mathbf{S}_{\mathbf{Y}}^{-1}\mathbf{G}(\mathbf{G}^T\mathbf{S}_{\mathbf{Y}}^{-1}\mathbf{G})^{-1}\}. \tag{A.16}$$

A.7 Miscellaneous Results

Lemma A.13 justifies replacing $\log|\mathbf{G}_0^T(\mathbf{M} + \mathbf{U})\mathbf{G}_0|$ with $\log|\mathbf{G}^T(\mathbf{M} + \mathbf{U})^{-1}\mathbf{G}|$ during the minimization to obtain $\mathbf{\Gamma}$. Lemma A.16 is used when studying properties of a sequential algorithm in Chapter 6.

Lemma A.13 *Suppose that $\mathbf{A} \in \mathbb{R}^{t\times t}$ is nonsingular and that the column-partitioned matrix $(\boldsymbol{\alpha}, \boldsymbol{\alpha}_0) \in \mathbb{R}^{t\times t}$ is orthogonal. Then* $|\boldsymbol{\alpha}_0^T\mathbf{A}\boldsymbol{\alpha}_0| = |\mathbf{A}| \times |\boldsymbol{\alpha}^T\mathbf{A}^{-1}\boldsymbol{\alpha}|$.

Proof: Define the $t \times t$ matrix

$$\mathbf{K} = \begin{pmatrix} \mathbf{I}_d, \boldsymbol{\alpha}^T\mathbf{A}\boldsymbol{\alpha}_0 \\ 0, \boldsymbol{\alpha}_0^T\mathbf{A}\boldsymbol{\alpha}_0 \end{pmatrix}.$$

Since $(\boldsymbol{\alpha}, \boldsymbol{\alpha}_0)$ is an orthogonal matrix,

$$\begin{aligned}|\boldsymbol{\alpha}_0^T \mathbf{A}\boldsymbol{\alpha}_0| &= |(\boldsymbol{\alpha}, \boldsymbol{\alpha}_0)\mathbf{K}(\boldsymbol{\alpha}, \boldsymbol{\alpha}_0)^T| = |\boldsymbol{\alpha}\boldsymbol{\alpha}^T + \boldsymbol{\alpha}\boldsymbol{\alpha}^T \mathbf{A}\boldsymbol{\alpha}_0\boldsymbol{\alpha}_0^T + \boldsymbol{\alpha}_0\boldsymbol{\alpha}_0^T \mathbf{A}\boldsymbol{\alpha}_0\boldsymbol{\alpha}_0^T|\\ &= |\mathbf{A} - (\mathbf{A} - \mathbf{I}_p)\boldsymbol{\alpha}\boldsymbol{\alpha}^T| = |\mathbf{A}||\mathbf{I}_d - \boldsymbol{\alpha}^T(\mathbf{I}_p - \mathbf{A}^{-1})\boldsymbol{\alpha}|\\ &= |\mathbf{A}||\boldsymbol{\alpha}^T \mathbf{A}^{-1}\boldsymbol{\alpha}|.\end{aligned}$$

□

Lemma A.14 *Let* $\mathbf{O} = (\mathbf{O}_1, \mathbf{O}_2) \in \mathbb{R}^{r\times r}$ *be a column partitioned orthogonal matrix and let* $\mathbf{A} \in \mathbb{S}^{r\times r}$ *be positive definite. Then* $|\mathbf{A}| \le |\mathbf{O}_1^T\mathbf{A}\mathbf{O}_1| \times |\mathbf{O}_2^T\mathbf{A}\mathbf{O}_2|$ *with equality if and only if* $\text{span}(\mathbf{O}_1)$ *reduces* $\mathbf{A}$.

Proof:

$$\begin{aligned}|\mathbf{A}| = |\mathbf{O}^T\mathbf{A}\mathbf{O}| &= \begin{vmatrix} \mathbf{O}_1^T\mathbf{A}\mathbf{O}_1 & \mathbf{O}_1^T\mathbf{A}\mathbf{O}_2 \\ \mathbf{O}_2^T\mathbf{A}\mathbf{O}_1 & \mathbf{O}_2^T\mathbf{A}\mathbf{O}_2 \end{vmatrix}\\ &= |\mathbf{O}_1^T\mathbf{A}\mathbf{O}_1| \times |\mathbf{O}_2^T\mathbf{A}\mathbf{O}_2 - \mathbf{O}_2^T\mathbf{A}\mathbf{O}_1(\mathbf{O}_1^T\mathbf{A}\mathbf{O}_1)^{-1}\mathbf{O}_1^T\mathbf{A}\mathbf{O}_2|\\ &\le |\mathbf{O}_1^T\mathbf{A}\mathbf{O}_1| \times |\mathbf{O}_2^T\mathbf{A}\mathbf{O}_2|.\end{aligned}$$

□

Lemma A.15 *The function* $f(\mathbf{G}) := \log|\mathbf{G}^T\mathbf{M}\mathbf{G}| + \log|\mathbf{G}^T\mathbf{M}^{-1}\mathbf{G}|$ *is minimized over semi-orthogonal matrices* $\mathbf{G}$ *by choosing* $\mathbf{G}$ *to span any reducing subspace of* $\mathbf{M}$*, in which case* $f(\mathbf{G}) = 0$.

Proof: Let $(\mathbf{G}, \mathbf{G}_0)$ be an orthogonal matrix, where span($\mathbf{G}$) reduces $\mathbf{M}$. Using Lemma A.13, we have

$$f(\mathbf{G}) = \log|\mathbf{G}^T\mathbf{M}\mathbf{G}| + \log|\mathbf{G}_0^T\mathbf{M}\mathbf{G}_0| - \log|\mathbf{M}|.$$

It follows from conclusion 3 of Corollary A.1 that $f(\mathbf{G})$ is constant for all $\mathbf{G}$ that reduce $\mathbf{M}$; that is, $f(\mathbf{G}) = 0$ for all $\mathbf{G}$ that reduce $\mathbf{M}$. It follows from Lemma A.14 that this is the minimum value for $f(\mathbf{G})$, and thus the conclusion follows. □

The next lemma was used in the justification of the sequential moment-based envelope algorithm discussed in Section 6.5.

Lemma A.16 *Let* $\mathbf{U} \in \mathbb{S}^{r\times r}$ *and* $\mathbf{V} \in \mathbb{S}^{r\times r}$ *be positive semi-definite matrices, let* $\mathcal{S}$ *be a* u*-dimensional subspace of* $\mathbb{R}^r$ *with semi-orthogonal basis matrix* $\boldsymbol{\Gamma} \in \mathbb{R}^{r\times u}$*, let* $\mathcal{T}$ *be a* u_1*-dimensional subspace of* $\mathbb{R}^r$ *that is orthogonal to* $\mathcal{S}$*,* $u_1 = \dim(\mathcal{T})$*, and let* $u_2 = \dim\{(\mathcal{S}+\mathcal{T})^{\perp}\}$*. Assume that* $\boldsymbol{\Gamma}^T\mathbf{V}\boldsymbol{\Gamma} > 0$*. Then*

$$\begin{aligned}\mathbf{w}_{\max} &= \arg\max_{D_1} \mathbf{w}^T\mathbf{P}_{\mathcal{S}}\mathbf{U}\mathbf{P}_{\mathcal{S}}\mathbf{w}\\ &= \arg\max_{D_2} \mathbf{w}^T\mathbf{P}_{\mathcal{S}}\mathbf{U}\mathbf{P}_{\mathcal{S}}\mathbf{w}\\ &= \boldsymbol{\Gamma}(\boldsymbol{\Gamma}^T\mathbf{V}\boldsymbol{\Gamma})^{-1/2}\ell_1\{(\boldsymbol{\Gamma}^T\mathbf{V}\boldsymbol{\Gamma})^{-1/2}\boldsymbol{\Gamma}^T\mathbf{U}\boldsymbol{\Gamma}(\boldsymbol{\Gamma}^T\mathbf{V}\boldsymbol{\Gamma})^{-1/2}\},\end{aligned}$$

where

$$D_1 = \{\mathbf{w}|\mathbf{w}^T\mathbf{P}_S\mathbf{V}\mathbf{P}_S\mathbf{w} + \mathbf{w}^T\mathbf{P}_{\mathcal{T}}\mathbf{V}\mathbf{P}_{\mathcal{T}}\mathbf{w} = 1\},$$
$$D_2 = \{\mathbf{w}|\mathbf{w}^T\mathbf{P}_S\mathbf{V}\mathbf{P}_S\mathbf{w} = 1\},$$

and $\ell_1(\mathbf{A})$ *is any eigenvector in the first eigenspace of* $\mathbf{A}$*. Clearly,* $\mathbf{w}_{\max} \in \mathcal{S}$*, although it is not necessarily unique.*

Proof: Let $\mathbf{\Gamma}_1$ be a semi-orthogonal basis matrix for $\mathcal{T}$, and let $\mathbf{\Gamma}_2$ be a semi-orthogonal basis matrix for $(\mathcal{S}+\mathcal{T})^{\perp}$ so that $(\mathbf{\Gamma}, \mathbf{\Gamma}_1, \mathbf{\Gamma}_2) \in \mathbb{R}^{r\times r}$ is an orthogonal matrix. Let $\mathbf{s} = \mathbf{\Gamma}^T\mathbf{w}$, $\mathbf{t} = \mathbf{\Gamma}_1^T\mathbf{w}$, and $\mathbf{v} = \mathbf{\Gamma}_2^T\mathbf{w}$. Then expressing $\mathbf{w}$ in the coordinates of $(\mathbf{\Gamma}, \mathbf{\Gamma}_1, \mathbf{\Gamma}_2)$, we have $\mathbf{w} = \mathbf{\Gamma}\mathbf{s} + \mathbf{\Gamma}_1\mathbf{t} + \mathbf{\Gamma}_2\mathbf{v}$ and $\mathbf{w}_{\max} = \mathbf{\Gamma}\mathbf{s}_{\max} + \mathbf{\Gamma}_1\mathbf{t}_{\max} + \mathbf{\Gamma}_2\mathbf{v}_{\max}$, where $\mathbf{s}_{\max} = \arg\max \mathbf{s}^T\mathbf{\Gamma}^T\mathbf{U}\mathbf{\Gamma}\mathbf{s}$ is now over all vectors $\mathbf{s} \in \mathbb{R}^u$, $\mathbf{t} \in \mathbb{R}^{u_1}$, and $\mathbf{v} \in \mathbb{R}^{u_2}$ such that $\mathbf{s}^T\mathbf{\Gamma}^T\mathbf{V}\mathbf{\Gamma}\mathbf{s} + \mathbf{t}^T\mathbf{\Gamma}_1^T\mathbf{V}\mathbf{\Gamma}_1\mathbf{t} = 1$, and $(\mathbf{s}_{\max}, \mathbf{t}_{\max}, \mathbf{v}_{\max})$ is the triplet of values at which the maximum occurs. Since $\mathbf{\Gamma}^T\mathbf{V}\mathbf{\Gamma} > 0$, we can make a change of variable in $\mathbf{s}$ without affecting $\mathbf{t}$ or $\mathbf{v}$. Let $\mathbf{d} = (\mathbf{\Gamma}^T\mathbf{V}\mathbf{\Gamma})^{1/2}\mathbf{s}$. Then $\mathbf{s}_{\max} = (\mathbf{\Gamma}^T\mathbf{V}\mathbf{\Gamma})^{-1/2}\mathbf{d}_{\max}$, where

$$\mathbf{d}_{\max} = \arg\max \mathbf{d}^T(\mathbf{\Gamma}^T\mathbf{V}\mathbf{\Gamma})^{-1/2}\mathbf{\Gamma}^T\mathbf{U}\mathbf{\Gamma}(\mathbf{\Gamma}^T\mathbf{V}\mathbf{\Gamma})^{-1/2}\mathbf{d}$$

and the maximum is computed over all vectors $(\mathbf{d}, \mathbf{t}, \mathbf{v})$ such that $\mathbf{d}^T\mathbf{d} + \mathbf{t}^T\mathbf{\Gamma}_1^T\mathbf{V}\mathbf{\Gamma}_1\mathbf{t} = 1$. We know that the inverse $(\mathbf{\Gamma}^T\mathbf{V}\mathbf{\Gamma})^{-1}$ exists because of the condition $\mathbf{\Gamma}^T\mathbf{V}\mathbf{\Gamma} > 0$. Clearly, the maximum is achieved when $\mathbf{t} = 0$ and $\mathbf{v} = 0$ and then $\mathbf{d}_{\max}$ is the first eigenvector of $(\mathbf{\Gamma}^T\mathbf{V}\mathbf{\Gamma})^{-1/2}\mathbf{\Gamma}^T\mathbf{U}\mathbf{\Gamma}(\mathbf{\Gamma}^T\mathbf{V}\mathbf{\Gamma})^{-1/2}$ and $\mathbf{w}_{\max} = \mathbf{\Gamma}\mathbf{s}_{\max} = \mathbf{\Gamma}(\mathbf{\Gamma}^T\mathbf{V}\mathbf{\Gamma})^{-1/2}\mathbf{d}_{\max}$. □

Lemmas A.17–A.20 contain optimization results. In Lemmas A.17–A.19, $\mathbf{V} \in \mathbb{S}^{p\times p}$ is positive definite with eigenvalues $\varphi_1 \geq \cdots \geq \varphi_d > \varphi_{d+1} \geq \cdots \geq \varphi_p > 0$ and corresponding eigenvectors $\ell_1, \ldots, \ell_p$. Also, $\mathbf{G} \in \mathbb{R}^{p\times d}$ is a semi-orthogonal matrix with $d \leq p$ and $\mathcal{G} := \mathrm{span}(\mathbf{G})$.

Lemma A.17 *The objective function* $\mathrm{tr}(\mathbf{G}^T\mathbf{V}\mathbf{G})$ *is maximized over* $\mathbf{G}$ *subject to* $\mathbf{G}^T\mathbf{G} = \mathbf{I}_d$ *by any orthogonal basis for* $\mathrm{span}(\ell_1, \ldots, \ell_d)$*.*

Proof:

$$\mathrm{tr}(\mathbf{G}^T\mathbf{V}\mathbf{G}) = \sum_{i=1}^{p} \varphi_i \|\mathbf{P}_{\mathcal{G}}\ell_i\|^2 = d\sum_{i=1}^{p} \varphi_i(\|\mathbf{P}_{\mathcal{G}}\ell_i\|^2/d).$$

This is a weighted average of the φ_i's because $\sum_{i=1}^{p} \|\mathbf{P}_{\mathcal{G}}\ell_i\|^2 = \mathrm{tr}(\mathbf{P}_{\mathcal{G}}) = d$. Consequently, to maximize we put all the weight on the first d eigenvalues, and this is done by choosing $\mathcal{G}$ to be the span of the first d eigenvectors of $\mathbf{V}$. □

Lemma A.18 *Let* $k_1 \geq k_2 \geq \cdots \geq k_d > 0$ *and* $\mathbf{D} = \mathrm{diag}(k_1, k_2, \cdots, k_d)$*. Then the objective function* $\mathrm{tr}(\mathbf{G}^T\mathbf{V}\mathbf{G}\mathbf{D})$ *is maximized over* $\mathbf{G} \in \mathbb{R}^{p\times d}$ *subject to* $\mathbf{G}^T\mathbf{G} = \mathbf{I}_d$ *by any orthogonal basis for* $\mathrm{span}(\ell_1, \ldots, \ell_d)$*.*

Proof: Let $\mathbf{G}_i = [\mathbf{g}_1, \ldots, \mathbf{g}_{d-i}]$ for $i = 1, \ldots, d-1$. Then

$$\begin{aligned}\mathrm{tr}(\mathbf{G}^T\mathbf{V}\mathbf{G}\mathbf{D}) &= k_d\mathrm{tr}(\mathbf{G}^T\mathbf{V}\mathbf{G}) + (k_{d-1} - k_d)\mathrm{tr}(\mathbf{G}_1^T\mathbf{V}\mathbf{G}_1) \\ &\quad + \cdots + (k_1 - k_2)\mathrm{tr}(\mathbf{G}_{d-1}^T\mathbf{V}\mathbf{G}_{d-1}).\end{aligned}$$

From Lemma A.17, we know $\mathbf{G} = (\boldsymbol{\ell}_1, \ldots, \boldsymbol{\ell}_d)$ maximizes every term on the right side of the formula above. The global maximum value of $\mathrm{tr}(\mathbf{G}^T\mathbf{V}\mathbf{G}\mathbf{D})$ equals $\sum_{i=1}^d \varphi_i k_i$. If $k_1, k_2, \ldots, k_d$ are not in descending order, then we need to permute the columns of $(\boldsymbol{\ell}_1, \ldots, \boldsymbol{\ell}_d)$ so that they correspond to the order of the k_i's. However, $\mathrm{span}(\boldsymbol{\ell}_1, \ldots, \boldsymbol{\ell}_d)$ does not change. □

Lemma A.19 *Let* $\boldsymbol{\Delta} \in \mathbb{S}^{d\times d}$ *be positive definite. Then* $\mathrm{tr}(\mathbf{G}^T\mathbf{V}\mathbf{G}\boldsymbol{\Delta})$ *is maximized over* $\mathbf{G}$ *subject to* $\mathbf{G}^T\mathbf{G} = \mathbf{I}_d$ *by any orthogonal basis for* $(\boldsymbol{\ell}_1, \ldots, \boldsymbol{\ell}_d)$.

Proof: Let $\mathbf{A}\mathbf{D}\mathbf{A}^T$ be the spectral decomposition of $\boldsymbol{\Delta}$. Then

$$\mathrm{tr}(\mathbf{G}^T\mathbf{V}\mathbf{G}\boldsymbol{\Delta}) = \mathrm{tr}(\mathbf{A}^T\mathbf{G}^T\mathbf{V}\mathbf{G}\mathbf{A}\mathbf{D}).$$

The conclusion follows from Lemma A.18 since $(\mathbf{G}\mathbf{A})^T\mathbf{G}\mathbf{A} = \mathbf{I}_d$ and $\mathrm{span}(\mathbf{G}\mathbf{A}) = \mathrm{span}(\mathbf{G})$. □

Lemma A.20 *Let the real function* $f(x) = \log(x) + C\log(K - x)$ *defined on the interval* $[a, b]$, $0 < a < K/(1 + C) < b < K$, *then* $f(x)$ *reaches its maximum at* $K/(1 + C)$ *and reaches its minimum at either* a *or* b.

Proof: The first derivative of $f(x)$ is $f'(x) = \{K - (1 + C)x\}/\{x(K - x)\}$ and the second derivative is $f''(x) = -x^{-2} - C/(K - x)^2 < 0$. Consequently, $f(x)$ is concave with the only stationary point $K/(1 + C)$. So we can conclude that $f(x)$ reaches its maximum at $K/(1 + C)$ and reaches its minimum at a boundary point, either a or b. □

A.8 Matrix Normal Distribution

This section contains a cursory introduction to the matrix normal distribution, which was used in Section 7.2.

A matrix normal distribution is a generalization of the multivariate normal to a matrix-valued random variable $\mathbf{Z} \in \mathbb{R}^{p_r\times p_c}$. Its mean $\boldsymbol{\mu}$ is just the corresponding matrix of element-wise means. The variances and covariances of its elements can be represented by using the vec operator: $var(\mathrm{vec}(\mathbf{Z})) = \boldsymbol{\Delta}$. Then $\mathbf{Z}$ is said to have a matrix normal distribution if $\mathrm{vec}(\mathbf{Z})$ is normally distributed and $\boldsymbol{\Delta}$ can be decomposed as the Kronecker product of two positive definite matrices $\boldsymbol{\Delta}_r \in \mathbb{R}^{p_r\times p_r}$ and $\boldsymbol{\Delta}_c \in \mathbb{R}^{p_c\times p_c}$, so $\boldsymbol{\Delta} = \boldsymbol{\Delta}_r \otimes \boldsymbol{\Delta}_c$ and $\mathrm{vec}(\mathbf{Z}) \sim N_{p_r\times p_c}(\mathrm{vec}(\boldsymbol{\mu}), \boldsymbol{\Delta}_r \otimes \boldsymbol{\Delta}_c)$. Its density is defined via the density of $\mathrm{vec}(\mathbf{Z})$:

$$\begin{aligned}f_{\mathrm{vec}(\mathbf{Z})}(\mathbf{z}) &= (2\pi)^{-p_r p_c/2}|\boldsymbol{\Delta}_r|^{-p_r/2}|\boldsymbol{\Delta}_c|^{-p_c/2} \\ &\quad \times \exp\{-\mathrm{tr}[\boldsymbol{\Delta}_r^{-1}(\mathbf{z} - \boldsymbol{\mu})^T\boldsymbol{\Delta}_c^{-1}(\mathbf{z} - \boldsymbol{\mu})/2]\}.\end{aligned}$$

The second moments of $\mathbf{Z}$ are $\mathrm{E}[(\mathbf{Z}-\boldsymbol{\mu})^T(\mathbf{Z}-\boldsymbol{\mu})] = \boldsymbol{\Delta}_r\mathrm{tr}(\boldsymbol{\Delta}_c)$ and $\mathrm{E}[(\mathbf{Z}-\boldsymbol{\mu})(\mathbf{Z}-\boldsymbol{\mu})^T] = \boldsymbol{\Delta}_c\mathrm{tr}(\boldsymbol{\Delta}_r)$. The matrices $\boldsymbol{\Delta}_r$ and $\boldsymbol{\Delta}_c$ are often called the row (left) and column (right) covariance matrices. The rows or columns of $\mathbf{Z}$ are independent if and only if $\boldsymbol{\Delta}_r$ or $\boldsymbol{\Delta}_c$ is diagonal. If both are diagonal, then $\boldsymbol{\Delta}_r \otimes \boldsymbol{\Delta}_c$ and the elements of $\mathbf{Z}$ are mutually independent.

Given a sample $\mathbf{Z}_1, \dots, Z_n$, the maximum likelihood estimator of $\boldsymbol{\mu}$ is just the average matrix $\bar{\mathbf{Z}}$. Given $\boldsymbol{\Delta}_c$, the maximum likelihood estimator of $\boldsymbol{\Delta}_r$ is

$$\widehat{\boldsymbol{\Delta}}_r = \frac{1}{np_r}\sum_{i=1}^{n}(\mathbf{Z}_i - \bar{\mathbf{Z}})^T\boldsymbol{\Delta}_c^{-1}(\mathbf{Z}_i - \bar{\mathbf{Z}})$$

and given $\boldsymbol{\Delta}_r$ the maximum likelihood estimator of $\boldsymbol{\Delta}_c$ is

$$\widehat{\boldsymbol{\Delta}}_c = \frac{1}{np_c}\sum_{i=1}^{n}(\mathbf{Z}_i - \bar{\mathbf{Z}})\boldsymbol{\Delta}_r^{-1}(\mathbf{Z}_i - \bar{\mathbf{Z}})^T.$$

An iterative algorithm is needed to determine the maximum likelihood estimator of $\boldsymbol{\Delta}$.

It follows from the review of the multivariate linear model given in Section 1.1 that the maximum likelihood estimator $\mathbf{B}$ of the coefficient matrix $\boldsymbol{\beta}$ has a matrix normal distribution with $\boldsymbol{\mu} = \boldsymbol{\beta}$ and $\boldsymbol{\Delta} = (\mathbb{X}^T\mathbb{X})^{-1} \otimes \boldsymbol{\Sigma}$, so $\boldsymbol{\Delta}_r = (\mathbb{X}^T\mathbb{X})^{-1}$ and $\boldsymbol{\Delta}_c = \boldsymbol{\Sigma}$.

A.9 Literature Notes

The material in Sections A.1, A.2, A.3.1, and A.6 comes largely from Cook, Li, and Chiaromonte (2010). Lemma A.6 is from Cook and Zhang (2015b). Propositions A.6 and A.7 are from Cook and Zhang (2016). Lemma A.16 is from Cook, Helland, and Su (2013). The material in Section A.4 is standard. Lemmas A.18–A.20 are from Chen (2010).

For additional properties and proofs on commutation, expansion, and contraction matrices, Section A.5, see Henderson and Searle (1979), Magnus and Neudecker (1979), and Neudecker and Wansbeek (1983).

In reference to the matrix normal distribution, Section A.8, see Dawid (1981) for a discussion of relationships between the matrix normal distribution and other distributions and Dutilleul (1999) for an introduction to estimative algorithms. Hypothesis tests for a Kronecker covariance structure can be found in Shitan and Brockwell (1995), Lu and Zimmerman (2005), and Roy and Khattree (2005).

Appendix B

Proofs for Envelope Algorithms

B.1 The 1D Algorithm

The argument in this section contains two major parts (see Section 6.4). The first, stated in Proposition B.1, shows that the algorithm returns $\mathcal{E}_{\mathbf{M}}(\mathcal{U})$ in the population. Root-n consistency is then demonstrated in the second part given in Proposition B.2.

Proposition B.1 *The sequential likelihood-based algorithm returns $\mathcal{E}_{\mathbf{M}}(\mathcal{U})$ in the population; that is,* $\text{span}(\mathbf{G}_u) = \mathcal{E}_{\mathbf{M}}(\mathcal{U})$.

Proof: Let $\boldsymbol{\Gamma} \in \mathbb{R}^{r\times u}$ be a semi-orthogonal basis matrix for $\mathcal{E}_{\mathbf{M}}(\mathcal{U})$. Then we can write $\mathbf{M} = \boldsymbol{\Gamma}\boldsymbol{\Omega}\boldsymbol{\Gamma}^T + \boldsymbol{\Gamma}_0\boldsymbol{\Omega}_0\boldsymbol{\Gamma}_0^T$ by Proposition A.1, where $\boldsymbol{\Omega} > 0$, $\boldsymbol{\Omega}_0 > 0$ and $(\boldsymbol{\Gamma}, \boldsymbol{\Gamma}_0) \in \mathbb{R}^r$ is orthogonal basis for $\mathbb{R}^r$. Since $\mathcal{U} \subseteq \mathcal{E}_{\mathbf{M}}(\mathcal{U})$, we also have the decomposition $\mathbf{M} + \mathbf{U} = \boldsymbol{\Gamma}\boldsymbol{\Phi}\boldsymbol{\Gamma}^T + \boldsymbol{\Gamma}\boldsymbol{\Omega}_0\boldsymbol{\Gamma}_0^T$, where $\boldsymbol{\Phi} = \boldsymbol{\Omega} + \mathbf{U}_1 > 0$ and $\mathbf{U} = \boldsymbol{\Gamma}\mathbf{U}_1\boldsymbol{\Gamma}^T$, as defined in Section 6.2.1. We first need to solve the optimization for the first direction $\mathbf{g}_1 = \arg\min_{\mathbf{g}\in\mathbb{R}^r} \phi_0(\mathbf{g})$, where $\phi_0(\mathbf{g}) = \log(\mathbf{g}^T\mathbf{M}\mathbf{g}) + \log(\mathbf{g}^T(\mathbf{M}+\mathbf{U})^{-1}\mathbf{g})$ and the minimization is subject to constraint $\mathbf{g}^T\mathbf{g} = 1$. Let $\mathbf{g} = \boldsymbol{\Gamma}\mathbf{h} + \boldsymbol{\Gamma}_0\mathbf{h}_0$ for some $\mathbf{h} \in \mathbb{R}^u$ and $\mathbf{h}_0 \in \mathbb{R}^{(r-u)}$. Consider the equivalent unconstrained optimization,

$$\mathbf{g}_1 = \arg\min_{\mathbf{g}\in\mathbb{R}^r}\{\log(\mathbf{g}^T\mathbf{M}\mathbf{g}) + \log(\mathbf{g}^T(\mathbf{M}+\mathbf{U})^{-1}\mathbf{g}) - 2\log(\mathbf{g}^T\mathbf{g})\}.$$

Then we will have the same solution as the original problem up to an arbitrary scaling constant. Next, we plug-in these expressions for $\mathbf{g}$, $\mathbf{M}+\mathbf{U}$ and $\mathbf{M}$, to get

$$\begin{aligned} f(\mathbf{h}, \mathbf{h}_0) :&= \log(\mathbf{g}^T\mathbf{M}\mathbf{g}) + \log(\mathbf{g}^T(\mathbf{M}+\mathbf{U})^{-1}\mathbf{g}) - 2\log(\mathbf{g}^T\mathbf{g}) \\ &= \log\{\mathbf{h}^T\boldsymbol{\Omega}\mathbf{h} + \mathbf{h}_0^T\boldsymbol{\Omega}_0\mathbf{h}_0\} + \log\{\mathbf{h}^T\boldsymbol{\Phi}^{-1}\mathbf{h} + \mathbf{h}_0^T\boldsymbol{\Omega}_0^{-1}\mathbf{h}_0\} \\ &\quad - 2\log\{\mathbf{h}^T\mathbf{h} + \mathbf{h}_0^T\mathbf{h}_0\}. \end{aligned}$$

An Introduction to Envelopes: Dimension Reduction for Efficient Estimation in Multivariate Statistics,
First Edition. R. Dennis Cook.

Taking partial derivative with respect to $\mathbf{h}_0$, we have

$$\frac{\partial}{\partial \mathbf{h}_0} f(\mathbf{h}, \mathbf{h}_0) = \frac{2\boldsymbol{\Omega}_0 \mathbf{h}_0}{\mathbf{h}^T \boldsymbol{\Omega} \mathbf{h} + \mathbf{h}_0^T \boldsymbol{\Omega}_0 \mathbf{h}_0} + \frac{2\boldsymbol{\Omega}_0^{-1} \mathbf{h}_0}{\mathbf{h}^T \boldsymbol{\Phi}^{-1} \mathbf{h} + \mathbf{h}_0^T \boldsymbol{\Omega}_0^{-1} \mathbf{h}_0} - \frac{4\mathbf{h}_0}{\mathbf{h}^T \mathbf{h} + \mathbf{h}_0^T \mathbf{h}_0}.$$

To get a stationary point, we set this derivative to zero, obtaining

$$\left\{ \frac{2\boldsymbol{\Omega}_0}{\mathbf{h}^T \boldsymbol{\Omega} \mathbf{h} + \mathbf{h}_0^T \boldsymbol{\Omega}_0 \mathbf{h}_0} + \frac{2\boldsymbol{\Omega}_0^{-1}}{\mathbf{h}^T \boldsymbol{\Phi}^{-1} \mathbf{h} + \mathbf{h}_0^T \boldsymbol{\Omega}_0^{-1} \mathbf{h}_0} \right\} \mathbf{h}_0 = \left\{ \frac{4}{\mathbf{h}^T \mathbf{h} + \mathbf{h}_0^T \mathbf{h}_0} \right\} \mathbf{h}_0.$$

Define the positive definite matrix $\mathbf{A}_0$ as

$$\mathbf{A}_0 = \left\{ \frac{2\boldsymbol{\Omega}_0}{\mathbf{h}^T \boldsymbol{\Omega} \mathbf{h} + \mathbf{h}_0^T \boldsymbol{\Omega}_0 \mathbf{h}_0} + \frac{2\boldsymbol{\Omega}_0^{-1}}{\mathbf{h}^T \boldsymbol{\Phi}^{-1} \mathbf{h} + \mathbf{h}_0^T \boldsymbol{\Omega}_0^{-1} \mathbf{h}_0} \right\} \Big/ \left\{ \frac{4}{\mathbf{h}^T \mathbf{h} + \mathbf{h}_0^T \mathbf{h}_0} \right\}.$$

Then $\mathbf{A}_0 \mathbf{h}_0 = \mathbf{h}_0$ has a solution only as an eigenvector of $\mathbf{A}_0$. The eigenvectors of $\mathbf{A}_0$ are the same as those of $\boldsymbol{\Omega}_0$. Hence, we have $\mathbf{h}_0$ equals 0 or any eigenvector $\ell_k(\boldsymbol{\Omega}_0)$ of $\boldsymbol{\Omega}_0$. Therefore, the minimum value of $f(\mathbf{h}, \mathbf{h}_0)$ has to be obtained by 0 or $\ell_k(\boldsymbol{\Omega}_0)$. Restricting $\boldsymbol{\Omega}_0 \mathbf{h}_0 = \lambda \mathbf{h}_0$, we have

$$\begin{aligned} f(\mathbf{h}, \mathbf{h}_0) &= \log \left\{ \frac{\mathbf{h}^T \boldsymbol{\Omega} \mathbf{h} + \lambda \mathbf{h}_0^T \mathbf{h}_0}{\mathbf{h}^T \mathbf{h} + \mathbf{h}_0^T \mathbf{h}_0} \right\} + \log \left\{ \frac{\mathbf{h}^T \boldsymbol{\Phi}^{-1} \mathbf{h} + \lambda^{-1} \mathbf{h}_0^T \mathbf{h}_0}{\mathbf{h}^T \mathbf{h} + \mathbf{h}_0^T \mathbf{h}_0} \right\} \\ &= \log \left\{ \frac{\mathbf{h}^T \boldsymbol{\Omega} \mathbf{h}}{\mathbf{h}^T \mathbf{h}} W_h + \lambda (1 - W_h) \right\} + \log \left\{ \frac{\mathbf{h}^T \boldsymbol{\Phi}^{-1} \mathbf{h}}{\mathbf{h}^T \mathbf{h}} W_h + \lambda^{-1} (1 - W_h) \right\} \end{aligned}$$

where $W_h = \mathbf{h}^T \mathbf{h} / (\mathbf{h}^T \mathbf{h} + \mathbf{h}_0^T \mathbf{h}_0)$ is a weight in the interval [0, 1]. Because $\log(\cdot)$ is concave, we have $\log(aW_h + b(1 - W_h)) \geq W_h \log(a) + (1 - W_h) \log(b)$, with strict inequality when $W_h \in (0, 1)$ and $a \neq b$. Hence,

$$\begin{aligned} f(\mathbf{h}, \mathbf{h}_0) &\geq W_h \left\{ \log \frac{\mathbf{h}^T \boldsymbol{\Omega} \mathbf{h}}{\mathbf{h}^T \mathbf{h}} + \log \frac{\mathbf{h}^T \boldsymbol{\Phi}^{-1} \mathbf{h}}{\mathbf{h}^T \mathbf{h}} \right\} + (1 - W_h) \{ \log(\lambda) + \log(\lambda^{-1}) \} \\ &= W_h f(\mathbf{h}, 0). \end{aligned}$$

Let $\mathbf{h}_{\min} = \arg\min_{\mathbf{h} \in \mathbb{R}^u} f(\mathbf{h}, 0)$. Then, repeating the last result for clarity,

$$f(\mathbf{h}, \mathbf{h}_0) \geq W_h f(\mathbf{h}, 0) \geq W_h f(\mathbf{h}_{\min}, 0) \geq f(\mathbf{h}_{\min}, 0),$$

where last inequality holds because

$$\min_{\mathbf{h} \in \mathbb{R}^u} f(\mathbf{h}, 0) < 0. \tag{B.1}$$

We defer the proof of this inequality to the end of this proof. This lower bound on $f(\mathbf{h}, \mathbf{h}_0)$, which is negative, is attained when $\mathbf{h}_0 = 0$, so $W_h = 1$, and $\mathbf{h} = \mathbf{h}_{\min}$.

So we have the minimum found at $W_h = \mathbf{h}^T\mathbf{h}/(\mathbf{h}^T\mathbf{h} + \mathbf{h}_0^T\mathbf{h}_0) = 1$, or equivalently, $\mathbf{g} = \mathbf{\Gamma h} \in \text{span}(\mathbf{\Gamma})$. This setting is unique since the log is a strictly concave function.

For the $(k+1)$th direction, $\mathbf{g}_{k+1} = \mathbf{G}_{0k}\mathbf{w}_{k+1}$, where $\mathbf{w}_{k+1} = \arg\min_{\mathbf{w}\in\mathbb{R}^{r-k}} \phi_k(\mathbf{w})$, subject to $\mathbf{w}^T\mathbf{w} = 1$. Because

$$\phi_k(\mathbf{w}) = \log(\mathbf{w}^T\mathbf{G}_{0k}^T\mathbf{M}\mathbf{G}_{0k}\mathbf{w}) + \log(\mathbf{w}^T\{\mathbf{G}_{0k}^T(\mathbf{M}+\mathbf{U})\mathbf{G}_{0k}\}^{-1}\mathbf{w})$$

has the same form as $\phi_0(\mathbf{g})$, we have

$$\mathbf{w}_{k+1} \in \mathcal{E}_{\mathbf{G}_{0k}^T\mathbf{M}\mathbf{G}_{0k}}(\mathbf{G}_{0k}^T\mathcal{U}).$$

Therefore, $\mathbf{g}_{k+1} = \mathbf{G}_{0k}\mathbf{w}_{k+1} \in \mathcal{E}_{\mathbf{M}}(\mathcal{U})$ by Proposition A.7.

We conclude the proof by demonstrating (B.1). We first show that $\min_{\mathbf{h}\in\mathbb{R}^u} f(\mathbf{h}, 0) \leq 0$ and then we assume the equality to conclude the proof by contradiction. Define the following two functions,

$$F(\mathbf{h}, \mathbf{\Omega}, \mathbf{\Phi}^{-1}) = \log\frac{\mathbf{h}^T\mathbf{\Omega}\mathbf{h}}{\mathbf{h}^T\mathbf{h}} + \log\frac{\mathbf{h}^T\mathbf{\Phi}^{-1}\mathbf{h}}{\mathbf{h}^T\mathbf{h}} = f(\mathbf{h}, 0),$$
$$F(\mathbf{h}, \mathbf{\Omega}, \mathbf{\Omega}^{-1}) = \log\frac{\mathbf{h}^T\mathbf{\Omega}\mathbf{h}}{\mathbf{h}^T\mathbf{h}} + \log\frac{\mathbf{h}^T\mathbf{\Omega}^{-1}\mathbf{h}}{\mathbf{h}^T\mathbf{h}},$$

where $F(\mathbf{h}, \mathbf{\Omega}, \mathbf{\Phi}^{-1})$ is a reexpression of $f(\mathbf{h}, 0)$ to display $\mathbf{\Omega}$ and $\mathbf{\Phi}^{-1}$ as arguments. Recall that $\mathbf{\Phi} - \mathbf{\Omega} \geq 0$, and hence $F(\mathbf{h}, \mathbf{\Omega}, \mathbf{\Phi}^{-1}) \leq F(\mathbf{h}, \mathbf{\Omega}, \mathbf{\Omega}^{-1})$ for any $\mathbf{h}$. Consider the minimum of both $F(\mathbf{h}, \mathbf{\Omega}, \mathbf{\Phi}^{-1})$ and $F(\mathbf{h}, \mathbf{\Omega}, \mathbf{\Omega}^{-1})$, we have

$$\min_{\mathbf{h}} f(\mathbf{h}, 0) = \min_{\mathbf{h}} F(\mathbf{h}, \mathbf{\Omega}, \mathbf{\Phi}^{-1}) \leq \min_{\mathbf{h}} F(\mathbf{h}, \mathbf{\Omega}, \mathbf{\Omega}^{-1}) = 0.$$

From Lemma A.15, the minimum of the right-hand side is zero by taking $\mathbf{h}$ equals to any eigenvector of $\mathbf{\Omega}$.

Now we assume that $\min_{\mathbf{h}} F(\mathbf{h}, \mathbf{\Omega}, \mathbf{\Phi}^{-1}) = 0$. Then for an arbitrary $\mathbf{h}$,

$$0 \leq F(\mathbf{h}, \mathbf{\Omega}, \mathbf{\Phi}^{-1}) \leq F(\mathbf{h}, \mathbf{\Omega}, \mathbf{\Omega}^{-1}).$$

Let $\mathbf{h}_i = \ell_i(\mathbf{\Omega})$, $i = 1, \dots, u$, be the ith unit eigenvector of $\mathbf{\Omega}$ and plug $\mathbf{h}_i$ into the above inequalities, we have

$$0 \leq F(\mathbf{h}_i; \mathbf{\Omega}, \mathbf{\Phi}^{-1}) \leq F(\mathbf{h}_i; \mathbf{\Omega}, \mathbf{\Omega}^{-1}) = 0, \quad i = 1, \dots, u,$$

which implies

$$0 = F(\mathbf{h}_i; \mathbf{\Omega}, \mathbf{\Phi}^{-1}) = F(\mathbf{h}_i; \mathbf{\Omega}, \mathbf{\Omega}^{-1}) = 0, \quad i = 1, \dots, u.$$

More explicitly,

$$\mathbf{h}_i^T\mathbf{\Phi}^{-1}\mathbf{h}_i = \mathbf{h}_i^T\mathbf{\Omega}^{-1}\mathbf{h}_i, \quad i = 1, \dots, u,$$

which implies $\mathbf{\Phi} = \mathbf{\Omega}$ because $\mathbf{\Phi}, \mathbf{\Omega} \in \mathbb{R}^{u\times u}$ and $\mathbf{h}_i$, $i = 1, \dots, u$, are u linear independent vectors. Then by definition $\mathbf{U} = \mathbf{\Gamma}(\mathbf{\Phi} - \mathbf{\Omega})\mathbf{\Gamma}^T = 0$, which is a contradiction. □

The previous proposition shows that the sequential likelihood-based methods give the desired answer in the population, and it an essential ingredient in showing $\sqrt{n}$-consistency in the next proposition.

Proposition B.2 *Let $\widehat{\mathbf{M}}$ and $\widehat{\mathbf{U}}$ be $\sqrt{n}$-consistent estimators of $\mathbf{M}$ and $\mathbf{U}$, and let $\widehat{\mathbf{G}}_u$ denote the estimated basis for $\mathcal{E}_{\mathbf{M}}(\mathcal{U})$ from the sequential likelihood-based algorithm using $\widehat{\mathbf{M}}$ and $\widehat{\mathbf{U}}$. Then $\mathbf{P}_{\widehat{\mathbf{G}}_u}$ is a $\sqrt{n}$-consistent estimator of $\mathbf{P}_{\mathcal{E}}$, the projection onto $\mathcal{E}_{\mathbf{M}}(\mathcal{U})$.*

Proof: This proof that the sample algorithm is $\sqrt{n}$-consistent makes use of classical results in the theory of extremum estimators (see, for example, Amemiya (1985, Theorems 4.1.1 and 4.1.3)). We review these results briefly and then sketch how they can be used to prove the $\sqrt{n}$-consistency for the algorithm.

Let $Q(\mathbf{Y}, \boldsymbol{\theta})$ be a real-valued function of the random variables $\mathbf{Y} = (Y_1, \dots, Y_n)^T$ and the parameters $\boldsymbol{\theta} = (\theta_1, \dots, \theta_K)^T$. For notational convenience, let $Q_n(\boldsymbol{\theta}) = Q(\mathbf{Y}, \boldsymbol{\theta})$. Let the parameter space be $\boldsymbol{\Theta}$, and let the true value of $\boldsymbol{\theta}$ be $\boldsymbol{\theta}_t \in \boldsymbol{\Theta}$. The following assumptions (A)–(F) are used to establish asymptotic properties of the extremum estimator, $\widehat{\boldsymbol{\theta}}_n = \arg\max_{\boldsymbol{\theta}\in\boldsymbol{\Theta}} Q_n(\boldsymbol{\theta})$:

(A) The parameter space $\boldsymbol{\Theta}$ is a compact subset of $\mathbb{R}^K$.
(B) $Q_n(\boldsymbol{\theta})$ is continuous in $\boldsymbol{\theta} \in \boldsymbol{\Theta}$ for all $\mathbf{Y}$ and is a measurable function of $\mathbf{Y}$ for all $\boldsymbol{\theta} \in \boldsymbol{\Theta}$.
(C) $n^{-1}Q_n(\boldsymbol{\theta})$ converges to a nonstochastic function $Q(\boldsymbol{\theta})$ in probability uniformly in $\boldsymbol{\theta} \in \boldsymbol{\Theta}$ as n goes to infinity, and $Q(\boldsymbol{\theta})$ attains a unique global maximum at $\boldsymbol{\theta}_t$.
(D) $\partial^2 Q_n(\boldsymbol{\theta})/\partial\boldsymbol{\theta}\partial\boldsymbol{\theta}^T$ exists and is continuous in an open, convex neighborhood of $\boldsymbol{\theta}_t$.
(E) $n^{-1}\{\partial^2 Q_n(\boldsymbol{\theta})/\partial\boldsymbol{\theta}\partial\boldsymbol{\theta}^T\}_{\boldsymbol{\theta}=\boldsymbol{\theta}_n^*}$ converges to a finite nonsingular matrix

$$\mathbf{A}(\boldsymbol{\theta}) = \lim_{n\to\infty} \mathrm{E}_{\boldsymbol{\theta}_t}\{n^{-1}\{\partial^2 Q_n(\boldsymbol{\theta})/\partial\boldsymbol{\theta}\partial\boldsymbol{\theta}^T\}\},$$

for any random sequences $\boldsymbol{\theta}_n^*$ such that $\mathrm{plim}(\boldsymbol{\theta}_n^*) = \boldsymbol{\theta}$.
(F) $n^{-1/2}\{\partial Q_n(\boldsymbol{\theta})/\partial\boldsymbol{\theta}\}_{\boldsymbol{\theta}=\boldsymbol{\theta}_t} \to N(0, \mathbf{B}(\boldsymbol{\theta}_t))$, where

$$\mathbf{B}(\boldsymbol{\theta}) = \lim_{n\to\infty} \mathrm{E}_{\boldsymbol{\theta}_t}\{n^{-1}\{\partial Q_n(\boldsymbol{\theta})/\partial\boldsymbol{\theta}\}\{\partial Q_n(\boldsymbol{\theta})/\partial\boldsymbol{\theta}^T\}\}.$$

Under assumptions (A)–(D), $\widehat{\boldsymbol{\theta}}_n$ converges to $\boldsymbol{\theta}_t$ in probability. Under assumptions (A)–(F), then $\sqrt{n}(\widehat{\boldsymbol{\theta}}_n - \boldsymbol{\theta}_t) \to N(0, \mathbf{A}^{-1}(\boldsymbol{\theta}_t)\mathbf{B}(\boldsymbol{\theta}_t)\mathbf{A}^{-1}(\boldsymbol{\theta}_t))$ (Amemiya, 1985).

To adapt these results for the first step of the algorithm, let $\boldsymbol{\theta} = \mathbf{g}$ whose true value is denoted by $\mathbf{g}_t$. The parameter space is the 1D manifold $\boldsymbol{\Theta} = \mathcal{G}_{(p,1)}$, which is a compact set of $\mathbb{R}^{p\times 1}$, so condition (A) is satisfied. The objective function depends on the random variables $\mathbf{Y}$ only through $(\widehat{\mathbf{M}}, \widehat{\mathbf{U}})$. The specific function to be maximized is

$$Q_n(\mathbf{g}) = -n/2\log(\mathbf{g}^T\widehat{\mathbf{M}}\mathbf{g}) - n/2\log(\mathbf{g}^T(\widehat{\mathbf{M}}+\widehat{\mathbf{U}})^{-1}\mathbf{g}) + n\log(\mathbf{g}^T\mathbf{g}).$$

Condition (B) then holds. Although arg max$Q_n(\mathbf{g})$ is not unique, span(arg max $Q_n(\mathbf{g})$) is unique, and we can choose one unique value for the maximizing argument by, say, setting the norm equal to one (Jennrich 1969).

We next verify condition (C) that $n^{-1}Q_n(\mathbf{g})$ converges uniformly to

$$Q(\mathbf{g}) = -1/2\log(\mathbf{g}^T\mathbf{M}\mathbf{g}) - 1/2\log(\mathbf{g}^T(\mathbf{M}+\mathbf{U})^{-1}\mathbf{g}) + \log(\mathbf{g}^T\mathbf{g}).$$

We have shown in Proposition B.1 that the population objective function $Q(\mathbf{g})$ attains the unique global maximum at $\mathbf{g}_t$. For simplicity, we assume $\mathbf{M}$ and $\mathbf{M}+\mathbf{U}$ have distinct eigenvalues so that $\mathbf{g}_t$ is the unique maximum of $Q(\mathbf{g})$ in the 1D manifold $\boldsymbol{\Theta}$. For the case where there are multiple local maxima of $Q(\mathbf{g})$, we can obtain similar results by applying Theorem 4.1.2 in Amemiya (1985). Since $\widehat{\mathbf{M}}$ and $\widehat{\mathbf{U}}$ are $\sqrt{n}$-consistent for $\mathbf{M}$ and $\mathbf{U}$, the eigenvectors and eigenvalues of $\widehat{\mathbf{M}}$, $\widehat{\mathbf{M}}$ and $\widehat{\mathbf{M}}+\widehat{\mathbf{U}}$ are $\sqrt{n}$-consistent for the eigenvectors and eigenvalues of their population counterparts. Then $n^{-1}Q_n(\mathbf{g})$ converges in probability to $Q(\mathbf{g})$ uniformly in $\mathbf{g}$, as can be seen from the following argument.

$$\begin{aligned} n^{-1}Q_n(\mathbf{g}) - Q(\mathbf{g}) &= -1/2(\log(\mathbf{g}^T\widehat{\mathbf{M}}\mathbf{g}) - \log(\mathbf{g}^T\mathbf{M}\mathbf{g})) \\ &\quad -1/2(\log(\mathbf{g}^T(\widehat{\mathbf{M}}+\widehat{\mathbf{U}})^{-1}\mathbf{g}) - \log(\mathbf{g}^T(\mathbf{M}+\mathbf{U})^{-1}\mathbf{g})) \\ &= -1/2\log\left[\frac{\mathbf{g}^T\widehat{\mathbf{M}}\mathbf{g}}{\mathbf{g}^T\mathbf{M}\mathbf{g}}\right] - 1/2\log\left[\frac{\mathbf{g}^T(\widehat{\mathbf{M}}+\widehat{\mathbf{U}})^{-1}\mathbf{g}}{\mathbf{g}^T(\mathbf{M}+\mathbf{U})^{-1}\mathbf{g}}\right]. \end{aligned}$$

Hence, $\sup_{\mathbf{g}\in\boldsymbol{\Theta}}\log(\mathbf{g}^T\widehat{\mathbf{M}}\mathbf{g}/\mathbf{g}^T\mathbf{M}\mathbf{g}) = \sup_{\mathbf{g}\in\boldsymbol{\Theta}}\log(\mathbf{g}^T\mathbf{M}^{-1/2}\widehat{\mathbf{M}}\mathbf{M}^{-1/2}\mathbf{g}/\mathbf{g}^T\mathbf{g})$, which equals to the logarithm of the largest eigenvalue of $\mathbf{M}^{-1/2}\widehat{\mathbf{M}}\mathbf{M}^{-1/2}$ and converges to 0 in probability. Similarly, $\sup_{\mathbf{g}\in\boldsymbol{\Theta}}\log[\mathbf{g}^T(\widehat{\mathbf{M}}+\widehat{\mathbf{U}})^{-1}\mathbf{g}/\mathbf{g}^T(\mathbf{M}+\mathbf{U})^{-1}\mathbf{g}]$ converges to zero in probability. Therefore, $n^{-1}Q_n(\mathbf{g})$ converges to $Q(\mathbf{g})$ in probability uniformly in $\mathbf{g}\in\boldsymbol{\Theta}$.

By straightforward calculation, condition (D) follows from the second derivative matrix

$$\begin{aligned} n^{-1}\frac{\partial^2 Q_n(\mathbf{g})}{\partial\mathbf{g}\partial\mathbf{g}^T} &= 2(\mathbf{g}^T\widehat{\mathbf{M}}\mathbf{g})^{-2}(\widehat{\mathbf{M}}\mathbf{g}\mathbf{g}^T\widehat{\mathbf{M}}) - (\mathbf{g}^T\widehat{\mathbf{M}}\mathbf{g})^{-1}\widehat{\mathbf{M}} \\ &\quad + 2[\mathbf{g}^T(\widehat{\mathbf{M}}+\widehat{\mathbf{U}})^{-1}\mathbf{g}]^{-2}[(\widehat{\mathbf{M}}+\widehat{\mathbf{U}})^{-1}\mathbf{g}\mathbf{g}^T(\widehat{\mathbf{M}}+\widehat{\mathbf{U}})^{-1}] \\ &\quad - [\mathbf{g}^T(\widehat{\mathbf{M}}+\widehat{\mathbf{U}})^{-1}\mathbf{g}]^{-1}(\widehat{\mathbf{M}}+\widehat{\mathbf{U}})^{-1} - 2(\mathbf{g}^T\mathbf{g})^{-2}\mathbf{P}_{\mathbf{g}} + (\mathbf{g}^T\mathbf{g})^{-1}\mathbf{I}_p. \end{aligned}$$

Condition (E) holds because the above quantity is a smooth function of $\mathbf{g}$, $\widehat{\mathbf{M}}$, and $\widehat{\mathbf{M}}+\widehat{\mathbf{U}}$.

Last, we need to verify condition (F). To demonstrate $\sqrt{n}$-consistency of the first step, we need to show only that

$$n^{-1}\{\partial Q_n(\mathbf{g})/\partial \mathbf{g}\}_{\mathbf{g}=\theta}$$
$$= -(\theta^T\widehat{\mathbf{M}}\theta)^{-1}\widehat{\mathbf{M}}\theta - (\theta^T(\widehat{\mathbf{M}}+\widehat{\mathbf{U}})^{-1}\theta)^{-1}(\widehat{\mathbf{M}}+\widehat{\mathbf{U}})^{-1}\theta + 2\theta$$

has order $O_p(1/\sqrt{n})$. From the proof of Proposition B.1, we know that $\{\partial Q(\mathbf{g})/\partial \mathbf{g}\}_{\mathbf{g}=\theta} = 0$. Then the result follows from the fact that $n^{-1}\partial Q_n(\mathbf{g})/\partial \mathbf{g}$ is a smooth function of $\widehat{\mathbf{M}}$ and $\widehat{\mathbf{M}}+\widehat{\mathbf{U}}$, which are $\sqrt{n}$-consistent estimators.

So far, we have verified the conditions (A)–(F) so that the sample estimator $\text{span}(\widehat{\mathbf{G}}_1)$ will be $\sqrt{n}$-consistent for the population estimator. For the $(k+1)$th direction, $k < u$, let $\widehat{\mathbf{G}}_k$ denote a $\sqrt{n}$-consistent estimator of the first k directions and let $(\widehat{\mathbf{G}}_k, \widehat{\mathbf{G}}_{0k})$ be an orthogonal matrix. The $(k+1)$th direction is defined by $\mathbf{g}_{k+1} = \widehat{\mathbf{G}}_{0k}\mathbf{w}_{k+1}$, where the parameters are $\mathbf{w}_{k+1} \in \mathbf{\Theta}_{k+1} \subset \mathbb{R}^{p-k}$, and the parameter space is $\mathbf{\Theta}_{k+1} = \mathcal{G}_{p-k,1}$. We show that we can obtain a $\sqrt{n}$-consistent estimator $\widehat{\mathbf{w}}_{k+1}$, so the $\sqrt{n}$-consistency of $\widehat{\mathbf{G}}_{k+1} = \widehat{\mathbf{G}}_{0k}\widehat{\mathbf{w}}_{k+1}$ then follows. We define our objective function $Q_n(\mathbf{w})$ and $Q(\mathbf{w})$ as

$$Q_n(\mathbf{w}) = -n/2\log(\mathbf{w}^T(\widehat{\mathbf{G}}_{0k}^T(\widehat{\mathbf{M}}+\widehat{\mathbf{U}})\widehat{\mathbf{G}}_{0k})^{-1}\mathbf{w}) - n/2\log(\mathbf{w}^T\widehat{\mathbf{G}}_{0k}^T\widehat{\mathbf{M}}\widehat{\mathbf{G}}_{0k}\mathbf{w})$$
$$+n\log(\mathbf{w}^T\mathbf{w})$$
$$Q(\mathbf{w}) = -1/2\log(\mathbf{w}^T(\mathbf{G}_{0k}^T(\mathbf{M}+\mathbf{U})\mathbf{G}_{0k})^{-1}\mathbf{w}) - 1/2\log(\mathbf{w}^T\mathbf{G}_{0k}^T\mathbf{M}\mathbf{G}_{0k}\mathbf{w})$$
$$+\log(\mathbf{w}^T\mathbf{w}).$$

Following the same logic as verifying the conditions for the first direction, we can see that $\widehat{\mathbf{w}} = \arg\max Q_n(\mathbf{w})$ will be $\sqrt{n}$-consistent for $\mathbf{v}_t = \arg\max Q(\mathbf{w})$ by noticing that $(\widehat{\mathbf{G}}_{0k}^T(\widehat{\mathbf{M}}+\widehat{\mathbf{U}})\widehat{\mathbf{G}}_{0k})^{-1}$ and $\widehat{\mathbf{G}}_{0k}^T\widehat{\mathbf{M}}\widehat{\mathbf{G}}_{0k}$ are $\sqrt{n}$-consistent estimators for $(\mathbf{G}_{0k}^T(\mathbf{M}+\mathbf{U})\mathbf{G}_{0k})^{-1}$ and $\mathbf{G}_{0k}^T\mathbf{M}\mathbf{G}_{0k}$. Since all the u directions will be $\sqrt{n}$-consistent, the projection onto $\widehat{\mathbf{G}}_u = (\widehat{\mathbf{G}}_1, \dots, \widehat{\mathbf{G}}_u)$ will be a $\sqrt{n}$-consistent estimator for the projection onto the envelope $\mathcal{E}_{\mathbf{M}}(\mathcal{U})$.

B.2 Sequential Moment-Based Algorithm

The first step in demonstrating that the sequence constructed by (6.6)–(6.8) has property (6.5) is to incorporate $\mathcal{E}_{\mathbf{M}}(\mathcal{U})$ into the algorithm (see Section 6.5). For notational convenience, we shorten $\mathcal{E}_{\mathbf{M}}(\mathcal{U})$ to $\mathcal{E}$ when used as a subscript. Since $\mathcal{U}$ is contained in $\mathcal{E}_{\mathbf{M}}(\mathcal{U})$, $\mathcal{U} = \text{span}(\mathbf{U}) \subseteq \mathcal{E}_{\mathbf{M}}(\mathcal{U})$, we can substitute into (6.6) $\mathbf{w}^T\mathbf{U}\mathbf{w} = \mathbf{w}^T\mathbf{P}_{\mathcal{E}}\mathbf{U}\mathbf{P}_{\mathcal{E}}\mathbf{w}$. We know from Proposition A.1 that $\mathbf{M} = \mathbf{P}_{\mathcal{E}}\mathbf{M}\mathbf{P}_{\mathcal{E}} + \mathbf{Q}_{\mathcal{E}}\mathbf{M}\mathbf{Q}_{\mathcal{E}}$. Substituting this into (6.7), we have $\mathbf{w}^T\mathbf{M}\mathbf{W}_k = 0$ if and only if $\mathbf{w}^T\mathbf{P}_{\mathcal{E}}\mathbf{M}\mathbf{P}_{\mathcal{E}}\mathbf{W}_k + \mathbf{w}^T\mathbf{Q}_{\mathcal{E}}\mathbf{M}\mathbf{Q}_{\mathcal{E}}\mathbf{W}_k = 0$. These considerations led to the following equivalent statement of the algorithm. For $k = 0, 1, \ldots, u - 1$ find

$$\mathbf{w}_{k+1} = \arg\max_{\mathbf{w}} \mathbf{w}^T \mathbf{P}_{\mathcal{E}} \mathbf{U} \mathbf{P}_{\mathcal{E}} \mathbf{w}, \text{ subject to} \tag{B.2}$$

$$\mathbf{w}^T \mathbf{P}_{\mathcal{E}} \mathbf{M} \mathbf{P}_{\mathcal{E}} \mathbf{W}_k + \mathbf{w}^T \mathbf{Q}_{\mathcal{E}} \mathbf{M} \mathbf{Q}_{\mathcal{E}} \mathbf{W}_k = 0 \tag{B.3}$$

$$\mathbf{w}^T \mathbf{P}_{\mathcal{E}} \mathbf{w} + \mathbf{w}^T \mathbf{Q}_{\mathcal{E}} \mathbf{w} = 1. \tag{B.4}$$

We next establish (6.5) by induction, starting with an analysis of (B.2)–(B.4) for $k = 0$.

B.2.1 First Direction Vector $\mathbf{w}_1$

For the first vector $\mathbf{w}_1$, only the length constraint (B.4) is operational since $\mathbf{w}_0 = 0$. It follows from Lemma A.16 with $\mathbf{V} = \mathbf{I}_r$ and $\mathcal{T} = \mathcal{S}^{\perp}$ that

$$\mathbf{w}_1 = \boldsymbol{\Gamma} \ell_1(\boldsymbol{\Gamma}^T \mathbf{U} \boldsymbol{\Gamma}) = \ell_1(\mathbf{P}_{\mathcal{E}} \mathbf{U} \mathbf{P}_{\mathcal{E}}) = \ell_1(\mathbf{U}) \in \operatorname{span}(\mathbf{M}),$$

where $\boldsymbol{\Gamma}$ is a semi-orthogonal basis matrix for $\mathcal{E}_{\mathbf{M}}(\mathcal{U})$. Clearly, $\mathbf{w}_1 \in \mathcal{S} \subseteq \mathcal{E}_{\mathbf{M}}(\mathcal{U})$, so trivially $\mathcal{W}_0 \subset \mathcal{W}_1 \subseteq \mathcal{E}_{\mathbf{M}}(\mathcal{U})$ with equality if and only if $u = 1$.

B.2.2 Second Direction Vector $\mathbf{w}_2$

Next, assume that $u \geq 2$ and consider the second vector $\mathbf{w}_2$. In that case $\mathcal{W}_1 \subset \mathcal{E}_{\mathbf{M}}(\mathcal{U})$ and so the second addend on the left of (B.3) is 0. Consequently,

$$\mathbf{w}_2 = \arg\max_{\mathbf{w}} \mathbf{w}^T \mathbf{P}_{\mathcal{E}} \mathbf{U} \mathbf{P}_{\mathcal{E}} \mathbf{w}, \text{ subject to}$$

$$\mathbf{w}^T \mathbf{P}_{\mathcal{E}} \mathbf{M} \mathbf{P}_{\mathcal{E}} \mathbf{w}_1 = 0 \tag{B.5}$$

$$\mathbf{w}^T \mathbf{P}_{\mathcal{E}} \mathbf{w} + \mathbf{w}^T \mathbf{Q}_{\mathcal{E}} \mathbf{w} = 1. \tag{B.6}$$

Condition (B.5) holds if and only if $\mathbf{w}$ is orthogonal to $\mathbf{P}_{\mathcal{E}} \mathbf{M} \mathbf{P}_{\mathcal{E}} \mathbf{w}_1$. Letting $\mathcal{E}_1 = \operatorname{span}(\mathbf{P}_{\mathcal{E}} \mathbf{M} \mathbf{P}_{\mathcal{E}} \mathbf{w}_1)$ for notational convenience, $\mathbf{w}$ is then constrained to be of the form $\mathbf{w} = \mathbf{Q}_{\mathcal{E}_1} \widetilde{\mathbf{w}}$ for $\widetilde{\mathbf{w}} \in \mathbb{R}^r$. This form for $\mathbf{w}$ satisfies (B.5) by construction, leaving only the length constraint.

We need to deal with $\mathbf{P}_{\mathcal{E}} \mathbf{w} = \mathbf{P}_{\mathcal{E}} \mathbf{Q}_{\mathcal{E}_1} \widetilde{\mathbf{w}}$ and $\mathbf{Q}_{\mathcal{E}} \mathbf{w} = \mathbf{Q}_{\mathcal{E}} \mathbf{Q}_{\mathcal{E}_1} \widetilde{\mathbf{w}}$ to write the algorithm in terms of $\widetilde{\mathbf{w}}$: $\mathbf{Q}_{\mathcal{E}} \mathbf{Q}_{\mathcal{E}_1} = \mathbf{Q}_{\mathcal{E}}$, $\mathcal{E}_1 = \operatorname{span}(\mathbf{P}_{\mathcal{E}} \mathbf{M} \mathbf{P}_{\mathcal{E}} \mathbf{w}_1) \subset \mathcal{E}_{\mathbf{M}}(\mathcal{U})$, (the strict containment holds because $u \geq 2$), and thus $\mathbf{Q}_{\mathcal{E}_1} \mathbf{P}_{\mathcal{E}} = \mathbf{P}_{\mathcal{E}} - \mathbf{P}_{\mathcal{E}_1}$ is the projection onto $\mathcal{E}_{\mathbf{M}}(\mathcal{U}) \backslash \mathcal{E}_1$, the part of $\mathcal{E}_{\mathbf{M}}(\mathcal{U})$ that is orthogonal to $\mathcal{E}_1$. Let $\mathcal{D}_1 = \mathcal{E}_{\mathbf{M}}(\mathcal{U}) \backslash \mathcal{E}_1$ and $\mathbf{P}_{\mathcal{D}_1} = \mathbf{P}_{\mathcal{E}} - \mathbf{P}_{\mathcal{E}_1}$. It follows that the length condition (B.6) can be written as

$$\begin{aligned}
\mathbf{w}^T \mathbf{P}_{\mathcal{E}} \mathbf{w} + \mathbf{w}^T \mathbf{Q}_{\mathcal{E}} \mathbf{w} &= \widetilde{\mathbf{w}}^T \mathbf{Q}_{\mathcal{E}_1} \mathbf{P}_{\mathcal{E}} \mathbf{Q}_{\mathcal{E}_1} \widetilde{\mathbf{w}} + \widetilde{\mathbf{w}}^T \mathbf{Q}_{\mathcal{E}_1} \mathbf{Q}_{\mathcal{E}} \mathbf{Q}_{\mathcal{E}_1} \widetilde{\mathbf{w}} \\
&= \widetilde{\mathbf{w}}^T \mathbf{P}_{\mathcal{D}_1} \widetilde{\mathbf{w}} + \widetilde{\mathbf{w}}^T \mathbf{Q}_{\mathcal{E}} \widetilde{\mathbf{w}} = 1.
\end{aligned}$$

Then the algorithm for $\widetilde{\mathbf{w}}_2$ becomes

$$\widetilde{\mathbf{w}}_2 = \arg\max_{\mathbf{w}} \mathbf{w}^T \mathbf{P}_{\mathcal{D}_1} \mathbf{U} \mathbf{P}_{\mathcal{D}_1} \mathbf{w}, \text{ subject to}$$
$$\mathbf{w}^T \mathbf{P}_{\mathcal{D}_1} \mathbf{w} + \mathbf{w}^T \mathbf{Q}_{\mathcal{E}} \mathbf{w} = 1.$$

Let $\boldsymbol{\Gamma}_1$ be a semi-orthogonal basis matrix for $\mathcal{D}_1$. It follows from Lemma A.16 with $\mathbf{V} = \mathbf{I}_r$, $\mathcal{U} = \mathcal{D}_1$, and $\mathcal{T} = \mathcal{E}_{\mathbf{M}}^{\perp}(\mathcal{U})$ that

$$\widetilde{\mathbf{w}}_2 = \boldsymbol{\Gamma}_1 \ell_1(\boldsymbol{\Gamma}_1^T \mathbf{U} \boldsymbol{\Gamma}_1) = \ell_1(\mathbf{P}_{\mathcal{D}_1} \mathbf{U} \mathbf{P}_{\mathcal{D}_1}) = \ell_1(\mathbf{Q}_{\mathcal{E}_1} \mathbf{U} \mathbf{Q}_{\mathcal{E}_1}) \in \text{span}(\mathbf{M}),$$

and thus that $\mathbf{w}_2 = \mathbf{Q}_{\mathcal{E}_1} \widetilde{\mathbf{w}}_2 = \widetilde{\mathbf{w}}_2$.

If $\mathbf{P}_{\mathcal{D}_1} \mathbf{U} = 0$, the algorithm will terminate and then $\text{span}(\mathbf{w}_1) = \mathcal{E}_{\mathbf{M}}(\mathcal{U})$. To see why this conclusion holds, we have $\mathbf{P}_{\mathcal{D}_1} \mathbf{U} = 0$ is equivalent to $\mathbf{P}_{\mathcal{E}_1} \mathbf{U} = \mathbf{U}$, so $\mathcal{U} = \text{span}(\mathbf{U}) \subseteq \mathcal{E}_1$. The conclusion will follow if, when $\mathbf{P}_{\mathcal{E}_1} \mathbf{U} = \mathbf{U}$, we can show that $\mathcal{E}_1 = \text{span}(\mathbf{w}_1)$ and that $\mathcal{E}_1$ reduces $\mathbf{M}$, because then $\mathcal{E}_1$ will be a reducing subspace of $\mathbf{M}$ that contains $\mathcal{U}$. Now, we know that

$$\mathcal{E}_1 = \text{span}(\mathbf{P}_{\mathcal{E}} \mathbf{M} \mathbf{P}_{\mathcal{E}} \mathbf{w}_1) = \text{span}(\mathbf{M} \mathbf{w}_1) = \mathbf{M}\text{span}(\ell_1(\mathbf{U})).$$

Assuming that $\mathbf{P}_{\mathcal{E}_1} \mathbf{U} = \mathbf{U}$ gives

$$\mathcal{E}_1 = \mathbf{M}\text{span}(\ell_1(\mathbf{U})) = \mathbf{M}\text{span}(\ell_1(\mathbf{P}_{\mathcal{E}_1} \mathbf{U} \mathbf{P}_{\mathcal{E}_1})).$$

The subspace $\mathcal{E}_1$ has dimension 1 and thus $\mathcal{U}$, which is assumed to be nontrivial, must have dimension 1. It follows that $\text{span}(\mathbf{w}_1) = \text{span}(\ell_1(\mathbf{P}_{\mathcal{E}_1} \mathbf{U} \mathbf{P}_{\mathcal{E}_1})) = \mathcal{E}_1$ and thus $\mathcal{E}_1 = \mathbf{M}\mathcal{E}_1$, so $\mathcal{E}_1$ reduces $\mathbf{M}$.

In sum, $\mathbf{w}_1 \in \mathcal{E}_{\mathbf{M}}(\mathcal{U})$, $\mathbf{w}_2 \in \mathcal{D}_1 = \mathcal{E}_{\mathbf{M}}(\mathcal{U}) \backslash \mathcal{E}_1 \subset \mathcal{E}_{\mathbf{M}}(\mathcal{U})$. If $\mathbf{w}_1$ and $\mathbf{w}_2$ are linearly independent, then $\mathcal{W}_0 \subset \mathcal{W}_1 \subset \mathcal{W}_2 \subseteq \mathcal{E}_{\mathbf{M}}(\mathcal{U})$, with equality if and only if $u = 2$.

B.2.3 $(q+1)$st Direction Vector $\mathbf{w}_{q+1}$, $q < u$

The reasoning here parallels that for $\mathbf{w}_2$. Consider weight matrices $\mathbf{W}_k = (\mathbf{w}_0, \dots, \mathbf{w}_k)$, with $\mathbf{w}_k \in \mathcal{E}_{\mathbf{M}}(\mathcal{U})$, $k = 0, 1, \dots, q$. Assuming that the columns of $\mathbf{W}_k$ are linearly independent, the weight matrices must satisfy

$$\mathcal{W}_0 \subset \mathcal{W}_1 \subset \cdots \subset \mathcal{W}_{q-1} \subset \mathcal{W}_q \subset \mathcal{E}_{\mathbf{M}}(\mathcal{U})$$

by construction. For $k = 0, 1, \dots, q-1$, let $\mathcal{E}_k = \text{span}(\mathbf{P}_{\mathcal{E}} \mathbf{M} \mathbf{P}_{\mathcal{E}} \mathbf{W}_k)$ and $\mathcal{D}_k = \mathcal{E}_{\mathbf{M}}(\mathcal{U}) \backslash \mathcal{E}_k$ so that $\mathbf{P}_{\mathcal{D}_k} = \mathbf{P}_{\mathcal{E}} - \mathbf{P}_{\mathcal{E}_k}$. As a consequence of the structure assumed for the $\mathbf{W}_k$'s, the $\mathcal{D}_k$'s are properly nested subspaces

$$\mathcal{D}_{q-1} \subset \cdots \mathcal{D}_1 \subset \mathcal{D}_0 = \mathcal{E}_{\mathbf{M}}(\mathcal{U}).$$

To use induction, we now must show that $\mathcal{W}_q \subset \mathcal{W}_{q+1} \subseteq \mathcal{E}_{\mathbf{M}}(\mathcal{U})$, with equality in the last containment if and only if $q + 1 = u$, starting from representations

(B.2)–(B.4). Since $\mathcal{W}_q \subset \mathcal{E}_{\mathbf{M}}(\mathcal{U})$ the second addend on the left of (B.3) is again 0, and the $q+1$st step of the algorithm becomes

$$\mathbf{w}_{q+1} = \arg\max_{\mathbf{w}} \mathbf{w}^T \mathbf{P}_{\mathcal{E}} \mathbf{U} \mathbf{P}_{\mathcal{E}} \mathbf{w}, \text{ subject to}$$

$$\mathbf{w}^T \mathbf{P}_{\mathcal{E}} \mathbf{M} \mathbf{P}_{\mathcal{E}} \mathbf{W}_q = 0 \tag{B.7}$$

$$\mathbf{w}^T \mathbf{P}_{\mathcal{E}} \mathbf{w} + \mathbf{w}^T \mathbf{Q}_{\mathcal{E}} \mathbf{w} = 1. \tag{B.8}$$

Condition (B.7) holds if and only if $\mathbf{w}$ is orthogonal to $\mathbf{P}_{\mathcal{E}} \mathbf{M} \mathbf{P}_{\mathcal{E}} \mathbf{W}_q$, so $\mathbf{w}$ is then constrained to be of the form $\mathbf{w} = \mathbf{Q}_{\mathcal{E}_q} \widetilde{\mathbf{w}}$ for $\widetilde{\mathbf{w}} \in \mathbb{R}^r$, where $\mathcal{E}_q = \text{span}(\mathbf{P}_{\mathcal{E}} \mathbf{M} \mathbf{P}_{\mathcal{E}} \mathbf{W}_q)$. This form for $\mathbf{w}$ satisfies (B.7) by construction, leaving only the length constraint (B.8). Straightforwardly,

$$\mathcal{E}_{q-1} = \text{span}(\mathbf{P}_{\mathcal{E}} \mathbf{M} \mathbf{P}_{\mathcal{E}} \mathbf{W}_{q-1}) \subset \mathcal{E}_q = \text{span}(\mathbf{P}_{\mathcal{E}} \mathbf{M} \mathbf{P}_{\mathcal{E}} \mathbf{W}_q) \subset \mathcal{E}_{\mathbf{M}}(\mathcal{U}),$$

and $\mathbf{Q}_{\mathcal{E}_q} \mathbf{P}_{\mathcal{E}} = \mathbf{P}_{\mathcal{E}} - \mathbf{P}_{\mathcal{E}_q}$ is a projection onto the part of $\mathcal{E}_{\mathbf{M}}(\mathcal{U})$ that is orthogonal to $\mathcal{E}_q$. Let $\mathcal{D}_q = \mathcal{E}_{\mathbf{M}}(\mathcal{U}) \backslash \mathcal{E}_q \subset \mathcal{D}_{q-1}$. Then we can use $\widetilde{\mathbf{w}}$ to rewrite the algorithm in terms of only the criterion and the length constraint leading to

$$\widetilde{\mathbf{w}}_{q+1} = \arg\max_{\mathbf{w}} \mathbf{w}^T \mathbf{P}_{\mathcal{D}_q} \mathbf{U} \mathbf{P}_{\mathcal{D}_q} \mathbf{w}, \text{ subject to}$$

$$\mathbf{w}^T \mathbf{P}_{\mathcal{D}_q} \mathbf{w} + \mathbf{w}^T \mathbf{Q}_{\mathcal{E}} \mathbf{w} = 1.$$

Let $\mathbf{\Gamma}_q$ be a semi-orthogonal basis matrix for $\mathcal{D}_q$. It follows from Lemma A.16 with $\mathbf{V} = \mathbf{I}_r$, $\mathcal{U} = \mathcal{D}_q$, and $\mathcal{T} = \mathcal{E}_{\mathbf{M}}^{\perp}(\mathcal{U})$ that

$$\widetilde{\mathbf{w}}_{q+1} = \mathbf{\Gamma}_q \ell_1(\mathbf{\Gamma}_q^T \mathbf{U} \mathbf{\Gamma}_q) = \ell_1(\mathbf{P}_{\mathcal{D}_q} \mathbf{U} \mathbf{P}_{\mathcal{D}_q}) = \ell_1(\mathbf{Q}_{\mathcal{E}_q} \mathbf{U} \mathbf{Q}_{\mathcal{E}_q}) \in \text{span}(\mathbf{M}),$$

and thus that $\mathbf{w}_{q+1} = \mathbf{Q}_{\mathcal{E}_q} \widetilde{\mathbf{w}}_{q+1} = \widetilde{\mathbf{w}}_{q+1}$. With $\mathbf{W}_{q+1} = (\mathbf{W}_q, \mathbf{w}_{q+1})$, it follows that $\mathcal{W}_q \subset \text{span}(\mathbf{P}_{\mathcal{E}} \mathbf{M} \mathbf{P}_{\mathcal{E}} \mathbf{W}_{q+1}) = \mathcal{W}_{q+1} \subseteq \mathcal{E}_{\mathbf{M}}(\mathcal{U})$.

B.2.4 Termination

The algorithm will terminate the first time $\mathbf{P}_{\mathcal{D}_q} \mathbf{U} = 0$ or, equivalently, $\mathbf{P}_{\mathcal{E}_q} \mathbf{U} = \mathbf{U}$ so $\mathcal{S} \subseteq \mathcal{E}_q$. We need to show that $\mathcal{E}_{\mathbf{M}}(\mathcal{U}) = \text{span}(\mathbf{W}_q) = \mathcal{E}_q$ to insure that the algorithm produces the envelope at termination. It is sufficient to show that $\text{span}(\mathbf{W}_q) = \mathcal{E}_q$ and that $\mathcal{E}_q$ reduces $\mathbf{M}$. Now,

$$\begin{aligned} \mathcal{E}_q &= \text{span}(\mathbf{M}\mathbf{W}_q) = \text{span}(\mathbf{P}_{\mathcal{E}} \mathbf{M} \mathbf{P}_{\mathcal{E}} \mathbf{W}_q) \\ &= \mathbf{M}\text{span}\{\ell_1(\mathbf{U}), \ell_1(\mathbf{P}_{\mathcal{D}_1} \mathbf{U} \mathbf{P}_{\mathcal{D}_1}), \dots, \ell_1(\mathbf{P}_{\mathcal{D}_{q-1}} \mathbf{U} \mathbf{P}_{\mathcal{D}_{q-1}})\}. \end{aligned}$$

Substituting the condition $\mathbf{P}_{\mathcal{E}_q} \mathbf{U} = \mathbf{U}$, we have

$$\begin{aligned} \mathcal{E}_q &= \mathbf{M}\text{span}\{\ell_1(\mathbf{P}_{\mathcal{E}_q} \mathbf{U} \mathbf{P}_{\mathcal{E}_q}), \ell_1(\mathbf{P}_{\mathcal{D}_1} \mathbf{P}_{\mathcal{E}_q} \mathbf{U} \mathbf{P}_{\mathcal{E}_q} \mathbf{P}_{\mathcal{D}_1}), \dots, \ell_1(\mathbf{P}_{\mathcal{D}_{q-1}} \mathbf{P}_{\mathcal{E}_q} \mathbf{U} \mathbf{P}_{\mathcal{E}_q} \mathbf{P}_{\mathcal{D}_{q-1}})\} \\ &= \mathbf{M}\text{span}\{\ell_1(\mathbf{P}_{\mathcal{E}_q} \mathbf{U} \mathbf{P}_{\mathcal{E}_q}), \ell_1[(\mathbf{P}_{\mathcal{E}_q} - \mathbf{P}_{\mathcal{E}_1}) \mathbf{U} (\mathbf{P}_{\mathcal{E}_q} - \mathbf{P}_{\mathcal{E}_1})], \\ &\quad \dots, \ell_1[(\mathbf{P}_{\mathcal{E}_q} - \mathbf{P}_{\mathcal{E}_{q-1}}) \mathbf{U} (\mathbf{P}_{\mathcal{E}_q} - \mathbf{P}_{\mathcal{E}_{q-1}})]\}. \end{aligned}$$

The argument to the span in the right-hand side of the last expression contains q linearly independent terms, each of which is in $\mathcal{E}_q$, and thus these terms must span $\mathcal{E}_q$. Thus,

$$\begin{aligned}\mathcal{E}_q = \text{span}\{&\boldsymbol{\ell}_1(\mathbf{P}_{\mathcal{E}_q}\mathbf{U}\mathbf{P}_{\mathcal{E}_q}), \boldsymbol{\ell}_1[(\mathbf{P}_{\mathcal{E}_q} - \mathbf{P}_{\mathcal{E}_1})\mathbf{U}(\mathbf{P}_{\mathcal{E}_q} - \mathbf{P}_{\mathcal{E}_1})],\\ &\ldots, \boldsymbol{\ell}_1[(\mathbf{P}_{\mathcal{E}_q} - \mathbf{P}_{\mathcal{E}_{q-1}})\mathbf{U}(\mathbf{P}_{\mathcal{E}_q} - \mathbf{P}_{\mathcal{E}_{q-1}})]\},\end{aligned}$$

and

$$\mathcal{E}_q = \mathbf{M}\text{span}(\mathbf{W}_q) = \mathbf{M}\mathcal{E}_q.$$

It follows that $\mathcal{E}_q = \text{span}(\mathbf{W}_q)$, that $\mathcal{E}_q$ reduces $\mathcal{E}_{\mathbf{M}}(\mathcal{U})$, and thus that $q = u$. We have therefore verified that the sequence generated by the algorithm satisfies (6.5).

Appendix C

Grassmann Manifold Optimization

The Grassmann manifold $\mathcal{G}(d,p)$, also called a Grassmannian, is the set of all d-dimensional subspaces in $\mathbb{R}^p$, which has algebraic dimension $d(p-d)$ and is compact and connected. An element of $\mathcal{G}(d,p)$ can be represented uniquely as a projection matrix or nonuniquely as a semi-orthogonal basis matrix. Optimization over a Grassmann manifold is involved in many problems in dimension reduction, or other fields where subspace estimation is needed. Here are two examples:

1. Suppose that we have a multivariate sample $\mathbf{X}_1, \ldots, \mathbf{X}_n$, and we wish to find the d-dimensional subspace $\hat{\mathcal{S}}$ that provides the best approximation of the sample in the sense that

$$\hat{\mathcal{S}} = \arg \min_{\mathcal{S} \in \mathcal{G}(d,p)} \sum_{i=1}^{n} \| \mathbf{X}_i - \bar{\mathbf{X}} - \mathbf{P}_{\mathcal{S}}(\mathbf{X}_i - \bar{\mathbf{X}}) \|^2 = \arg \max \mathrm{tr}(\mathbf{P}_{\mathcal{S}} \mathbf{S}_{\mathbf{X}}),$$

   where $\mathbf{S}_{\mathbf{X}}$ is the sample covariance matrix of the $\mathbf{X}_i'$s. For later reference, let $L_1(\mathcal{S}) = \mathrm{tr}(\mathbf{P}_{\mathcal{S}} \mathbf{S}_{\mathbf{X}})$.
2. In envelope model (1.20), $\mathbf{Y} = \boldsymbol{\alpha} + \boldsymbol{\Gamma}\boldsymbol{\eta}\mathbf{X} + \boldsymbol{\varepsilon}$, $\boldsymbol{\Sigma} = \boldsymbol{\Gamma}\boldsymbol{\Omega}\boldsymbol{\Gamma}^T + \boldsymbol{\Gamma}_0\boldsymbol{\Omega}_0\boldsymbol{\Gamma}_0^T$, the maximum likelihood estimator of the envelope subspace $\hat{\mathcal{E}}_{\boldsymbol{\Sigma}}(\mathcal{B})$ is found by minimizing the objective function $L_2(\mathbf{G}) = \log |\mathbf{G}^T \mathbf{S}_{\mathbf{Y}|\mathbf{X}} \mathbf{G}| + \log |\mathbf{G}_0^T \mathbf{S}_{\mathbf{Y}} \mathbf{G}_0|$, where $\mathbf{G} \in \mathbb{R}^{p \times d}$ is semi-orthogonal, $(\mathbf{G}, \mathbf{G}_0)$ is an orthogonal matrix, $\mathbf{S}_{\mathbf{Y}}$ is the sample covariance matrix of $\mathbf{Y}$, and $\mathbf{S}_{\mathbf{Y}|\mathbf{X}}$ is the sample covariance matrix of the residual vectors from the linear regression of $\mathbf{Y}$ on $\mathbf{X}$.

Sometimes it is possible to get an analytic solution for a Grassmann optimization. For example, $L_1(\mathcal{S}) = \mathrm{tr}(\mathbf{P}_{\mathcal{S}} \mathbf{S}_{\mathbf{X}}) = \mathrm{tr}(\mathbf{P}_{\mathcal{S}} \mathbf{S}_{\mathbf{X}} \mathbf{P}_{\mathcal{S}}) = \sum_{i=1}^{p} \varphi_i \| \mathbf{P}_{\mathcal{S}} \boldsymbol{\ell}_i \|^2$, where φ_i and $\boldsymbol{\ell}_i$, $i = 1, \ldots, p$ are ordered eigenvalues (descending) and corresponding eigenvectors of $\mathbf{S}_{\mathbf{X}}$. So $\hat{\mathcal{S}} = \mathrm{span}(\boldsymbol{\ell}_1, \ldots, \boldsymbol{\ell}_d)$ maximizes $L_1(\mathcal{S})$. However, in many problems, analytic solutions are very hard or impossible to get and numerical algorithms are needed. In this appendix, we give a cursory introduction to numerical optimization over a Grassmann manifold, describing a basic gradient algorithm. The description is adapted from Liu et al. (2004) and

An Introduction to Envelopes: Dimension Reduction for Efficient Estimation in Multivariate Statistics,
First Edition. R. Dennis Cook.

Edelman et al. (1998), and it makes use of matrix exponential functions and skew-symmetric matrices.

Matrix exponential. The matrix exponential is defined for square matrices $\mathbf{C} \in \mathbb{R}^{p\times p}$ as

$$\exp\{\mathbf{C}\} = \sum_{k=0}^{\infty} \frac{\mathbf{C}^k}{k!} = \mathbf{I} + \mathbf{C} + \frac{\mathbf{C}\times\mathbf{C}}{2} + \frac{\mathbf{C}\times\mathbf{C}\times\mathbf{C}}{3\cdot 2} + \cdots$$

Here are a few basic properties of matrix exponentials; a and b denote scalars.

1) $\exp\{\mathbf{0}\} = \mathbf{I}$.
2) $\exp\{a\mathbf{C}\}\exp\{b\mathbf{C}\} = \exp\{(a+b)\mathbf{C}\}$, so $\exp\{\mathbf{C}\}\exp\{-\mathbf{C}\} = \mathbf{I}$.
3) If $\mathbf{D}$ is nonsingular then $\exp\{\mathbf{D}\mathbf{C}\mathbf{D}^{-1}\} = \mathbf{D}\exp\{\mathbf{C}\}\mathbf{D}^{-1}$. Consequently, if $\mathbf{C}$ is symmetric and has spectral decomposition $\mathbf{C} = \mathbf{H}\boldsymbol{\Theta}\mathbf{H}^T$, then $\exp\{\mathbf{C}\} = \mathbf{H}\exp\{\boldsymbol{\Theta}\}\mathbf{H}^T$.
4) $|\exp\{\mathbf{C}\}| = \exp\{\mathrm{tr}(\mathbf{C})\}$.
5) $\exp\{\mathbf{C}^T\} = (\exp\{\mathbf{C}\})^T$.
6) If $\mathbf{C}$ is skew-symmetric, then $\exp\{\mathbf{C}\}$ is orthogonal.

Skew symmetric matrix. A matrix $\mathbf{A} \in \mathbb{R}^{p\times p}$ is skew-symmetric if $\mathbf{A}^T = -\mathbf{A}$.

C.1 Gradient Algorithm

Consider optimizing an objective function $L(\mathbf{G})$ over $\mathbf{G} \in \mathbb{R}^{p\times d}$, $d \le p$, where L is smooth and $L(\mathbf{G}\mathbf{O}) = L(\mathbf{G})$ for any orthogonal matrix $\mathbf{O} \in \mathbb{R}^{d\times d}$. In consequence, the value of $L(\mathbf{G})$ depends only on span($\mathbf{G}$) and not on a particular basis, and the essential domain of L is $\mathcal{G}(p,d)$. Without loss of generality, we restrict $\mathbf{G}$ to $p\times d$ semi-orthogonal matrices as representations of bases for subspaces in $\mathcal{G}(p,d)$. Any argument $\hat{\mathbf{G}}$ that optimizes L then stands for $\mathbb{G} = \{\hat{\mathbf{G}}\mathbf{O} \mid \mathbf{O} \in \mathbb{R}^{p\times d}, \mathbf{O}^T\mathbf{O} = \mathbf{I}_d\}$, the set of all semi-orthogonal bases for span($\hat{\mathbf{G}}$). For example, we could then write $\mathbb{G} = \arg\max L(\mathbf{G})$.

Let

$$\mathbf{J}_1 = \begin{pmatrix} \mathbf{I}_{d\times d} \\ 0_{(p-d)\times d} \end{pmatrix} \quad \text{and} \quad \mathbf{J}_2 = \begin{pmatrix} 0_{d\times(p-d)} \\ \mathbf{I}_{(p-d)\times(p-d)} \end{pmatrix},$$

so that $(\mathbf{J}_1, \mathbf{J}_2) = \mathbf{I}_p$. Let $\mathbf{e}_{p,j} \in \mathbb{R}^p$ denote the vector with a 1 in position i and zeros elsewhere, and let $\mathbf{E}_{ij} \in \mathbb{R}^{p\times p}$ be a skew-symmetric matrix with a 1 in position (i,j) and a -1 in position (j,i), $1 \le i \le d$, $d < j \le p$; $\mathbf{E}_{ij} = \mathbf{e}_{p,i}\mathbf{e}_{p,j}^T - \mathbf{e}_{p,j}\mathbf{e}_{p,i}^T$.

The vector space that is tangent to $\mathcal{G}(p,d)$ at $\mathcal{J}_1 = \mathrm{span}(\mathbf{J}_1)$ is

$$T(\mathcal{J}_1) = \mathrm{span}\{\mathbf{E}_{i,j}\mathbf{J}_1 \mid i = 1,\dots,d, j = d+1,\dots,p\}.$$

This tangent space is spanned by the $d(p-d)$ matrices $\mathbf{E}_{i,j}\mathbf{J}_1$, which form an orthonormal basis for $T(\mathcal{J}_1)$ relative to the usual trace inner product:

$$\begin{aligned}\mathrm{tr}(\mathbf{E}_{i,j}\mathbf{J}_1\mathbf{J}_1^T\mathbf{E}_{k,l}^T) &= \mathrm{tr}(\mathbf{e}_{p,j}\mathbf{e}_{p,i}^T\mathbf{e}_{p,k}\mathbf{e}_{p,l}^T)\\ &= 1 \text{ if } i=k, j=l\\ &= 0 \text{ otherwise.}\end{aligned}$$

Next, consider the vector space $T(\mathcal{U})$ that is tangent to $\mathcal{G}(p,d)$ at $\mathcal{U} = \mathrm{span}(\mathbf{U})$, where $\mathbf{U} \in \mathbb{R}^{p\times d}$ is semi-orthogonal. Select $\mathbf{U}_0 \in \mathbb{R}^{p\times(p-d)}$ so that $\mathbf{W} = (\mathbf{U}, \mathbf{U}_0) \in \mathbb{R}^{p\times p}$ is orthogonal. With this we can write the tangent space for $\mathcal{U}$ as

$$T(\mathcal{U}) = \mathrm{span}\{\mathbf{W}\mathbf{J} : \mathbf{J} \in T(\mathcal{J}_1)\}.$$

An orthonormal basis for this space is $\{\mathbf{W}\mathbf{E}_{i,j}\mathbf{J}_1 : i = 1,\dots,d, j = d+1,\dots,p\}$.

The *gradient* $L'(\mathcal{U})$ of L at $\mathcal{U}$ is then

$$L'(\mathcal{U}) = \mathbf{W}\mathbf{A}(\mathbf{B})\mathbf{J}_1,$$

where $\mathbf{A}(\mathbf{B})$, which is sometimes called the *gradient direction* of L, is a skew-symmetric matrix of the form

$$\mathbf{A}(\mathbf{B}) = \begin{pmatrix} 0_{d\times d} & \mathbf{B}_{d\times(p-d)} \\ -\mathbf{B}^T_{(p-d)\times d} & 0_{(p-d)\times(p-d)} \end{pmatrix} \in \mathbb{R}^{p\times p}.$$

The matrix $\mathbf{B}$ is sometimes called the *gradient direction matrix* and is defined later.

For step size $\delta \in \mathbb{R}^1$, a single step of the gradient algorithm is

$$\mathbf{W}_{+1} = (\mathbf{G}_{+1}, \mathbf{G}_{0,+1}) = \mathbf{W}\exp\{\delta\mathbf{A}(\mathbf{B})\},$$

where $\mathbf{G}_{+1} = \mathbf{W}\exp\{\delta\mathbf{A}(\mathbf{B})\}\mathbf{J}_1$, $\mathbf{G}_{0,+1} = \mathbf{W}\exp\{\delta\mathbf{A}(\mathbf{B})\}\mathbf{J}_2$, and exp is the matrix exponential. Iteration continues until a stopping criterion is met. Because $\delta\mathbf{A}(\mathbf{B})$ is a skew-symmetric matrix, $\exp\{\delta\mathbf{A}(\mathbf{B})\}$ is orthogonal. Consequently, $\mathbf{W}_{+1}$ is also an orthogonal matrix. The algorithm thus works by transforming the starting orthonormal basis $\mathbf{W}$ to a new basis $\mathbf{W}_{+1}$ by right multiplication by an orthogonal matrix. This basic algorithm holds for any differentiable function that maps $\mathcal{G}(p,d)$ to $\mathbb{R}^1$; only $\mathbf{B}$ changes with the choice of L.

Computation of $\mathbf{B}$ and $\mathbf{A}(\mathbf{B})$ are necessary to implement the algorithm. These are considered in the next two subsections.

C.2 Construction of B

In this section, we demonstrate that $\mathbf{B}$ can be constructed as

$$\mathbf{B} = \{\partial L(\mathbf{G})/\partial\mathbf{G}\}^T\mathbf{G}_0.$$

The elements of $\mathbf{B}$ are

$$(\mathbf{B})_{ij} = \lim_{\epsilon \to 0} \frac{L(\mathbf{W}\exp\{-\epsilon \mathbf{E}_{i,j+d}\}\mathbf{J}_1) - L(\mathbf{G})}{\epsilon}, \; i = 1, \ldots, d, \; j = 1, \ldots, (p-d).$$

The matrix $\exp\{-\epsilon \mathbf{E}_{i,j+d}\}$ is the $p \times p$ identity matrix except for the following four elements:

$$(i,i) \text{ and } (j+d, j+d) : \cos(\epsilon), \quad (i, j+d) : -\sin(\epsilon), \quad (j+d, i) : \sin(\epsilon).$$

The submatrix consisting of the first d rows of $\exp\{-\epsilon \mathbf{E}_{i,j+d}\}\mathbf{J}_1 \in \mathbb{R}^{p\times d}$ is $\mathbf{I}_d + \mathbf{e}_{d,i}\mathbf{e}_{d,i}^T \cos(\epsilon)$. The last $p-d$ rows of $\exp\{-\epsilon \mathbf{E}_{i,j+d}\}\mathbf{J}_1$ are 0 except for element $(j+d, i)$, which is $\sin(\epsilon)$. Thus,

$$\begin{aligned}
\exp\{-\epsilon \mathbf{E}_{i,j+d}\}\mathbf{J}_1 &= \begin{pmatrix} \mathbf{I}_d + \mathbf{e}_{d,i}\mathbf{e}_{d,i}^T \cos(\epsilon) \\ \mathbf{e}_{p-d,j}\mathbf{e}_{d,i}^T \sin(\epsilon) \end{pmatrix} \\
&= \mathbf{J}_1 + \mathbf{e}_{p,i}\mathbf{e}_{p,i}^T(\cos(\epsilon) - 1) + \mathbf{e}_{p,j+d}\mathbf{e}_{p,i}^T \sin(\epsilon) \\
&:= \mathbf{J}_1 + \mathbf{A}_{i,j+d}.
\end{aligned}$$

In short, $\exp\{-\epsilon \mathbf{E}_{i,j+d}\}\mathbf{J}_1$ can be written as $\mathbf{J}_1$ plus a $p \times d$ matrix $\mathbf{A}_{i,j+d}$ of all zeros expect for element (i, i) that is equal to $\cos(\epsilon) - 1$ and element $(j+d, i)$ that is equal to $\sin(\epsilon)$. Therefore,

$$\begin{aligned}
(\mathbf{G}, \mathbf{G}_0)\exp\{-\epsilon \mathbf{E}_{i,j+d}\}\mathbf{J}_1 &= (\mathbf{G}, \mathbf{G}_0)(\mathbf{J}_1 + \mathbf{A}_{i,j+d}) \\
&= \mathbf{G} + \mathbf{T},
\end{aligned}$$

where $\mathbf{T}$ is a matrix $p \times d$ with all elements 0 except for column i that is equal to the ith column of $\mathbf{G}$ multiplied by $\cos(\epsilon) - 1$ plus the jth column of $\mathbf{G}_0$ multiplied by $\sin(\epsilon)$. Using this result that $\lim_{\epsilon\to 0}(\cos(\epsilon) - 1)/\epsilon = 0$ and $\lim_{\epsilon \to 0}(\sin(\epsilon))/\epsilon = 1$, we get

$$(\mathbf{B})_{ij} = \sum_{k=1}^{p} \frac{\partial L}{\partial \mathbf{G}_{ki}} \mathbf{G}_{0,kj},$$

which is the ijth element of

$$\mathbf{B} = \{\partial L(\mathbf{G})/\partial \mathbf{G}\}^T \mathbf{G}_0. \tag{C.1}$$

For example, consider $L_2(\mathbf{G})$ as in the second example of this appendix.

$$\begin{aligned}
L_2(\mathbf{G}) &= \log|\mathbf{G}^T \mathbf{S}_{\mathbf{Y}|\mathbf{X}} \mathbf{G}| + \log|\mathbf{G}_0^T \mathbf{S}_{\mathbf{Y}} \mathbf{G}_0| \\
&= \log|\mathbf{G}^T \mathbf{S}_{\mathbf{Y}|\mathbf{X}} \mathbf{G}| + \log|\mathbf{G}^T \mathbf{S}_{\mathbf{Y}}^{-1} \mathbf{G}| + \log|\mathbf{S}_{\mathbf{Y}}|.
\end{aligned}$$

Since

$$\begin{aligned}
\partial\{\log|\mathbf{G}^T \mathbf{S}_{\mathbf{Y}|\mathbf{X}} \mathbf{G}|\}/\partial \mathbf{G} &= 2\mathbf{S}_{\mathbf{Y}|\mathbf{X}}\mathbf{G}(\mathbf{G}^T \mathbf{S}_{\mathbf{Y}|\mathbf{X}} \mathbf{G})^{-1} \\
\partial\{\log|\mathbf{G}^T \mathbf{S}_{\mathbf{Y}}^{-1} \mathbf{G}|\}/\partial \mathbf{G} &= 2\mathbf{S}_{\mathbf{Y}}^{-1}\mathbf{G}(\mathbf{G}^T \mathbf{S}_{\mathbf{Y}}^{-1} \mathbf{G})^{-1} \\
\partial\{\log|\mathbf{S}_{\mathbf{Y}}|\}/\partial \mathbf{G} &= 0,
\end{aligned}$$

we have

$$\begin{aligned}\partial\{L_2(\mathbf{G})\}/\partial\mathbf{G} &= 2\{\mathbf{S}_{\mathbf{Y}|\mathbf{X}}\mathbf{G}(\mathbf{G}^T\mathbf{S}_{\mathbf{Y}|\mathbf{X}}\mathbf{G})^{-1} + \mathbf{S}_{\mathbf{Y}}^{-1}\mathbf{G}(\mathbf{G}^T\mathbf{S}_{\mathbf{Y}}^{-1}\mathbf{G})^{-1}\}\\ &= 2\{\mathbf{S}_{\mathbf{Y}|\mathbf{X}}\mathbf{G}(\mathbf{G}^T\mathbf{S}_{\mathbf{Y}|\mathbf{X}}\mathbf{G})^{-1} + \mathbf{G} - \mathbf{G}_0(\mathbf{G}_0^T\mathbf{S}_{\mathbf{Y}}\mathbf{G}_0)^{-1}\mathbf{G}_0^T\mathbf{S}_{\mathbf{Y}}\mathbf{G}\},\end{aligned}$$

where the last equality is from Appendix A of Cook and Forzani (2009). Since $\mathbf{B} = \{\partial L_2(\mathbf{G})/\partial\mathbf{G}\}^T\mathbf{G}_0$,

$$\mathbf{B} = 2\{(\mathbf{G}^T\mathbf{S}_{\mathbf{Y}|\mathbf{X}}\mathbf{G})^{-1}\mathbf{G}^T\mathbf{S}_{\mathbf{Y}|\mathbf{X}}\mathbf{G}_0 - \mathbf{G}^T\mathbf{S}_{\mathbf{Y}}\mathbf{G}_0(\mathbf{G}_0^T\mathbf{S}_{\mathbf{Y}}\mathbf{G}_0)^{-1}\}.$$

See also (A.15)and (A.16).

C.3 Construction of exp{δA(B)}

To describe the construction of $\mathbf{A}(\mathbf{B})$, we assume that $2d \le p$. Corresponding results for $2d > p$ can be shown similarly and are stated at the end of this section.

Construct the full singular value decomposition of $\mathbf{B} = \mathbf{H}_1\boldsymbol{\theta}\mathbf{H}_2^T$, where $\mathbf{H}_1 \in \mathbb{R}^{d\times d}$, $\mathbf{H}_2 \in \mathbb{R}^{(p-d)\times(p-d)}$, both being orthogonal matrices, and $\boldsymbol{\theta} = (\boldsymbol{\theta}_1, 0)$ where $\boldsymbol{\theta}_1 \in \mathbb{R}^{d\times d}$ is a diagonal matrix. $\boldsymbol{\theta}$ has this shape because we have assumed that $2d \le p$. Then

$$\begin{aligned}\mathbf{A}(\mathbf{B}) &= \begin{pmatrix} 0 & \mathbf{H}_1\boldsymbol{\theta}\mathbf{H}_2^T \\ -\mathbf{H}_2\boldsymbol{\theta}^T\mathbf{H}_1^T & 0 \end{pmatrix}\\ &= \begin{pmatrix} \mathbf{H}_1 & 0 \\ 0 & \mathbf{H}_2 \end{pmatrix}\begin{pmatrix} 0 & \boldsymbol{\theta} \\ -\boldsymbol{\theta}^T & 0 \end{pmatrix}\begin{pmatrix} \mathbf{H}_1^T & 0 \\ 0 & \mathbf{H}_2^T \end{pmatrix}\\ &:= \mathbf{H}\boldsymbol{\Theta}\mathbf{H}^T.\end{aligned}$$

Since $\mathbf{H}$ is an orthogonal matrix, we have

$$\exp\{\delta\mathbf{A}(\mathbf{B})\} = \mathbf{H}\exp\{\delta\boldsymbol{\Theta}\}\mathbf{H}^T.$$

Next, let the (ii)th diagonal of $\boldsymbol{\theta}_1$ be denoted as θ_{1ii}, and let $\mathbf{S}$ and $\mathbf{C}$ be diagonal matrices with diagonal elements $\sin(\theta_{1ii})$ and $\cos(\theta_{1ii})$, respectively, $i = 1, \dots, d$. Then it can be shown that, for $2d \le p$,

$$\exp\{\delta\boldsymbol{\Theta}\} = \begin{pmatrix} \mathbf{C} & \mathbf{S} & 0 \\ -\mathbf{S} & \mathbf{C} & 0 \\ 0 & 0 & \mathbf{I}_{p-2d} \end{pmatrix}.$$

For $2d > p$,

$$\exp\{\delta\boldsymbol{\Theta}\} = \begin{pmatrix} \mathbf{C} & 0 & \mathbf{S} \\ 0 & \mathbf{I}_{2d-p} & 0 \\ -\mathbf{S} & 0 & \mathbf{C} \end{pmatrix}.$$

C.4 Starting and Stopping

Like all nonlinear optimization problems, the starting value of $\mathbf{G}$ is important. If possible, it is best to avoid random starts for Grassmann optimization. Instead, inspect the objective function for recognizable special cases or for matrices whose eigenvalues may furnish useful starting values, as we did in Section 6.2 for $L_2(\mathbf{G})$.

Starting values are particularly important in Grassmann optimization because the objective function typically has multiple stationary points. For example, consider $L_1(\mathbf{G}) = \text{tr}(\mathbf{G}\mathbf{G}^T\mathbf{S}_\mathbf{X})$ written terms of a semi-orthogonal basis matrix $\mathbf{G}$ for $\mathcal{S}$. The unconstrained derivative with respect to $\mathbf{G}$ is $\partial L(\mathbf{G})/\partial\mathbf{G} = 2\mathbf{S}_\mathbf{X}\mathbf{G}$, and thus from (C.1) the Grassmann derivative is $\{\partial L(\mathbf{G})/\partial\mathbf{G}\}^T\mathbf{G}_0 = 2\mathbf{G}^T\mathbf{S}_\mathbf{X}\mathbf{G}_0$, where $(\mathbf{G}, \mathbf{G}_0)$ is an orthogonal matrix. Clearly, $\{\partial L(\mathbf{G})/\partial\mathbf{G}\}^T\mathbf{G}_0 = 0$ when evaluated at any subset of d eigenvectors of $\mathbf{S}_\mathbf{X}$.

Stopping can be on the usual criteria: (a) a fixed number of iterations and (b) when the difference $|L_2(\mathbf{G}) - L_2(\mathbf{G})_{+1}|$ is small for a sufficient number of iterations and, most strict, when a norm of $\mathbf{B}$ is sufficiently small.

Bibliography

Adcock, R. J. (1878). A problem in least squares. *The Analyst* 5: 53–54.
Adragni, K. P. (2009). Some basis functions for principal fitted components. http://userpages.umbc.edu/~kofi/reprints/BasisFunctions.pdf.
Aldrich, J. (2005). Fisher and regression. *Statistical Science* 20 (4): 401–417.
Alter, O., Brown, P., and Botstein, D. (2000). Singular value decomposition for genome-wide expression data processing and modeling. *Proceedings of the National Academy of Sciences* 97 (18): 10101–10106.
Amemiya, T. (1985). *Advanced Econometrics*. Harvard University Press.
Anderson, T. W. (1999). Asymptotic distribution of the reduced-rank regression estimator under general conditions. *Annals of Statistics* 27 (4): 1141–1154.
Andrews, D. W. K. (2002). Higher-order improvements of a computationally attractive k-step bootstrap for extremum estimators. *Econometrica* 70 (1): 119–162.
Artemiou, A. and Li, B. (2009). On principal components and regression: A statistical explanation of a natural phenomenon. *Statistica Sinica* 19: 1557–1566.
Bickel, P. J. and Doksum, K. A. (1981). An analysis of transformations revisited. *Journal of the American Statistical Association* 76 (374): 296–311.
Boik, R. J. (2002). Spectral models for covariance matrices. *Biometrika* 89 (1): 159–182.
Boulesteix, A.-L. and Strimmer, K. (2006). Partial least squares: a versatile tool for the analysis of high-dimensional genomic data. *Briefings in Bioinformatics* 8 (1): 32–44.
Box, G. E. P. and Cox, D. R. (1982). An analysis of transformations revisited, rebutted. *Journal of the American Statistical Association* 77 (377): 209–210.
Buckland, S. T., Burnham, K. P., and Augustin, N. H. (1997). Model selection: An integral part of inference. *Biometrics* 53 (2): 603–618.
Bura, E. and Cook, R. D. (2003). Rank estimation in reduced-rank regression. *Journal of Multivariate Analysis* 87 (1): 159–176.
Burges, C. J. C. (2009). Dimension reduction: A guided tour. *Foundations and Trends in Machine Learning* 2 (4): 275–365.

An Introduction to Envelopes: Dimension Reduction for Efficient Estimation in Multivariate Statistics,
First Edition. R. Dennis Cook.

Burnham, K. P. and Anderson, D. R. (2004). Multimodal inference. *Sociological and Methods Research* 33 (2): 261–304.

Candés, E. and Recht, B. (2009). Exact matrix completion via convex optimization. *Foundations of Computational Mathematics* 9: 717–772.

Cavalli-Sforza, L., Menozzi, P., and Piazza, A. (1994). The History and Geography of Human Genes. New Jersey: Princeton University Press.

Chen, X. (2010). Sufficient Dimension Reduction and Variable Selection. Ph. D. thesis, School of Statistics, University of Minnesota, School of Statistics, 313 Ford Hall, 224 Church St SE, Minneapolis, MN 55455.

Chun, H. and Keleş, S. (2010). Sparse partial least squares regression for simultaneous dimension reduction and predictor selection. *Journal of the Royal Statistical Society B* 72 (1): 3–25.

Conway, J. (1990). *A Course in Functional Analysis. Second edition.* New York: Springer.

Cook, R. D. (1977). Detection of influential observations in linear regression. *Technometrics* 19: 15–18.

Cook, R. D. (1986). Assessment of local influence (with discussion). *Journal of the Royal Statistical Society B* 48 (2): 133–169.

Cook, R. D. (1994). Using dimension-reduction subspaces to identify important inputs in models of physical systems. In *Proceedings of the Section on Engineering and Physical Sciences*, pp. 18–25. Alexandria, VA: American Statistical Association.

Cook, R. D. (1998). *Regression Graphics*. New York: Wiley.

Cook, R. D. (2007). Fisher lecture: dimension reduction in regression. *Statistical Science* 22 (1): 1–26.

Cook, R. D. and Forzani, L. (2008a). Covariance reducing models: An alternative to spectral modelling of covariance matrices. *Biometrika* 95 (4): 799–812.

Cook, R. D. and Forzani, L. (2008b). Principal fitted components for dimension reduction in regression. *Statistical Science* 23 (4): 485–501.

Cook, R. D. and Forzani, L. (2009). Likelihood-based sufficient dimension reduction. *Journal of the American Statistical Association* 104 (485): 197–208.

Cook, R. D. and Forzani, L. (2017). Big data and partial least squares prediction. *The Canadian Journal of Statistics/La Revue Canadienne de Statistique to appear*.

Cook, R. D., Forzani, L., and Rothman, A. J. (2012). Estimating sufficient reductions of the predictors in abundant high-dimensional regressions. *The Annals of Statistics* 40 (1): 353–384.

Cook, R. D., Forzani, L., and Su, Z. (2016). A note on fast envelope estimation. *Journal of Multivariate Analysis* 150: 42–54.

Cook, R. D., Forzani, L., and Yao, A.-F. (2010). Necessary and sufficient conditions for consistency of a method for smoothed functional inverse regression. *Statistica Sinica* 20 (1): 235–238.

Cook, R. D., Forzani, L., and Zhang, X. (2015). Envelopes and reduced-rank regression. *Biometrika* 102 (2): 439–456.

Cook, R. D., Helland, I. S., and Su, Z. (2013). Envelopes and partial least squares regression. *Journal of the Royal Statistical Society B* 75 (5): 851–877.

Cook, R. D. and Li, B. (2002). Dimension reduction for the conditional mean in regression. *Annals of Statistics* 30 (2): 455–474.

Cook, R. D., Li, B., and Chiaromonte, F. (2007). Dimension reduction in regression without matrix inversion. *Biometrika* 94 (3): 569–584.

Cook, R. D., Li, B., and Chiaromonte, F. (2010). Envelope models for parsimonious and efficient multivariate linear regression. *Statistica Sinica* 20 (3): 927–960.

Cook, R. D. and Su, Z. (2013). Scaled envelopes: scale-invariant and efficient estimation in multivariate linear regression. *Biometrika* 100 (4): 939–954.

Cook, R. D. and Su, Z. (2016). Scaled predictor envelopes and partial least-squares regression. *Technometrics* 58 (2): 155–165.

Cook, R. D. and Weisberg, S. (1982). *Residuals and influence in regression.* London: Chapman and Hall.

Cook, R. D. and Weisberg, S. (1991). Sliced inverse regression for dimension reduction: Comment. *Journal of the American Statistical Association* 86 (414): 328–332.

Cook, R. D. and Weisberg, S. (1998). *Applied Regression Including Computing and Graphics.* New York: Wiley.

Cook, R. D. and Zhang, X. (2015a). Foundations for envelope models and methods. *Journal of the American Statistical Association* 110 (510): 599–611.

Cook, R. D. and Zhang, X. (2015b). Simultaneous envelopes for multivariate linear regression. *Technometrics* 57 (1): 11–25.

Cook, R. D. and Zhang, X. (2016). Algorithms for envelope estimation. *Journal of Computational and Graphical Statistics* 25 (1): 284–300.

Cook, R. D. and Zhang, X. (2017). Fast envelope algorithms. *Statistica Sinica,* 28 (3): 1179–1197. DOI:10.5705/ss.202016.0037.

Dawid, A. P. (1981). Some matrix-variate distribution theory: Notational considerations and a bayesian application. *Biometrika* 68 (1): 265–274.

de Jong, S. (1993). Simpls: An alternative approach to partial least squares regression. *Chemometrics and Intelligent Laboratory Systems* 18 (3): 251–263.

Delaigle, A. and Hall, P. (2012). Methodology and theory for partial least squares applied to functional data. *Annals of Statistics* 40 (1): 322–352.

Diaconis, P. and Freedman, D. (1984). Asymptotics of graphical projection pursuit. *Annals of Statistics* 12 (3): 793–815.

Ding, S. and Cook, R. D. (2017). Matrix-variate regressions and envelope models. *Journal of the Royal Statistical Society B* 80 (2): 387–408. doi: 10.1111/rssb.12247. Working paper: https://arxiv.org/abs/1605.01485.

Dutilleul, P. (1999). The MLE algorithm for the matrix normal distribution. *Journal of Statistical Computation and Simulation* 64 (2): 105–123.

Eck, D. J. (2017). Bootstrapping for multivariate linear regression models. *Statistics & Probability Letters* 134: 141–149.

Eck, D. J. and Cook, R. D. (2017). Weighted envelope estimation to handle variability in model selection. *Biometrica* 104 (3): 743–749.

Eck, D. J., Geyer, C. J., and Cook, R. D. (2016). Supporting data analysis for "an application of envelope methodology and aster models". http://hdl.handle.net/11299/178384.

Eck, D. J., Geyer, C. J., and Cook, R. D. (2017). An application of envelope and aster models. Submitted. arXiv:1701.07910.

Edelman, A., Arias, T., and Smith, S. (1998). The geometry of algorithms with orthogonality constraints. *SIAM Journal on Matrix Analysis and Applications* 20 (2): 303–353.

Edgeworth, F. Y. (1884). On the reduction of observations. *Philosophical Magazine* 17 (104): 135–141.

Efron, B. (2014). Estimation and accuracy after model selection. *Journal of the American Statistical Association* 109 (507): 991–1007.

Fearn, T. (1983). A misuse of ridge regression in the calibration of a near infrared reflectance instrument. *Journal of the Royal Statistical Society C* 32 (1): 73–79.

Fisher, R. A. (1922). On the mathematical foundations of theoretical statistics. *Philosophical Transactions of the Royal Society A: Mathematical, Physical and Engineering Sciences* 222 (594-604): 309–368.

Fitzmaurice, F., Larid, N., and Ware, J. (2004). *Applied Longitudinal Analysis*. New York: Wiley.

Flury, B. (1987). Two generalizations of the common principal component model. *Biometrika* 74 (1): 59–69.

Flury, B. (1988a). *Common Principal Components and Related Multivariate Models*. New York: Wiley.

Flury, B. (1988b). Common principal components in k groups. *Journal of the American Statistical Association* 79 (388): 892–898.

Flury, B. and Riedwyl, H. (1988). *Multivariate Statistics: A Practical Approach*. London: Chapman and Hall.

Frank, I. E. and Friedman, J. H. (1993). A statistical view of some chemometrics regression tools. *Technometrics* 35 (2): 102–246.

Freedman, D. A. (1981). Bootstrapping regression models. *Annals of Statistics* 9 (2): 1218–1228.

Geyer, C. J., Wagenius, S., and Shaw, R. G. (2007). Aster models for life history analysis. *Biometrika* 94 (2): 415–426.

Guo, Z., Li, L., Lu, W., and Li, B. (2015). Groupwise dimension reduction via the envelope method. *Journal of the American Statistical Association* 110 (512): 1515–1527.

Haenlein, M. and Kaplan, A. M. (2004). A beginner's guide to partial least squares analysis. *Understanding Statistics* 3 (4): 283–297.

Hall, P. and Li, K. C. (1993). On almost linearity of low dimensional projections from high dimensional data. *Annals of Statistics* 21: 867–889.

Hand, D. J., Daly, F., Lunn, A. D., McConway, K. J., and Ostrowski, E. (1994). *A Handbook of Small Data Sets*. New York: Chapman and Hall.

Harville, D. A. (2008). *Matrix Algebra from a Statistician's Perspective*. New York: Springer-Verlag.

Helland, I. S. (1990). Partial least squares regression and statistical models. *Scandinavian Journal of Statistics* 17 (2): 97–114.

Helland, I. S. (1992). Maximum likelihood regression on relevant components. *Journal of the Royal Statistical Society B* 54 (2): 637–647.

Henderson, H. and Searle, S. (1979). Vec and vech operators for matrices, with some uses in Jacobians and multivariate statistics. *The Canadian Journal of Statistics/La Revue Canadienne de Statistique* 7 (1): 65–81.

Hinkley, D. V. and Runger, G. (1984). The analysis of transformed data (with discussion). *Journal of the American Statistical Association* 79 (386): 302–309.

Hjort, N. L. and Claeskens, G. (2003). Frequentist model average estimators. *Journal of the American Statistical Association* 98 (464): 879–899.

Hotelling, H. (1933). Analysis of a complex statistical variable into principal components. *Journal of Educational Psychology* 24 (6): 417–441.

Hsing, T. and Ren, H. (2009). An RKHS formulation of the inverse regression dimension-reduction problem. *Annals of Statistics* 27 (2): 726–755.

Jennrich, R. I. (1969). Asymptotic properties of non-linear least squares estimators. *Annals of Mathematical Statistics* 40 (2): 633–643.

Jensen, S. T. and Madsen, J. (2004). Estimation of proportional covariance matrices in the presene of certain linear restrictions. *Annals of Statistics* 32 (1): 219–232.

Jiang, C.-R., Yu, W., and Wang, J.-L. (2014). Inverse regression for longitudinal data. *Annals of Statistics* 42 (2): 563–591.

Johnson, O. (2008). Theoretical properties of cook's PFC dimension reduction algrithm for linear regression. *Electronic Journal of Statistics* 2 (3): 807–827.

Johnson, R. A. and Wichern, D. W. (2007). *Applied Multivariate Analysis*. New Jersey: Pearson Prentice Hall.

Johnstone, I. M. and Lu, A. Y. (2009). On consistency and sparsity for principal components analysis in high dimensions. *Journal of the American Statistical Association* 104: 689–693.

Jolliffe, I. T. (2002). *Principal Component Analysis*. New York: Springer.

Kenward, M. G. (1987). A method for comparing profiles of repeated measurements. *ournal of the Royal Statistical Society C* 36 (3): 296–308.

Khare, K., Pal, S., and Su, Z. (2016). A bayesian approach for envelope models. *Annals of Statistics* 45 (1): 196–222.

Konstantinides, K., Natarajan, B., and Yovanof, G. (1997). Noise estimation and filtering using block-based singular value decomposition. *IEEE Transactions on Image Processing* 6 (3): 478–483.

Krishnan, A., Williams, L. J., McIntosh, A. R., and Abdi, H. (2011). Partial least squares (pls) nethods for neuroimaging: A tutorial and review. *NeuroImage* 56 (2): 455–475.

Lee, K.-Y., Li, B., and Chiaromonte, F. (2013). A general theory for non-linear sufficient dimension reduction: formulation and estimation. *Annals of Statistics* 41: 221–249.

Li, K. C. (1991). Sliced inverse regression for dimension reduction (with discussion). *Journal of the American Statistical Association* 86 (414): 316–342.

Li, L., Cook, R. D., and Tsai, C.-L. (2007). Partial inverse regression. *Biometrika* 94 (3): 615–625.

Li, B. (2018). *Sufficient Dimension Reduction. Methods and Applications with R.* London: Chapman and Hall/CRC.

Li, B., Kim, M. K., and Altman, N. (2010). On dimension folding of matrix or array valued statistical objects. *The Annals of Statistics* 38 (2): 1094–1121.

Li, B. and Song, J. (2017). Nonlinear sufficient dimension for functional data. *Annals of Statistics* 45 (3): 1059–1095.

Li, B. and Wang, S. (2007). On directional regression for dimension reduction. *Journal of the American Statistical Association* 102: 997–1008.

Li, G., Yang, D., Nobel, A. B., and Shen, H. (2016). Supervised singular value decomposition and its asymptotic properties. *Journal of Multivariate Analysis* 146: 7–17.

Li, L. and Zhang, X. (2017). Parsimonious tensor response regression. *Journal of the American Statistical Association* 112 (519): 1131–1146.

Liland, K. H., Høy, M., Martens, H., and Sæbø, S. (2013). Distribution based truncation for variable selection in subspace methods for multivariate regression. *Chemometrics and Intelligent Laboratory Systems* 122: 103–111.

Liu, X., Srivastiva, A., and Gallivan, K. (2004). Optimal linear representations of images for object recognition. *IEEE Transactions on Pattern Analysis and Machine Intelligence* 26 (5): 662–666.

Lu, N. and Zimmerman, D. L. (2005). The likelihood ratio test for a separable covariance matrix. *Statistics and Probability Letters* 73 (4): 449 –457.

Ma, Y. and Zhu, L. (2014). On estimation efficiency of the central mean subspace. *Journal of the Royal Statistical Society B* 76 (5): 885–901.

Ma, Y. and Zhu, L. A. (2012). Semiparametric approach to dimension reduction. *Journal of the American Statistical Association* 107 (497): 168–179.

Magnus, J. R. and Neudecker, H. (1979). The commutation matrix: some properties and applications. *Annals of Statistics* 7 (2): 381–394.

Mallik, A. (2017). Topics in Functional Data Analysis. Ph. D. thesis, University of Minnesota, School of Statistics.

Manly, B. J. F. (1986). *Multivariate Statistical Methods.* New York: Chapman and Hall.

Martens, H. and Næs, T. (1992). *Multivariate Calibration.* New York: Wiley.

Muirhead, R. J. (2005). *Aspects of Multivariate Statistical Theory*. New York: Wiley.

Næs, T. and Helland, I. S. (1993). Relevant components in regression. *Scandinavian Journal of Statistics* 20 (3): 239–250.

Næs, T. and Martens, H. (1985). Comparison of prediction methods for multicolliear data. *Communications in Statistics – Simulation and Computation* 14 (3): 545–576.

Neudecker, H. and Wansbeek, T. (1983). Some results on commutation matrices, with statistical applications. *Canadian Journal of Statistics* 11 (3): 221–231.

Nguefack-Tsague, G. (2014). On optimal weighting scheme in model averaging. *American Journal of Applied Mathematics and Statistics* 2 (3): 150–156.

Nguyen, D. V. and Rocke, D. M. (2002). Tumor classification by partial least squares using microarray gene expression data. *Bioinformatics* 18 (1): 39–50.

Nguyen, D. V. and Rocke, D. M. (2004). On partial least squares dimension reduction for microarray-based classification: A simulation study. *Computational Statistics and Data Analysis* 46 (3): 407–425.

Park, J.-H., Sriram, T. N., and Yin, X. (2009). Central mean subspace for time series. *Journal of Computational and Graphical Statistics* 18 (3): 717–730.

Park, Y., Su, Z., and Zhu, H. (2017). Groupwise envelope models for imaging genetic analysis. *Biometrics to appear*.

Pearson, K. (1901). On lines and planes of closest fit to systems of points in space. *Philosophical Magazine* 2 (11): 559–572.

Rao, A., Aljabar, P., and Rueckert, D. (2008). Hierarchical statistical shape analysis and prediction of sub-cortical brain structures. *Medical Image Analysis* 12 (1): 55–68.

Reinsel, G. C. and Velu, R. P. (1998). *Multivariate Reduced-rank Regression: Theory and Applications*. New York: Springer.

Rekabdarkolaee, H. M. and Wang, Q. (2017). New parsimonious multivariate spatial model: Spatial envelope. https://arxiv.org/abs/1706.06703.

Rönkkö, M., McIntosh, C. N., Antonakis, J., and Edwards, J. R. (2016). Partial least squares path modeling: Time for some serious second thoughts. *Journal of Operations Management* 47–48: 9–27.

Rothman, A. J., Bickel, P. J., Levina, E., and Zhu, J. (2008). Sparse permutation invariant covaraince estimation. *Electronic Journal of Statistics* 2: 495–515.

Roy, A. and Khattree, R. (2005). On implementation of a test for kronecker product covariance structure for multivariate repeated measures data. *Statistical Methodology* 2 (4): 297–306.

Sæbø, S., Almøy, T., and Aaastveit, A. (2007). St-pls: a multi-directional nearest shrunken centroid classifier via plsst-pls: a multi-directional nearest shrunken centroid classifier via. *Journal of Chemometrics* 22: 54–62.

Schott, J. R. (1991). Some tests for common principal component subspaces in several groups. *Biometrika* 78 (4): 771–777.

Schott, J. R. (1999). Partial common principal component subspaces. *Biometrika* 86 (4): 899–908.

Schott, J. R. (2003). Weighted chi-squared test for partial common principal component subspaces. *Biometrika* 90 (2): 411–421.

Schott, J. R. (2013). On the likelihood ratio test for envelope models in multivariate linear regression. *Biometrika* 100 (2): 531–537.

Shapiro, A. (1986). Asymptotic theory of overparameterized structural models. *Journal of the American Statistical Association* 81 (393): 142–149.

Shaw, R. G. and Geyer, C. J. (2010). Inferring fitness landscapes. *Evolution* 64 (9): 2510–2520.

Shaw, R. G., Geyer, C. J., Wagenius, S. et al. (2008). Unifying life-history analyses for inference of fitness and population growth. *The American Naturalist* 172 (1): E35–E47.

Shitan, M. and Brockwell, P. (1995). An asymptotic test for separability of a spatial model. *Communications in Statistics – Theory and Methods* 24 (8): 2027 –2040.

Small, C. G., Wang, J., and Yang, Z. (2000). Eliminating multiple root problems in estimation. *Statistical Science* 15 (4): 313–341.

Smith, H., Gnanadesikan, R., and Hughes, J. B. (1962). Multivariate analysis of variance (MANOVA). *Biometrika* 18 (1): 22–41.

Stoica, P. and Viberg, M. (1996). Maximum likelihood parameter and rank estimation in reduced-rank multivariate linear regressions. *IEEE Transactions on Signal Processing* 44 (12): 3069–3079.

Su, Z. and Cook, R. D. (2011). Partial envelopes for efficient estimation in multivariate linear regression. *Biometrika* 98 (1): 133–146.

Su, Z. and Cook, R. D. (2012). Inner envelopes: efficient estimation in multivariate linear regression. *Biometrika* 99 (3): 687–702.

Su, Z. and Cook, R. D. (2013). Estimation of multivariate means with heteroscedastic errors using envelope models. *Statistica Sinica* 23 (1): 213–230.

Su, Z., Zhu, G., Chen, X., and Yang, Y. (2016). Sparse envelope model: estimation and response variable selection in multivariate linear regression. *Biometrika* 103 (3): 579–593.

Tenenhaus, M., Vinzi, V. E., Chatelin, Y.-M., and Lauro, C. (2005). PLS. path modeling. *Computational Statistics & Data Analysis* 48 (1): 159–205.

Thomson, A. and Randall-Maciver, R. (1905). *Ancient Races of the Thebaid.* Oxford: Oxford University Press.

Tipping, M. E. and Bishop, C. M. (1999). Probabilistic principal component analysis. *Journal of the Royal Statistical Society B* 61 (3): 611–622.

Tuddenham, R. D. and Snyder, M. M. (1954). Physical growth of california boys and girls from birth to age 18. *University of California Publications in Child Development* 1 (2): 183–364.

Vølund, A. (1980). Multivariate bioassay. *Biometrics* 36 (2): 225–236.

von Rosen, D. (1988). Moments for the inverted wishart distribution. *Scandinavian Journal of Statistics* 15 (2): 97–109.

Wang, C. (2017). A Study of Sufficient Dimension Reduction Methods. Ph. D. thesis, North Carolina State University, Department of Statsitics.

Weiss, R. E. (2005). *Modeling Longitudinal Data*. New York: Springer.

Welling, M., Williams, C., and Agakov, F. (2004). Extreme component analysis. In S. Thrun, S. K. Saul, and B. Schölkopf (Eds.), *Advances in Neural Information Processing Systems 16*, pp. 137–144. MIT Press.

Wold, H. (1975). Path models with latent variables: The nipals approach. In H. M. Blalock, A. Aganbegian, F. M. Borodkin, R. Boudon, and V. Capecchi (Eds.), *Quantitative Sociology: International Perspectives on Mathematical and Statistical Model Building*, pp. 307–357. Academic Press.

Wold, H. (1982). Soft modeling, the basic design and some extensions. In H. Wold and K.-G. Jöreskog (Eds.), *Systems Under Indirect Observation: Causality-Structure-Prediction. Part II*, pp. 1–54. Amsterdam: North-Holland.

Wold, S., Ketteneh, N., and Tjessem, K. (1996). Hierarchical multiblock pls and pc models for easier model interpretation and as an alternative to variable selection. *Journal of Chemometrics* 10 (5–6): 463–482.

Wold, S., Martens, H., and Wold, H. (1983). The multivariate calibration problem in chemistry solved by the pls method. In A. Ruhe and B. Kågström (Eds.), *Proceedings of the Conference on Matrix Pencils*, Lecture Notes in Mathematics, Volume 973: pp. 286–293. Heidelberg: Springer Verlag.

Wold, S., Sjöström, M., and Eriksson, L. (2001). PLS-regression: a basic tool of chemometrics. *Chemometrics and Intelligent Laboratory Systems* 58 (2): 109–130.

Yang, Y. (2005). Can the strengths of aic and bic be shared? a conflict between model indentification and regression estimation. *Biometrika* 92 (4): 937–950.

Yao, F., Lei, E., and Wu, Y. (2015). Effective dimension reduction for sparse functional data. *Biometrika* 102 (2): 421–437.

Yin, X., Li, B., and Cook, R. D. (2008). Successive direction extraction for estimating the central subspace in a multiple-index regression. *Journal of Multivariate Analysis* 99 (8): 1733–1757.

Yoo, J. K. (2013). Advances in seeded dimension reduction: Bootstrap criteria and extensions. *Computational Statistics & Data Analysis* 60: 70–79.

Yoo, J. K. and Im, Y. (2014). Multivariate seeded dimension reduction. *Journal of the Korean Statistical Society* 43: 559–566.

Zhang, H. (2007). Maximum-likelihood estimation for multivariate spatial linear coregionalization models. *Environmetrics* 18 (2): 125–139.

Zhang, X. (2017, November). Discussion on model selection uncertainty. Penn State University; personal communication.

Zhang, X. and Mai, Q. (2017). Model-free envelope dimension selection. https://arxiv.org/abs/1709.03945.

Zhang, X., Wang, C., and Wu, Y. (2017). Functional envelope for model-free sufficient dimension reduction. *Journal of Multivariate Analysis* 163 (Suppl. C): 37–50.

Zhu, H., Khondker, Z., Lu, Z., and Ibrahim, J. G. (2014). Bayesian generalized low rank regression models for neuroimaging phenotypes and genetic markers. *Journal of the American Statistical Association* 109 (507): 977–990.

Zhu, L. and Wei, Z. (2015). Estimation and inference on central mean subspace for multivariate response data. *Computational Statistics & Data Analysis* 92 (C): 68–83.

Zhu, Y. and Zeng, P. (2006). Fourier methods for estimating the central subspace and the central mean subspace in regression. *Journal of the American Statistical Association* 101 (476): 1638–1651.

Author Index

Adcock, R. J. 217
Adragni, K. P. 229, 232
Aldrich, J. 95
Alter, O. 199, 219
Amemiya, T. 260, 261
Anderson, D. R. 48
Anderson, T. W. 195
Andrews, D. W. K. 46
Artemiou, A. 219

Bickel, P. J. 40
Bishop, C. M. 85, 200, 218–220
Boik, R. J. 213
Boulesteix, A. -L. 84
Box, G. E. P. 40
Brockwell, P. 256
Buckland, S. T. 48
Bura, E. 38
Burges, C. J. C. xv
Burnham, K. P. 48

Candés, E. 199
Cavalli-Sforza, L. 219
Chen, X. 221, 224, 256
Chun, H. 91, 92, 94, 103
Claeskens, G. 48
Conway, J. 235
Cook, R. D. 1, 5, 7, 8, 13, 22, 25, 35, 37, 38, 48, 49, 58, 61, 65, 69, 72, 75, 78, 84, 88, 89, 91, 92, 95, 98, 101, 102, 106, 109, 115, 121, 122, 127–129, 135, 139–144, 149, 150, 154, 155, 159, 161, 164, 167, 173, 178, 183, 185–189, 194–199, 202, 203, 205, 207–211, 213–219, 229, 230, 232, 233, 256, 271
Cox, D. R. 40

Dawid, A. P. 256
de Jong, S. 84, 88
Delaigle, A. 84
Diaconis, P. 203
Ding, S. 161, 164, 167
Doksum, K. A. 40
Dutilleul, P. 256

Eck, D. J. 46, 48, 49, 157
Edelman, A. 20, 252, 268
Edgeworth, F. Y. xv
Efron, B. 48, 157

Fearn, T. 13
Fisher, R. A. xv
Fitzmaurice, F. 67
Flury, B. 54, 213, 215
Forzani, L. 91, 92, 178, 213–218, 229, 230, 232, 233, 271
Frank, I. E. 84
Freedman, D. 203
Freedman, D. A. 36
Friedman, J. H. 84

An Introduction to Envelopes: Dimension Reduction for Efficient Estimation in Multivariate Statistics, First Edition. R. Dennis Cook.

Geyer, C. J. 156, 157
Guo, Z. 206, 207

Haenlein, M. 85
Hall, P. 84, 203
Hand, D. J. 55
Harville, D. A. 183
Helland, I. S. 84, 90, 93, 101, 102
Henderson, H. xix, 256
Hinkley, D. V. 41
Hjort, N. L. 48
Hotelling, H. 217
Hsing, T. 211

Im, Y. 207

Jennrich, R. I. 261
Jensen, S. T. 213
Jiang, C. -R. 211
Johnson, O. 230
Johnson, R. A. 59, 78
Johnstone, I. M. 226, 227, 229
Jolliffe, I. T. 179, 217, 218, 220

Kaplan, A. M. 85
Keleş, S. 91, 92, 94, 103
Kenward, M. G. 14
Khare, K. 149, 171, 172
Khattree, R. 163, 256
Konstantinides, K. 199
Krishnan, A. 84

Lee, K. -Y. 203
Li, B. 202–204, 206–210, 219
Li, G. 199–200
Li, K. C. 203–205
Li, L. 161, 207
Liland, K. H. 92, 103
Liu, X. 267
Lu, A. Y. 226, 227, 229
Lu, N. 163

Ma, Y. 203, 204, 211
Madsen, J. 213
Magnus, J. R. xix, 256
Mai, Q. 141
Mallik, A. 126
Manly, B. J. F. 55
Martens, H. 84, 101
Muirhead, R. J. 1, 5, 213

Næs, T. 84, 101
Neudecker, H. xix, 256
Nguefack-Tsague, G. 48
Nguyen, D. V. 84

Park, J. -H. 210
Park, Y. 130, 131
Pearson, K. 217

Randall-Maciver, R. 55
Rao, A. 85
Recht, B. 199
Reinsel, G. C. 195
Rekabdarkolaee, H. M. 149, 166, 167
Ren, H. 211
Riedwyl, H. 54, 213
Rocke, D. M. 84
Rönkkö, M. 84
Rothman, A. J. 170, 233
Roy, A. 163, 256
Runger, G. 41

Sæbø, S. 109
Schott, J. R. 36, 213
Searle, S. xix, 256
Shapiro, A. 25, 26, 35, 100
Shaw, R. G. 156
Shitan, M. 256
Small, C. G. 141
Smith, H. 74, 75
Snyder, M. M. 51
Song, J. 204
Stoica, P. 195
Strimmer, K. 84
Su, Z. 5, 33–35, 37, 69, 72, 78, 127–129, 149, 168–171, 173, 183, 185–189

Tenenhaus, M. 85
Thomson, A. 55
Tipping, M. E. 85, 200, 218–220
Tuddenham, R. D. 51

Velu, R. P. 195
Viberg, M. 195
Vølund, A. 63
von Rosen, D. 87

Wang, C. 149, 166, 167, 211, 212
Wang, Q. 149, 166, 167
Wang, S. 203
Wansbeek, T. 256
Wei, Z. 211
Weisberg, S. 5, 58, 61, 75, 203
Weiss, R. E. 6, 13, 14
Welling, M. 225
Wichern, D. W. 59, 78
Wold, H. 84, 88
Wold, S. 84, 94

Yang, Y. 37
Yao, F. 211
Yin, X. 203
Yoo, J. K. 207

Zeng, P. 211
Zhang, H. 167
Zhang, X. 39, 109, 115, 141–144, 149, 150, 154, 155, 159, 161, 194, 211, 212, 256
Zhu, H. 65
Zhu, L. 211
Zhu, L. A. 203, 204
Zhu, Y. 211
Zimmerman, D. L. 163

Subject Index

A

Added variable plot 5, 61, 76
Akaike's information criterion, AIC 10, 37
Algorithms 133–148
  Gaussian elimination 140
  likelihood-based 133–135
    1D algorithm 141, 257–262
    component screening (ECS) 142–145
    coordinate decent (ECD) 142
    non-Grassmann 139–141
    sequential 141–145
    starting values 135–139
  moment-based 145–148, 262–266
  nested envelope estimators 102, 141, 145
  selecting constituent estimators 135
  SIMPLS, *see* Partial least squares
Asymptotic standard error 27

B

Bayes information criterion, BIC 10, 37
Bingham distribution 172
Bonferroni inequality 17, 66
Box's M test 129
Box–Cox power transformation 40

C

Canonical correlations 66, 191–194
  connection to envelopes 194
  review 191–194
Canonical variates 191
Canonical vectors 191
Central subspace, *see* Sufficient dimension reduction
Compound symmetry 13
CORE, *see* Sufficient dimension reduction for covariance matrices
Covariance matrices, comparison of
  and envelopes 215
  SDR methods 213–215
  spectral methods 212–213

D

Dimension reduction subspace, DRS, *see* Sufficient dimension reduction
Dimension selection
  asymptotic behavior 38–41
  bounding the rank 38
  model free 141
  overestimation 41–43
  underestimation 41–43
  via cross validation 37
  via information criteria 37

An Introduction to Envelopes: Dimension Reduction for Efficient Estimation in Multivariate Statistics,
First Edition. R. Dennis Cook.

Dimension selection (*Continued*)
via likelihood ratio testing 36
via likelihood ratio testing, LRT(α) 36, 37

E
Eigenspace, *see* Matrix algebra, eigenspace
Envelope algebra 240–246
coordinate reductions 244–246
direct sums 244
envelope relationships 241–244
Envelopes
asymptotic, definition of 150
basic 7
Bayesian 171–172
characterization 8
for a matrix-valued parameter 157
for a matrix-valued response 160–166
for a vector-valued parameter 149–157
heteroscedastic 126–131
definition 127
regression 130–131
inner 173–182
definition 174
inner response envelopes 175
relationship to response envelopes 175
predictor, *see* Predictor envelopes
reduced-rank predictor 199
reduced-rank response 197–199
response, *see* Response envelopes
scaled predictor 187–190
scaled response 182–187
sparse 33, 103, 145, 168–171
when $r > n$ 168–170
when $r \ll n$ 170–171
spatial 166–167
tensor 157
definition 159
Extreme components 225

F
Functional data 126, 211–212

G
Grassmannian 20, 24, 71, 142, 171
algebraic dimension 20
derivatives 251–252
optimization 267–272

I
Illustrations
Air pollution 59–63
added variable plots 62
envelope estimate of $\boldsymbol{\Sigma}$ 63
immaterial variation, plot of 63
response envelopes 61–62
Aster models 156–157
Australian Institute of Sport
predictor envelopes 103–105
response envelopes 58–59
Banknotes
comparing covariance matrices 216
response envelopes 54–55
Berkeley Guidance Study
bootstrap performance 53
response envelopes 51–54
Birds-planes-cars
comparing covariance matrices 216–217
Brain volumes 65–67
plots of canonical variates 66
sparse fit 169
Cattle weights
bootstrap smoothing 49
bootstrapping u 47–48
heteroscedastic envelope fit 129
influence of u 43
initial envelope fit 14
partial response envelopes 74
sparse fit 169
X-invariant part of Y 23
Egyptian skulls 55–58
response envelopes 56

Meat properties 109
Mens' urine
added variable plot 76
partial response envelope 76
response transformations 75
Minneapolis schools
transformed responses 123
untransformed responses 124
Multivariate bioassay 63–65
Mussels' muscles
predictor envelopes 106–109
sliced inverse regression 205–206
Pulp fibers
partial response envelopes 78
partial response envelopes for prediction 78
response envelopes 78
Race times
inner envelopes 179–182
scaled response envelopes 185–187
Wheat protein
bootstrap 46
full data response envelope 51
introductory illustration 13
predicting protein content 105
Invariant subspace 235–236, 238
definition 235
Inverse Gaussian distribution 172

K

Krylov matrices 90, 101, 147, 207, 210

L

Longitudinal data 6, 13, 79, 117
LRT(α), *see* Dimension selection

M

Majorization-minimization principle 169
Matrix algebra
vec and *vec h* 246–247
commutation matrices 248–249
derivatives 249–251
determinants 252–253
eigenspace 237–238, 241–244
definition 237
expansion and contraction matrices 248–249
Kronecker products 246–247
matrix exponential 268
similar matrices 183
skew symmetric matrix 268
Matrix normal distribution 159, 163, 166, 255–256
Multivariate linear model 2
asymptotic properties 5
estimation 2–4
Fisher information 5
OLS estimator 2
standardized OLS estimator 3
Z-score 4, 14, 15
Multivariate mean envelope 151
Multivariate mean envelopes 117–131
multiple means 126–130
single mean 117–126

N

NIPALS algorithm, *see* Partial least squares
Non-linear least squares 154

P

Parallel coordinate plot 15
Partial envelopes for prediction 77–78
Partial least squares 81
latent variable formulation 84–86
NIPALS algorithm 88
scaled SIMPLS algorithm 189–190
SIMPLS algorithm 88, 90, 105, 109, 145, 173, 210
Krylov matrices and 90
properties of when $n < p$ 91–94

Partial least squares (*contd.*)
response scale, importance of 98
sparse version 91
vs OLS 90
sparse versions 92, 94
Partial predictor envelopes 152–153
Partial response envelopes
asymptotic distribution 72
estimation 71–72
model 69
rationale 69, 152
schematic illustration 70
selecting u 73
PCR, *see* Principal component regression
Penalized objective functions 168–170
PFC, *see* Principal fitted components
PLS, *see* Partial least squares
Prediction
Wheat protein data 105
with partial response envelopes 77–78
with response envelopes 28–29
Predictor envelopes
dimension selection 101
functional 211–212
likelihood based estimation 95–96
asymptotic properties 98–100
fitted values and predictions 100
vs principal components 98
vs SIMPLS 98
model 83
motivation for 81–83, 151
parameter count 83
potential advantages 86–88
reduced-rank, *see* Envelopes, reduced-rank
scaling the predictors 187–190
Predictor-response envelopes 109–115
asymptotic properties 115
estimation 113–115
model 109
potential gain 110–113
Principal component regression 25, 82, 98
Principal components 62, 85
Principal fitted components 229–233
and SIR 233
envelopes and 230, 231
high-dimensional 233
model 229
model choices 232–233
non-normal errors 232
with anisotropic errors 231
with isotropic errors 230–231
Probabilistic principal components 200, 219–228
asymptotic behavior 226–228
envelope formulations 220–225
dimension selection 225
intrinsic/extrinsic variation 220
with isotropic intrinsic variation 223–225
with isotropic variation 222–223
fixed latent model 225
isotropic variation 219
random latent model 219
Profile plots 15, 16, 18, 44, 45

R

Reduced-rank regression 195–199
contrast with envelopes 196
Reducing subspace 7–9, 235–240
definition 7, 235
Reflexive indicator 85
Relevant components 101
Response envelope
partial, *see* Partial response envelopes
Response envelopes
asymptotic distributions 25–28

bootstrap 45–50
bootstrap smoothing 48–50
clues to effectiveness 14
definition 8
diagnostic plots 44, 63
dimension selection, *see* Dimension selection
fitted values 28
heteroscedastic regressions 130–131
immaterial variation and X-invariants 7
inner, *see* Envelopes, inner
introduction 6–10
introductory illustrations 10–18
maximum likelihood estimation 21–23
maximum likelihood estimators 23
model 8
motivation 7, 152
multivariate mean, *see* Multivariate mean envelopes
non-normal errors 34
parameter count 19
partial, *see* Partial response envelopes
potential advantages of 21
prediction, *see* Prediction
reduced-rank, *see* Envelopes, reduced-rank
scaled, *see* Envelopes, scaled
scaling the responses 25, 182–187
sparse, *see* Envelopes, sparse
sufficiency and 19
testing responses 29–34
X-invariants 7
Z-score 27

S

SDR, *see* Sufficient dimension reduction
SIMPLS algorithm, *see* Partial least squares
SIR, *see* Sliced inverse regression
Sliced inverse regression 204–207
  envelopes and 207
Sparse permutation invariant covariance estimation 170–171
Stiefel manifold 171
Sufficient dimension reduction 155, 202–204
  central mean subspace, definition 207–208
  central subspace 202–203, 231
    definition 202
  conditional mean reduction 207–211
    and envelopes 209–211
    and SIMPLS 210
    envelopes and 209
    iterative Hessian transformations 210
  connection to envelopes 204
  constant covariance condition 203
  dimension reduction subspace 202
  for covariance matrices 212–217
    and envelopes 215
  functional predictor 211
  linearity condition 155, 203
Supervised singular value decomposition 199–201

W

Weighted least squares 153–154